# The Bird Name Book

# The Bird Name Book

A History
of
English Bird Names

Susan Myers

Princeton University Press
Princeton and Oxford

Published by Princeton University Press
41 William Street, Princeton, New Jersey 08540
99 Banbury Road, Oxford OX2 6JX

press.princeton.edu

All Rights Reserved

Library of Congress Cataloging-in-Publication Data

Names: Myers, Susan, 1962– author.
Title: The bird name book: a history of English bird names / Susan Myers.
Description: Princeton : Princeton University Press, [2022] | Includes
   bibliographical references.
Identifiers: LCCN 2022000490 (print) | LCCN 2022000491 (ebook) |
   ISBN 9780691235691 (hardback) | ISBN 9780691236858 (ebook)
Subjects: LCSH: Birds–Identification.
Classification: LCC QL673 .M94 2022  (print) | LCC QL673  (ebook) |
   DDC  598–dc23/eng/20220113
LC record available at https://lccn.loc.gov/2022000490
LC ebook record available at https://lccn.loc.gov/2022000491

British Library Cataloging-in-Publication Data is available

Editorial: Robert Kirk and Megan Mendonça
Production Editorial: Ellen Foos
Text Design: Julie Dando
Jacket Design: Wanda España
Production: Steven Sears
Publicity: Matthew Taylor and Caitlyn Robson

Jacket image: Slaty-headed Parakeet by H. C. Richter in *Birds of Asia, Vol. VI* by John Gould, 1850–1883.
   Courtesy of Dragan Jelic / Alamy Stock Photo. Photography by Susan Myers.
Text photos: Susan Myers

This book has been composed in Adobe Garamond Pro and Myriad Pro

Printed on acid-free paper. ∞

Printed in Italy

10 9 8 7 6 5 4 3 2 1

*Right: from Wilson, Bonaparte, and Jardine's* American Ornithology *(1870)*

# Contents

Acknowledgments      7

Author's Note      7

Introduction      9

The Namers      12

Bird Names A to Z      23

Glossary      387

Illustrations References      389

Notes      393

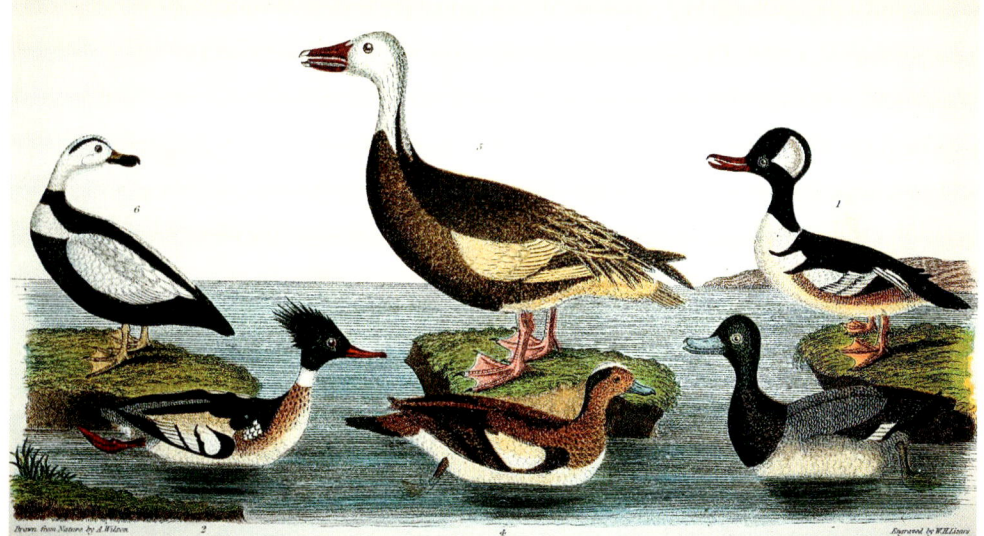

Drawn from Nature by A Wilson    2      4      Engraved by W.H.Lizars

1. Hooded or Crested Merganser. 2.Red-breasted M. 3.Blue Bill or Scaup Duck. 4.American Widgeon Male. 5.Female Snow Goose. 6.Pied Duck.

I dedicate this book to my father, Ken Myers, a man larger than life, with an insatiable curiosity and a love for life that was infectious.
Thank you, Dad . . .

# Acknowledgments

I'm deeply indebted to Steve Howell for his exceptionally helpful comments and suggestions on the manuscript. My thanks go to Som GC and Aasheesh Pittie for their comments on Nepali and Hindi bird names and to Murray Lord and Hans Breuer for providing access to references. I'd like to express my heartfelt appreciation for the hard work, sharp eyes and enthusiastic support of the wonderful PUP staff. I'd like to recognize the hard work of the administrators of the Biodiversity Heritage Library, without which this book could never have happened. Lastly, my thanks go to all my colleagues and friends at WINGS Birding Tours, for their support and friendship.

# Author's Note

Names matter—simply put, they allow us to communicate. But lately, bird names have been receiving more scrutiny than ever. It seems that rather belatedly it's come to our attention that many of the names we take for granted have a checkered background. You'll note when reading this book that nearly all the people involved in the naming of birds are or were privileged, white, and male. Even when bird names were derived from local names, they were often misspelled, misinterpreted, or just plain wrong. We can't rewrite history, and, face facts, we are talking about English names here so it's hardly surprising that European figures predominate. This book is an attempt to consolidate and summarize the etymology and history of all the common generic names of birds. Discussion of the merits or otherwise of these names is not within the scope of this work, but I hope that it may provide a basis for informed discussion on the subject. Let's hope that moving forward, the human world of ornithology and birding will reflect the wonderful diversity of the birds themselves.

Many of the quotes in this book contain terms or views that were acceptable within the mainstream cultures in the period in which they were written but may no longer be considered appropriate. These quotes do not necessarily reflect the views of the author.

As an Australian, I would like to pay my respects to the First Nations peoples, not just of Australia but worldwide. Aboriginal and Torres Strait Islander people should be aware that this book contains the names of deceased persons.

Any and all factual errors or omissions that I may have inadvertently committed are mine and mine alone.

*Susan Myers,* 2021

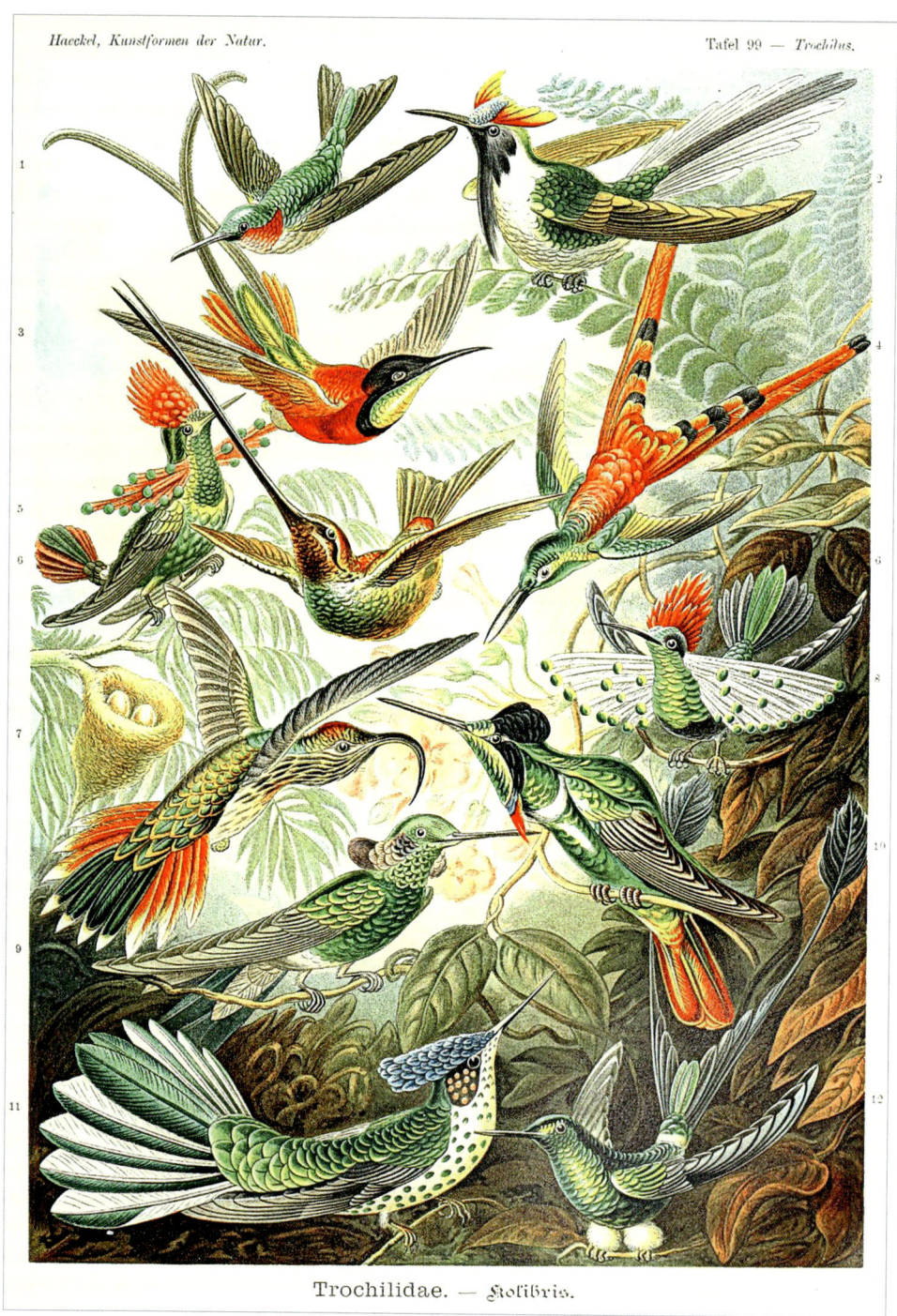

Trochilidae. — Kolibris.

*Hummingbirds in* Kunstformen der Natur *by Ernst Haeckel (1904)*
(1. Ruby-throated Hummingbird, 2. Horned Sungem, 3. Crimson Topaz, 4. Tufted Coquette, 5. Red-tailed
Comet, 6. Sword-billed Hummingbird, 7. Buff-tailed Sicklebill, 8. Dot-eared Coquette, 9. White-vented
Violetear, 10. Hooded Visorbearer, 11. Juan Fernandez Firecrown, 12. Booted Racket-tail)

# Introduction

Why do we name birds? Without a common nomenclature, our conversations about the birds we love would be nonsensical. An understanding of these names will surely make the birder's experiences even more rewarding. The scientific field of ornithology is one of the few in which common names are widely used alongside the scientific names in the literature. In other words, these names—whether they be dubbed common, vernacular, colloquial, popular, or simply English names—are not trivial, as they are also sometimes known. Birds are arguably the most visible and beloved group of animals that humans engage with. We have given them names since language began. A myriad of names for most species of birds has existed throughout human history in all languages. Even in English, there was a multitude of local names first in the British Isles, and then farther afield, as the language spread along with the colonial aspirations of the British Empire. Not surprisingly, much confusion has arisen from this natural human proclivity toward regionalism. This lexical mayhem has been described as "a swirling sea of polylexy and polysemy, where one bird may have many names and one name may apply to many birds."[288]

Only in recent years has an attempt been made, by various players, to standardize the common English names of the birds of the world. In 1883, the British Ornithologists' Union published the first edition of their *A List of British Birds*, but, while common English names were listed, no mention was made of any attempt to standardize these. The American Ornithologists' Union was arguably the first of its ilk to undertake the task of standardizing the English names of birds in 1886, but their efforts applied only to the avifauna of the United States and Canada. In this publication, common English names were given more weight, with this point in the introduction:[585]

> Every technical name [will] be followed by a vernacular name, selected with due regard to its desirability . . . Vernacular names, though having no standing in scientific nomenclature, and being not strictly subject to the law of priority, have still an importance that demands the due exercise of care in their selection, especially with reference to their fitness and desirability.

Then, in 1896, Newton published his important *Dictionary of Birds*, and it remains a useful and relevant work. In fact, his comment regarding confusion over common English names may be as pertinent today as it was in the late 1800s:

> Readers who in most respects are certainly not ignorant of things in general, frequently find in works of all sorts, but especially in books of travel, mention of Birds by names which no ordinary dictionary will explain; and, on meeting with a *Caracara*, a *Koel* or a *Paauw*, a *Leatherhead*, a *Mollymawk* or a *Tom-fool*, are at a loss to know what kind of bird is intended by the author.

In 1926, the Royal Australasian Ornithologists Union published the first edition of the *Official Checklist of the Birds of Australia* with the comment:

> Vernacular names have also been closely examined. Some indefinite names like Ground-bird have been replaced by more appropriate names, such as Quail-thrush; some inelegant names such as Black-vented Ground-Bird, have been displaced in favour of more suitable names; and some long formal names such as Great Brown Kingfisher, White-shouldered Caterpillar-eater and Rose-breasted Cockatoo, have been replaced by Laughing Kookaburra, White-winged Triller, and Galah respectively.

A few long names such as White-throated Grass-Wren and Black-faced Cuckoo-Shrike have so far defied efforts for improvement.* Appropriate, descriptive, children's, and poets' names are still required for many Australian birds.

*and still do.

In 2006, Gill and Wright published their important work *Birds of the World: Recommended English Names*, at the same time acknowledging that reaching a complete global consensus on names and spelling was unlikely. The authors took on the unenviable task of balancing the competing American and British names and spellings, as well as continuing taxonomic changes. Gill and Wright's work was endorsed by the International Ornithologists' Union in the International Ornithological Committee's (IOC) World Bird List but was not adopted by other bird species lists/taxonomies, such as Clements/eBird. And there remains widespread, albeit relatively minor, disagreement regarding common names among the various authorities.

The full common names of birds should, ideally, be descriptive so that they are not only more easily learned and remembered, but also facilitate identification. Another important function of common names for animals is that they very often prompt beginners and laypeople to take an interest in the wildlife around them—scientific names can be intimidating to many. As Skutch noted in 1954, "when one writes or speaks about birds in English, the advantage of having English names is obvious."[116]

This book is not about the full English names of the over ten thousand species of birds, but, rather, an exploration of the etymology and history of the common group names of birds, sometimes referred to, for better or worse, as the last names, primary names, collective names, generic names, or family names of birds. The history of these myriad appellations is often complex, quirky, and fascinating. The origins of many are straightforward, but, equally, many are nebulous. Unlike scientific binomials, no rules apply to the designation of common names, and most have evolved over time or were given by early European naturalists in an ad hoc manner. Skutch wrote on the subject in 1954:[116]

> In the absence of truly indigenous names for the birds they treated, Gould, Sclater, Sharpe, Seebohm, and other authors of great illustrated monographs of bird families wholly or in part American, of which England during the Nineteenth Century was so prolific, undertook to supply the deficiency with English names of their own invention.

In order to begin this conversation, though, we should start at the start—what is "bird"? The original English word used to communicate about birds, in general, was the Old English word *fugol* or *fugel* (spelling in Old and Middle English was often very inconsistent[401]), which came from a Proto-Germanic word, *fuglaz*, the reconstructed early Germanic word for "bird." It is thought the word is an alteration of another word meaning, literally, "flier," from the word *pleuk-* (from the even older theorized common ancestor language, Proto-Indo-European) from the root *pleu-*, meaning "to flow." From this original beginning, the word evolved into the better-known English word *fowl*. (This is also the source of the Dutch and German words *vogel*, both meaning *bird* in those respective languages.)

*Fugol* and *fowl* were displaced in their original sense by *bird* in the later 1400s so that now the latter is used in a narrower sense of a domestic hen or rooster; but, of course, "fowl" is often still used in names for gallinaceous bird species, such as junglefowl and guineafowl.

*Bird* comes from the Old English *bridd* or *brid*, which originally referred to a chick or fledgling. Up to the 1400s, the word *bridd* was most often used in the specific sense of the young of a bird—a fledgling, nestling, or chick, and of the young of other animals (bees, fish, snakes) and even for human children.[2] The transposition of letters, or metathesis, from *bridd* to *bird* was complete by the late 1500s. By the late 1700s *bird* had come to mean any type of bird.

Many of the common generic names we use for the ten thousand or more bird species have their origins in ancient languages and non-English languages. This is a book about the origins, meanings, and histories of all of these names. Of great interest is not only the etymology of the names, but the why and how and who of them as well.

According to Jobling's *Dictionary of Scientific Bird Names,* there are nine categories of bird names.[14] In that book, these categories are applied to scientific names but can equally be applied to common generic names.

1. eponym—named after a person or persons, as in Bernieria.
2. morphonym—named after morphological characteristics, like plumage, as in Cacholote.
3. toponym—named in reference to a geographic place, as in Madanga.
4. autochthonym—names from other languages, as in Jabiru.
5. taxonym—named in relation to other taxa, as in Catbird.
6. bionym—named after habitat or environmental conditions, as in Mountaineer.
7. ergonym—named after behavioral characteristics, like breeding or display behaviors, as in Accentor.
8. phagonym—named after diet or prey type, as in Bee-eater.
9. phononym—named after vocal characteristics, as in Babax.

There is a further category of common bird names—namely, nicknames—wherein familiar birds were affectionately given a person's name. These could be said to be a subset of the eponyms. Examples include jackdaw, magpie, and robin. Newton, who appears frequently in these pages, discussed them in his *Dictionary of Birds* from 1896 when he wrote:[7]

> Perhaps the earliest instance of nicknaming birds is to be found in Langland's *Piers the Plowman,* written soon after 1400, where the Sparrow is called "Philip"; but the practice, as all know, extended, and Swift in his *Description of a Salamander* thus mentions it:
>
> > "As mastiff-dogs in modern phrase are
> > Call'd Pompey, Scipio, and Caesar;
> > As pyes and daws are often stil'd
> > With Christian nicknames like a child."

The spelling and grammar in much of the historical literature quoted in this book was very different to what is deemed to be correct in Modern English. Note that these are not errors, but the way these passages appear in the books and papers I have quoted from.

This volume follows the taxonomy of the eBird/Clements Checklist, with some minor exceptions. The taxonomic arrangement of birds is in a constant state of flux, so that some of these names, numbers, and associations will change over time. That said, the etymology and stories of these common names are now a matter of historical fact and, to that extent, are immutable.

# The Namers

Certain names and languages appear over and over in this book. What follows is by no means an exhaustive list, but it provides a brief background to some of the most frequently mentioned ornithological figures, who helped us arrive where we are today when we talk about birds.

Born in 1522, Ulisse **Aldrovandi** was an Italian Renaissance naturalist who studied medicine at the University of Padua before returning to his native Bologna in 1561 as a professor at the University of Bologna. Later in his academic career, he developed an interest in natural history. Carl Linnaeus and the Comte de Buffon reckoned him the father of natural history studies. He assembled a huge collection of animals, plants, minerals, and fossils, said to contain more than eighteen thousand specimens, which in 1547 he opened to the public, making it the first natural history museum. He was prolific and wrote several hundred books and essays, including three influential volumes entitled *Ornithologiae*.

The French traveler, naturalist, writer, and diplomat Pierre **Belon** lived from 1517 to 1564, during which time he studied medicine at the University of Wittenberg and traveled around Germany with the German botanist Valerius Cordus, then independently through Flanders to England,

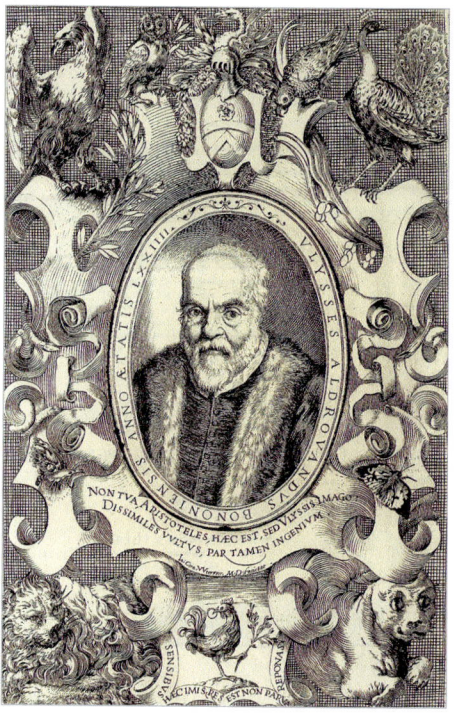

*The great Aldrovandi, the first ornithologist?*

before taking a position as apothecary to Cardinal François de Tournon. Starting in 1546 and with the backing of the cardinal, Belon then traveled extensively through Greece, Crete, Asia Minor, Egypt, Arabia, and Palestine before returning to France in 1549. He was considered a "Renaissance scholar," publishing extensively on a wide range of subjects, including *L'Histoire de la Nature des Oyseaux*. He was in the process of translating the works of the Greek scientists Dioscorides and Theophrastus when, in 1564, he was murdered in the Bois de Boulogne while returning from Paris.

Thomas **Bewick** (1753–1828) was an English wood engraver and natural history author who wrote and illustrated numerous books but is probably best known for his *A History of British Birds*. The book is illustrated with Bewick's superb wood engravings and was the first popular guide to the identification of different species of the birds of Britain. It is widely regarded as the forerunner of all modern field guides.

Edward **Blyth** (1810–1873) was born in London but traveled to India at the age of thirty-one to take a position as curator of zoology at the museum of the Asiatic Society of India in Calcutta. One of his tasks was to update the museum's catalogs, and he published a *Catalogue* *of the Birds in the Museum Asiatic Society* in 1849. He was not a field naturalist as such and received and described many bird specimens from A. O. Hume, Samuel Tickell, Robert Swinhoe, and others. Blyth published numerous reports, articles, and monographs, with particular emphasis on mammals and birds. John Gould declared that Blyth was "one of the first zoologists of his time, and the founder of the study of that science in India" while A. O. Hume described him as "the greatest of Indian naturalists," stating that "It is impossible to over-rate the extent and importance of Blyth's many-sided labours."[392]

Born in 1803, Charles Lucien Jules Laurent **Bonaparte**, Prince of Canino and Musignano, was a French biologist and ornithologist. In 1822 he married the daughter of his uncle, Joseph Bonaparte, the king of Spain, and moved with the family to Philadelphia. During his residence in the United States, he studied the ornithology of the country. While there, he published a supplement to Wilson's *Ornithology*, entitled *American Ornithology, or History of the Birds of the United States*, describing more than 100 new species. He returned to Europe in the 1840s but was expelled from France by order of Louis Napoleon. When he was finally permitted to return to Paris in 1850, he became director of the Jardin des Plantes. He later wrote extensively on American and European ornithology and other branches of natural history.

Mathurin Jacques **Brisson** was a French naturalist who lived from 1723 to 1806. Thanks to the mentorship of the French entomologist René-Antoine Ferchault de Réaumur, he embarked on a study of natural history. His published works included *Le Règne Animal* (1756) and the highly regarded six-volume *Ornithologie* (1760), one of the largest ornithological catalogs ever written. The work was written in Latin and French and contained descriptions of over 1,500 species of birds, grouped into 115 genera, twenty-six orders, and two classes distinguished by the presence or absence of webbed feet. The volumes were illustrated by François-Nicolas Martinet with 220 plates of 500 birds, many of which had never before been illustrated. Brisson's work was one of the most complete treatises in ornithology before Buffon's *Histoire des Oiseaux*.

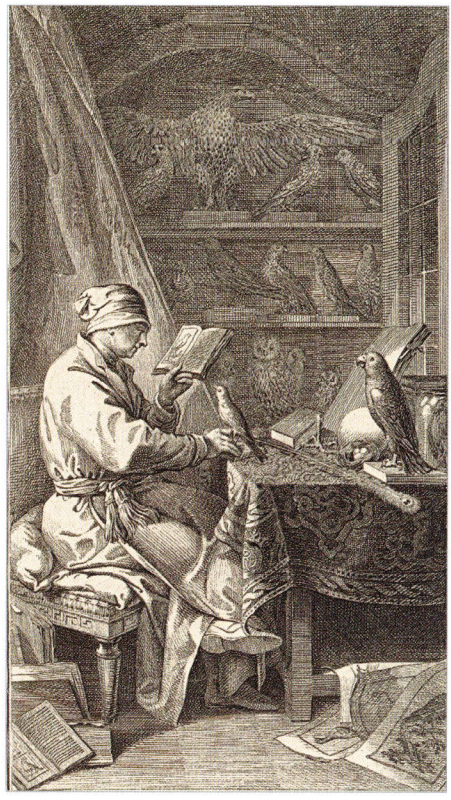

The French naturalist, mathematician, cosmologist, and encyclopedist Georges-Louis Leclerc, Comte de **Buffon**, lived from 1707 to 1788. As his title suggests, he was born into the wealth and prestige of the French aristocracy, allowing a privileged education in law and medicine, but from a young age his real interest was natural history. He was one of the first to question the prevailing status quo regarding the origins of life. Buffon built his career on the encyclopedic *Histoire*

Ornithologist in his Study *from Buffon's Naturgeschichte der Vögel (1772)*

*Naturelle* in which he planned to describe every aspect of the natural world known at the time. Sadly, he managed to publish only thirty-six out of his projected fifty volumes before his death.

Walter Lawry **Buller** (1838–1906) was a New Zealand lawyer and naturalist who was a dominant figure in New Zealand ornithology in the late 1800s. When he was only nineteen, he was admitted as a fellow of the Linnean Society of London. His book, *A History of the Birds of New Zealand*, first published in 1873, remains one of the most important in the history of ornithology of the nation. Like many of his era, he was not without flaws, holding some contentious views. He believed that the Māori, "are dying out and nothing can save them. Our plain duty as good compassionate colonists is to smooth down their dying pillow."[572] He generally rejected ideas for the conservation of both native forests and birds, although he did support plans for the protection of endangered birds, such as the huia, as well as the creation of sanctuaries at Resolution and Little Barrier Islands.

Jean Louis **Cabanis** (1816–1906) was a German ornithologist, born in Berlin to an old Huguenot family who had moved from France. He studied at the University of Berlin before traveling to North America, returning to Germany in 1841 with a large collection of specimens. He was instrumental in the collation and description of the large collection of ornithological specimens in the Museum Heineanum in Halberstadt. He started his career as an assistant in the Natural History Museum, Berlin, later becoming director. In 1853 he founded the *Journal für Ornithologie* in 1853, editing it for the next forty-one years.

John **Cassin** was a major figure of nineteenth-century ornithology. Born in Pennsylvania in 1813, he took an interest in natural history from a young age and at twenty helped found the Delaware County Institute of Science. In 1842, he was promoted to curator of the Philadelphia Academy of Natural Sciences, which at the time possessed the largest ornithological collection in the United States with over twenty-five thousand specimens, including many from Australia, India, and Africa. By that stage, Cassin was considered to be the leading ornithological taxonomist in the world. He described almost 200 new species of birds and revised a number of families in the academy's publications. Like so many of the early naturalists, he sacrificed much for his passion, writing that "It is hard work, this studying foreign birds. It would do very well was there no arrangement to be made for ensuring the supply of bread and butter"[573] and eventually dying in 1869 of arsenic poisoning thought to be caused by his handling of bird skins preserved with the toxin.

Mark **Catesby**, who lived from 1683 to 1749, was an English naturalist who was an early student of the flora and fauna of the New World. Between 1729 and 1747, Catesby published his *Natural History of Carolina, Florida and the Bahama Islands*, the first published account of the flora and fauna of North America. His friendship with John Ray led to his interest in natural history, and he first visited Virginia in 1712, returning to England in 1719, only to return to South Carolina in 1722. Catesby read a paper entitled "Of Birds of Passage" at a meeting of the Royal Society in London in 1747, and he is now recognized as one of the first people to describe bird migration.

Georges **Cuvier** was a French naturalist and zoologist who lived from 1769 to 1832. He is sometimes referred to as the "founding father of paleontology." Through his work in comparative anatomy, comparing living animals with fossils he established the legitimacy of the science of paleontology. Although not strictly an ornithologist, his major work from 1830, the five-volume *Le Règne Animal*, was highly influential. In it, Cuvier presented the results of his research into the structure of living and fossil animals. The French paleontologist Philippe Taquet noted that "Cuvier introduced clarity into natural history, accurately reproducing the actual ordering of animals."[574]

Father Armand **David**, who was also known by his French name Père David, was a Lazarist missionary Catholic priest with a keen interest in zoology and botany. He lived from 1826 to 1900 and spent an important part of his life in China as a missionary, during which time he made numerous scientific journeys throughout China acquiring many biological specimens of great importance.

The so-called father of British ornithology, English naturalist George **Edwards** was born in 1694 in Stratford, Essex. Early on, he traveled extensively throughout mainland Europe, studying natural history, and gained a reputation for the quality of his colored drawings of birds. In 1743 he published the first of four volumes of his *A Natural History of Uncommon Birds* followed by three supplementary volumes, entitled *Gleanings of Natural History*. The works featured engravings and descriptions of more than 600 subjects in natural history never before described. He was a correspondent of Linnaeus, who would later supply binomial names for those listed in Edwards's general index.

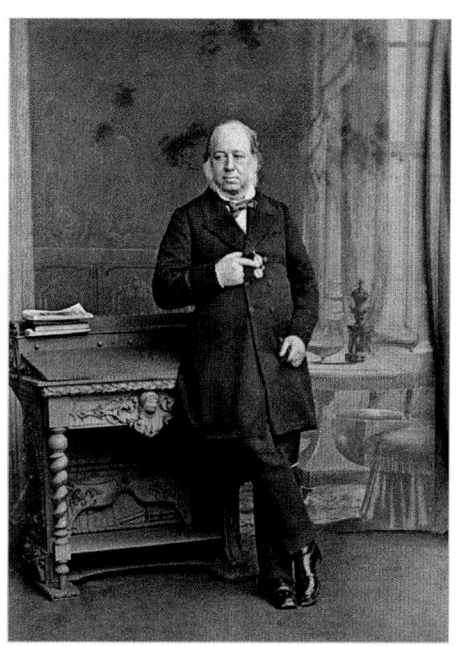

John Gould in 1860, aged 56

John **Gould** was an English ornithologist who lived from 1804 to 1881. He was, and to a degree still is, one of the most influential figures in birding history. He published numerous monographs on birds, illustrated by plates produced by his wife, Elizabeth Gould, and several other artists, including Edward Lear—perhaps better known for his whimsical limericks. Gould worked closely with Darwin, assisting him with the identification of many of the specimens Darwin brought home to England from his famous voyage on the *Beagle*. Gould is credited with being highly instrumental in Darwin's formation of his theory of natural selection. Subsequently, he and Elizabeth traveled to Australia, where he traveled and collected extensively for two years. He also took a special interest in the hummingbirds, accumulating a collection of over 320 specimens. He described and named many species, despite having never seen a living hummingbird. Gould quite probably gave more English common names to birds than any other figure in ornithology.

John Henry **Gurney** Jr. (1848–1922) was a British ornithologist from the influential family of English Quakers and bankers, which had a major role in the development of Norwich. His politician father, John Henry Gurney Sr., was an accomplished ornithologist who fostered in his son, from an early age, a love birds. A wealthy landowner and country gentleman, Gurney soon gained an international reputation as an ornithologist, and in 1868 he was elected as a fellow of the Zoological Society. He was then promoted as a member of the British

Ornithologists' Union in 1870, and a fellow of the Linnean Society in 1885. He published many books on various ornithological subjects, arguably the most important of which was the *Early Annals of Ornithology*, which appeared in 1921 shortly before his death.

The American physician and naturalist Thomas **Horsfield** lived from 1775 to 1859. In 1799, he took up a post as surgeon on the vessel *China*, a merchant vessel that took him to Java where he immediately became enamored with the region. And in 1801 he applied for a post as a surgeon with the Dutch colonial army in Batavia. While there, he took a keen interest in the flora, fauna, and geology of the region. Horsfield made a large collection of plants and animals on behalf of the governor of the Dutch East Indies, Sir Thomas Stamford Raffles. Due to ill health, he reluctantly left the island in 1819 for London, where he took a post as curator of the East India Company Museum in London. He wrote several books and papers, one of the most important of which was the *Zoological Researches in Java and the Neighbouring Islands*. Another important work was his "A Description of the Australian Birds in the Collection of the Linnean Society; with an Attempt at arranging them according to their natural Affinities,"[575] coauthored with the Irish zoologist Nicholas Aylward Vigors.

Allan Octavian **Hume** (1829–1912) was a British member of the Imperial Civil Service, a political reformer, and an ornithologist and botanist who worked in British India. Hume has been called "the father of Indian ornithology." He founded the journal *Stray Feathers* and built a vast collection of bird specimens, making expeditions and obtaining specimens through a large network of contacts. The largest single collection of birds ever obtained by the British Natural History Museum was donated by Hume in 1885. The collection contains more than sixty thousand bird skins and nearly twenty thousand eggs from the Indian subcontinent. The museum notes that despite Hume's astounding achievements, birds were just a hobby for him and that his main focus was his job as a high-ranking official of the British Raj. He was also involved in politics, playing a key role in the founding of the Indian National Congress. He retired in the early 1880s when his superiors dis-

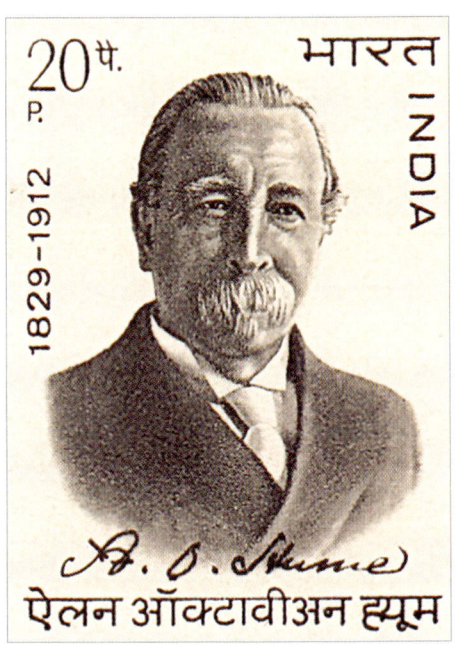

*Allan Octavian Hume, a 1973 stamp of India*

approved of his egalitarian attitudes toward the Indian people, after which the manuscript material for his planned magnum opus, *Birds of the British Indian Empire*, was stolen. This prompted him to donate his bird collection to the Natural History Museum and effectively give up ornithology.

Ernest **Ingersoll** (1852–1946) was an American naturalist, writer, and explorer. He was an early advocate for the protection of wildlife and natural habitats and preferred field notes and photographs to taking specimens. For thirty-eight years, Ingersoll wrote a weekly column entitled "The Natural History Club" for a Montreal newspaper. The column, consisting of answers to readers' questions and presentations of new information about nature, had a readership of an estimated half a million people a week throughout Canada and the

Caribbean. During this time, Ingersoll prepared a classification list of nearly all the birds of Canada, giving descriptions of each species. Ingersoll believed his mission in life should be to make science popular and interesting to the public.

William **Jardine**'s formal title was Sir William Jardine, 7th Baronet of Applegarth. He was born in 1800 into a wealthy family but didn't waste his privilege, studying medicine at Edinburgh University while pursuing his interest in natural history. At twenty-five he became a member of the Royal Society of Edinburgh; he cofounded the Berwickshire Naturalists' Club and the Ray Society, going on to become a fellow of the Linnean Society of London and of the Society of Antiquaries of London. His major contribution, though, was his role as editor, and author of many of the forty volumes, of the highly influential *Naturalist's Library* published from 1833 to 1843.

Thomas Caverhill **Jerdon** was a British physician, zoologist, and botanist whose pioneering work in India led to the first scientific descriptions of many birds of India. Born in 1811, he had an interest in natural history from a young age but trained as a physician and surgeon. He was appointed as an assistant surgeon in the East India Company in 1835, pursuing his interest in ornithology during his thirty-four years on the subcontinent. His most important publication, *The Birds of India*, included descriptions of over 1,000 species, many of them described for the first time, in two volumes. Importantly, Jerdon documented the local names of many birds, although he did not always follow a consistent transliteration of Hindi and Urdu words.

John **Latham** was the author of two influential books on the subject of ornithology in the late 1700s and the early 1800s. In these works, *A General Synopsis of Birds* and *General History of Birds*, he described many new species he had discovered in various museums and collections. Although he never visited Australia, he described so many of the continent's birds from specimens for the first time that he is still often referred to as the grandfather of Australian ornithology. He was also the first to coin common English names for many species, including the Emu, Sulphur-crested Cockatoo, Wedge-tailed Eagle, Superb Lyrebird, and Australian Magpie.

François **Le Vaillant** was a French author, explorer, naturalist, zoological collector, and noted ornithologist. He was born in French Guiana and went to southern Africa in 1781 at the age of twenty-eight. He collected over two thousand specimens of birds, describing many new species, and subsequently published the six-volume *Histoire Naturelle des Oiseaux d'Afrique* between 1799 and 1808. This work, among the first to use color plates for illustrating birds, established his

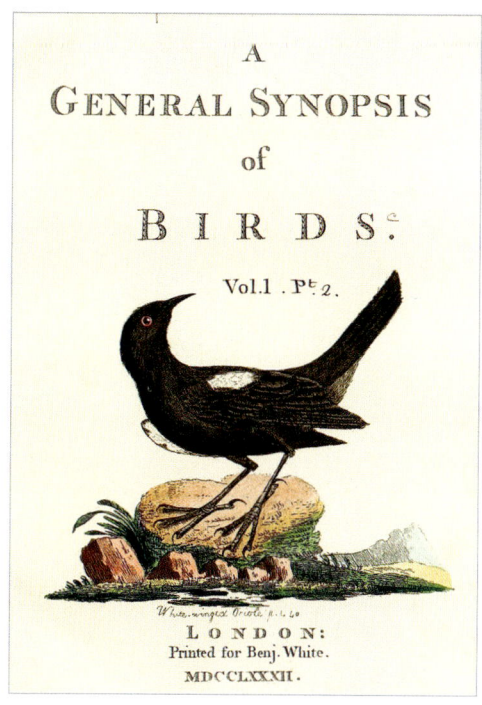

*The cover of one of six issues of Latham's* General Synopsis of Birds, *published from 1781 to 1785*

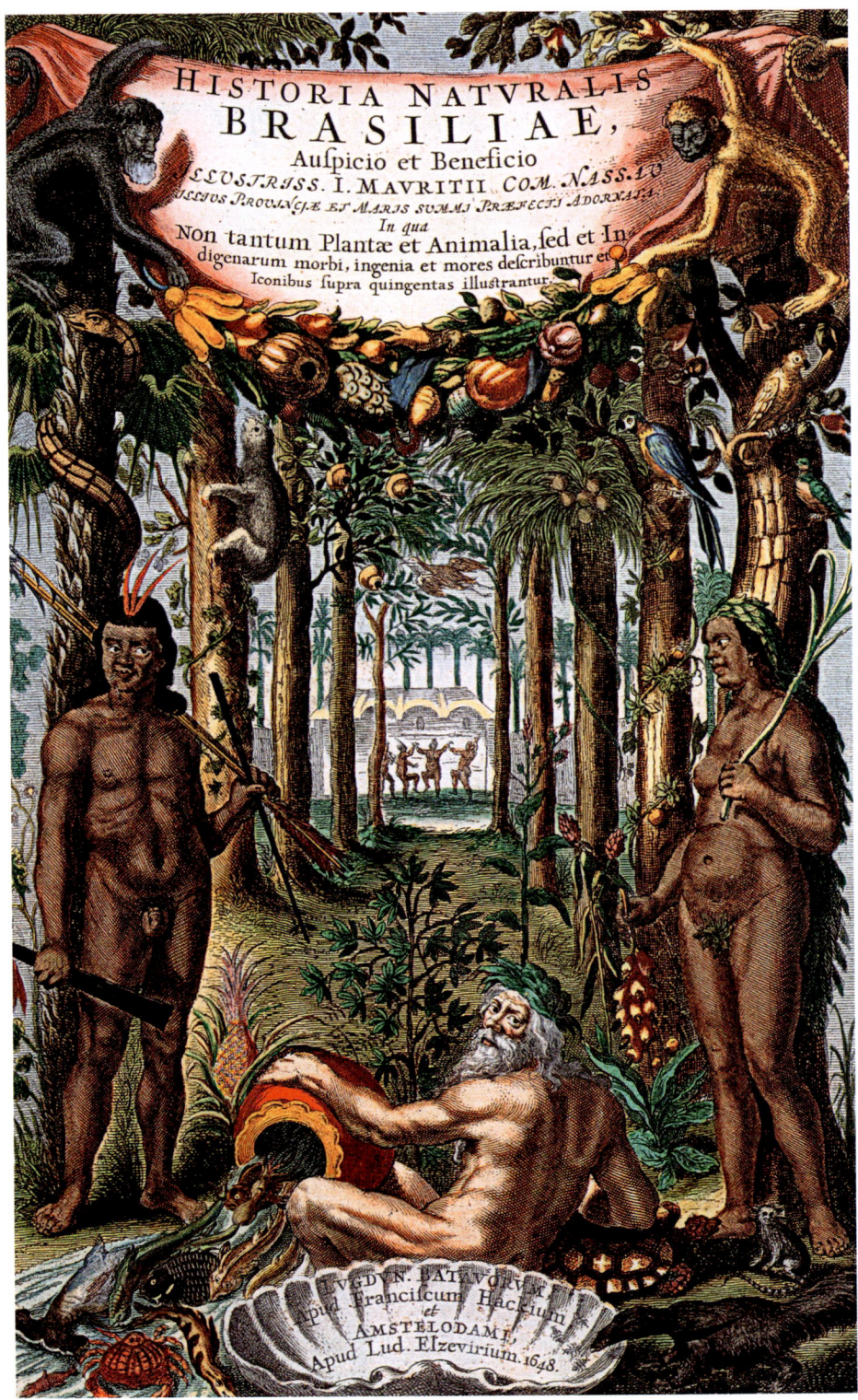

*Marcgrave's* Historia Naturalis Brasiliae

reputation as the so-called founder of African ornithology. Le Vaillant lacked formal training and opposed the use of the binomial nomenclature introduced by Linnaeus, instead championing the use of descriptive French names such as the Bateleur. He is reputed to have been the most famous ornithologist of the 1700s.

René Primevère **Lesson** (1794–1849) was a French surgeon and naturalist. He entered the Naval Medical School in Rochefort at the age of sixteen and served in the French Navy during the Napoleonic Wars. He joined a round-the-world voyage on *La Coquille* (1822–25), amassing a large collection of natural history specimens. On the voyage he visited the Moluccas and New Guinea, becoming the first European naturalist to see birds-of-paradise in the wild. On returning to Paris, he wrote the section on vertebrates for the official account of the expedition. He published numerous important ornithological works, including the seminal *Manuel d'Ornithologie* in 1828, several monographs on hummingbirds, and a book on the birds-of-paradise.

William Burley **Lockwood** (1917–2012) was a professor of Germanic and Indo-European philology at the University of Reading from 1968 until his retirement in 1982. Although he did not coin names for birds, he wrote one of the few linguistic examinations of English bird names as well as numerous articles.[84]

Georg **Marcgrave** (1610–1644) was a German naturalist and astronomer whose posthumously published *Historia Naturalis Brasiliae* was a major contribution to early biological sciences. Born in Germany, Marcgrave was a polymath who studied botany, astronomy, mathematics, and medicine before being appointed astronomer of a company that dispatched him to Dutch Brazil. He entered the service of the governor of Dutch Brazil and undertook the first zoological, botanical, and astronomical expeditions there. Marcgrave went on to describe the natural history of numerous remote countries during the sixteenth and seventeenth centuries, and was considered by many authorities to have been one of the most capable and most precise naturalists of his era. In 1644 he fell ill with a fever and was preparing to sail for home when he was ordered to travel to Angola, where, on arrival, he died from the illness.

Alfred **Newton** was Cambridge University's first professor of the Department of Zoology ("Zoology and Comparative Anatomy" at the founding of the chair), serving for over forty years. Newton came from a wealthy family that prospered from investments in West Indian sugar plantations. He was born near Geneva in 1829 when his family was making an extended trip to the Continent. An accomplished and dedicated ornithologist since boyhood, Newton approached Charles Darwin, among others, for support in his application for the new Chair of Zoology, but many of them felt he might be too specialized for that role. He was a prolific author, but arguably his most important work was his four-volume *Dictionary of Birds* published in the 1890s. He was one of the founders of the British Ornithologists' Union and its journal *Ibis* (*The Ibis* at its founding).

Alfred John **North** was an important figure in early Australian ornithology. He was born in Melbourne in 1855 and was appointed to the Australian Museum in Sydney in 1886, where he worked as a permanent assistant in Ornithology until 1917. As a founding member of the Field Naturalists Club of Victoria, he described a number of species of birds for the first time, many in the club's journal, *The Victorian Naturalist*. His publications, *A List of the Insectivorous Birds of New South Wales* and *Descriptive Catalogue of the Nests and Eggs of Birds Found Breeding in Australia and Tasmania*, were groundbreaking works at the time.

The Welsh naturalist Thomas **Pennant** never traveled farther afield than continental Europe, yet was the author of several important volumes, including *British Zoology*, *History of Quadrupeds*, *Arctic Zoology*, and *Indian Zoology*. By his own admission, he first acquired a taste for ornithology at the age of twelve, when John Ray and Francis Willughby's book on birds was presented to him by a relative. Although he never completed his degree from Oxford University, he received many honors and marks of distinction during his lifetime, including his election in 1757 as a member of the Royal Society of Upsala.

One of America's greatest ornithologists, Robert **Ridgway** lived from 1850 to 1929. As a specialist in systematics, he was appointed in 1880 as the first full-time curator of birds at the United States National Museum, a title he held until his death. In 1883 he was one of the founding members of the American Ornithologists' Union, where he served as an officer and journal editor. No other ornithologist described as many new species of North American birds, and his eight-volume publication *The Birds of North and Middle America*, published from 1901 to 1919, is certainly one of the most important works of the era and beyond.

Walter **Rothschild** was born in 1868 into the prominent Rothschild banking family. Despite this, he had little interest in finance and instead pursued his childhood passion for collecting natural history specimens. As a child he announced to his parents that he was going to "make a museum," which he did at the age of ten when he used his parents' garden shed to house the specimens he had already collected. As his collection of both local and exotic specimens grew, it became apparent that the shed would no longer suffice and, despite the family's disapproval, his father, the first Lord Rothschild, built him a museum on the edge of Tring Park as a twenty-first birthday present. Three years later, in 1892, Walter's Zoological Museum opened to the public. From this modest start, Rothschild's collection became the now world-famous Natural History Museum at Tring. Rothschild was also an outspoken Zionist who was active in the formulation of a plan to establish a Jewish homeland in Palestine.

The English lawyer and zoologist Philip Lutley **Sclater** lived from 1829 to 1913. He was secretary of the Zoological Society of London for forty-two years, from 1860 to 1902. Specializing in the field of ornithology, Sclater was one of the first to identify the main zoogeographic regions of the world. He was a highly productive author with over 1,400 publications on ornithology, but he is best remembered for his 1858 paper setting out the classification of the faunal regions of zoogeography, later adopted by Alfred Russel Wallace. The system is still in common use today. His most lasting ornithological work is probably the four volumes he contributed to the *Catalogue of the Birds in the British Museum*. He was the founder and editor of *The Ibis*, a council member of the Zoological Society of London, and a council member of the Royal Society of London.

Richard Bowdler **Sharpe** (1847–1909) was an English zoologist and ornithologist. He took an early interest in ornithology and, even at the age of sixteen, expressed an interest in writing a monograph on the kingfishers. At nineteen he became a librarian at the Zoological Society of London and completed his *Monograph of the Kingfishers* during this period. Sharpe took charge of the British Museum's bird collection when he was appointed as a senior assistant in the Department of Zoology in 1872. He founded the British Ornithologists' Club in 1892 where he worked as editor of the club's bulletin. As well as describing many new bird species, Sharpe published several monographs on bird groups as well as a multivolume catalog of the specimens in the collection of the museum. As Sharpe himself wrote in his preface, his two-volume *Birds of Paradise* contained "a great number of the species [that were] here figured for the first time."

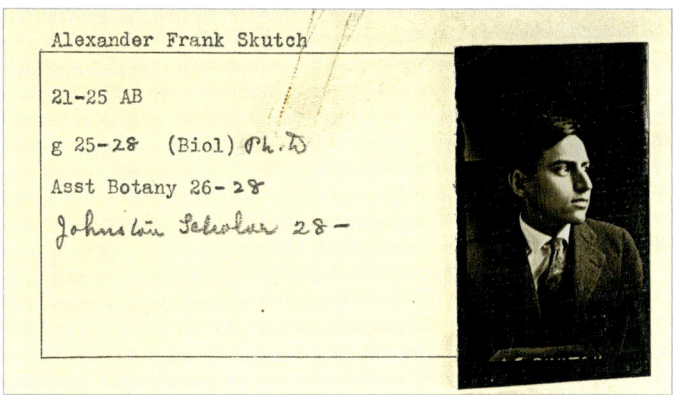

*Skutch's ID card from Johns Hopkins University, issued around 1921*

Alexander Frank **Skutch** (1904–2004) was an American naturalist and writer, a prolific author of numerous scientific papers and books about birds as well as several books on philosophy. He graduated from Johns Hopkins University in 1928, initially working as a botanist, before travels in Central America sparked a deep interest in birds. He is still universally regarded as one of the world's greatest ornithologists.

Living from 1789 to 1855, William John **Swainson** was an English ornithologist, malacologist, conchologist, entomologist, and artist. He started his working life as a junior clerk for Liverpool Customs at the age of fourteen, but his adventures started in 1806 when he accompanied the English explorer Henry Koster to Brazil. From then on he traveled and collected widely, including to Australia and New Zealand. Swainson's father was an original fellow of the Linnean Society, and his son followed in his footsteps in 1815. This was followed by his election as a fellow of the Royal Society after his return from Brazil in 1820. He published prolifically, but his *Zoological Illustrations* is arguably his most important work. His work was lauded not only for the establishment of common and scientific names for many species, but also for the quality of his illustrations, which represented the first use of lithography in the literature.

Harry Kirke **Swann** (1871–1926) was an English ornithologist who began his ornithological career with a journey to Nova Scotia at the age of twenty. He published an account of his Canadian travels under the title *Nature in Acadie* in 1895. After returning to England in 1892, he published numerous journals and books, including the excellent *Dictionary of English Folk-Names of British Birds*, which showcased his extensive knowledge of the literature and bibliography of ornithology. He died at the age of only fifty-five during surgery, and his important work *A Monograph of the Birds of Prey* was published posthumously.

Louis Pierre **Vieillot** (1748–1830) was a French ornithologist, the author of the first scientific descriptions and binomial names of several birds, including species he collected in the West Indies and North and South America. At least twenty-six genera erected by Vieillot are still in use. He was a pioneer in the study of molt and plumage, and one of the first to study live birds in the field, as opposed to specimens. Very little is known of his life as he left no journals or detailed accounts of his voyages. It is known he fled Haiti during the French Revolution. Arriving in the United States, he studied the birds of the country and collected material for his *Histoire Naturelle des Oiseaux de l'Amérique Septentrionale*, published in 1808. On returning to France in 1800, he produced *Histoire naturelle et générale des colibris,* *oiseaux-mouches, jacamars et promerops* in 1802, followed by *Histoire Naturelle des plus Beaux Oiseaux Chanteurs de la Zone Torride* in 1806. Other volumes followed, including the *Analyse d'une Nouvelle Ornithologie Elémentaire*, in which he set out a proposed system of ornithological classification, and *Ornithologie Française*. In 1831 he died in poverty and almost unremarked.

The German herpetologist and ornithologist Johann Georg **Wagler** was born in 1800, living a highly productive short life before dying at the age of only thirty-two from an accidental self-inflicted gunshot wound while out collecting in Bavaria. He was an assistant to Johann Baptist von Spix (of Spix's Macaw fame) and was also a lecturer in zoology at the Ludwig Maximilian University in Munich.

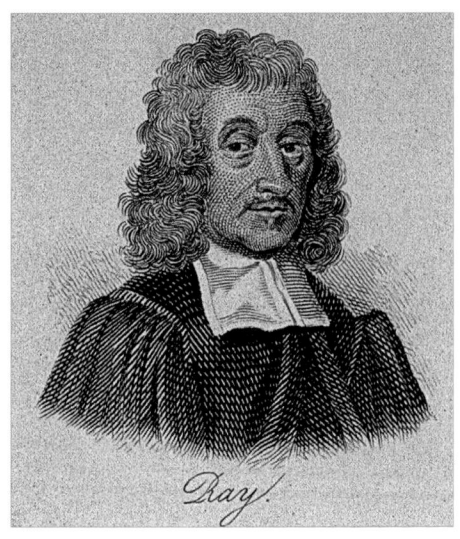

Line engraving of John Ray by Mary Beale

He published his first larger work in 1824, a description of the amphibians collected by Spix in Brazil, after which he was appointed a member of the Royal Academy of Sciences. This was followed by expeditions to France, England, and the Netherlands and numerous other publications in the fields of herpetology and ornithology.

Francis **Willughby** and John **Ray**: Willughby was born into a wealthy family and was a student at Trinity College, Cambridge, before pursuing an interest in ornithology and ichthyology. While there, he was tutored by the mathematician and naturalist John Ray, who became a lifelong friend and colleague. Ray, on the other hand, was the son of a blacksmith but also studied at Trinity, later holding college offices there. The two men worked together closely until Willughby's death at the age of thirty-six from pleurisy. They toured Europe together in the 1660s, amassing a huge collection of specimens. Based on these collections and more, *The Ornithology of Francis Willughby of Middleton in the county of Warwick* was published by Ray after the titular author's death; Ray saw it as his duty to complete and publish his colleague's work on animals. The relative contributions of the two authors have long been disputed. Some saw Ray as the brains behind their collaboration, with the younger and wealthier Willughby demoted to a mere financial backer. Others believe that both men were talented naturalists and point out that decades after Willughby's death, Ray's approach to zoology was shaped by the two men's joint investigations in the 1650s and 1660s. In this book, the work will be referred to as "Ray and Willughby."

Born in 1865, Scott Barchard **Wilson** was a British ornithologist and explorer. In 1887, sponsored by the University of Cambridge, he spent eighteen months in the Hawaiian Islands, where he described more than a dozen new bird species. On his return, he wrote, with Arthur Humble Evans, the important *Aves Hawaiienses*. Although his work was highly acclaimed, he struggled to maintain his success and reputation following criticism from both Newton and Rothschild. Wilson died at his home in England in 1923 when he put a gun to his head and shot himself. At the inquest, his sister revealed that "he was very subject to extremes, was easily depressed over quite trivial matters and was always abnormally sensitive."[576] He was apparently greatly troubled by financial issues and had lost money due to failed investments.

# A IS FOR ACCENTOR

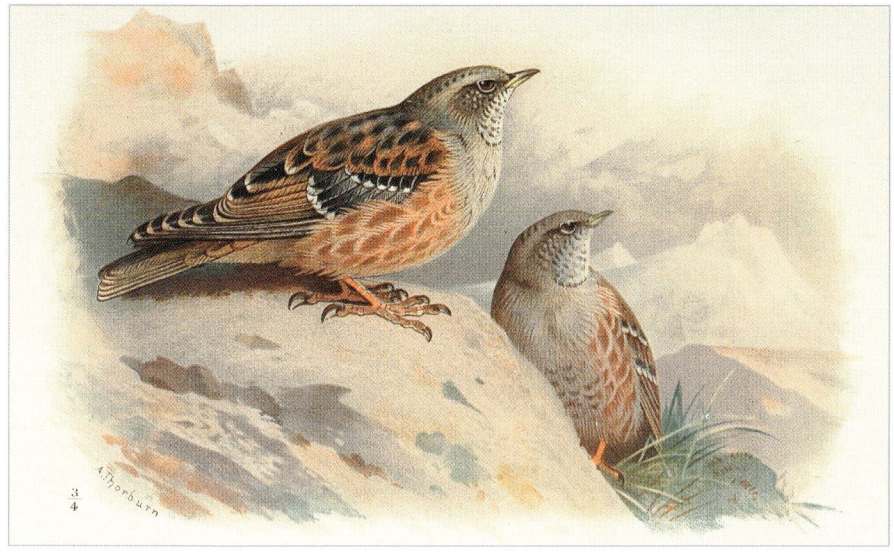

*Alpine Accentor in* Coloured figures of the birds of the British Islands *(1885)*

## Accentor

The Prunellidae is a monogeneric family with thirteen members in the genus *Prunella* bearing the common name accentor. The German naturalist Bechstein first created the now-disused genus name in 1793, albeit for the "Water Ouzel" (White-throated Dipper):[469]

> I am . . . creating a separate genus for it under the Latin name *Accentor*, or German water singer, because he sings well, and in winter no matter how cold it, as long as the sun is shining.

This name derives from the Latin "*accentor*," meaning "one who sings with another" from *ad*, the preposition "to," and *cantor*, "singer."

## Adjutant

Two species of storks are named for their supposed stiff military bearing. Latham called the Greater Adjutant the "Giant Crane" but gave a hint of the origin of the present-day name:[166]

> This singular species is not unfrequent at *Bengal*, where it arrives before the rainy season comes on, and is called *Argala*, or *Adjutant* . . . I have been told, that the bird has obtained this last name from its appearing, when looked on in front at a distance, like a man having a white waistcoat and breeches.

According to the same author, it had other local names, although it's unclear if these were names given by the English in India or if they were translations of local names:

> It has also, from its immense gape, gained the name of *Large Throat*; and, from its swallowing bones, the *Bone-eater*, or *Bone-taker*.

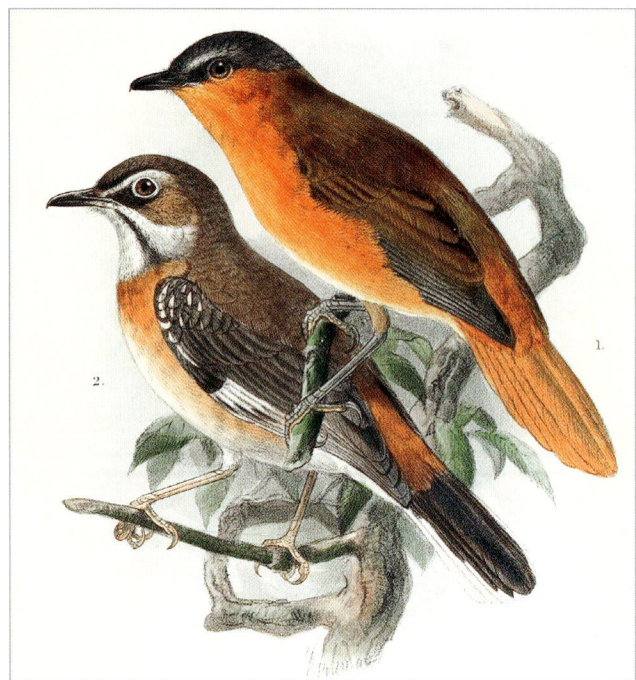

*Bocage's Akalat and Miombo Scrub-Robin in Barbosa's* Ornithologie d'Angola *(1881)*

# Akalat

Nine species of akalats in the Muscicapidae (Old World flycatchers) are found in Central Africa. The name derives from the languages of Bulu and Fang (both Bantu languages), probably as a general term for small birds. In 1908 the British collector George L. Bates was quoted by Sharpe as corresponding:[243]

> The little members of the genus *Turdinus* [as they were considered at the time], which are called in Fang and Bulu "Akalat," are among the most secretive of birds, keeping to the dark thickets of the forest.

# Akeke'e

A single species of Hawaiian honeycreeper in the Fringillidae bears the name Akeke'e. It's likely it stems from the Hawaiian word *ke'e*, which is defined in the Hawaiian-English dictionary as meaning "deformity, crooked, bent," as a reference to the bird's asymmetric bill shape, like that of a crossbill. In the early 1900s, Perkins wrote in *Fauna Hawaiiensis*:[244]

> The native names of the forest-birds are themselves of some interest, showing as they sometimes do the rudiments, as it were, of a crude, and often erroneous, classification. The names are certainly very aptly chosen, and their meaning is in most cases apparent to anyone with some knowledge of the language.

He included the Akeke'e in the category of

> names given from peculiarities of structure or plumage, e.g. Akihialoa (*Hemignathus*) from its long, sharply-pointed beak; Nukupuu (*Heterorhynchus*) from its hill-like (i.e. strongly rounded) bill; Palila from its aberrant grey plumage. Such names are often compounded with *a*- beak (lit. jaw) e.g. Akekee, Amakihi, Akohekohe.

# Akepa

Three species of Hawaiian honeycreepers in the genus *Loxops* bear the common name akepa. The name is derived from a Hawaiian word meaning "agile, active, or quick" due to the birds' habit of very actively foraging at the tips of branches. Perkins noted:[244]

> The various species of *Loxops* are amongst the most active of native birds and their name Akepa, signifying "sprightly" "turning this way and that," is singularly appropriate. This name is applied by well-informed natives of the present day to both the species inhabiting Maui and Hawaii, and Bloxam gave the same name for the Oahuan form three-quarters of a century ago.

# Akialoa

Four species in the genus *Akialoa* possess the same common name. All four are now very sadly extinct, but all possessed remarkably long decurved bills. Hawaiians named it for its *a-loa* or very long bill, *a-* for the bill and *loa* meaning "long." The *Akia* element of the name is from the Hawaiian for "bird," *akihi*. Perkins wrote that they are notable for "the excessive elongation of the beak" and:[244]

> In their habits the species of this genus are quite intermediate between *Chlorodrepanis* and *Heterorhynchus*, since they are greater nectar-eaters than the members of the latter genus and hunt more persistently, creeper-like, on the limbs of forest trees for wood- and bark-eating insects than does the Amakihi (*Chlorodrepanis*).

# Akiapolaau

Closely related to the above species, this Hawaiian honeycreeper is still extant on the island of Hawaii. The name comes from three Hawaiian words—*akihi*, "bird," *po'o*, "dig," and *la'au*, "branch," which is indeed what the bird does with its superbly adapted bill. Henshaw wrote in his checklist of the birds of Hawaii in 1902:[245]

> In the akiapolaau we have another of the interesting and extraordinary bird forms with which Nature has favored the Hawaiian Islands . . . [it] resembles the akialoa . . . but the yellow belly and the short, blunt mandible, in contrast with the long, delicate maxilla, serve at once to distinguish the two apart . . . the short, blunt mandible of the akiapolaau has conferred new powers upon it . . . By means of it, when the maxilla is agape, it can flake off lichens and even pound off small knobs and excrescences under which it suspects larvae to be concealed . . . [Nature] has given to others long bills and brush tipped tongues for probing hidden cavities and seizing the insect prey; and she has equipped the akiapolaau with a special device in the shape of a more or less effective hammer to expose the hidden retreats of larvae.

# Akikiki

Another unique Hawaiian honeycreeper, this small, rather plain bird behaves somewhat like a nuthatch, but the name, although of the Hawaiian languages, is said to be onomatopoeic. However, it's likely that the first part of the name is *akihi*, "bird" (see above). Perkins, who called the bird *Akikeke*, wrote:[244]

> Both the Kauai and Molokai [the Kakawahie] species no doubt received their names from the characteristic cry which distinguishes *Oreomyza* from all other native birds—the reiterated 'chip,' with which they resent the presence of an intruder.

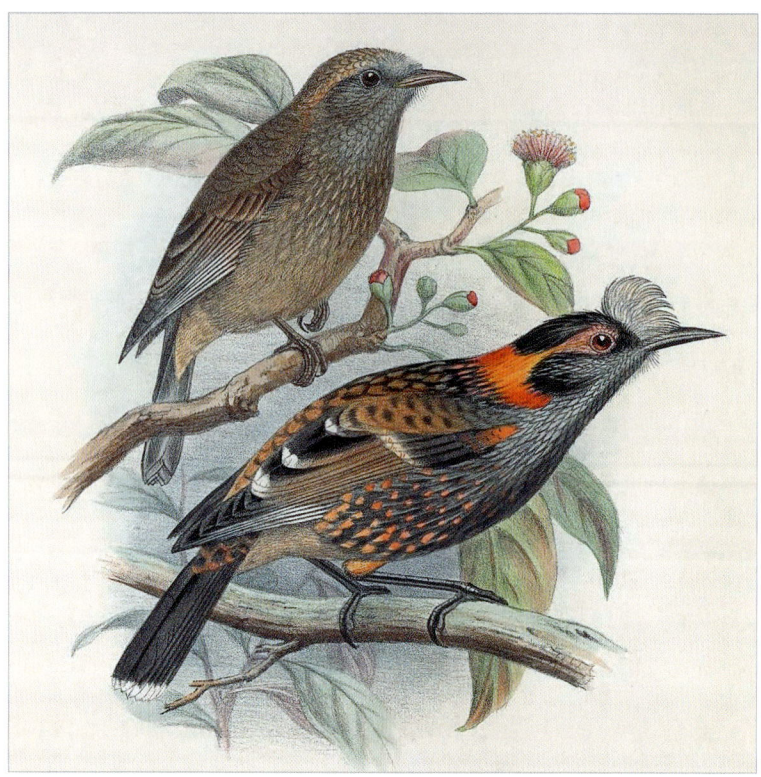

*Akohekohe from Wilson's* Aves Hawaiienses: the birds of the Sandwich Islands *(1890)*

# Akohekohe

This very unusual Hawaiian honeycreeper resides in a monotypic genus, *Palmeria*. It's said that its Hawaiian name most likely originated from a variation of its low guttural song, "AH-kohay-kohay." And in 1895 Perkins wrote:[242]

> The genus *Palmeria* contains but a single species (*P. dolii*), which inhabits the higher forests of both Molokai and Maui, especially the wetter portions, where fog and rain are of constant occurrence. On the latter island the natives call it "akohekohe," but on Molokai several of them gave it the name of "hoe," and by repetition of this word gave a very recognizable imitation of its song, showing thereby that they were well acquainted with the bird.

However, Perkins claimed that it is one of the "Names given from peculiarities of structure or plumage . . . Such names are often compounded with *a-* beak (literally jaw) e.g. Akekee, Amakihi, Akohekohe."[244] According to the *A Dictionary of the Hawaiian Language*, *kohe-kohe* is the name for a type of sedge grass, so the name would mean *a-* for "beak" and *kohe-kohe* for "grass," which certainly describes the brushy feather crest that curves forward over the bill.[246] Rothschild recounted the diary of Henry Palmer, who collected on Hawaii from 1891 to 1893, who spoke of the name but gave no hint as to its meaning:[37]

> I was talking to an old Kanaka, and he described to me a bird somewhat like that of a farmyard-cock. In Keanei they told me about a bird with a comb, but as I could get no definite information from them I thought it was a myth, but this native assures me it was often seen in Kipalmlu Valley years ago. He calls it "Akohekohe."

# Alauahio

Two species of Hawaiian honeycreepers in the genus *Paroreomyza* sport the name alauahio; one is now extinct and the other is endangered. Perkins wrote in the early 1900s:[244]

> The three most nearly resembling each other, *O. maculata* of Oahu, *O. newtoni* [no longer valid] of Maui, and *O. montana* of Lanai are called Alauwahio, or by the shorter name Alauwi or Lauwi, indiscriminately in each case.

The exact derivation of the name is unknown, given in Andrews's *A Dictionary of the Hawaiian Language* from 1865 as:[246]

> Alauwahio (ā-lǎ'u-wǎ-hǐ'o), n. A small yellow bird (*Oreomyza montana*) resembling the canary. Also known as alauhiio. See lauwi.

The name is thought to be echoic of the bird's pleasant song.

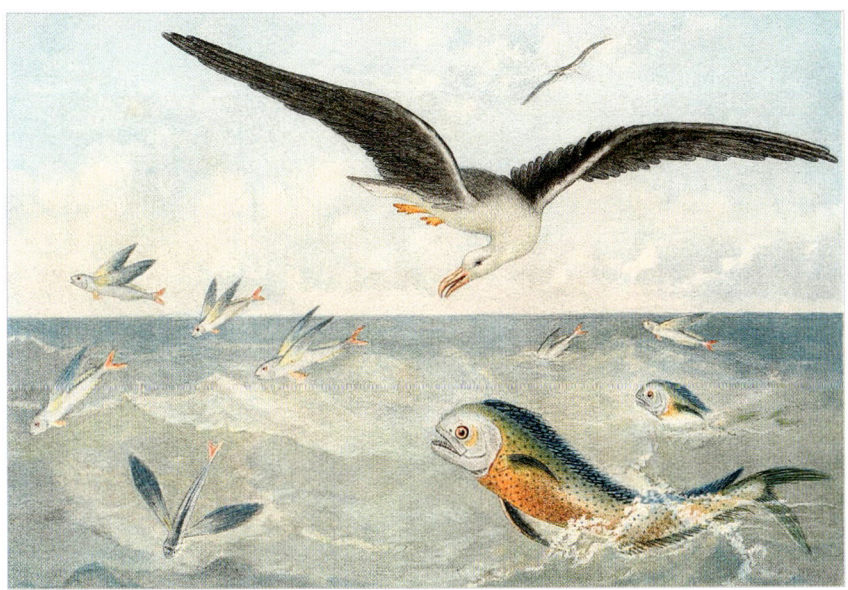

*Wandering Albatross in Alfred Frédol's (Moquin-Tandon's)* Le Monde de la Mer *(1866)*

# Albatross

The fifteen species in the Diomedeidae all go by the vernacular name albatross. The story of the word is a convoluted one. The word came into the English language from the Portuguese form *alcatraz* for the pelican. The Portuguese acquired their word from the Arabic الغَطَّاس *al-ġaṭṭās*, "the diver," also their name for the pelican. The Arabic name is an etymon of القَادُوس *al-qādūs*, "the bucket," which in turn is from the Ancient Greek κάδος *kádos*, meaning a "pail or jar," a reference to the pelican's voluminous gular pouch in which, as it was believed at the time, they carried water to their young. Newton wrote albatross is[7]

> a corruption of the Spanish and Portuguese *Alcatraz* or *Alcaduz* . . . The word is Arabic, *al-câdous*, adopted from the Greek κάδος, water-pot or bucket, and especially signifying the leathern bucket of an irrigating machine. Thence it was applied to the Pelican, from the resemblance of that bird's pouch, in which it was believed to carry water to its young in the wilderness.

The name, with various spellings such as *alcatraza, algatross* and *albitross*, was used fairly indiscriminately for all sorts of seabirds in the past, but over time came to be used for the members of the Diomedeidae.[7] It's thought the later modification of the spelling to albatross was perhaps influenced by the Latin word *albus,* feminine *alba,* meaning "white," in contrast to frigatebirds and other waterbirds, which are black.

# Alethe

There are six species of alethes, two in the genus *Alethe* and four in the genus *Chamaetylas,* all in the Old World flycatcher family, the Muscicapidae, and all found in sub-Saharan Africa. The etymology of this unusual common name is unclear. The genus was erected by Cassin in 1859, but the same year du Chaillu appeared to be surprised at the choice of name when he wrote in *Journal für Ornithologie*:[89]

> The new genus *Alethe* (!?), the somewhat mysterious etymology of which we have unfortunately not yet managed to find, is described by Cassin.

The word *alitheia* is the Greek for "truth or reality," and Alethe is presented as the Priestess of Isis in a novel, *The Epicurean,* by Thomas Moore, the seventeenth-century Irish poet. In the tale, apparently, Alethe was given "the most honourable of the minor ministries . . . to wait upon the sacred birds of the Moon, to feed them daily with eggs from the Nile, of which they were fond, and provide for their use purist water." She is the most beautiful of the priestesses and the hero of the story, Alciphron, endures an initiation into Christianity in order to stay with her. In the end, they are both persecuted and killed by the Romans. But Alethe's birds were Sacred Ibis, so why this moniker was attached to such dissimilar birds is unclear. (There is also the goddess of truth, Alithea, who was apparently Apollo's nurse; and Alethes, son of Hippotas and the king of Corinth.)

# Amakihi

Four species (one now extinct) of Hawaiian honeycreepers sport the name amakihi. Although Perkins includes the Amakihi in his group of "names given from peculiarities of structure or plumage . . . [which] are often compounded with *a-* beak (lit. jaw)," this doesn't account for the *ma* syllable.[244] The name is properly spelled 'Ama-kihi. The word *ama* means "talkative," while *kihi* means "curved or corner." It seems likely that these common (formerly, in some cases) and vocal birds were dubbed "talkative curve bills" by the original inhabitants of the islands.

# Amaui

This Hawaiian thrush, which was sadly the first of the genus to go extinct, got its name from the name 'Āmaui for all the thrushes from Maui, Molokai, Lanai, and Oahu that were considered by the Hawaiians to be one species. See Hawaiian Thrushes for more details.

# Anhinga

Although there are four species in the *Anhinga* genus, found on four different continents, only one bears the common name Anhinga. The name was first used by Marcgrave in the scientific literature in 1648 when he referred to the *Anhinga Brasiliensibus Tupinambæ,* with the meaning "the Anhinga of the Brazilian Tupinamba."[90] Garcia wrote that the etymology is[133]

> é nombre antiguo tupi, que equivale á *cabeza pequeña.* [an old Tupi name, which is equivalent to *small head.*]

The *añãgá*, *ajíŋa*, or *ayinga* was in local beliefs a jungle spirit or demon (the word is still used in present-day Brazilian-Portuguese, for example in the Brazilian municipality Itanhangá "itá + añãgá": devil's rock). No doubt the bird's hieratic stance with wings outstretched and snake-like appearance in the water contributed to the Tupi superstitions. As Wilson noted:[98]

> In those countries where noxious animals abound, we may readily conceive that the appearance of this bird, extending its slender neck through the foliage of a tree, would tend to startle the wary traveller, whose imagination had portrayed objects of danger lurking in every thicket.

Nash wrote that in the mid-1500s, Aspilcueta, the Spanish missionary[91]

> would go among them [the Tupi] adopting the technique of the Indian pagés, singing out the mysteries of the Roman faith, running round his auditors, stamping his feet, clapping his hands, making the easy substitution of Hell for Anhangá, copying the very tones and gestures of the medicine men by whom they were wont to be affected.

Over time the name changed to *anhinga* in the Tupi-Portuguese Língua Geral, the simplified version of the Tupi languages.

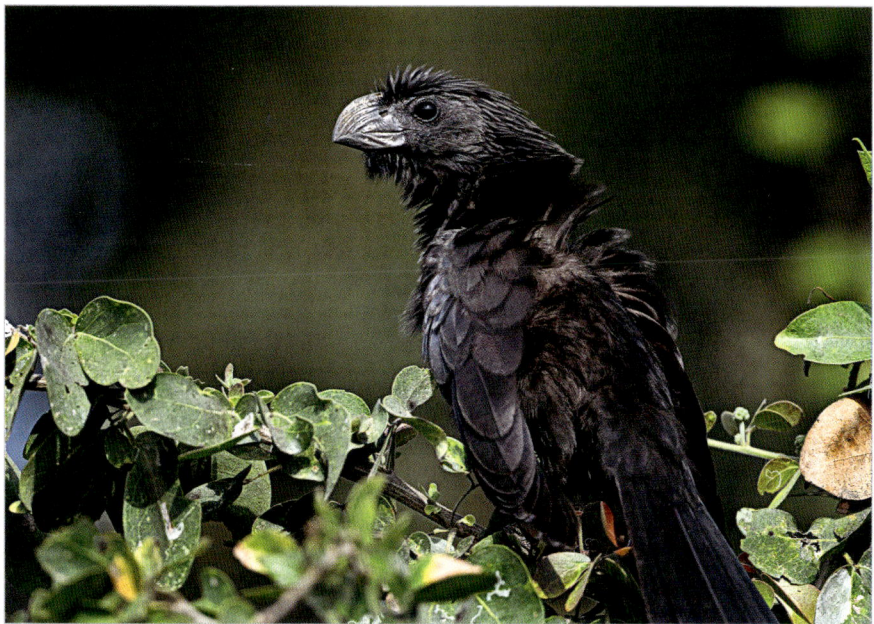

*Groove-billed Ani (Mexico)*

# Ani

The anis are three species of large, all-black cuckoos found in the Caribbean and Middle and South America. The name was coined in the early nineteenth century from the Spanish *ani* and Portuguese *anum*, which both came from the Tupi *a'nū*, the name given specifically to these three species in the genus *Crotophaga*.[25] According to Garcia in his *Nomes de Aves em Lingua Tupi (Names of Birds in the Tupi Language)* from 1913, it is not an onomatopoeic name, but rather it derives from the Tupi radical *anã* for "related, kin" followed by the suffix *úm* for "black or dark."[133] He wrote that:

The interpretation of Batista Caetano seems to me all the more true as the ornithologists say that this species is mainly characterized by their communal customs, living in bands and making large colonial nests, where the females lay the eggs jointly.

# Anianiau

Another Hawaiian honeycreeper with a unique name deriving from the Hawaiian language. Like many of the Hawaiian bird names beginning with *a-*, this bird is named for the shape of its beak. Perkins wrote in the early 1900s that the name[244]

is admirably adapted to the bird in question, A-nianiau meaning simply straight-beak.

# Antbird

Most members of the Thamnophilidae, the typical antbirds, bear compound names based on the ant theme. Ninety-seven, and counting, different species possess the antbird designation. The etymology is very straightforward—many, but not all, of these forest-dwelling understory birds regularly attend army-ant swarms feeding on fleeing insects as they go.

# Anteater-Chat

These two very closely related birds are members of the *Myrmecocichla* genus in the Muscicapidae from Africa. Both of them feed on all sorts of invertebrates, but mainly ants, which they forage for by hopping and jumping on the ground. (See Chat.)

# Antpecker

Three species of antpeckers in the Estrildidae occur in West and Central Africa. Most of the estrildids feed primarily on seeds, but the antpeckers, in the genus *Parmoptila*, as the name suggests, feed on ants, and sometimes other small invertebrates, by pecking them from stems or off the ground.

# Antpipit

The antpipits are two species of *Corythopis* in the Tyrannnidae, the tyrant flycatchers of the New World. As is often the case with compound bird names, they superficially resemble their Old World counterparts, the pipits (see entry). The first element of the name suggests their feeding habits, although studies have shown beetles constitute 40 percent of their diet, with ants only 20 percent.[1]

# Antpitta

The seventy antpittas (two in the Conopophagidae and sixty-eight in the Grallariidae) superficially resemble the pittas (see entry) of the Old World, thus explaining the latter part of the compound word. Surprisingly, they show no obvious predilection for eating ants but do favor a mixed diet of invertebrates.

# Antshrike

Fifty-three species belonging to fourteen genera in the Thamnophilidae bear a superficial resemblance to the Old World shrikes, with strong, hooked bills. As with so many birds, the early naturalists or settlers of English descent misnamed this group due to a similarity to species from Europe with which they were more familiar. They regularly join mixed feeding flocks of insectivores, and some are known to occasionally follow army-ant swarms.

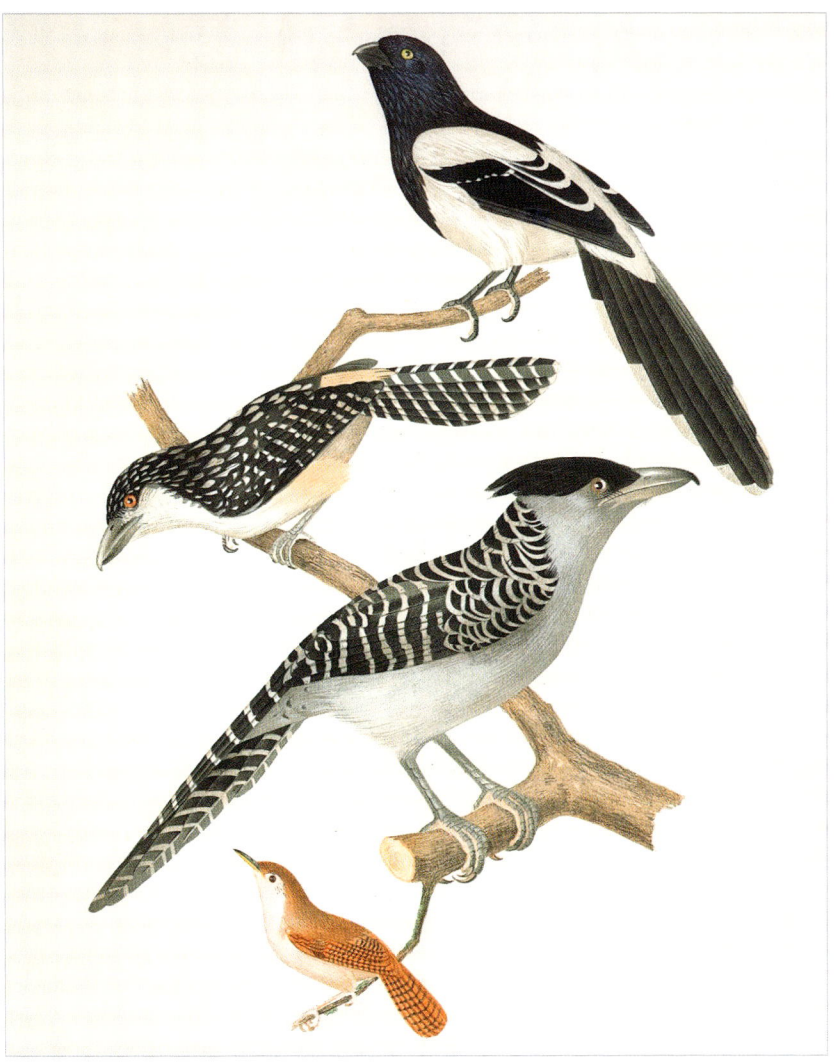

*Spot-backed Antshrike and Giant Antshrike (with Magpie Tanager and House Wren) in Descourtilz's* Ornithologie Brésilienne *(1854)*

## Antthrush

The antthrushes are a family, the Formicariidae, of twelve species in two genera from the Neotropics. They bear a (very) superficial resemblance to thrushes (family Turdidae), but, despite the moniker, they infrequently attend ant swarms and more typically forage on the forest floor, flipping leaf litter for invertebrate prey.

## Ant-Thrush

The capitalized word "Thrush" in the hyphenated name denotes that these are two species of true thrushes. They are members of the *Neocossyphus* genus in the Turdidae found in Central and West Africa. Although they will feed on all types of invertebrates, they attend army-ant swarms, dominating smaller bird species. One species in particular, the White-tailed Ant-Thrush, is rarely seen away from the swarms and is heavily dependent on them.

## Antvireo

Eight species of thamnophilids possess the name antvireo. As with all the members of the antbird family, they feed on ants, although, as foliage gleaners, a variety of invertebrates comprise the diet. Presumably, the early English-speaking naturalists who encountered these birds thought they looked a lot like vireos. (See Vireo.)

## Antwren

Sixty-one species belonging to nine genera in the Thamnophilidae, the typical antbirds, are somewhat wren-like but, of course, unrelated to the wrens of the Troglodytidae. They generally do not follow ant swarms, but feed by actively gleaning insects from foliage.[3]

## Apalis

There are twenty-six species of apalis of two genera, *Oreolais* and the synonymous *Apalis*, in the Cisticolidae. All occur in Africa. In 1838, Swainson erected the genus name *Apalis* from the Ancient Greek απαλος *hapalos*, meaning "delicate, tender, or gentle," an apt description of these small birds.[50]

## Apapane

This Hawaiian fringilid (finch) is the most abundant of the honeycreepers; it is a vocal gymnast with at least ten different songs and six calls. The name is the Hawaiian name for the bird; Perkins included it in his category of "Names derived from the nature of the sounds uttered by the bird," describing it as[244]

> a most untiring songster and its song though short is pleasing, but from constant repetition becomes wearisome. Its call note is a plaintive whistle.

## Apostlebird

The single species of Apostlebird, in the Corcoracidae, is endemic to Australia and is supposedly always found in groups of twelve, hence the reference to apostles, but of course, the numbers vary. The name was given to it by local settlers in the late 1800s. Robert Hall, who called it the "Grey Jumper" noted the popular name:[170]

> It is a noticeable feature in winter to see about a dozen together, from which the common name "Twelve Apostles," or Apostle-bird, has been derived.

## Aracari

Eleven species in the Ramphastidae, the toucans, are known as aracaris. All are in the genus *Pteroglossus*. The name was coined in the early nineteenth century via Portuguese from the Tupi *arasa'ri*. Marcgrave, who first introduced the name to European scholars in the mid-1600s, suggested that the name is echoic of the vocalizations, writing in Latin:[90]

> Avis haec quasi suum nomen profert, clamando acuto sono sed non admodum clangoro *Aracari*. [This brings us to the name of the bird, calling with a sharp tone and not a very loud *Aracari*.]

And Garcia wrote in 1913:[133]

> O nome *Araçari* por que tambem se designam essas aves, é onomatopaico do grito qne ellas soltam. [The name *Araçari*, as these birds are also called, is onomatopoeic of the cry that they release.]

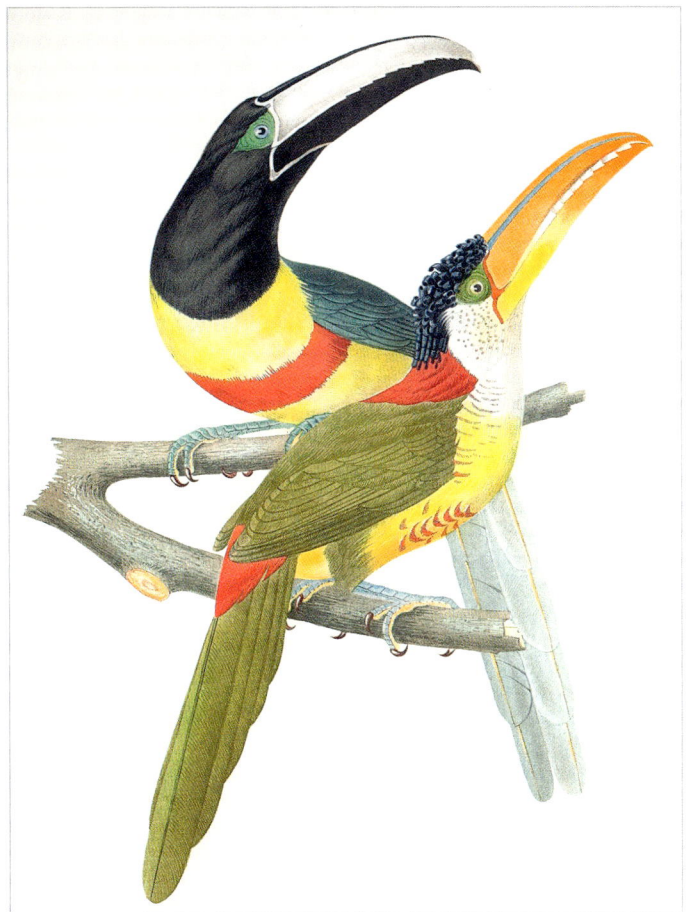

*Black-necked Aracari and Curl-crested Aracari in Descourtilz's* Ornithologie Brésilienne *(1854)*

WikiAves[247], a website aimed at the Brazilian community of birdwatchers, claims that the name means "little bright bird (like the day)"; the word *'ara* meaning "day" in Tupi.

## Argus

The three argus pheasants found in Southeast Asia have hundreds or thousands of ocelli, eye-like spots in their plumage pattern, and thus the name refers to the mythical hundred-eyed giant Argus Panoptes of Greek mythology, whose murder by Hermes so upset Hera that she transformed his eyes into the beautiful jewels of a peacock's tail. Although he had only ever seen a specimen of the bird, Edwards wrote an accurate description of the plumage in 1751:[78]

> What is most extraordinary in these feathers is that each of them has on the outer web, close adjoining to the shaft, a row of very distinct spots like eyes, so shaded as to appear imbost; they are larger and smaller as the feathers to the outer quills; they are from twelve to fifteen on each feather; the largest eyes are an inch diameter; they are incircled first with black, and without that with light brown, their shafts are white; the eyes, in the two or three innermost quills, are not so regularly marked, they lose their roundness and become confused. These beautiful eyes are not seen unless the wings are a little spread.

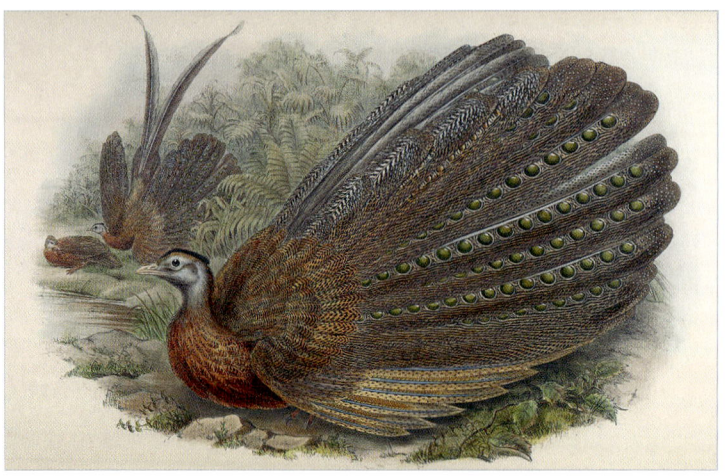

*Great Argus from Gould's* Birds of Asia *(1850)*

# Asity

The etymology of the name for these two species of Madagascar endemics in the Philepittidae lies in the local Malagasy name for the birds. In G. E. Shelley's discussion of the Velvet Asity in *The Birds of Africa, comprising all the species which occur in the Ethiopian region* (1900), he stated:[139]

> Owing to the numerous native dialects spoken in Madagascar, the present species is not only known as "Asity," but according to the Rev. J. Sibree as, "Variamanangana" in the Betsileo country, and as "Tsoitsoy" by the Betsimisaraka people.

He also applied the name to the bird he called the *Yellow-breasted "Asity,"* the present-day Schlegl's Asity. In his *New Malagasy-English Dictionary* from 1885, Richardson listed "Asi'ty: A bird allied to the ground thrushes [Pittae]" referring to the Philepittidae.[231] (See also Sunbird-Asity.)

# Astrapia

These remarkable birds are five species of birds-of-paradise from New Guinea in the synonymous genus *Astrapia*. The common name derives from the genus name, which is said to be from the Greek ἀστραπαῖος *astrapaios* for "lightning or flashing." The genus name was erected by Vieillot in 1816 without elaboration.[248] It is a reference to the male's iridescent plumage and long, flowing tail feathers. Vieillot may have been thinking of the ἀστραπίας *astrăpĭas*, the lightning stone, that is mentioned in Pliny as "a precious stone, black in color, with gleams of light crossing the middle of it," a description that would perfectly fit the iridescent velvety-black plumage of these spectacular birds-of-paradise.[470]

# Attila

The genus *Attila* is also the common name given to these seven members of the Tyrannidae of the New World. Lesson erected the genus in 1831, calling Bright-rumped Attila by the colloquial name "Le Tyran olive."[241] Although he gave no specific reason for his choice of the appellation, we can be fairly certain that the use of the *tyran(t)* and the name Attila are correlated. Attila the Hun was indeed a tyrant who terrorized large swathes of Europe in the fifth century. He was described in the fifth century by Priscus, the Greek historian, as being "short of stature with a broad chest and a large head . . . and a swarthy complexion," a description that could possibly fit for the gray or even the typical morphs of the Bright-rumped Attila.[471] It may seem odd that a New World bird should be named after an Old World despot, but the hegemony of Europe was overarching in the early 1800s, and many names were given to species from the study of specimens that had never been seen in the wild by the students of natural history whose world view 200 years ago was very different from our own.

## Auk and Auklet

Seven species of small Alcidae, the auks, murres, and puffins, get their names from the late-seventeenth-century Old Norse *alka*, and the Proto-Germanic *alkō*, usually claimed as probably originally imitative of a waterbird cry. The name *Alka* was first mentioned in the literature, possibly for the Razorbill, of Clusius in 1605 from correspondence he received from Høyer, the Norwegian physician who collected in the Faroe Islands in the 1600s, thus accounting for the Scandinavian origins.[249, 338] It's likely the name *Alka* was used in Scandinavia for many species of black and white alcids. The first use of the English name auk is in Ray and Willughby, where they used it to refer to a related bird, "the Razor-bill, *Auk* or *Murre. Alca Hoieri*":[16]

> *Hoiers Alka* and our *Awk* . . . hatcheth its young ones in holes and chinks of high Promontories. That Hoier was not mistaken in the name of this Bird I conclude, because it is called by the very same name, *viz. Auk*, in the North of *England*; so that it is manifest either our Northern men borrowed it of the *Ferroese*, or the *Ferroese* of them, it being very unlikely that by chance they should impose the same name upon it. But that ours borrowed this name of the *Ferroese* seems to me more probable because in other parts of *England*, farther distant from the *Ferroyer* Islands this Bird is called by other names.

Lockwood argued that the Old Norse *alka* had the original meaning of *neck*, in reference to the Razorbill's habit of bending its head over it back.[337]

> Such characteristic behavior has played a part in the nomenclature, witness one of the Faroese names *nakkalanga* lit. "long neck." Indeed, the very name we are dealing with appears to be no more than a secondary application of an unrecorded sense of ON *alka*, namely "neck," preserved in the Icelandic idiom *teygja álkuna* "crane the neck."

Most authorities don't appear to have accepted this explanation, a fact that Lockwood referred to:

> The etymologists have been rather cavalier in their treatment of this item, and the circumstances of its transmission into English and subsequent fate await clarification.

The addition of the diminutive suffix *-let*, from the Latin diminutive *-ettus*, is applied to smaller auks.

## Avadavat

There are two avadavats, small estrildid finches found in Asia. The name is a variant of the earlier *amadavat*, an old spelling of Ahmedabad, a city in Gujarat, India, from which the bird was imported to Europe. Another old name for the bird was Amandava. In *Hobson-Jobson: being a glossary of Anglo-Indian colloquial words and phrases and of kindred terms etymological, historical, geographical and discursive* by Henry Yule in 1886 an entry reads:[250]

**Avadavat**: Improperly for Amadavat. The name given to a certain pretty little cage-bird (*Estrelda amandava*, L. or "Red Wax-Bill") found throughout India, but originally brought to Europe from Ahmadābād in Guzerat, of which the name is a corruption.

One of the earliest references to the bird is in Fryer's *New Account of East India and Persia* from 1673:[331]

> From Amidavad, small Birds, who, besides that they are spotted with white and Red no bigger than Measles, the principal Chorister beginning, the rest in Consort, Fifty in a Cage, make an admirable Chorus.

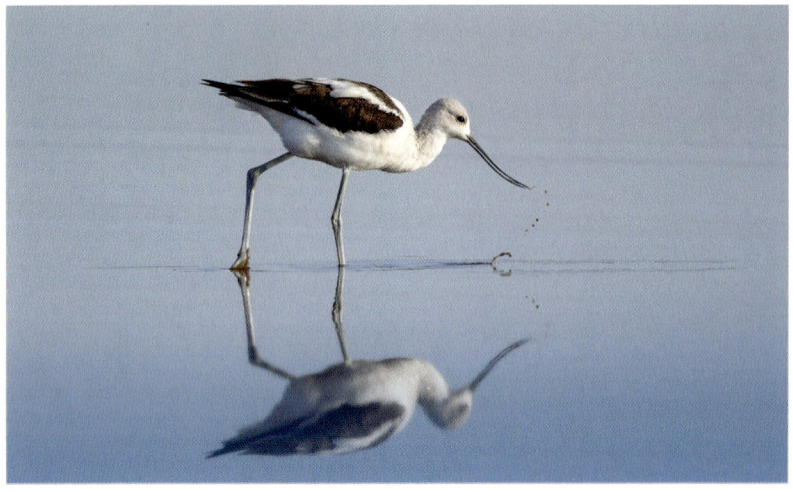

*American Avocet (Arizona, USA)*

# Avocet

The common name was originally assigned to the Pied Avocet and is said to derive from the Venetian name *avosetta*. It appeared first in the Italian naturalist Aldrovandi's *Ornithologiae* in 1637 under the heading "Avosetta Italis Dicta."[101] Ray and Willughby referred to "The Avosetta of the Italians" in 1678,[16] while Pennant popularized the name in English in *British Zoology* in 1776.[177] The etymology of the Italian word is unrecorded, but it seems likely that it came from the Latin *avis* for "bird" and the diminutive *-etta*. Newton stated that the word comes "from the Ferrarese *Avosetta*" and goes on to say:[7]

> This word is considered to be derived from the Latin *avis*—the termination expressing a diminutive of a graceful or delicate kind, as *donnetta* from *donna* . . . but it is spelt *Avocetta* by Prof. Giglioli [in Avifauna Italica].

# Avocetbill

One species of hummingbird, the Mountain Avocetbill, in the monotypic genus *Opisthoprora* carries this name. It is a clear reference to the shape of the bill, which is sharply upturned at the tip, in a very similar shape to that of an avocet's. In every other way, it bears no resemblance whatsoever to an avocet. Wood (1862) stated in *The Illustrated Natural History*:[85]

> This singular species is remarkable for the curious manner in which the bill is curved upwards at the extremity, after running nearly straight for the greater part of its length As this formation of beak bears some resemblance to that which is found in

the well-known Avocet, the present species has been named the Avocet Humming-bird. When the first specimen of this bird was brought to Europe, the peculiar shape of the beak was thought to be accidental, and owing to pressure against the side of the box in which the bird had been packed; but it is now clear that the structure is intentional, and that in all probability, it subserves some very important purpose. Some persons have suggested, with some show of reason, that the beak is recurved in order to enable the bird to feed upon the nectar and insects which reside in the deepest recesses of certain tubular flowers.

# Awlbill

The single species of awlbill, Fiery-tailed Awlbill in the genus *Avocettula*, is a hummingbird with a very unusually shaped bill. The end of the bill is sharply upturned in a shape resembling the tip of a stitching awl, a tool used to puncture holes when sewing heavy and thick materials. Gould noted the somewhat similar shapes of the bills of this species and the Avocetbill, calling the bird now known as an awlbill, the "Fiery-tailed Avocet":[6]

> *Avocettula* and *Avocettinus* are the generic terms applied to the two species rendered remarkable by the points of the mandibles being curved upwards in the shape of a hook: this extraordinary deviation from the usual structure is doubtless designed for some especial purpose; but what that may be, is at present unknown to us.

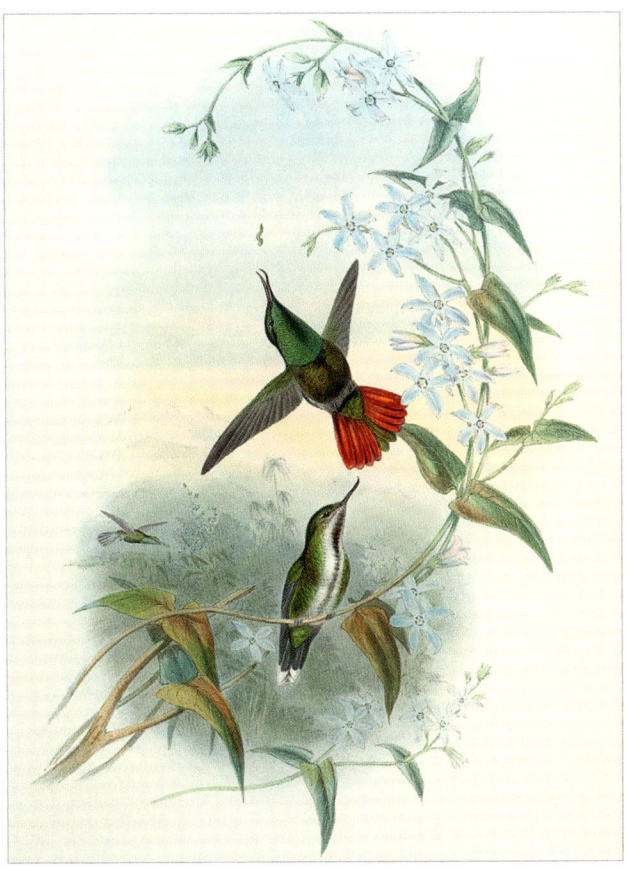

*Fiery-tailed Awlbill from Gould's* Monograph of the Trochilidæ *(1849)*

# B IS FOR BABAX

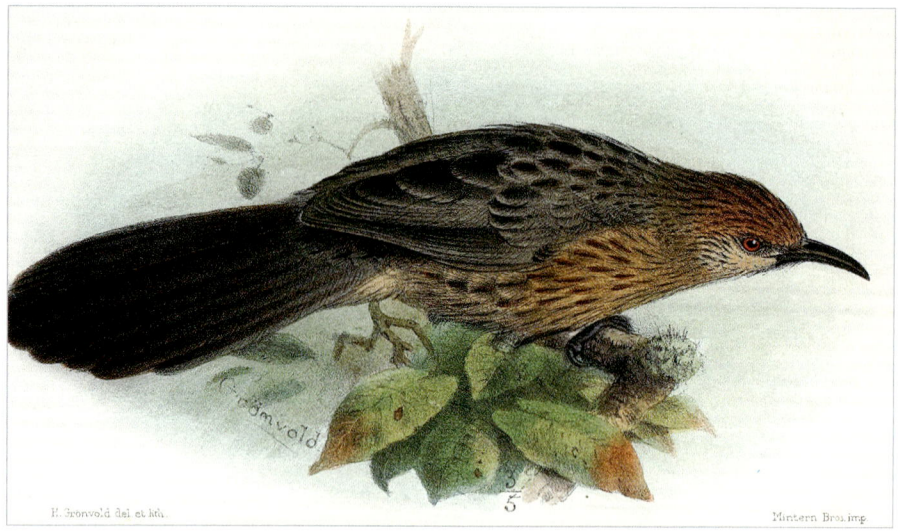

*Giant Babax in Dresser's* Descriptions of Three new Species of Birds obtained during the recent Expedition to Lhassa *(1905)*

## Babax

Four species of the Leiothrichidae, the laughingthrushes and allies, were previously in the genus *Babax*, which was erected by Père Armand David in 1875. In his account of the Chinese Babax he wrote:[92]

> And all year round, this bird is very talkative and makes its strange notes heard about everything, which sometimes are very sweet and sometimes seem to express anger. The new generic name we are proposing for this species, *Babax*, a synonym of *Garrulax*, alludes to this endless babble.

The name is derived from the Ancient Greek onomatopoeic word βαβάζω *babázō*, "to speak inarticulately," with related words such as βάβιον *bábion*, "child," and βάβακοι *bábakoi*, "cicadas," and clearly refers to the bird's garrulous, varied vocalizations.

## Babbler

It's thought that the name was first used in ornithology in 1837, but it derives from a word used from the sixteenth century to describe a foolish, chattering person. The word "babble" comes from the mid-thirteenth-century word *babeln*—to prattle and utter sounds indistinctly and talk like a baby, so probably originally imitative of baby talk. Certainly many of the birds given the epithet are vocal and active, as well as often very social, hence the designation. Most species boasting the name were confined to the Timaliidae, described in Newton as a family that[7]

> no systematist has yet been able to define satisfactorily, while many have not unjustly regarded it as a "refuge for the destitute."

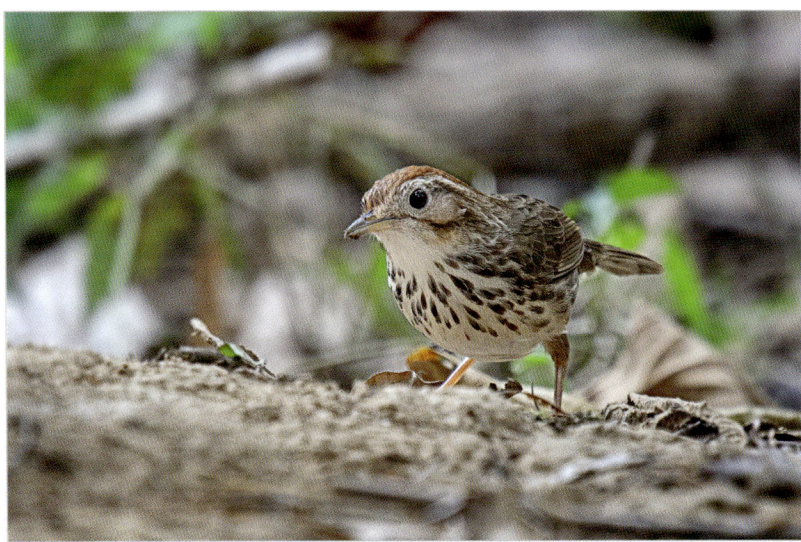

*Puff-throated Babbler (Vietnam)*

Indeed, he and others have been proven to be correct as the many babblers now spread across several different families. The name is used in many hyphenated forms, as well. These are striped-babbler, shrike-babbler, jewel-babbler, rail-babbler, pygmy-babbler, tit-babbler, wren-babbler, scimitar-babbler, pied-babbler, mountain-babbler, and thrush-babbler.

## Bamboo-Partridge

Three Asian members of the Phasianidae in the genus *Bambusicola* are called bamboo-partridge. Gould erected the genus in 1863 from the scientific name for "bamboo," *Bambusa*, and the *-cola*, "dweller":[33]

> The predilection the bird . . . evinces for bamboo forests suggested the term *Bambusicola*.

He also appears to have been the first to use the name "Bamboo Partridge," although Swinhoe may have foreshadowed it with the name "Bamboo-fowl."[226]

## Bamboowren

The Spotted Bamboowren is not a wren, but rather a member of the Rhinocryptidae, the tapaculos. Or at least, it now is—in the past, it has been placed in the Thamnophilidae, the antbirds; the Troglodytidae, the wrens; and the Polioptilidae, the gnatcatchers. With its cryptic coloring and cocked tail, it certainly looks like a wren. It favors bamboo stands and very dense vegetation, in which it skulks. The name appears to have first been used as late as the 1990s.

## Bananaquit

A small, nectarivorous songbird in the New World, the name comes from banana and *quit*, used for several species of small passerines dating back to the 1800s. The word *quit* is used in several other bird names (e.g., Grassquit, Orangequit) and derives from a Caribbean word applied to small birds. *Quit-quit* was the small bird in the "Anansi" Jamaican folk stories that originated in West Africa and were transmitted to the Caribbean by way of the transatlantic slave trade. It is onomatopoeic, from the call note described as *guit-guit*.[404] Newton wrote:[7]

> Quit [is] a name applied in Jamaica, and perhaps some others of the British Antilles, to several very different kinds of birds, probably from the note they utter.

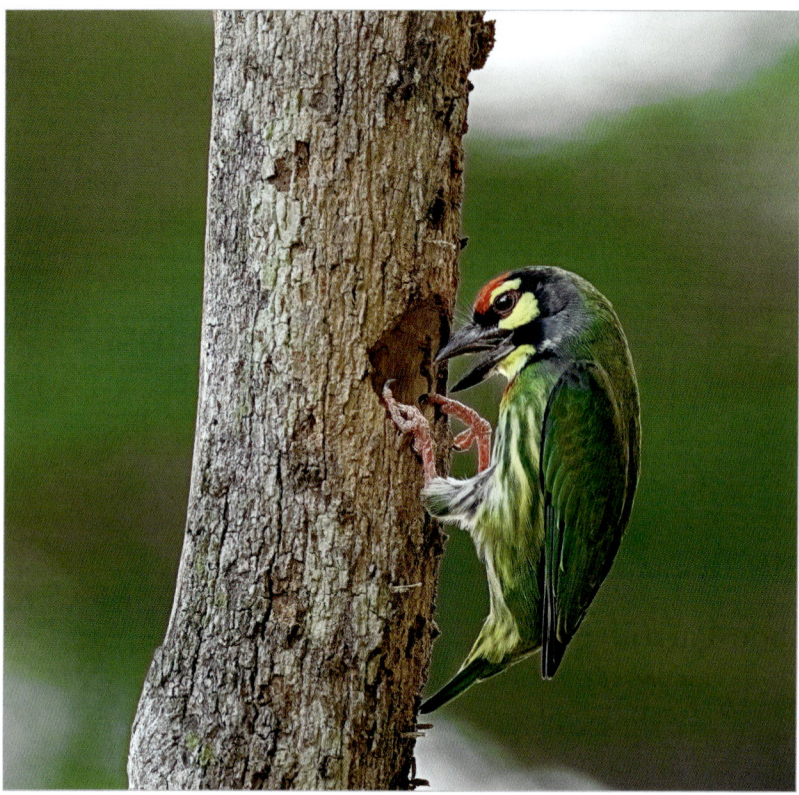

*Coppersmith Barbet (Malaysia)*

# Barbet

The barbets are birds of three families (previously one) found on three continents: Asia, Africa, and South America. Across the three families, all share the characteristics of short, thick bills; small, ovoid bodies; and rictal bristles. These canopy-dwelling fruit eaters derive their name from the French *barbe* meaning "beard; long hair of certain animals," which references the group's distinctive rictal bristles. Buffon justified the common name in 1780:[8]

> The Naturalists have given the name of barbet to several birds which have the base of the beak trimmed with slender feathers, long, stiff as bristles & all directed forward.

# Barbtail

This is a group of four birds in the Furnariidae, the ovenbirds and woodcreepers, of Central and South America. The name is descriptive—all have tails that are graduated with the central rectrices slightly stiffened, with the distal 3–6 mm of shafts lacking barbs, giving the appearance of spines. The tails are used for support as the birds forage by creeping along trunks and branches.

# Barbthroat

The barbthroats are three species of hummingbirds, all in the genus *Threnetes*. The name is descriptive, referring to the spiky feathers on the chin and throat. Gould was the first to give this group a common English name, calling them "Barbed-throats" in 1849.[6]

# Bare-eye

As the common generic name suggests, the three members of the Thamnophilidae in the genus *Phlegopsis* all have bare skin around the eyes. Bare orbital patches in birds can serve a number of functions, such as sexual communication for advertising status or quality, group signaling, thermoregulation, or the prevention of soiling of the feathers in species that feed on fruit or carcasses. In the case of these understory denizens of the Neotropical rainforests, the function is debated but is most likely associated with communication.

# Barwing

There are seven species of medium-sized Asian "babblers," all in the genus *Actinodura*, in the diverse Leiothrichid family. The name is descriptive; all have distinctive black barring on a chestnut background on the wings.

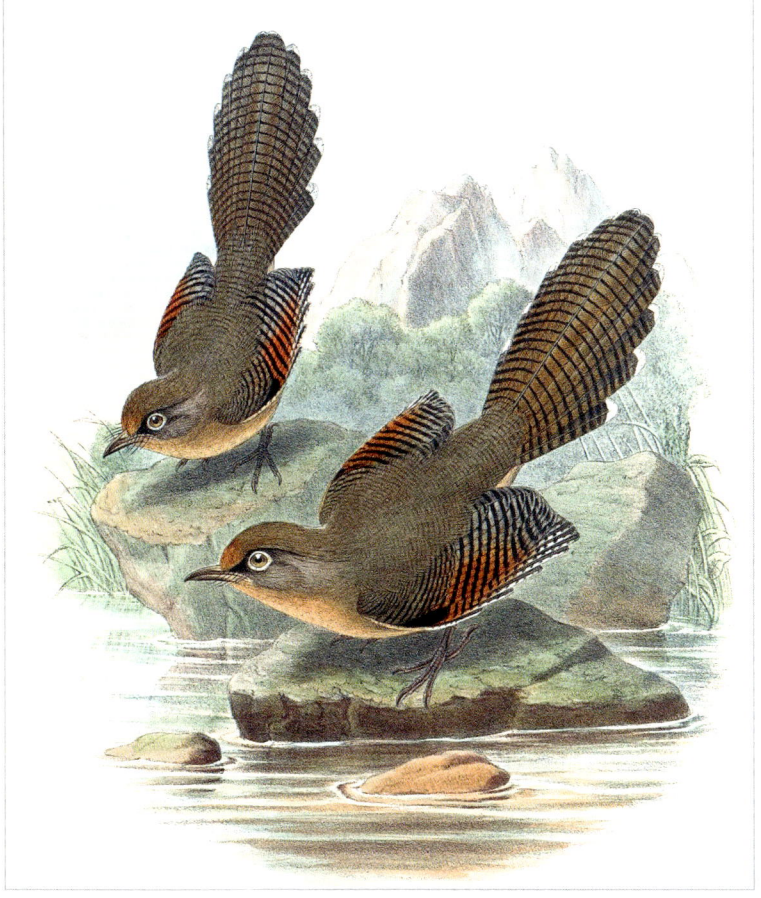

*Spectacled Barwing from Gould's* Birds of Asia *(1850)*

# Bateleur

The Bateleur is a striking and large raptor, widespread throughout the African continent. The unusual name, originally assigned by legendary eighteenth-century naturalist and explorer Le Vaillant, is borrowed from the French word *bateleur* meaning "juggler" or "acrobat" due to the bird's habit of rocking and tilting in gliding flight and from its habit of performing aerial somersaults. He described the bird in 1799:[93]

> The Bateleur soars in circles, and lets escape from time to time two very raucous sounds, one of which is waxing an octave higher than the other; often it suddenly folds down its flight, and descends to a certain distance, beating the air of his wings, so that one would think that he is injured and will fall to the ground. His female never fails to repeat the same game. We can hear these wingbeats at a very great distance; I cannot better compare the noise which results from it, and which is only a rustling in the air, than that made by a sail, one of the corners of which has come loose, and which a great wind blows violently. . . . I got the name of this bird from its way of playing in the air: one might say, in fact, a juggler who performs feats of strength to amuse the spectators.

# Batis

Small songbirds with a sub-Saharan distribution in Africa, the batises comprise nineteen species in the Platysteiridae. All are fairly similar in appearance, with predominantly black and white plumage and distinctive yellow to gold irides. The common name is also the genus name and derives from the Ancient Greek βατίς *batis* or *batidos*, an unidentified worm-eating bird mentioned by Aristotle:[109]

> Others feed on grubs, such as the chaffinch, the sparrow, the "batis," the green linnet, and the titmouse.

In the *Thesaurus linguæ Romanæ & Britannicæ*, printed in 1578, an entry reads:[108]

> *Batis, Plin. [Pliny].* . . . A little bird, also much like the bunting.

At the time, the use of the word was probably in reference to a variety of buntings, warblers, and chats or possibly the European Stonechat. The nineteenth-century German naturalist Boie, who coined the name for the genus in 1833, possibly thought that the description of the calls of the various species recalled those of the Stonechat.

# Bay-Owl

The genus *Phodilus*, containing three species of bay-owls, is characterized by the birds' unusually shaped facial disc. One of the earliest accounts of the Oriental Bay Owl is in Latham who confirmed the etymology of the name:[17]

> Plumage bay, spotted with black, paler beneath; front of the head, and chin whitish, variegated with bay; legs covered with down, colour pale chestnut.

Not related to the name but an amusing anecdote was related by Temminck:[94]

> This Owl is little known in Java, its favorite home away from homes and villages is always in the interior of the most dense forests, of which it rarely abandons the protective shade. It is believed that she prefers the den of the Royal Tiger to any other dwelling, and the people claim that she approaches this animal with impunity, in the same manner, says Mr. Horsfield, that the [myna] will pose without fear on

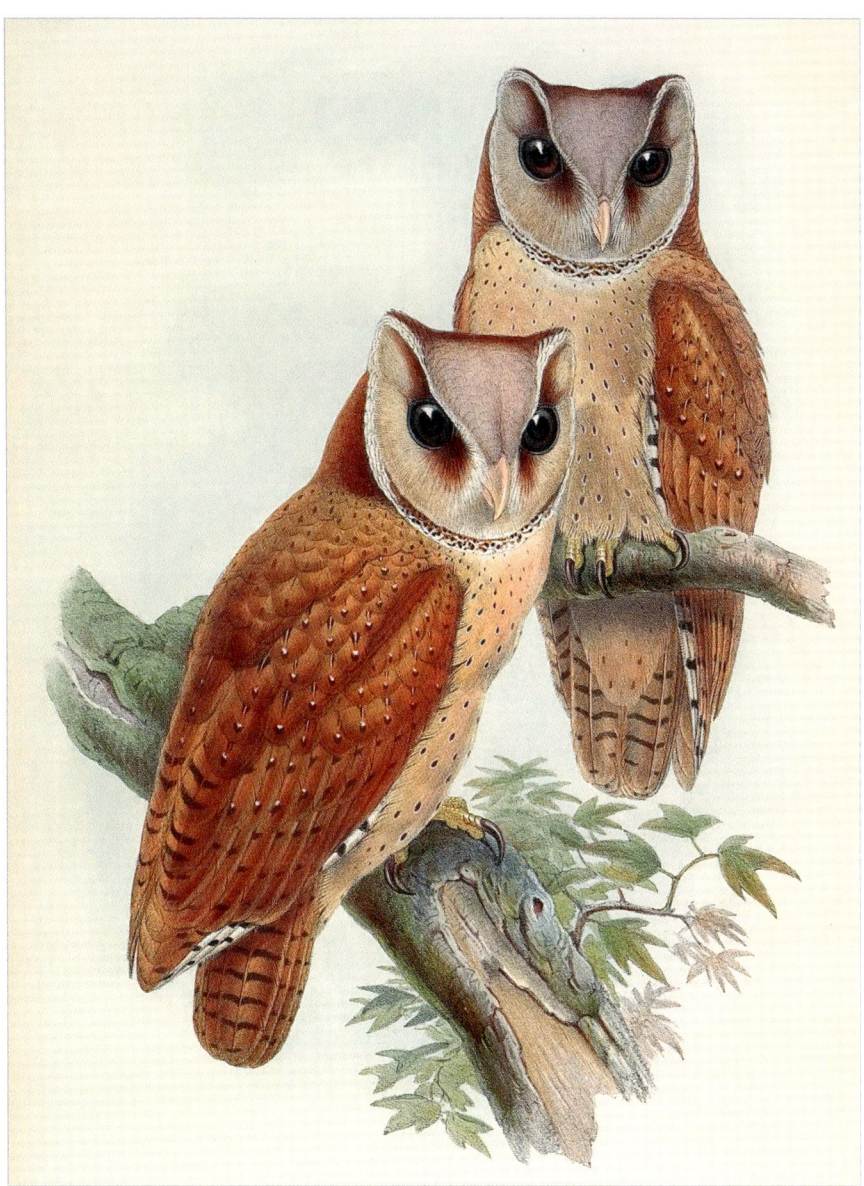

*Oriental Bay Owl from Gould's* Birds of Asia *(1850)*

the backs of the Oxen: our Owl would have no distrust of the Tiger, and would rest on its back. This popular opinion needs to be confirmed by observations so that it can be credited.

# Baywing

The baywings are two species of medium-sized icterids (troupials and allies) found in Brazil, Bolivia, Paraguay, Uraguay, and Argentina. The name is descriptive, as both species have rufous to chestnut wings. This name was formerly used for Vesper Sparrow in North America.[405]

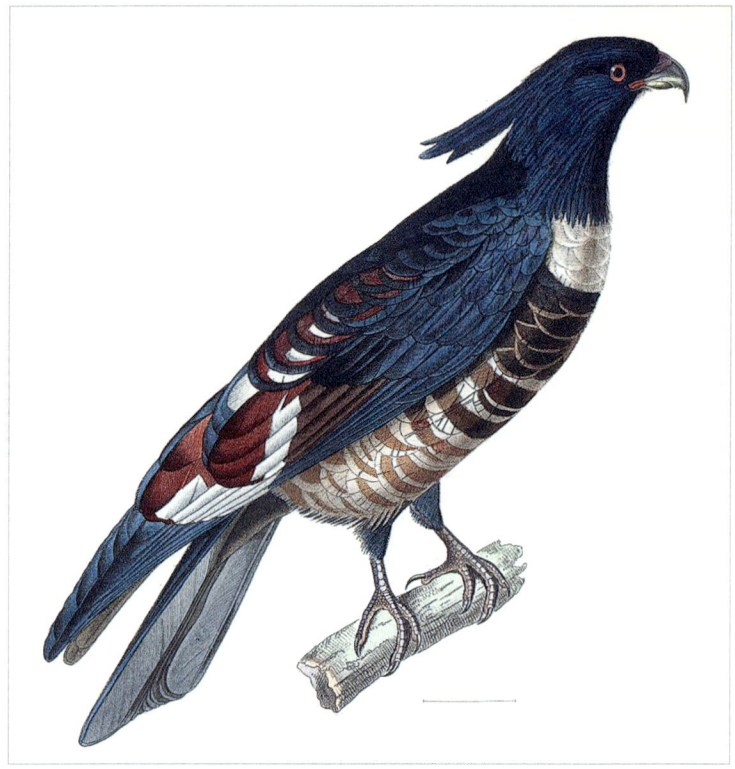

*Black Baza in Temminck's* Nouveau recueil de planches coloriées d'oiseaux *(1838)*

# Baza

A small genus, the *Aviceda* is three species of medium-sized, crested hawks found from India and China, through Southeast Asia to New Guinea and Australia. The name derives from the Hindi *baaz*, meaning "hawk." The origin is, however, from the Farsi word باز *bāz*. The name was originally erected as a genus without explanation by Hodgson in the 1836 issue of *The Journal of the Asiatic Society of Bengal*.[96] (See Besra.)

# Becard

The genus *Pachyramphus*, all known colloquially as becards, comprises sixteen species in the Tityridae, the tityras and allies. The name comes from the French *bécarde*, from *bec* meaning "beak," and the suffix *-arde*, a marker of "belonging." Although originally referring to Black-tailed Tityra, Buffon created the name in the eighteenth century in his *Histoire Naturelle des Oiseaux* in homage to the broad, slightly hooked bills:[8]

> Ainsi nommées à cause de leur gros & long bec rouge; ont le corps plus épais que nos pîe-grieschies; celles envoyées de Cayenne sous les noms de pie-grieche grise & de pie-grîesche tachetée, paroissent être le mâle & la femelle; notre bécarde à ventre jaune, est la pie-grîesche jaune de Cayenne. [So named because of their large & long red beaks; they have a thicker body than our shrikes; those sent from Cayenne under the names of gray shrike & spotted shrike, appear to be the male & female; our yellow-bellied becard is the yellow shrike of the Cayenne.]

The genus name similarly means broad bill.

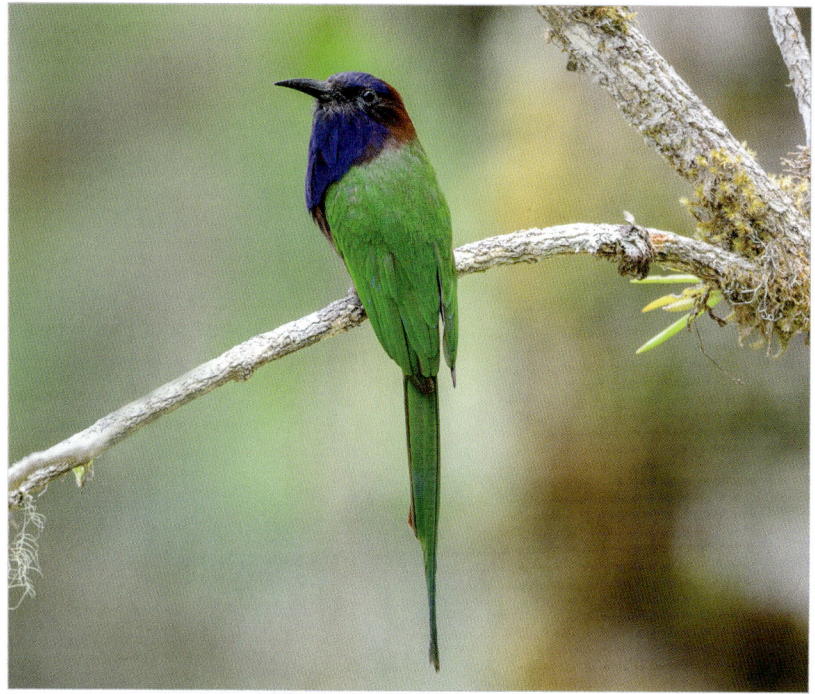

*Purple-bearded Bee-eater (Sulawesi, Indonesia)*

# Bee-eater

The Meropidae are a family of twenty-eight species found throughout much of Europe, Africa, Asia, and Australia. All bear the common name bee-eater despite belonging to three different genera. The name derives from their habits—they are sit-and-wait predators and are literally eaters of bees, favoring bees and wasps.[3] The name is first used in Charleton's *Onomasticon Zoicon* in 1668, which is written in Latin with the English "Bee-eater" a direct translation of *Apiaster* in the heading.[339] Ray and Willughby wrote in 1678:[16]

> Flying in the air it catches and preys upon Bees, as Swallows do upon flies. It flies not singly, but in flocks, and especially by the sides of those Mountains where the true Thyme grows. Its Voice is heard afar off, almost like to the whistling of a man.

# Bellbird

A number of unrelated species are endowed with the name bellbird: four species of cotingas in Central and South America, two species of honeyeaters in the New Zealand genus *Anthornis*, and three species of Oreoicidae (Australo-Papuan bellbirds) in New Guinea and Australia. In all cases, the name is descriptive of the bell-like vocalizations. Also note that throughout much of Australia the Bell Miner, an aggressive honeyeater, is known informally as "the Bellbird" due to its tinkling song. Latham described the Bearded Bellbird, a cotinga, in 1821:[17]

> This species inhabits Brasil, called there *Araponga*; and has a loud voice, which may be heard a great way off . . . : this cry is of two kinds, one like that of a hammer, striking on a wedge; the other similar to the noise of a cracked bell . . . Hence called by the English, the *Bell Bird*.

# Bentbill

Two species of unusual Tyrannidae belong to the genus *Oncostoma*, derived from the Ancient Greek ὄγκος *onkos* and στόμα *stoma,* which translates as "bulky mouth." This is probably a better description of the unusual shape of the bill than the common name conveys. Nevertheless, their name references the bird's short, broad, and distinctively downcurved bill with a notably thick upper mandible. An older name from the early 1900s for the bentbills was "bent-billed flycatcher."

# Bernieria

Alphonse Charles Joseph Bernier was a French medical doctor and naval surgeon who collected extensively in Madagascar in the early to mid-1800s. One of the birds he collected was a Malagasy Warbler in the Bernieridae now known as Long-billed Bernieria. Pucheran described the species from the specimen in 1855, naming the genus *Bernieria* in the doctor's honor.

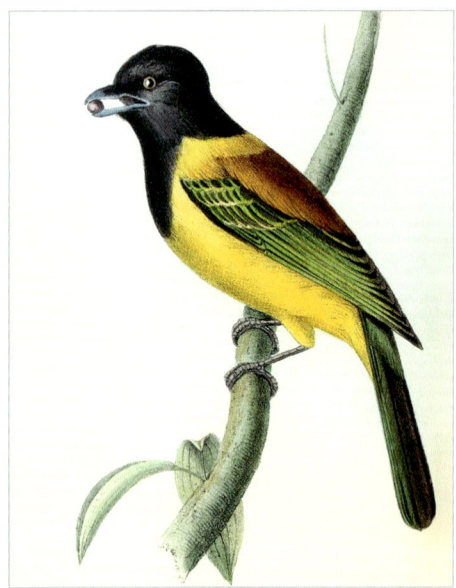

*Hooded Berryeater in Swainson's* Zoological Illustrations *(1820)*

# Berryeater

Two rather similar-looking species of cotingas in the genus *Carpornis* are known, somewhat unimaginatively, as berryeaters, due to their penchant for fruit.

# Berryhunter

The single species of berryhunter is a drab, poorly known bird from the mountains of New Guinea belonging to a monotypic family, the Rhagologidae. What is known is that it likes berries, hence the name, which is, of course, descriptive.

# Berrypecker

The common name berrypecker is assigned to two species in the Paramythiidae and seven species of Melanocharitidae, all from New Guinea. All eight species feed predominantly, if not entirely, on fruit, hence the descriptive name.

# Besra

This sleek avivore is a relatively small *Accipiter*. Its unique name comes from the Hindi word *Basra*, which is the diminutive of *Baaz*, meaning "hawk" or "goshawk," as it does in Arabic also (باز baz). The name in Hindi is used only for the female bird; the males are known as *Dhotee* or *Dharti*. Blyth explained the names in 1849:[283]

> *Básra* (diminutive of *Báz*, "Goshawk"), and the male—*Dharti* ("a handful," or "held in the hand"). Hindi.

Vikram Jit Singh wrote in the *Hindustan Times*:[102]

> The choice of English name [used for both sexes] underlines the female's precedence over the male because in raptors or birds of prey, the females are larger and pack more power than males.

(See Baza.)

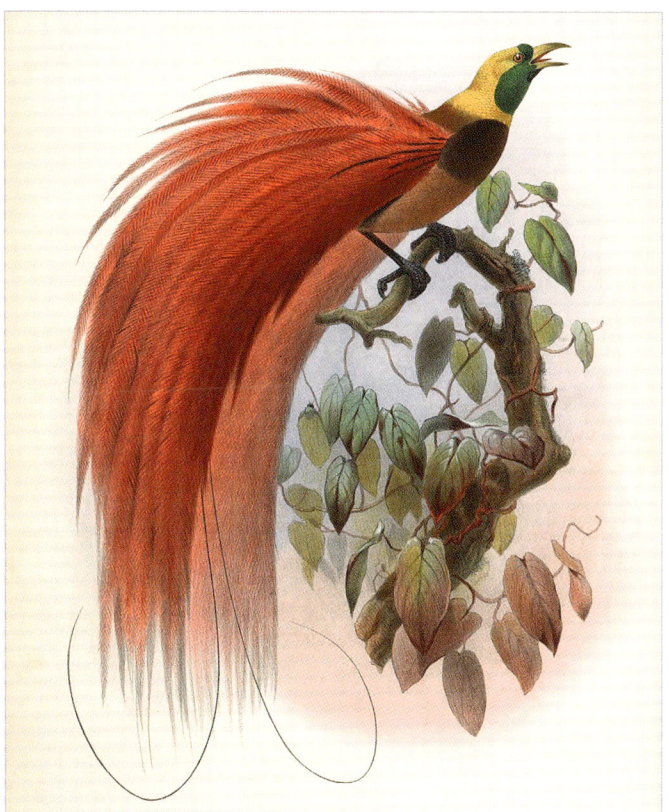

*Raggiana Bird-of-Paradise from Elliot's* A Monograph of the Paradiseidae *(1873)*

# Bird-of-Paradise

The birds-of-paradise, found in eastern Indonesia, New Guinea, and Australia, are among the most legendary and mysterious of birds. And the story of how they got their common name is no less intriguing. The first Europeans to encounter birds-of-paradise acquired their skins during the voyage of Ferdinand Magellan on his circumnavigation of the Earth. The Italian scholar Antonio Pigafetta, who was on the expedition, wrote:

The people told us that those birds came from the terrestrial paradise, and they call them *bolon diuata*, that is to say, "birds of God."

The voyagers brought the specimens back to their home countries in the early sixteenth century. Unknown to the explorers, however, the skins had been prepared by native traders who removed the wings and feet in order to use them as decorations. As a result, the Europeans came to believe that these exotic birds were without feet and never landed, some sort of mythical phoenix or *Garuda*. It was even believed that the females laid their eggs on the backs of the males.

In the 1678 publication *The Ornithology of Francis Willughby of Middleton in the county of Warwick, esq.* we find that:[16]

Birds of Paradise, so called, should want feet, as was not long since generally believed, not only by the Vulgar, but also by the greatest Naturalists themselves, even such as were most conversant in the History of Animals, because those brought out of the Indies were wont to be mutilated and bereaved of their feet, is now sufficiently convinced to be false by the testimony of eye-witnesses, and by the Birds themselves brought over intire, so that no man in his right wits can any longer doubt of that matter.

## Bishop

There are sixteen species of bishops, all in the genus *Euplectes*. They are a type of weaver belonging to the Ploceidae, the weavers and allies. These small, stocky seedeaters with their stout, conical bills all sport a variation on the theme of red, orange, or yellow on black plumage with ruff-like head feathering. Originally referred to as either cardinals or grosbeaks, the bishops were named for an analogy with the colors of the robes of bishops in the Catholic church. In 1794, Hayes wrote about the Southern Red Bishop under the heading of Grenadier Grosbeak:[148]

As I had an opportunity of making drawings from, and examining a cage of these birds which were sent from Lisbon as a present to the late Earl of Sandwich, when at the head of the admiralty; and although there were several of them, there were scarcely two alike: in a letter which accompanied them, they were called the Portugal Bishop.

Later, the vernacular name was sometimes used for a number of species of New World tanagers. Newton wrote that the names Bishop-bird or Bishop-Tanager were[17]

Latham's rendering of the French *l'Évêque* [the bishop], by which a species inhabiting Louisiana was, according to Dupratz, originally called . . . Dupratz's bird was probably the *Spiza cyanea* of modern ornithology, the Indigo-bird or Indigo-Bunting of the English in North America but Buffon confounded it with his *Organiste* [euphonias] of Santo Domingo, a very different species . . . while Brisson had already applied the French name (*l'Evesque*, as he wrote it) to a third species from Brazil, which subsequently became the *Tanagra episcopus* of Linnaeus, and this seems to be the only one now known (and that to few but "fanciers") as the "Bishop-Bird" or "Bishop-TANAGER"—the colour of its plumage suggesting, as in the original case, the appellation. Audubon, himself a Louisianian, makes no mention of the name "Bishop-Bird"; but says that it was known to his countrymen as the *Petit Papebleu* [little blue pope]. He adds that the first settlers called all the Buntings, Finches and "Orioles" *Papes* [popes].

While the birds now known as bishops are predominantly red and black, the birds discussed by Newton are either blue, purple, or blue and yellow, as in the case of the euphonias. Both colors are traditionally worn by bishops of the Catholic church—French canons of the cathedral wore red to distinguish themselves from other clergymen, while the use of purple attire by bishops was originally intended to show that they possessed religious and spiritual authority equal to the authority of princes and kings, who used it as a symbol of their status since it was among the most difficult of colors to produce.

# Bittern

There are many species of bitterns within the Ardeidae, all secretive denizens of reed beds and marshes. The name is based on the Latin *butio*, "bittern," and *taurus*, " bull" (because of its call). It then traveled into Old French *butor*, and then to the late Middle English *bitore*, appearing in Chaucer's "The Wife of Bath's Tale" in 1386:

> Til she came there, hir herte was a-fyre,
>
> And, as a bitore bombleth in the myre.
>
> [And then went there, her heart afire,
>
> And, as a bittern bumbled in the mire.]

The -*n* was added in the sixteenth century, perhaps by association with *hern*, which is an obsolete variant of heron. Ray and Willughby mentioned the bittern under the heading "The Bittour or Bittern or Mire-drum":[16]

> It is called by later Writers, *Butorius* and *Botaurus*, because it seems to imitate boatum tauri, the bellowing of a Bull. The Author of Philomela calls it *Butio*: But his mistakes are so many, that no account is to be made of his authority.

The local name "mire-drum," mentioned by Ray, displays similarities and is a reference to the bird's booming call heard in the marshes, the "mire."

# Blackbird

The origin of this name is straightforward: the Middle English was *blakebird* or *blacbrid*, the equivalent to *black* + *bird*. As Ray and Willughby succinctly noted:[16]

> Blackbirds, so called from their colour.

If we want to get into more detail, the word *black* comes from the Proto-Indo-European word *bhleg*, "to burn." The original blackbird was undoubtedly the Eurasian Blackbird *Turdus merula*, but there are now a number of other blackbirds—six within the genus *Turdus* and twenty-four in the New World Icteridae (troupials and allies), all of which share the common characteristic of being predominantly black. Lockwood noted that the name illustrates an interesting evolution of the word "bird," which, up to the 1400s, was used only to refer to small birds:[84]

> When this name was created, there was . . . no possibility of confusion with such birds as Crows or Ravens, which at the time would have been referred to as "black fowls." [See Introduction.]

# Blackcap

Both of these two species of Sylviidae, the sylviid warblers, parrotbills, and allies, have conspicuous black caps (though only in the male in the case of Eurasian Blackcap) that contrast with olive upperparts and gray throat. While Ray and Willughby referred to the Marsh Tit as the "Marsh Titmouse or Black-cap," there is also an entry for the bird now known as Eurasian Blackcap:[16]

> *The Black-cap*: Atricapilla seu Ficedula . . . called *by the Italians*, Capo Negro.

> This is a very small bird, not weighing above half an ounce: Its length from the tip of the Bill to the end of the Tail is six inches; its breadth between the ends of the Wings stretch out nine. The top of the Head is black, whence it took its name.

Swann pointed out in 1913:[121]

> The name is also applied to many other species which have the cap or summit of the head black, i.e. the COAL-TITMOUSE, MARSH-TITMOUSE, GREAT TITMOUSE. REED-BUNTING, STONECHAT, BULLFINCH and BLACK-HEADED GULL. The present species is the "Atricapilla seu Ficedula" of Aldrovandus.

## Black-Chat

This small chat, the White-fronted Black-Chat, is a representative of the *Oenanthe* genus, all black apart from a bright, white forehead patch in the male. Other members of the *Oenanthe* are wheatears, chats, and the Blackstart. (See Chat.)

## Blackstart

The single species of Blackstart, a small Old World flycatcher in the Muscicapidae, is found in northern Africa and the Arabian Peninsula. The second part of the name comes from the Middle English *stert* and *start*, in turn from the Old English *steort* or *stert*, and from Proto-Germanic *stertaz*, all meaning "tail." This is probably due to the conspicuous behavior of fanning, flexing, and wagging the tail when alighting on the ground. (Compare to redstart.)

## Blackthroat

The Blackthroat is a congener of the rubythroats and the Firethroat in the genus *Calliope*. It is obviously named for this morphological feature, its black throat, but in the past has been called by other names such as Black-throated Robin and Black-throated Blue-Robin.

## Bleeding-heart

The stunning bleeding-hearts are pigeons in the *Gallicolumba* genus, all endemic to the islands of the Philippines. They sport a neat red patch on the breast that contrasts conspicuously with the white plumage of the underparts, giving the appearance to the human eye of a bleeding wound. Beebe wrote in 1906:[100]

> What explanation can we give of the Blood-breasted Pigeon or Bleeding-heart Pigeon, which, as its name denotes, has a splash of blood-like scarlet in the centre of its breast? The remarkable and inexplicable resemblance is heightened by the stiffened vanes of the centre feathers, causing them to appear bedraggled and clotted, as if by an actual wound!

In fact, it's thought that the function is one of sexual selection, as the male will chase a female, displaying his inflated breast to show fully his vivid red marking or "heart," in order to gain her attentions.

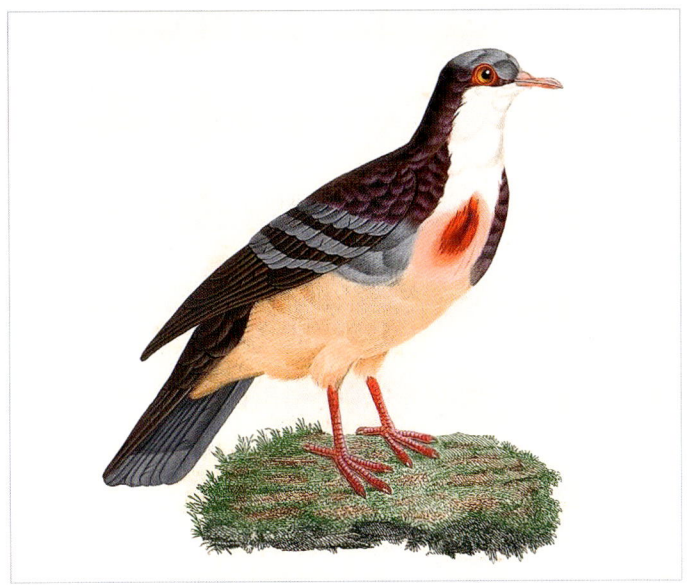

*Luzon Bleeding-heart in Temminck's* Les Pigeons *(1811)*

# Blossomcrown

Two species of blossomcrown hummingbirds in the *Anthocephala* (which also means "blossom crown") are found in Colombia. The name is descriptive of the male's white and orange crown feathers. Gould noted in his 1861 description of a specimen that had been sent to him:[6]

> This pretty little species, to which I have given the trivial name of Blossom-crown, is an inhabitant of the great country of Columbia, and is one of the most recent discoveries made in that rich region.

# Bluebill

Three species of *Spermophaga* finches from western and central Africa are known as bluebills. However, they have varying amounts of blue on the bills, with only two species, the Red-headed and Western Bluebills, possessing a predominantly blue bill. The exception, Grant's Bluebill, has a predominantly red bill.

# Bluebird

The bluebirds are three species of *Sialia* in the Turdidae, the thrushes and allies, native to North and Central America. One of the earliest accounts of the bird was in Vieillot's *Histoire Naturelle des Oiseaux* in 1807:[68]

> The name of *Blue-bird*, Oiseau bleu, is what the Americans have imposed on this small bird, one of the most common in North America. Although this denomination can also be applied to several other birds of the same region, it is always better suited to it than that of Robin, since it does not have a red throat, and that, if not a kind familiarity and a tail swing up and down, it bears no relation to our Robin, to which it has been compared as a very similar species and as its representative in the new Continent. It does not have the flight, the song, the manners, or the habits.

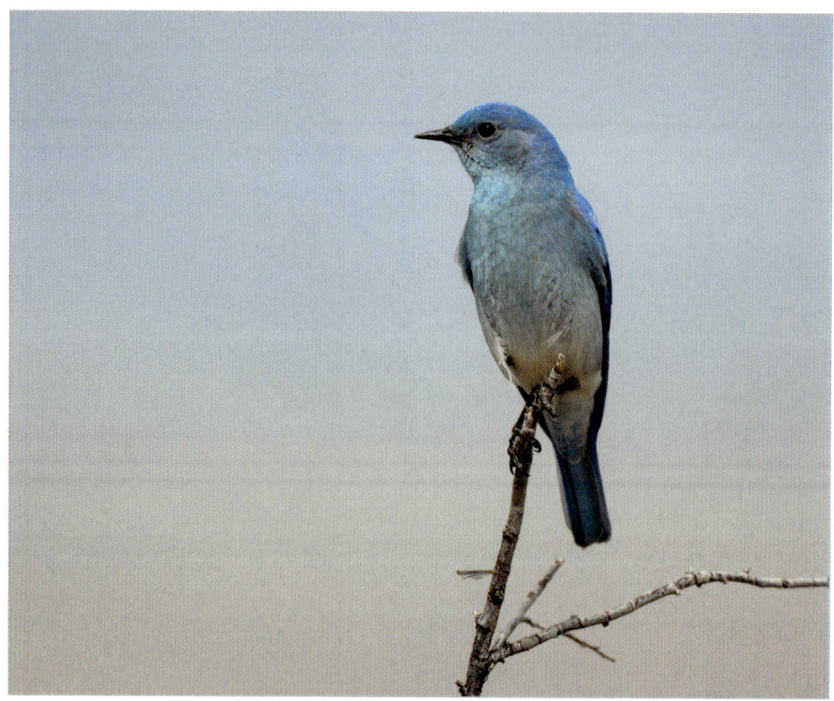

*Mountain Bluebird (Washington State, USA)*

## Bluebonnet

The bluebonnets are two species of small parrot found in outback southeastern Australia. The forehead and face are blue, giving the appearance of a blue face rather than a bonnet; nevertheless, the name is descriptive. Often rendered as "Blue Bonnet," the bird's name has gone through various iterations, as Leach illustrated in his comment about the species in 1912:[144]

> The "Bull-oak," or "Blue Bonnet," is a beautiful bird, and is common on inland plains. It has a brownish-olive back and a gray chest. The bright, blue cheeks, forehead, and shoulder, red abdomen, and light yellow base to tail present a striking appearance. Its vernacular name—Yellow-vented Parrakeet—has now been altered to Blue Bonnet Parrot.

## Bluetail

The two bluetails were previously considered to be conspecific; they are in the genus *Tarsiger* in the Muscicapidae, the Old World flycatchers. Both are blue above with rufous flanks, hence an alternative name, no longer favored, blue-robin. Both have blue tails, but this is more noticeable in the female as this is her only blue plumage.

## Bluethroat

The migratory Bluethroat, a *Luscinia* in the Muscicapidae, has a huge distribution over the Palearctic to southern Asia, central Africa, and southern Europe. The throat is indeed blue, but only in the male. Earlier names included Blue-breast, Blue-throated Warbler, and the cumbersome Blue-throat Redstart. Linnaeus wrote about the Bluethroat in 1776:[294]

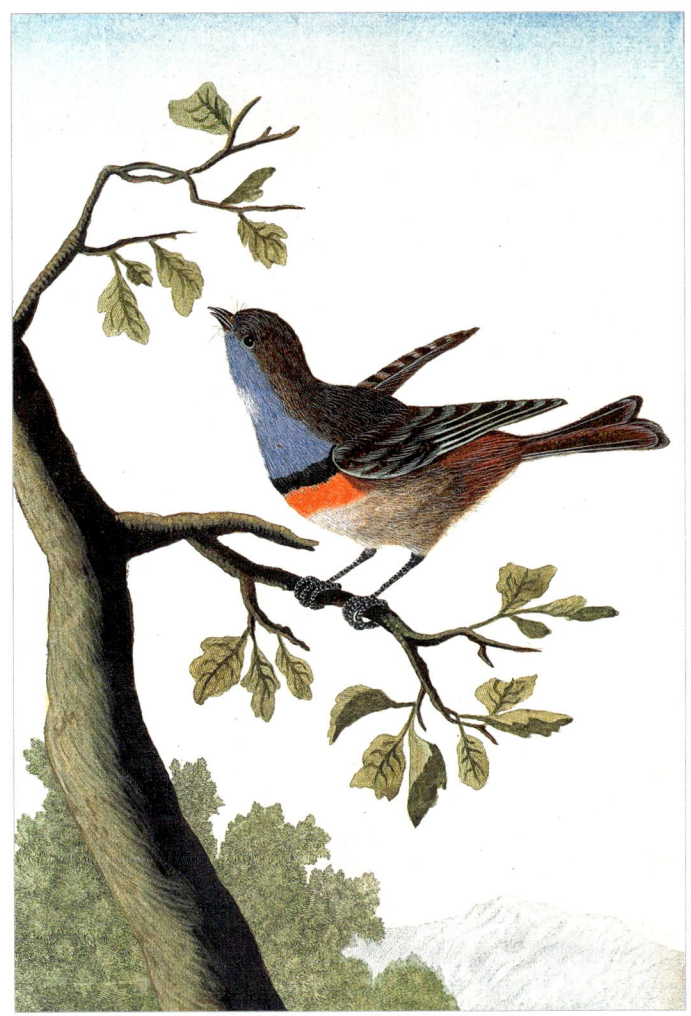

*Bluethroat in Huth's* Recueil de divers oiseaux étrangers et peu communs qui
se trouvent dans les ouvrages de Messieurs Edwards et Catesby *(1776)*

The Blue-throat Redstart . . . this Bird is bigger than the common *Redstart*; it is
of the Size here expressed; for Shape, like other small Birds of this Kind. I have
seen a Drawing of it from *Holland* or *Germany*, which was named *Blau-keckle*,
which is *Blue-throat*, and it being so near of Kin to the *Red-start*, I have given it
the above Name.

By the late 1700s, Buffon was calling it the "Blue-throat."

# Boatbill

The Machaerirhynchidae comprises just two species of boatbills, found in New Guinea and
northeastern Australia. The broad, dorsolaterally flattened bill accounts for the common
name, which was adopted by the Royal Australasian Ornithologists Union in the 1970s.
Wallace mentioned "the curious little boat-billed flycatchers (Machaerirhynchus)" in *The
Malay Archipelago* published in 1869.[205]

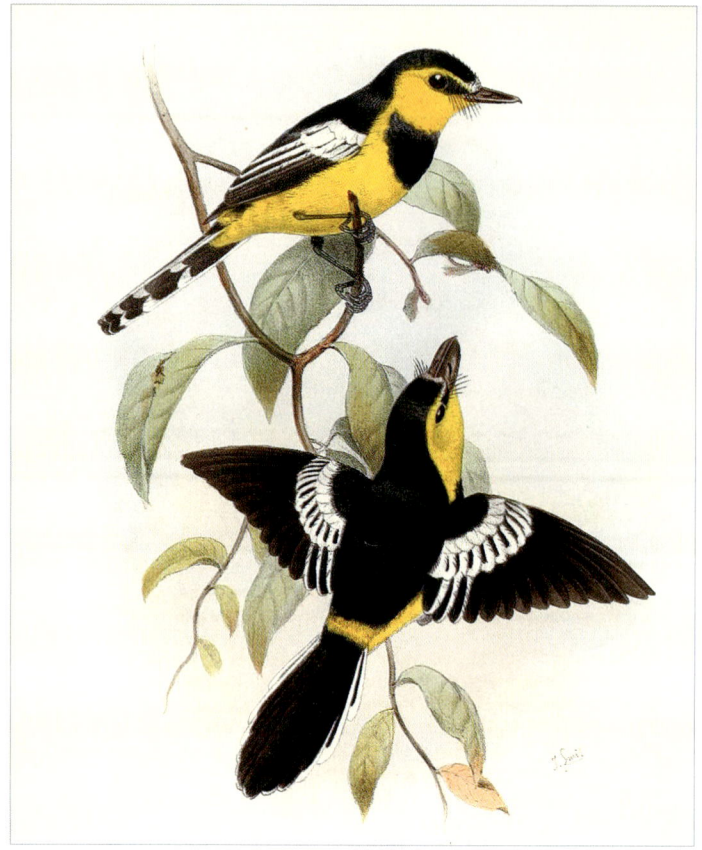

*Black-breasted Boatbill in Rowley's* Ornithological Miscellany *(1878)*

# Bobolink

The Bobolink is a single species within the Icteridae (troupials and allies), found throughout the Americas from Alaska to southern Argentina. The name is onomatopoeic, supposedly from the mnemonic "*bob o'lincoln*," which came into American English in the late 1700s. In 1937, Ingersoll wrote:[9]

> Some Puritan ear happily caught these syllables, and everybody adopted them as they do a "catchy" tune. Nuttall and other ornithologists have devoted pages of their books to imitations in print of the merry fellow's tinkling prattle, only to show that words and syllables can't do it. It is the jazz in the spring concert.

In the past, it has also been referred to as "Ricebird," reflecting its preference for grassland habitats.

# Bobwhite

There are three species of bobwhites, a type of New World quail in the Odontophoridae. The name is onomatopoeic, descriptive of one of their many loud, whistled calls and songs. Although Latham called it the "Maryland Partridge" he noted:[218]

> The note is a loud kind of whistle, twice quick repeated. Known by the natives by the name of *Ho-ouy*; the *New Englanders* call it *Bob White*.

# Bokmakierie

The Bokmakierie is one of a number of bushshrikes in the genus *Telophorus* found in Africa. Found in southern Africa, the name is borrowed from the Afrikaans language and is onomatopoeic. The species is known for its antiphonal duetting, that is, the male makes a *bok* call and the female immediately answers with *makierie.* The word was recorded first in English, as *beckbecary* in 1834 and then, closer to Afrikaans, as Bokmakary Thrush by the English bird artist Christopher Webb Smith in the later 1830s. Layard wrote in 1867:[292]

> Its loud call of "bacbakiri," its imitative powers, and bright plumage, render it one of the most conspicuous birds of the colony. I have not unfrequently heard two birds uttering their peculiar note for twenty or thirty minutes together—one bird giving out the harsh "*backbach*," the other the shrill "*kiri*," the two performers being at a considerable distance from each other.

In a later edition of the same book, revised by Sharpe, it was said:[293]

> At the Cape . . . it is very familiar, and is frequently seen perched on the garden walls, whilst it utters a succession of ringing calls which the Dutch liken to the word "Bacbakiri"; but its notes and calls are in fact very varied.

# Boobook

A number of species of *Ninox* owls are now known as boobooks, but the original boobook was the Southern Boobook, *Ninox boobook.* The common name came to us from an Australian Aboriginal word when William Dawes, an officer in the British Navy who came to Australia on the First Fleet recorded the name *bōkbōk* in 1790 or 1791 in a transcription of Dharuk, the language of the Aboriginal people of western Sydney.[104] Several different names in other Aboriginal languages are also onomatopoeic, for example, *goor-goor-da* (Western Australia), *koor-koo* (South Australia), and *guurrguurr* in the Gamilaraay language of southeastern

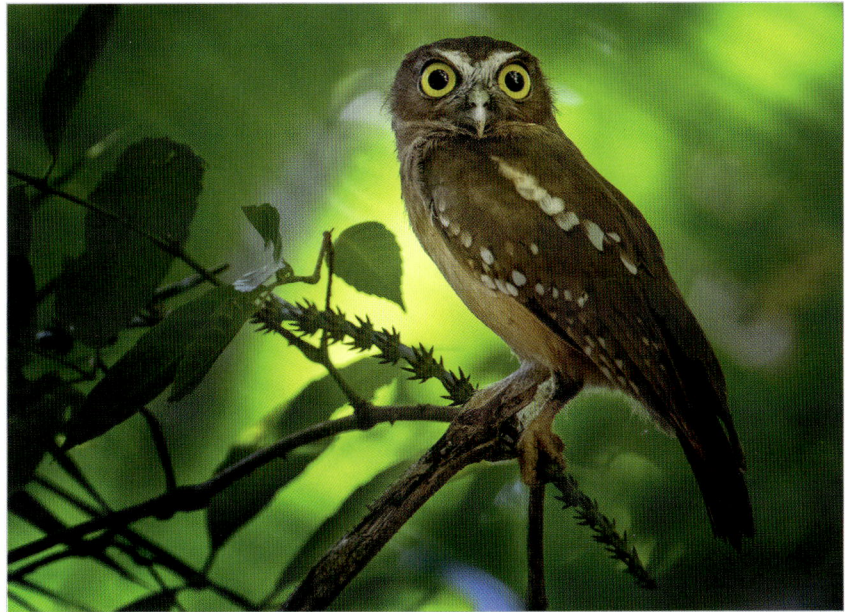

*Ochre-bellied Boobook (Sulawesi, Indonesia)*

Australia. A number of other *Ninox* species that were previously known as "hawk-owls" are now referred to as boobooks, despite the fact that the descriptor may not relate to their vocalizations. Latham coined the scientific name "Strix Boobook" [*sic*] in 1801 and used the common name "Boobook Owl" and stated:[17]

> This inhabits New Holland, where it is known by the name of Boobook.

# Booby

Seven species in the Sulidae have been blessed with the name booby. It comes from the Spanish word *bobo* meaning "fool" or "stupid." In the seventeenth century, Spanish sailors coined the name when birds would land on the masts of their ships and were then easily captured and eaten. It didn't help that they also appear clumsy on land and seem to be fearless of humans. Even Darwin was impressed by the supposed feeble-mindedness of the birds:[295]

> [They] are of a tame and stupid disposition, and are so unaccustomed to visitors, that I could have killed any number of them with my geological hammer.

# Boubou

The name boubou is attached to a number of species in the *Laniarus*, a genus of bushshrikes in the Malaconotidae. The name is onomatopoeic for the mellow, hooted vocalizations of some members of the group. That said, there's a wide vocal repertoire, and many species will engage in synchronous male-female duetting. Le Vaillant coined the common name in 1799's *Histoire Naturelle des Oiseaux d'Afrique* when he wrote:[93]

> L'espèce de pie-grièche dont il est question ici, et que j'ai nommée, comme la précédente, par onomatopée . . . Du reste, le ramage de ma pie-grièche se rapporte assez à ce qu'il nous dit de celui de son merle, car elle chante sans cesse *bou-bou—bou-bou*; ce qui imite assez bien le *cou-cou,* ou, si l'on veut, une horloge de bois: il est vrai que la femelle, qui est toujours près de son mâle, répond aussitôt par un autre cri qu'on exprime très-bien par *cou-ï*; et elle lui répond si à propos que très-longtems j'ai cru que c'étoit le mâle seul qui chantoit ces syllabes *bou-bou-cou-i.*

> [The species of shrike in question here, and that I have named, like the previous one, by onomatopoeia . . . Besides, the song of my shrike is quite similar to what he (Montbaillard) tells us about that of his blackbird, for she constantly sings *bou-bou-bou-bou*; which mimics the *cou-cou* quite well, or, if one wishes, a wooden clock: it is true that the female, who is always close to her male, responds immediately with another cry which one expresses very well by *cou-ï*; and she answers him so promptly that for a very long time I believed that it was the male alone who sang these *bou-bou-cou-i* syllables.]

# Bowerbird

The bowerbirds are a remarkable group of birds within the Ptilonorhynchidae. There are seventeen species with the name. It is thought that John Gould, the famous English ornithologist and artist, was the first to use the name after he spent time studying the birds in Australia in 1839 and first observed the often elaborate structures, known as bowers (a shelter made with tree boughs or vines), that the male birds build for the purposes of attracting a female. Gould was fascinated by the birds:[40]

> One point to which I more particularly allude,—a point of no ordinary interest, both to the naturalist and the general admirer of nature,—is the formation of a

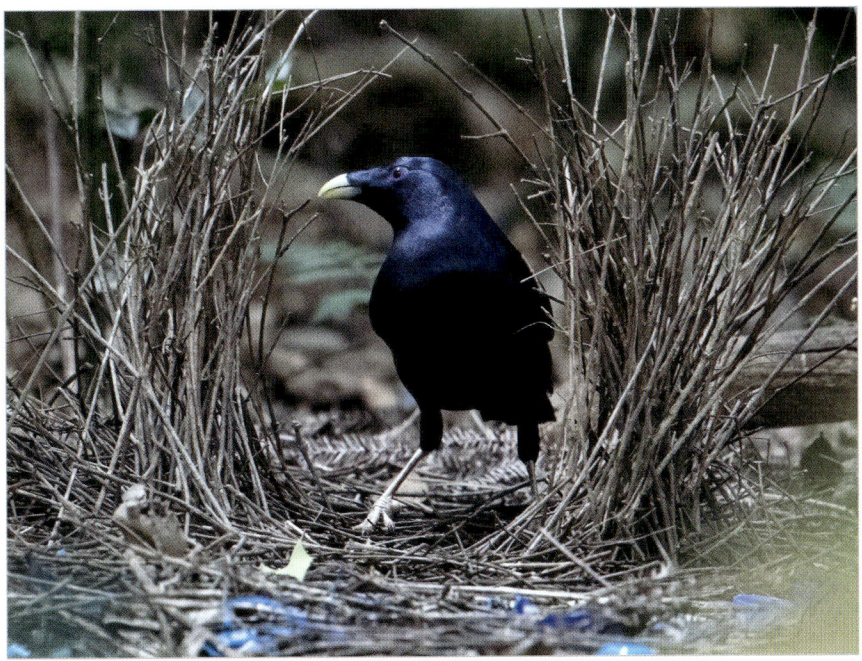

*Satin Bowerbird (Queensland, Australia)*

bower-like structure by this bird for the purpose of a playing-ground or hall of assembly, a circumstance in its economy which adds another to the many anomalies connected with the Fauna of Australia.

# Brambling

A member of the Fringillidae, the finches, euphonias, and allies, the Brambling owes its unique name to its familiarity to the inhabitants of the British Isles from antiquity. The name was originally *bramline* in Middle English, derived from the words *bramble* and *ling*. The word *bramble* is well known as referring to blackberries but would have been used in the past for many types of thorny bush. It came into English first as the Old English *bræmbel*, from the Proto-Germanic *brēmila-*, a diminutive of *brēm*, English *broom*. The *-ling* suffix is added in a similar sense to bird names such as Sanderling and Reedling, either as a diminutive or denoting "one concerned with."

Although it has not been recognized widely, Lockwood disagrees with the accepted etymology and relates the name to *Brandling*, which is used for young salmon and trout to denote their brindled markings.[100] Noting that other names for the fish include Bramling and Bramlin, he suggested that the name could equally apply to the bird with its

striking, brindled plumage [which] is calculated to evoke a name, as in local synonyms TARTAN BACK and FRENCH PIE FINCH.

He believed that the present-day name is a corruption of Brandling, going as far as to suggest that the name should be revived. Citing Ray and Willughby's use of the name as a caption for an illustration, he boldly stated:

Quite evidently then, the bird name too is properly Brandling, not Brambling, now seen to be a misnomer resulting from an erroneous normalization.

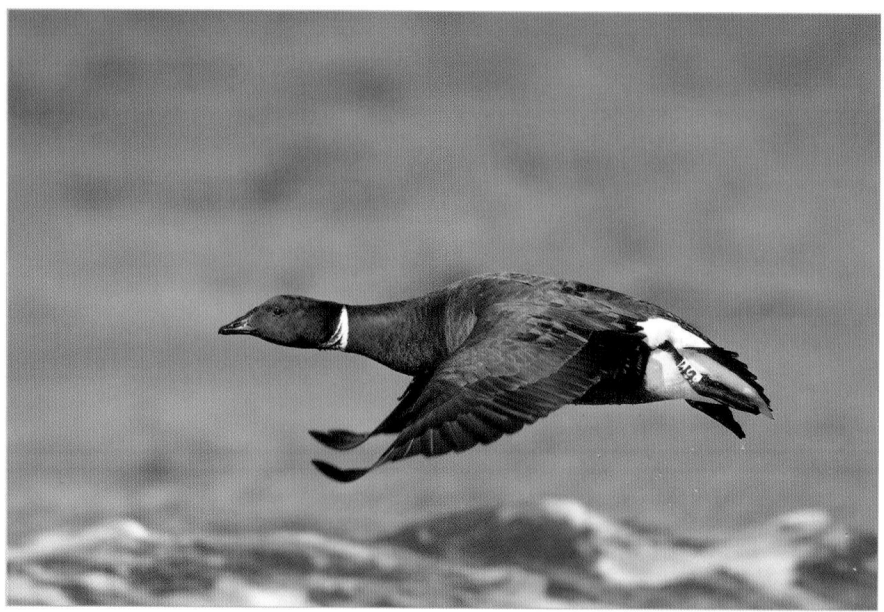

*Brant (Washington State, USA)*

# Brant

The Brant, or Brent Goose in Europe, is a large, dark waterfowl that is widespread across the northern Palearctic and America. The common name echoes the genus name (there are five other species in the genus) *Branta*. This is a Latinized form of the Old Norse *brandgás*, "burnt (black) goose," and came into Middle English as *brand gos*, morphing over time to the present appellation. The word appeared in Femina, one of the Trinity College Manuscripts from around 1400:[340, 341]

> Here comyþ fleynge a wylde goos, A brandgoos [F brallet] to hym ys felawet. [Here comes flying a wild goose, a Brant to him as companion.]

Pennant popularized the name in 1766, calling it the Brent Goose, in his influential publication, *The British Zoology*.[342]

# Brilliant

There are eight species of brilliants, all in the genus *Heliodoxa*. All of them are fairly uniformly glossy green (some with brown, red, or purple) with a glittering green forehead, crown, or throat in the males. The name brilliant, meaning "bright and saturated," undoubtedly stems from this plumage characteristic. It's interesting to note that the English word *brilliant*, borrowed from French, derives from the Latin *beryllus*—"a gem," the name of which in turn came from Pali and Sanskrit, after the name of an ancient city in Southern India.

# Bristlebill

Four species of bristlebills in the *Bleda* genus are found in West and Central Africa. They are members of the *Pycnonotidae*, the bulbul family and, as the name suggests, have well developed rictal bristles. Jardine wrote in 1854:[58]

> The bristles of the mouth are two-thirds the length of the bill and are very strong.

## Bristlebird

The three species of bristlebirds in the family Dasyornithidae are all endemic to Australia. The name comes from the two or three prominent rictal bristles (stiff hair-like feathers) that protrude on either side of the gape. Many insectivorous birds that forage in dark forested areas possess these types of bristles, and it's thought that this anatomical adaptation may assist with locating prey and navigating narrow spaces.

## Bristlefront

Two tapaculos in the genus *Merulaxis* in the Rhinocryptidae are characterized by a prominent tuft of short, stiff bristles on the forecrown above the bill in the males.

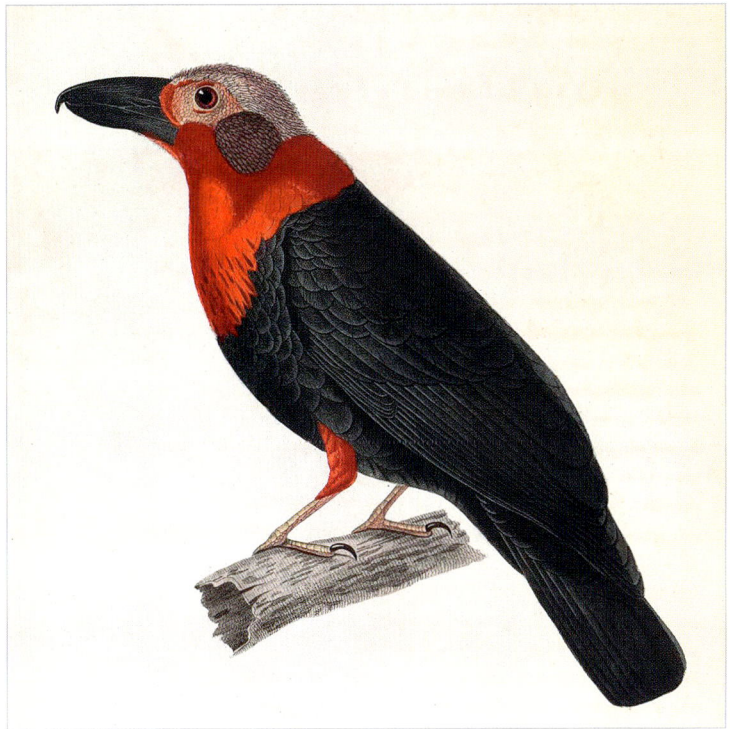

*Bornean Bristlehead in Temminck's* Tableau Méthodique *(1838)*

## Bristlehead

The Bornean Bristlehead is so unusual that it's been assigned its own family, thus it is the sole representative of the Pityriasidae. Very few birds have had such a checkered taxonomic history as the Bristlehead, having in the past been classified as a starling, a shrike, a helmet-shrike, a vanga, a babbler, and a woodshrike. And indeed, in the past, it was often called the "Bald-headed Wood Shrike." While the main body of plumage is black, the head and neck sport brilliant red plumage while the bare yellow-orange skin of the crown is topped with spiky bristles, which are actually short outgrowths of skin, 3–4 mm long, on the crown. It's from this unique feature that the common name is derived. Temminck, who first formally described the bird in 1838, called it the "Cassican Gymnocéphale," the bald-headed butcherbird. He described the head of this extraordinary bird thus:[94]

The bright and brilliant colors of this beautiful and rare bird, the very particular nature, somewhat heterogeneous, of certain parts of the plumage, the massive shape of the beak, and its more or less hairless head, contribute to the singular appearance . . . [The adult] has the top of his head covered with cartilaginous filaments in more or less rounded laminae: these filaments, very short, are rough to the touch, and present on the dried skin a yellowish tint which appears also be that of perfectly hairless skin with which the eye socket is surrounded; the lores are covered with a small brush of red feathers; we see [around the ears] an ample and large tuft in the shape of a brush, covering all this part; its small pinnae are cartilaginous in nature.

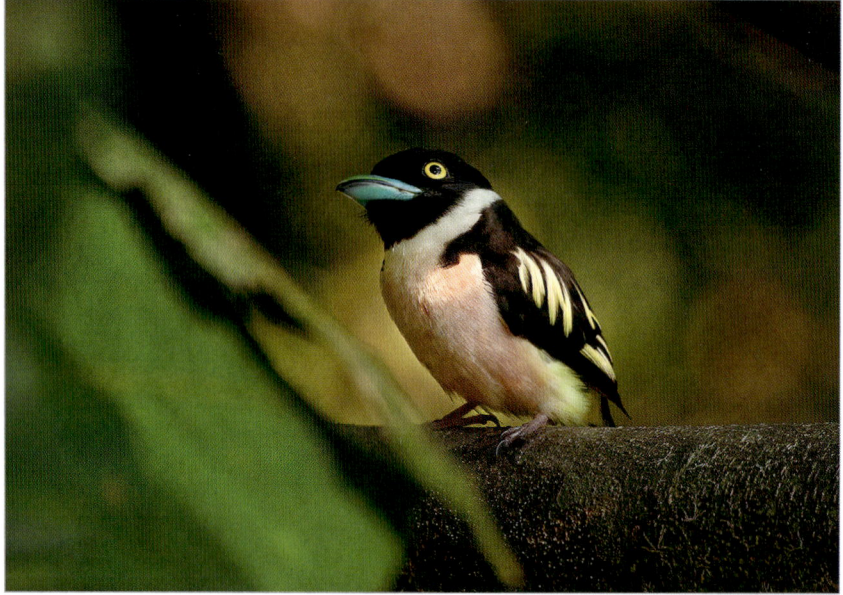

*Black-and-yellow Broadbill (Malaysia)*

# Broadbill

There are two closely related families of broadbills (Calyptomenidae and Eurylaimidae), all members of which are known by the common name broadbill. The reasons for the appellation are obvious—these colorful and characterful birds all possess large, wide-gaped bills. The bill structure is probably an adaptation to catching fast-moving insects on the wing in a sit-and-pounce hunting technique. Swainson was the first to adopt the name in 1836, writing that the birds are[64]

about the size of starlings; while the enormous breadth of their bills, and the peculiar brightness of their coloring, renders it impossible for the student to mistake them for any other genus. The bill is not only excessively broad, but the margins of the base are so dilated, that they often project over those of the lower mandible.

# Brolga

Initially called the Australian Crane, Brolga was made the official name for the bird in 1926 by the Royal Australasian Ornithologists Union, which used this popular name derived from the Gamilaraay or Kamilaroi word *burralga*.[472]

Gamilaraay is a now-endangered Aboriginal language from Australia's southeast. Mathews recorded the Wiradyuri name as *burolgang*, using an early colonial name, the Native Companion.[240] Leach gave alternative names for the "Australian Crane" as Native Companion and Brolga in 1911, writing that the former was "an inappropriate name."[144]

# Bronzewing

Five species of pigeons in two genera are known as bronzewings (*Phaps* and *Henicophaps*) due to the iridescent bronze, green, blue, and purple plumage on the upperwings. Phillip, the then-governor of New South Wales, wrote of the "Bronze-winged Pigeon" in 1789, describing the namesake feature:[110]

> The wing coverts are much the same colour as the back, but the greater ones, or lower series, have each of them a large oval spot of bronze on the outer webs near the ends, forming together, when the wings are closed, two bars of the most brilliant and beautiful bronze, changing into red, copper, and green, in different reflections of light.

(He goes on to say "This bird inhabits Norfolk Island; and is clearly a non-descript species," neither of which is correct, and his full description fits no known pigeon that has ever occurred on the island.) By the mid-1900s the "pigeon" part of the name had been dropped.

# Brownbul

These are two species of bulbuls, brown in color. The present-day name appears to have been adopted in the 1970s; prior to this it was referred to in the English literature by the French name *Le Jaboteur*, meaning "the jabberer." (See Bulbul.)

# Brubru

This monotypic genus (*Nilaus*) of bushshrikes from sub-Saharan Africa has quite a wide vocal repertoire, including a telephone-like, trilling *brrrru brrrru*, hence the name is onomatopoeic. The appellation was first attached to the bird by the French explorer and ornithologist François Le Vaillant, who collected and described many new species of birds in Africa.[93] He justified the name thus:

> The babbling of the male expresses the word *bru* very distinctly, repeated two and three times in succession, guttural and drawling a little on the *r*.

# Brushfinch

These somewhat finch-like birds of two genera (*Atlapetes* and *Arramon*) are in fact New World sparrows of the family Passerellidae. The brushfinches all forage on the ground, in a similar manner to finches, in dense forest and secondary growth—in other words, in the brush.

# Brushrunner

The lark-like Brushrunnner lives up to all aspects of its name—despite being a member of the Furnariidae, the ovenbirds and woodcreepers. This bird is a dweller of the South American *chaco* habitat, where it looks and behaves very much like an Old World lark, running rapidly on the ground among the brush of the tropical grasslands and savanna.

## Brushturkey

The brushturkeys are not turkeys, they are Megapodes with a superficial resemblance to the true turkeys. They are, however, denizens of the brush, in the Australian sense of the rainforest. This was a name originally given the Australian representative of the group, as Gould noted:[103]

> From the colonists of Australia the three species [of Megapode] inhabiting that country have received the trivial names of Brush-Turkey, Native Pheasant, and Jungle-Fowl; but to none of these birds are they in any way allied.

## Budgerigar

Not many bird names have had such a checkered history as the world-famous parrot, the Budgerigar. With regard to the European literature, the first reference to the name most similar to the present-day one is said to be in Leichhardt's 1847 work *Journal of an Overland Expedition in Australia*, in which he referred to "the little Betshiregah."[611] In his *Birds of Australia* in 1848 Gould described the "Warbling Grass-Parrakeet" with the alternative names Undulated Parrot, Undulated Parrakeet [*sic*], Canary Parrot of the Colonists, Scolloped Parrakeet [*sic*] per the explorer Captain Sturt, and, lastly, the Betcherrygah of the "Natives of Liverpool Plains."[40] Much later in a short paper intriguingly entitled "John Roach, the Budgerigar, and the Unfortunate Officer" in 1970, the authors mention so many names for the Budgerigar that it's hard to keep track.[473] Among the names they cite are *Betcherrygah*, *Bugernigang Parrot* or *Bugerrigang*, *Budgeree-gar* "Budgeree signifying handsome or good," *buggery-gong* in a flowery poem from 1861, "The Aboriginal names . . . *Budgeree-gar* and *Betcherrygah*, both referring to the beauty of their appearance,"* and, lastly, *bujirigar* and *bugirigar*.

The etymology has long been confused. There are numerous postulations as to the origins of the word. Some dictionaries believe it is a modification of the word *gijirrigaa* from Yuwaalaraay, an Australian Aboriginal language of northern New South Wales. Others state that it is from a different Indigenous Australian language, said to mean "good cockatoo," from *budgeri*, "good," and *gar*, "cockatoo." Hamilton & District Budgerigar Society claims without references:

> The usual definition of Betcherrygah is "good bird" but it is stated on reliable authority that it is an old aboriginal dialect word meaning "good food." According to eminent naturalists who have written descriptions of the Budgerigars [*sic*] native habits, these birds are migratory, the movements of the flocks being governed by the food supply. They stay in one district so long as the plains are green and luxuriant, but as soon as scarcity of water dries up the herbage, away they go to the proximity of streams and to the northern parts of Australia where a rich supply of ripening and ripe grass seeds awaits them.

This claim that the name means "good food" is repeated ad nauseam everywhere except in reliably researched literature. The thinking is that they were either tasty treats, which is very unlikely, or that their nomadic wanderings led the Gamilaraay to places of rainfall and abundant food. In other words, the birds would lead the people to "good food."

The most authoritative word on the matter comes from Nash in a 2010 paper amusingly titled "The smuggled budgie: case study of an Australian loanblend."** He concludes the name is a loanblend (a word constituted of a combination of native and foreign elements) of the Kamilaroi word *gijirragaa* for the bird and the former New South Wales pidgin word *budgery* or *boojery* meaning good. He dismisses the "good food" hypothesis, proposing:[612]

We should instead presume that it was speakers of English who formed the *budgery garr* (literally "beautiful parrot") loandblend in NSW Pidgin (perhaps even taking it to be faithful to the original language), and simultaneously adopted the combination into their own English.

* Rev. John Graham's 1861 "Night in the Australian Bush":

> *The forest's primal stillness breathes*
> *Its deep and unbroken rest.*
> *Unbroken—although a pulse of air*
> *The gum-leaves stirreth among,*
> *And the flying squirrel's sudden leap*
> *Awakes the buggery-gong.*

**"Budgie smuggler" is a somewhat rude Australian slang term for men's Speedo-type swimwear.

# Buffalo-Weaver

The buffalo-weavers are three members of the Ploceidae, the weavers and allies, that are often closely associated with cattle. The etymology of the name was elaborated in *Cassell's Book of Birds* in 1875:[106]

> The three species . . . alike frequent pasture land, keeping as much as possible in the immediate vicinity of the herds of buffaloes, upon whose backs they perch, to obtain the ticks that form their principal nourishment; they may constantly be seen around these huge creatures, to whom they not only render this service, but warn them of the approach of danger.

# Bufflehead

In the mid-1600s, the term *buffle head* was a pejorative meaning "stupid person, foolish, or big headed." There is no reason to conclude that the insult applies to this small and charming sea duck. Although Linnaeus put the Bufflehead in the genus *Anas* in 1758's *Systema Naturæ*, it was later placed in the genus *Bucephala*, from the Greek *bous*, "bull" and *kephale*, "head," a reference to the oddly bulbous head shape.[474] The common name is simply an anglicization of the genus name. There have been various common names over the years, though. In *Annals of the Lyceum of Natural History of New York* in 1824, the bird is called by the common name "Butter Ball."[475] And then in the *Reports of explorations and surveys, to ascertain the most practicable and economical route for a railroad from the Mississippi River to the Pacific Ocean* (1858) from the United States War Department, we see this heading for the species:[196]

BUCEPHALA ALBEOLA, Baird

Butter Ball; Dipper; Buffle Head.

# Bulbul

The Pycnonotidae, the bulbuls, are a large family of 151 species. Of these, 96 birds go by the name of bulbul, which comes from either the Persian بلبل *bolbol* or the Arabic بُلْبُل *bulbul*, meaning "nightingale." But the word nightingale seems to have been used in translations incorrectly many times for any melodious bird and thus the original word could well have referred to a bulbul of the Pycnonotidae that occurs in the Middle East. Its origins are

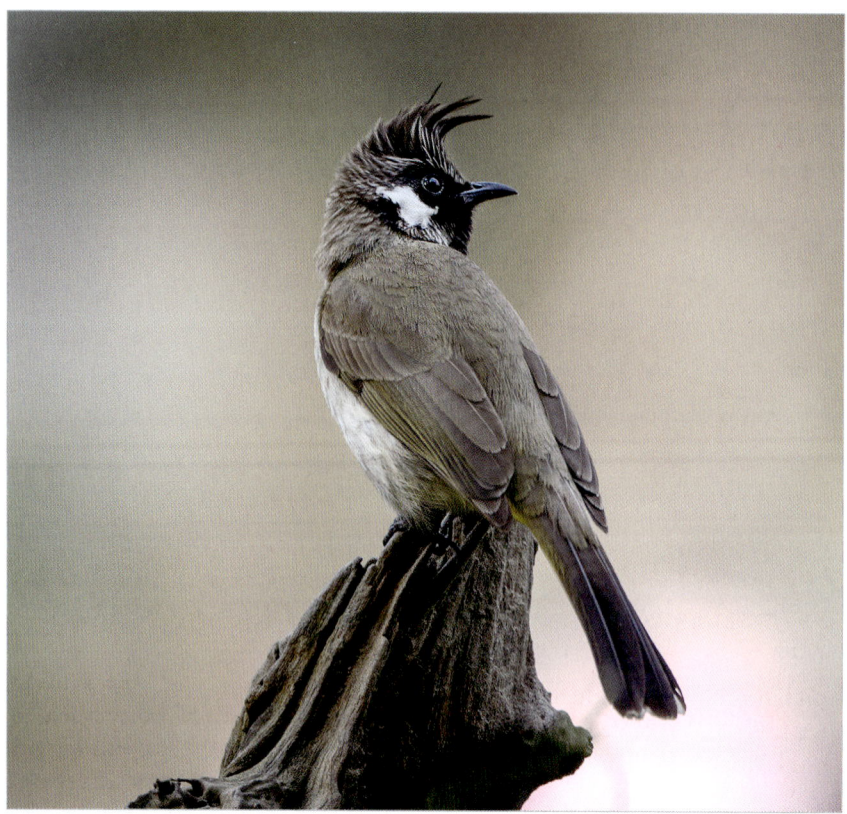

*Himalayan Bulbul (India)*

unclear, but it may be onomatopoeic. Other species are known as greenbuls and brownbuls, an interesting bilingual portmanteau. After the word entered English, in the mid-seventeenth century, it was frequently used as a synonym for "singer." Bulbul was the stage name of a famous Azerbaijani opera tenor, Murtuza Rza oglu Mammadov, of the early 1900s, where the Common Nightingale is called the Cənub bülbülü.

## Bullfinch

This name is used for a number of birds in two different families, eight species of *Pyrrhula* in the Fringillidae, the finches, euphonias, and allies, and six species of Thraupidae, the tanagers and allies. The original bird to bear the name bullfinch was, of course, the Eurasian Bullfinch. The etymology of this old common generic name, dating from the 1500s, is complicated and poorly understood. The name supposedly derives from the thick shape of the bird's head and neck, which would correlate with other old names such as bullfrog or bulldog (that said, there is some controversy over whether the bullfrog was named after its shape or its voice). Yarrell, however, asserted that the name simply referred to its relatively large size:[344]

> The names bulldog, bullfinch, bullhead, bulltrout, bullfrog, and bullrush, are applied to species of large size in Zoology and Botany.

On the other hand, one German name is *Blutfinnk* (or *Blödtfinck,* as Turner noted in 1544[26]), meaning "Blood Finch," and, with its bright pinkish-red plumage, maybe the English name derives from that.

# Bunting

The name for this group, consisting of three families—two species in the Calcariidae, longspurs and Snow Bunting; forty-three in the Emberizidae, the Old World buntings; and seven in the Cardinalidae, the cardinals and allies—came into Middle English as *bunting*, *bountyng*, or *buntynge*. The name has been in use for the bird since the 1300s, but bunting as a surname predates it. The bird name appeared in the Harley Manuscripts in a poem dating from 1350:[381]

> Ich wolde ich were a þrestelcok / a bountyng oþer a lauercok / swete bryd. [It would be a thrush, a bunting or a lark, sweet bird.]

The origin of the name is unclear, but it seems most likely that it stemmed from the Scottish *buntin* or English *bunty*, meaning "short and stout or thick." Newton rather uncharitably observed that the name is connected to[7]

> the Scottish word buntin = short and thick, or plump, which, however, seems as likely to have been derived from the bird, for the clumsy figure of the true [Corn] Bunting is very evident to any observer.

Lockwood pointed out:[100]

> The literal meaning [of the word] is plump or thick-set person or creature, an appropriate designation for the bird.

That said, other theories include a possible reference to the speckled plumage of the Corn Bunting from an unrecorded Middle English word *bunt*, meaning "spotted, speckled, pied" combined with the diminutive suffix *-ing*, but Newton thought this was unlikely.

# Bushbird

The Buff-banded Bushbird found in Timor and currently in the Locustellidae (the grassbirds and allies) has many alternative common generic names, including thicket-warbler, grassbird, warbler, babbler-warbler, grass-wren, and thicketbird. This reflects the changes in the bird's taxonomic history over time. The other bushbirds are three species of Thamnophilidae, typical antbirds from the Neotropics. All four bushbirds are secretive denizens of thick, tangled vegetation in tropical rainforest—in other words, birds of the bush.

# Bushchat

Thirteen species in the genus *Saxicola* sport the name bushchat. (See Chat.) Interestingly, while the common name means a small bird that lives in bushes, the genus name equates to rock dweller (*Saxi* for "rock," *cola* for "dweller").

# Bush-crow

Stresemann's Bush-Crow belongs to the monotypic genus *Zavattariornis* in the Corvidae. In 1962 Hall and Moreau described its habitat in a tiny area of southern Ethiopia:[111]

> The Abyssinian "Bush-crow" has a very restricted range, in park-like Acacia country.

Although not particularly crow-like, this corvid is named for its appearance and habitat—it is both a bionym and a morphonym. However, when one compares the habitats of the many birds with "bush" in the name, we can see that it is not a terribly reliable indicator of the respective species' preferred natural surroundings.

# Bush-hen

These are four somewhat secretive members of the Rallidae in the genus *Amaurornis*. Their forest-dwelling habits explain the first part of the name, while the second part hints at their chicken-like appearance. The word *hen* comes from the Old English *henn*, the feminine form of *hanna* for male chickens, the Proto-Indo-European (PIE) root word meaning "singer."[272]

# Bushlark

There are twenty-four species in the genus *Mirafra*, of which seven are called bushlarks. The remaining members all bear the appellation "lark." Many of the bushlarks are found in grasslands with scattered trees, but Oates points out Jerdon told him that the Bengal Bushlark[112]

> is a tolerably familiar bird, feeding in gardens and bushy places, squatting when watched and then taking a short flight; and it appears to have the propensity to hide itself more than any of the other Indian species. It frequently perches on bushes

and that Jerdon's Bushlark:

> is found in gardens and compounds, as well as on roadsides and patches of jungle.

(See Lark.)

# Bush-Quail

Four species of bush-quails in the *Perdicula* are closely related to the other quails in the Phasianidae. They can be found in bushy habitats, including dense grasslands, swamps, scrub jungle, and open plains with thorny shrubs. It was noted in the 1842 issue of *The Journal of the Asiatic Society of Bengal* that previously the Rock Bush-Quail went by a number of different names:[587]

> Java Partridge of Latham: Bush Quail of sports-men. Also termed Rock Quail in the peninsula.

(See Quail.)

# Bush-Robin

These robins are members of the Muscicapidae, the Old World flycatchers; they are four species in the genus *Tarsiger*. The first element of the name references the group's preferred forest habitats. (See Robin.)

# Bushshrike

Sixteen species of bushshrikes of three different genera are nestled in the family Malaconotidae, the bushshrikes and allies; they occur throughout sub-Saharan Africa. Their superficial resemblance to the true shrikes, the Laniidae, and the fact that they occur in woodlands explains the common name. The lack of a hyphen possibly makes this the only English word with the letters *shsh*.

# Bushtit

The only New World representative of the family of long-tailed tits, the Aegithalidae, the Bushtit is a familiar forest bird in large swathes of western North America and Middle

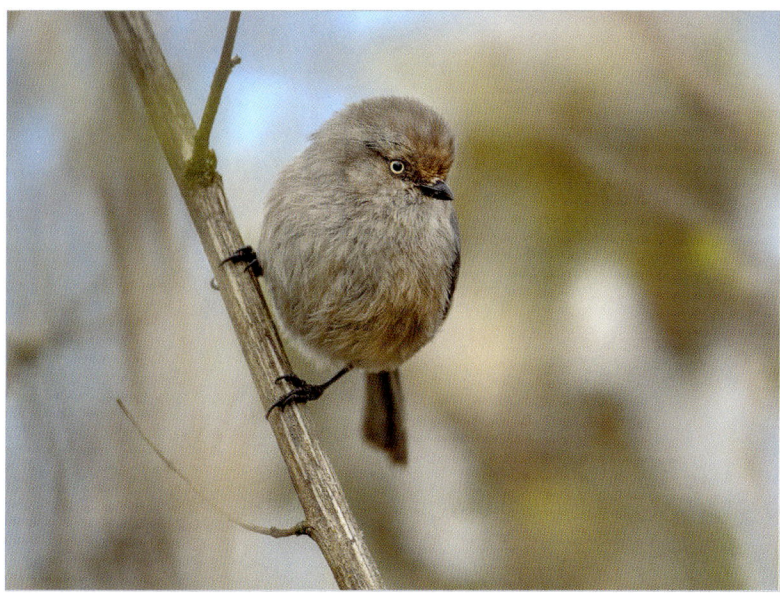

*Bushtit (Washington State, USA)*

America. Originally, the *Psaltriparus* bushtits were classified in the Paridae and called by the names "tit" and "titmouse." By the 1870s, Baird was calling it the "bush-titmouse." [251] Even though the Aegithalidae was introduced by the German naturalist Reichenbach in 1850, it wasn't until much later that they were moved into that family as the only New World representative. (See Tit.)

# Bustard

There are twenty-four species of bustards in the Otididae. The name originated in the fifteenth century from the Middle English *bustarde*, which in turn derives from an Anglo-Norman French blend of Old French *bistarde* and *oustarde*, from the Latin *avis tarda*, ostensibly meaning "slow bird." But the name is unexplained, as the bustards are actually swift runners. Pliny noted that in Spain, the bustard was known as *Tarda* in an unrecorded language, so it might be possible that the Romans, on first encountering the bird, referred to it as the "Tarda bird"—*avis Tarda*. Turner translated Pliny, writing:[26]

> Very near them [bustards] are those which Spaniards call "Aves tardæ" and Greece "Otides"; they are condemned as food.

The Great Bustard, for which the name was originally coined, previously occurred in southeastern England but became extinct there when the last bird was shot in 1832. The birds appeared in the *Boke of Nurture* from 1315:[252]

> *Pecok / Stork / Bustarde / & Shovellewre,*
>
> *[Peacock, Stork, Bustard & Shoveler]*
>
> *ye must vnlace þem in þe plite / of þe crane prest & pure,*
>
> *[you must carve them on the plate, like you do the crane]*
>
> *so þat vche of þem̄ haue þeyre feete aftur my cure*
>
> *[taking care to keep their feet on.]*

# Butcherbird

The butcherbirds are a group of carnivorous, shrike-like birds—both in looks and habit—found throughout Australia and New Guinea. And, in fact, the name "butcher bird" was colloquially used in England for the shrikes since the seventeenth century due to their habit of impaling prey items on thorns and the like. The name was imported to Australia by the early settlers and was later contracted to butcherbird.

# Buttonquail

There are sixteen species in the genus *Turnix* bearing the name buttonquail. These small and round ground dwellers superficially resemble true quails, and these features probably account for the use of the appellation button combined with quail. Newton stated that it is an[7]

> Anglo-Indian name for a little bird, *Turnix sykesi*, and one if not more of its congeners, which, though for a long while confounded with the true QUAILS, really belong to a very distinct group, Turnicidae, and may be more conveniently treated under the title of Hemipode.

# Buzzard

There are twenty buzzards in the Accipitridae, the hawks, eagles, and kites. The name came into use in the 1300s, meaning a "type of hawk not used in falconry," from the Old French *buisart* or *busard*, "harrier, inferior hawk," apparently from the Latin *Buteo*, a kind of hawk. *Buteo* is possibly based on the onomatopoeic syllable *būt*, which according to de Vaan, is probably onomatopoeic, rendering the call of a hawk or a buzzard.[476] The suffix *-art* or *-ard*, known as a "pejorative agent suffix," denotes someone who is in a specified condition (such as a "drunkard") and carries a derogatory connotation. As Newton wrote:[7]

> The Buzzards are fine-looking birds, but are slow and heavy of flight, so that in the old days of falconry they were regarded with infinite scorn, and hence in common English to call a man a "*buzzard*" is to denounce him as stupid.

In the New World, the word was extended to the American vultures in the 1800s.

# C IS FOR CACHOLOTE

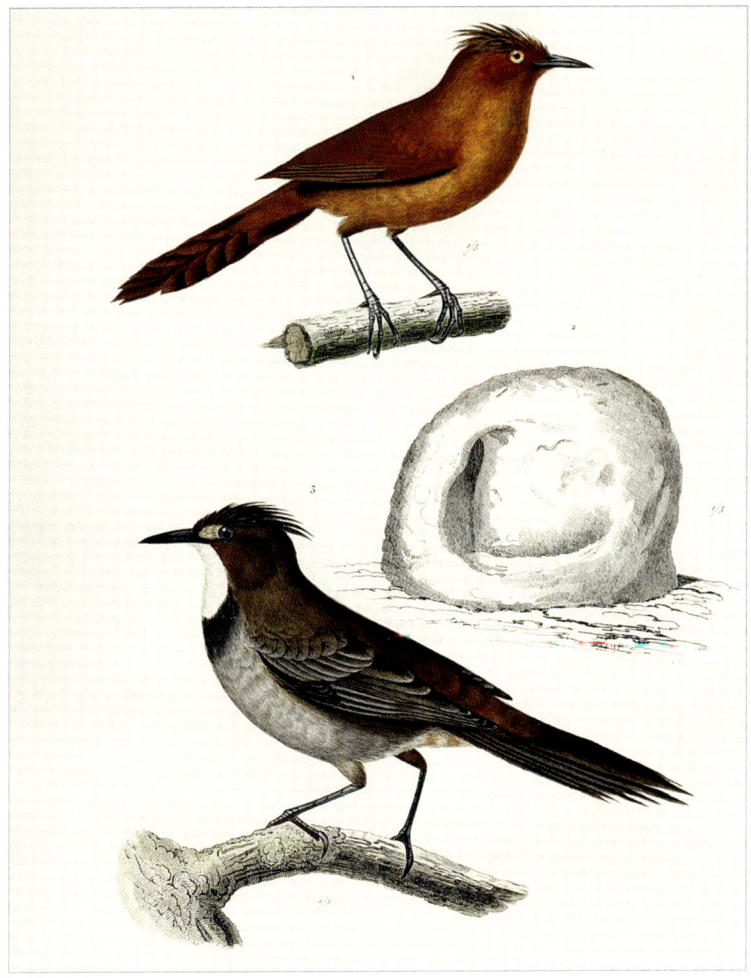

*Rufous and White-throated Cacholotes in Orbigny's* Voyage dans l'Amérique méridionale *(1846)*

## Cacholote

The four species of cacholotes in the ovenbird family (Furnariidae) are all somewhat similar in appearance; three of them have prominent, thick crests. They are found throughout Latin America, including Brazil, and the name most probably derives from the Portuguese word *cachola*, meaning "big head." The Spanish word for Sperm Whale is *Cachalote*, which also has a big head. In 1883, Barrows wrote in "Birds of the Lower Paraguay" that the Brown Cacholote has[117]

> a respectable crest, an outrageous disposition and voice, and a nest the size of a barrel, [it] is a bird that cannot be overlooked.

He then went on to tell us that the alternative names are

> *Copeton* (Big-crest) and *Casero* (House-builder).—The name *Cachalote* assigned by most writers to this bird I have never heard at Concepcion where it is well known by the name *Copeton*. . . . Like many a more pretentious creature, however, his house is more interesting than himself.

# Cacique

Twelve species of icterids (troupials, blackbirds, and allies) are known by the common name cacique. The name is borrowed from the Spanish and Portuguese word *cacique*, meaning a "king" or "prince" of an Indigenous group, derived from the Taino (a Carribean language) word *kasike* or the Arawak *kassequa* for the tribal chiefs or bosses in the Caribbean Islands. In the colonial era, the Europeans used the word as a title for the leaders of practically all Indigenous groups that they encountered in the Western Hemisphere. Why the name is applied to a covey of birds that are not particularly magisterial in appearance or manner is unclear, but it may relate to their aggressive behavior. The French naturalist Bernard Germain de Lacépède introduced the genus *Cacicus* in 1799 without explanation.[297]

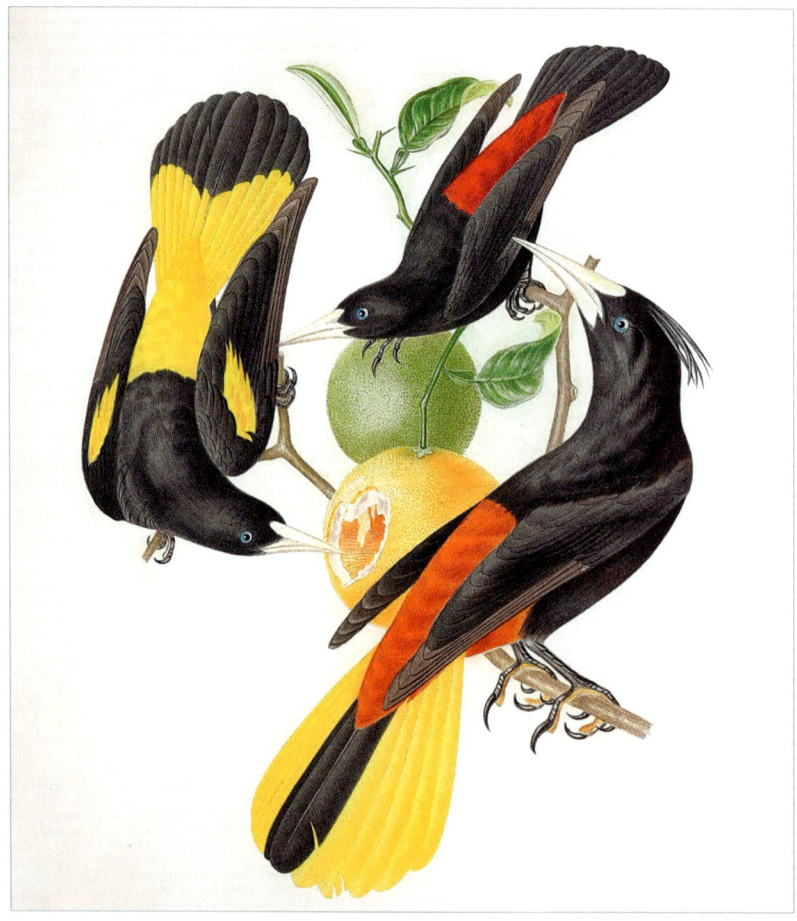

*Red-rumped and Yellow-rumped Caciques and Crested Oropendola in Descourtilz's* Ornithologie Brésilienne *(1854)*

# Calyptura

The tiny Kinglet Calyptura, though unrelated to kinglets, but rather a member of the Tyrannidae, looks remarkably similar and is found in a tiny area of woodland near Rio de Janeiro, Brazil. The common name, which is also the genus name, derives from the Ancient Greek *kalupto*, "to cover" and *oura*, "tail." This is a reference to the very short tail, which only just projects beyond the tail coverts. Strangely enough, when this bird was first described by Vieillot, he assigned the genus name *Pardalotus*, associating it with the Australian pardalotes, giving it the vernacular name Le Pardalote Huppé, the Crested Pardalote, and wrote:[39]

> We know only the specimens of the birds of which this division is composed. With the exception of one which is in Brazil, the others live in New Holland [Australia].

# Camaroptera

The common name follows the genus name in this group of four species of the Cisticolidae. Sundevall erected the genus in 1850 with the comment:[162]

> Alæ fornicatæ, breves: tantum basim caudæ tegunt. [Wings arched, short: only serving to cover the base of the tail.]

His explanation for his choice of name is "Καμάρα *fornix*; πτερόν *ala*." In other words, it derives from the Ancient Greek καμάρα *kamara* for "arch" and πτερά *ptera* for "wings" (plural of πτερόν *pteron*), referring to the short wings that reach only to the uppertail coverts. The wings in all species are variously snapped and quivered in social and breeding behaviors.

# Canary

The original canary, the Island Canary *Serinus canaria*, was named by Linnaeus as *Fringilla canaria* in 1758. It was named after the islands from which it came to Europe, the Canary Islands off the coast of Northern Africa. And in turn, the islands were so named after the populations of large dogs there, from the Latin word *canis*, "dog." Pliny the Elder first assigned the name to the islands in *Naturalis Historia*, completed in AD 77:[477]

> The one next to it is Canaria; it contains vast multitudes of dogs of very large size, two of which were brought home to Juba . . . While all these islands abound in fruit and birds of every kind, this one produces in great numbers the date palm which bears the caryota, also pine nuts. Honey too abounds here, and in the rivers papyrus, and the fish called silurus, are found. These islands, however, are greatly annoyed by the putrefying bodies of monsters, which are constantly thrown up by the sea.

Ray and Willughby explained the name in 1678:[16]

> All those Islands, which the Ancients called Fortunate, are nowadays called the *Canaries*. Out of which in our Age are wont to be brought certain singing birds, which from the place where they are bred they commonly call *Canary-birds*.

There are now sixteen species of canaries in various genera, all in the Fringillidae.

# Canary-Flycatcher

The canary-flycatchers are two members of the genus *Culicicapa* in the Stenostiridae, the fairy flycatchers, a family that was only established in 2005. Prior to this, the enigmatic members of the family were scattered across a number of different families. In the case of the canary-flycatchers, as the name suggests, they were classified with the Old World flycatchers.

In 1820, Swainson used the name "Ceylonese Flat-bill," but by the 1830s the name had been changed to "Grey-headed Flycatcher."[50] The word "canary" was added by the early 1900s, an obvious reference to the predominantly yellow plumage of both species. (See Canary and Flycatcher.)

# Canastero

Twenty-two species of canasteros in the Furnariidae are confined to South America. The name refers to the nests that are generally large, round stick-and-grass structures, usually with an opening at the top, situated off the ground among tree branches, the word *canastero* meaning "basket maker" in Spanish. In 1896 the canasteros were listed in the "Catálogo de las Aves Chilenas" with the note:[113]

> Debe su nombre comun al gran nido que hace de palitos. [Owes its common name to the large nest made of sticks.]

# Canvasback

This stately pochard was given its name by the English settlers on the Atlantic Coast based on the "the color of the back, which is white, very finely vermiculated with narrow, zigzag, blackish bars or rows of dots"; supposedly they likened the plumage to white canvas fabric.[114] Newton described the bird well:[7]

> The male Canvas-back has a darker head, and the black lines on the back and flanks are much broken up and further asunder, so that the effect is to give these parts a much lighter colour, and from this has arisen the bird's common though fanciful name.

# Capercaillie

The name for two of the four species of *Tetrao* grouse comes from a corruption of the Scottish Gaelic *capall-coille*, meaning "horse of the woods." Other early authors conjecture that it's more likely the name is derived from the Gaelic *cabhar*, an "old man" or "bird" and *coileach*, a "cock," drawing comparisons with *coileach-coille* for a "woodcock" and *coileach-oidhche* for an "owl," among other Gaelic bird names. The most complete discussion of this interesting bird name was given in a full volume about the bird from 1879, which can be summarized as follows:[394]

> About the second part of the word . . . there can be little room for doubt . . . but the first part of the word . . . is more difficult . . . *Cabhar* pronounced *Cavar*, means, according to our dictionaries, a hawk or old bird. It is not at all unlikely that it is the word spelled *Caper*. There is a similar word used in the name for a snipe, *Gabhar-athar*, thought by some to mean *the goat of the air*, from its bleating note. . . . I therefore lean to the idea that both in *Cabhar-athar* and *Cahhar-coille*—the one being the bird of the air, and the other bird of the woods—the original term is Cabhar.
>
> On the other hand, not a few Gaelic scholars consider that Capercaillie is derived from "*Capull, a horse*" . . . This reading gives "*Horse of the woods*." "It is called 'Horse of the woods,' because of its size, strength, and beauty, as compared with other wood birds." The first author of a Gaelic dictionary . . . thus renders it, and all subsequent authors of Gaelic dictionaries do so likewise. . . . (A)ll the Gaelic dictionaries [at the time] without exception, give *Capull coille*. "None have *caper, cabar*, or *cabher*."

The author finished with some words that remain prescient in modern times:

> In order to obtain the correct translations of Gaelic names, we must not, I believe, go to the Gaelic scholar alone, but first to the shepherd or crofter, whose family has for generations lived upon the same land, and whose father or grandfather was very likely the person who first applied the names, and which, being handed down from father to son, would preserve their purity of pronunciation, intonation, and significance, as well as, probably, a relation of the circumstances under which they were so named.

# Capuchinbird

Just like the Capuchin Monkey, the one species of Capuchinbird, in the Cotingidae, is said to resemble the Franciscan monks in the Order of St. Francis known colloquially as Capuchins. This name in turn derives from the Italian word *cappuccio*, "hood, cowl" for the long, pointed hoods on their cloak uniforms. The bird's most distinctive feature, its unfeathered head with its strange cowl of orange-brown feathers, draws comparisons with the Capuchin Monks' monastic crown, a tonsure where the top of the head is shaved, keeping only a narrow ring of short hair, as a sign of religious devotion or humility.

# Caracara

Ten birds in the Falconidae are known by the common name caracara. There are various genera, but at least one is the genus *Caracara*. This bird's name came to English via the Spanish and Portuguese, who in turn named the bird after the Tupi name. The word is *Caracaraí*, the name given by the Indigenous people to a small hawk, common in the region of Brazil now sharing the same name. Orbigny wrote about the Crested Caracara in his *Voyage dans l'Amérique Méridionale* in 1835:[138]

> Le nom du *carácará*, d'origine guarani, est le même chez toutes les tribus de cette grande nation. Ce nom est-il formé de la réduplication de *cará*, qui signifie adresse, astuce, curiosité, etc., toutes qualités que nous retrouvons chez l'oiseau qu'il désigne?

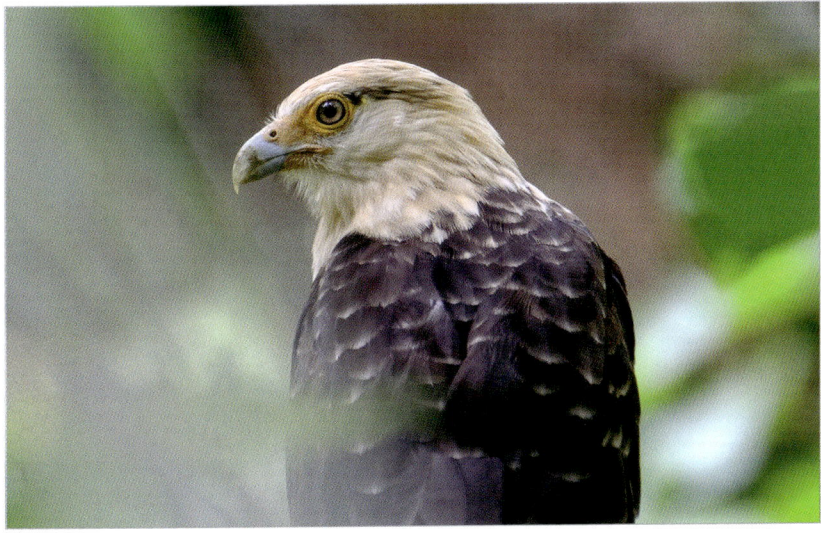

*Yellow-headed Caracara (Costa Rica)*

[The name of the *carácará*, of Guarani origin, is the same among all the tribes of this great nation. Is this name formed from the reduplication of *carâ*, which signifies skill, cleverness, curiosity, etc., all qualities that we find in the bird it designates?]

But Garcia wrote that while many believe the name is onomatopoeic, the more plausible etymology is that it is derived from the Tupi word *carãe*, meaning "to scratch or tear at something with claws or nails," becoming the frequentative word *carãe-carãe*, literally, "the scratcher," but possibly more accurately a "raptor."[133] The name of the bird gave rise to the name of the municipality, in the state of Roraima. It is also known to be the name for several species of hawks in a Carib language, Trio, that is usually transliterated as *karakara*.

# Cardinal

There are seven species of cardinals in two genera in the Thraupidae, the tanagers and allies, and two species in the Cardinalidae. The scientific name *Loxia cardinalis* was given to the Northern Cardinal *Cardinalis cardinalis* by Linnaeus in 1758. It's thought that the European colonists of North America coined the name, although Ray and Willughby noted in 1678:[16]

By the *Portugues* it is commonly called, The *Cardinal bird*, because it is of a scarlet [*purpurei*] colour, and seems to wear on its Head a red hat.

But there is no doubt that the name *cardinal* is an allusion to the brilliant red color and the crest, which reminded observers of similarly clad religious figures of the Catholic church with their tall miter caps. And although the seven cardinals in the Thraupidae are not fully red, all but one of them have conspicuous bright red in their plumage. The Yellow Cardinal is something of an anomaly, though.

# Carib

The two *Eulampis* caribs are hummingbirds, both found on the eastern Caribbean Islands, hence the name. The word Carib was the name of the original human inhabitants of the Lesser Antilles and the adjacent South American coast. The vernacular name appears to have first been used in Gould's *Monograph of the Trochilidæ* in 1849, but he gave no justification, although he did make much of his (correct) belief that both species are restricted to the Caribbean Islands.[6]

# Casiornis

This is another species whose common name derives from the genus name. In this case, the word is for the Ancient Greek κασία *kasia*, meaning "cinnamon tree," and *ornis*, for "bird." So, it's not surprising that the two Tyrranidae species from South America bearing this name should both be cinnamon colored.

# Cassowary

There are three species of cassowaries in the Casuariidae, a family that also contains the Emu. The first cassowary arrived in Europe in 1597 as a gift to Emperor Rudolf II, but the name was coined in the 1610s, via Dutch, according to many authorities, from the Indonesian (Malay) word for the bird, *kasuari*. As Boles observed in 1987, the name actually originated from two Papuan-language words, *kasu*, meaning "horned," and *were*, meaning "head," clearly a reference to the horny casque on the head.[119] Given the distribution of all three species, this makes more sense. The Papuan words were probably used by the Indonesians, who interacted with the Dutch and Portuguese sailors who carried the birds to Europe. In 1631, the Dutch physician and pioneer of tropical medicine, Jacobus Bontius wrote in *Historia naturalis & medicae Indiae Orientalis*:[332]

> De Emeu, vulgo Casoaris. In insula Ceram, aliisque Moluccensibus vicinis insulis, celebris haec avis reperitur. [The Emeu, generally Casoaris. On the island Seram, the Moluccas and other neighboring islands, this famous bird is found.]

The names Emu and Cassowary were used interchangeably up to the late 1700s, and, perhaps surprisingly, "New Holland Cassowary" was an early name for the Emu. Both the Emu and the Southern Cassowary were described by Phillip in 1789:[110]

> The whole of the head and neck is also covered with feathers, except the throat and fore part of the neck about half way, which are not so well feathered as the rest; whereas in the common Cassowary, the head and neck are bare and carunculated as in the turkey.

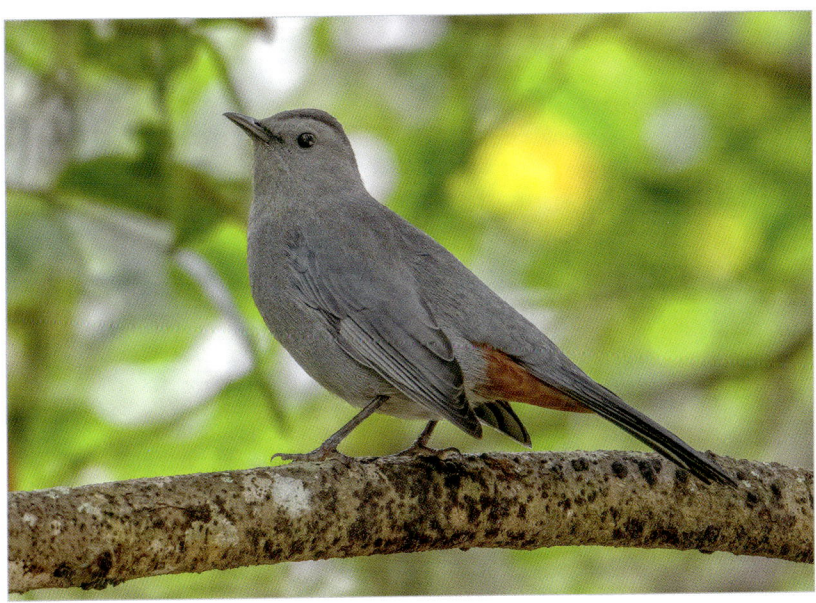

*Gray Catbird (Georgia, USA)*

# Catbird

There are thirteen species known as catbirds—ten of them in the bowerbird family, Ptilonorhynchidae, one in the Sylviidae (sylviid warblers), and two in the Mimidae (mockingbirds and thrashers). In the case of the ptilonorhynchids, all ten birds belong to the *Ailuroedus* genus, meaning "cat singer" from the Ancient Greek (*ailouros*, "cat," *ōdos*, "singer") and they do indeed often vocalize in a tone reminiscent of a cat's. In the case of the Abyssinian Catbird of the Sylviidae, it sounds nothing like a cat, but it does bear a superficial resemblance to the Gray Catbird of the Americas. The Mimidae catbirds are known to have one somewhat catlike mewing call, despite Catesby's mistaken assertion that the "Cat-Bird":[70]

> Has but one note, which resembles the mewing of a Cat, and which has given it its name.

On the other hand, the Australasian *Ailuroedus* catbirds' vocalizations are very catlike, prompting Gould's wonderful description in his *Handbook to the Birds of Australia* from 1865:[103]

> In its disposition it is neither a shy nor a wary bird, little caution being required to approach it, either when feeding or while quietly perched upon the lofty branches of the trees. It is at such times that its loud, harsh and extraordinary note is heard; a note which differs so much from that of all other birds, that having been once heard it can never be mistaken. In comparing it to the nightly concerts of the domestic cat, I conceive that I am conveying to my readers a more perfect idea of the note of this species than could be given by pages of description. This concert, is performed either by a pair or several individuals, and nothing more is required than for the hearer to shut his eyes to the neighbouring foliage to fancy himself surrounded by London grimalkins of house-top celebrity.

# Cave-Chat

The Angola Cave-Chat is a distinctive pied bird in the monotypic genus *Xenocopsychus* in the Muscicapidae, the Old World flycatchers. Its favored habitat is sandstone rocky hills, gorges, and outcrops with scatterings of forest patches, gullies, and caves. This accounts for the first element of the common name. See Chat for details of the second element of the name.

# Chachalaca

There are sixteen species of chachalacas, all in the *Ortalis* genus. The unusual name derives from the Nahuatl (an extinct Aztec language) verb *chachalaca*, meaning "to chatter," which is a good description of the vocalizations of all birds in the genus. In 1954, Skutch observed:[116]

> Only exceptionally, as in *quetzal* and *chachalaca*, have we a good indigenous name for some of the more conspicuous members of the avifauna; and these should be carefully preserved in modern usage.

# Chaco-Finch

The Many-colored Chaco-Finch is a finch-like tanager in the Thraupidae. Its name is a bionym combined with a taxonym. The first element of the name, the bionym, references the bird's habitat, which is the Chaco Plain, a hot semiarid region of eastern Bolivia, western Paraguay, northern Argentina, and parts of southwestern Brazil. The name Chaco comes from an Indigenous language of the Andes, Quechua, *chaqu*, meaning "hunting land," a nod to the rich biodiversity of the region.

# Chaffinch

This is a very old name for a common European bird that derives from the Old English *ceaffinc* (from *ceaf*, "chaff, husk" and *finc*), literally a "chaff finch" due to its habit of eating grain found among the chaff on farms in winter. Turner listed the name in Middle English in 1544:[26, 119]

> De Fringilla: Σπίζα,fringilla, Anglicé a chaffinche a sheld appel, a spink.

> [Of the Fringilla: Spiza, fringilla, in English a chaffinche, a sheld-appel, a spink.]

Kersey's dictionary in 1715 gave the definition of "Chaffinch" as[120]

> a Bird so call'd because it delights in Chaff.

Some authorities cite other possible etymologies, such as the Middle English *chaufen*, meaning "to warm," which could allude to the "reddish or "warm" breast of the male.[121] There is not a lot of support for this theory, however. The name was originally applied to the Common Chaffinch, but two other species in the Fringillidae, both found in the Canary Islands, now also bear the name.

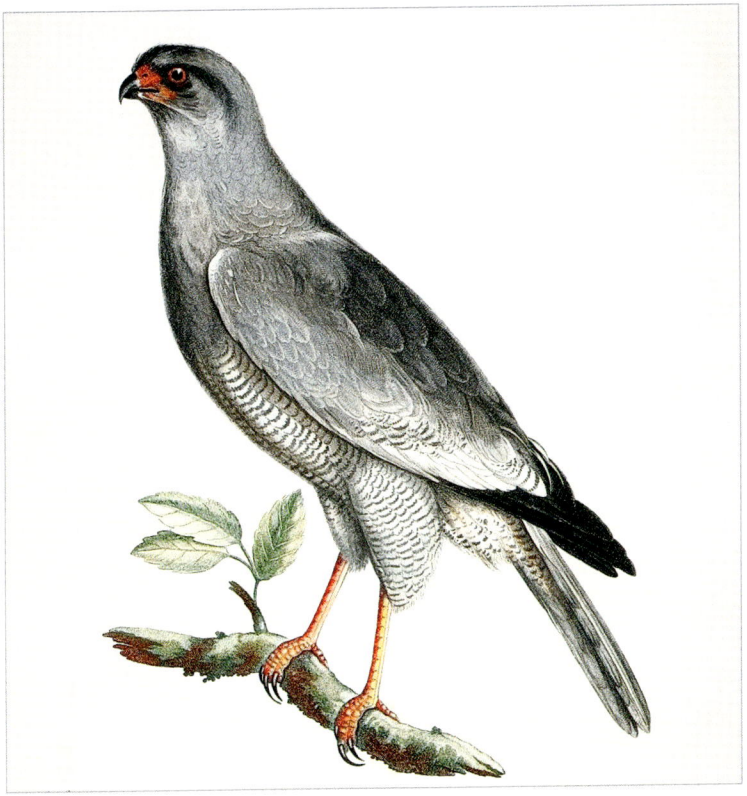

*Pale Chanting-Goshawk in Le Vaillant's* Histoire Naturelle des Oiseaux d'Afrique *(1799)*

# Chanting-Goshawk

(See Goshawk.) Three species of chanting-goshawks belong to the same genus, *Malierax*, in the Accipitridae. The chanting prefix refers to the birds' vocalizations, but this bird was originally dubbed "Le Faucon Chanteur," the Singing Falcon, by Le Vaillant, who was the first to describe the bird to European ornithology.[93] Although Buffon used the name "Singing Falcon," he evidently disagreed with the name:[8]

> Perched upon the summit of a tree, near its female, which it never quits all the year, or else in the vicinity of the nest where she is brooding, he sings for hours together, and in a particular manner . . . it is to [*Le Vocifer*'s*] resounding *ca-hou-cou-cou*, that we must approximate the equally boisterous cry of the singing falcon: it is to clamours a long time repeated, but monotonous, that all its musical art is reduced: so that this rapacious, bird ought rather be called the *crying* than the *singing*, falcon, and we must not censure nature so grossly as to imagine that she would unite delicate and touching sounds to ferocious tyranny and the revolting habits of carnage. [*African Fish-Eagle]

# Chat

The name chat is used for a wide range of small, active passerines. Most are members of the Muscicapidae, the Old World flycatchers, but others given the name are four chats in the Meliphagidae, the honeyeaters, one member of the Icteriidae, the Yellow-breasted Chat, and three in the Cardinalidae, the cardinals and allies. The epithet was first given to the European Stonechat, Turner calling it the "Stone-Chatter" in 1544.[119] The name refers to the vocalizations, which are quite "chatty," and is now used for many such vocal small birds, often in hyphenated forms as in chat-tyrant, robin-chat, cave-chat, cliff-chat, anteater-chat, black-chat, and chat-tanager. The English word "chat," meaning to make small talk, derives from Middle English *chateren*, from earlier *cheteren, chiteren* "to twitter, chatter, jabber," of imitative origin. Ray and Willughby wrote about the Whinchat, earlier known by the French name *Le Jaboteur*, the Jabberer (see Whinchat):[16]

> Lives about Rivers and Fens, especially in moist meadows and if it be driven away by Horses feeding there, it flies away with a certain chattering, wherein it seems after a fashion to imitate the neighing of a horse.

# Chat-Tanager

Two species of chat-tanagers are the sole members of the Calyptophilidae endemic to the island of Hispaniola in the central Caribbean. Of course, strictly speaking, they are neither chats nor tanagers, but they were classified in the Thraupidae with the tanagers until recent molecular studies showed their unique phylogeny. The first element of the name refers to their varied and cheery vocalizations.

# Chatterer

The three species of chatterers in the Leiothrichidae are found in northern Africa and the Middle East. They are so-called for their chattering vocalizations; all three species give loud six- to ten-note descending whistles or twitters and various other loud calls. The first use of the name in 1678 was for a very different bird when Ray and Willughby called the Bohemian Waxwing the "Bohemian Chatterer," a bird that they claim is "of a very hot temperament."[16] In the past, the name has been used for members of the Cotingidae as well, and as Newton pointed out the name is[7]

> a word that has been used by ornithologists in a very wide sense, and wholly irrespective of its meaning.

# Chickadee

Seven species of chickadees in the Paridae derive their names from their vocalizations. One of the most commonly heard calls is one that can be transliterated as *chicka-dee-dee-dee-dee*. These namesake vocalizations have been well studied and are said to be highly variable with multiple functions, one of which is to give the all-clear after a potential predator has left. Another function may be to alert congeners of a new food source. As Ingersoll noted in 1937:[9]

> It is best to . . . call the whole tribe Chickadees, which they themselves ask us to do in their cheery calls to one another.

# Chiffchaff

This is another onomatopoeic bird name. There are four species of warblers in the genus *Phylloscopus* that have chiffchaff in their common names. The "original" chiffchaff is the

Common Chiffchaff, which is found from Iceland all the way eastward to India, as well as the British Isles, home of the English language. The name Chiffchaff is an imitation of its song, as an entry in White's *A Naturalists' Calendar* in 1795 expounded:[123]

> The smallest uncrested or willow wren, or chiff chaf, is the next early summer bird which we have remarked; it utters two sharp piercing notes, so loud in hollow woods as to occasion an echo, and is usually first heard about the 20th of March.

# Chilia

The Crag Chilia is endemic to Chile and was previously placed in the monotypic genus *Chilia* but has now been moved to the *Ochetorhynchus*. Clearly, it was named for its country of origin, which, until at least 1900, was known by the older spelling "Chili" in English.

# Chlorophonia

The five species of chlorophonias in the Fringillidae (finch) family all take their common name from their genus name. The word is derived from the Ancient Greek χλωρός *khloros*, meaning "green" and from the genus *Euphonia*, also in the Fringillidae—in other words, the chlorophonias are "green euphonias." (See Euphonia.)

# Chlorospingus

The genus *Chlorospingus* consists of nine species of New World sparrows called chlorospingus in the Passerellidae. They are all basically green, hence the appellation, erected by Cabanis in 1851, which derives from the Ancient Greek χλωρός *khloros*, for "green," and σπίγγος *spingos*, for "finch."[63]

# Chough

There are three species with the common name chough, in two different families—one in the small Australian family the Corcoracidae, and two in the Corvidae. The Australian White-winged Chough was given the name by the early European settlers due to its supposed similarity to the two species of Palearctic chough in the Corvidae. The onomatopoeic name comes from the Old English word *ceo* or *ceahhe* that later became *choʒen, chawʒe,* or *choughe* in Middle English. The first known use of the word dates back to the early 1200s with the name *kaue* (another Middle English spelling for chough) mentioned in a cautionary tale in the *Ancrene Riwle or Guide for Anchoresses*, which was a guide for female religious ascetics.[301]

> Þe hen hwen ha haueð ileid, ne con bute cakelin; ah hwet biʒet ha þrof? kimeð þe *kaue* ananriht & reaueð hire hire eairen. When the hen has laid, she must needs cackle. And what does she get by it? Straightway comes the *chough* and robs her of her egg and devours all that of which she should have brought forth her live birds.

As can be seen in this spelling, the name has its roots in the Proto-West Germanic *kahwu* for a jackdaw or crow, which in turn comes from the imitative Proto-Indo-European *gewH-* "to crow, caw, shout." Originally, the name was used interchangeably for crows, jackdaws, and choughs. In 1552 an entry in Huloet's *Abcedarium* read:[364]

> Caddowe or choughe, byrde some call them Jacke dawe, *cornix, monedula*. [Hooded Crow, Eurasian Jackdaw.]

Later, in 1688, Holme wrote:[363]

> Jack-daw or Daw . . . in some places is called a Caddesse or Choff.

# Chowchilla

The charming Chowchilla is a member of the Orthonychidae, the logrunners. It has a tiny range in Australia's northeast, where it can be found foraging on the rainforest floor. The bird was first collected in 1889 by the naturalists Cairn and Grant, at which time it was commonly known as the Spalding's Orthonyx. Later, it went by a number of different names, including Auctioneer-bird, Black-headed Logrunner, and Spalding's Spinetail. The name appears to have first appeared in the literature in Alfred North's 1889 *Descriptive Catalogue of the Nests & Eggs of Birds Found Breeding in Australia and Tasmania*.[124] North wrote the name with the heading "ORTHONYX SPALDINGI, Ramsay Spalding's Orthonyx," and the subheading "'Chowchilla.' Aborigines of Cairns District." In a list of local Aboriginal names from 1909, the Australian ornithologist Sidney Jackson cited "Chowchilla" and described its call as "Chow-chilla, chow-chow, chow-chilla," confirming that the name is echoic of the bird's song.[125] The name is taken from the Dyirbal and Yidiny languages of North Queensland, who called the bird *jawujala*.[127]

# Chuck-will's-widow

This member of the Caprimulgidae is in the genus *Antrostomus,* the other members of which are variously called nightjars or whip-poor-wills. The name is onomatopoeic as Holder recounted in 1877:[65]

> The Chuck-will's-widow is another American species, the somewhat singular name applied to which has been derived from its note. This remarkable cry is said by the American writers to resemble the words Chuck-will's-widow, each syllable being slowly and distinctly pronounced, with the principal emphasis laid on the last word. It is so loud that in a still evening it may be heard at a distance of nearly a mile and in those districts where the birds are numerous, their incessant vociferation makes the mountains ring with echoes during the whole evening. In general the note is heard only in the morning and evening, but on moonlight nights it is continued throughout the whole night.

# Chukar

The Chukar is a partridge of the Palearctic region, found in rocky, scrubby hill country. The name comes from the onomatopoeic Hindi name चकोर *cakor* for the bird. Hume wrote in 1880:[127]

> The Chukor [*sic*] is a very noisy bird, repeating constantly in a sharp, clear tone, that may be heard for a mile or more through the pure mountain air, his own well-applied trivial name.

# Cicadabird

There are twelve species of cuckooshrikes in the Campephagidae known as cicadabirds. In 1926 the name appeared in the *Official Checklist of the Birds of Australia* with "Cicada-bird" given as an alternative name for the "Jardine Caterpillar-eater," which is now known as the Common Cicadabird.[128] In 1931, Neville Cayley turned the tables and gave Cicada-bird as the name, with the former listed as an alternative.[216] As he noted:

> "Cicada-bird" is an appropriate name; its call-notes resemble the buzzing sound of a large cicada, like "Kree-kree," uttered continuously.

# Cinclodes

Fifteen species of cinclodes in the similarly named *Cinclodes* genus are nested in the Furnariidae (ovenbirds and woodcreepers) family. The scientific name comes from κίγκλος *kinklos*, which was a type of small, tail-wagging waterbird in the Ancient Greek literature, maybe wagtails or dippers (cf. genus *Cinclus*), and *-oides*, "like or resembling," a fitting description for these terrestrial birds that are often found near water. According to Aristotle, the *Kinklos* was smaller than a thrush, lived near inland waters, and wagged its tail, but he later stated that it lives by the sea and, in character, had wicked tricks and was difficult to hunt, but once caught was very tame. Arnott believed the Greek *Kinklos* was the White Wagtail *Motacilla alba*, a very different bird to the cinclodes.[302]

# Cisticola

A large genus of fifty-one species in the Cisticolidae. The name *Cisticola* is from the Ancient Greek κίσθος *kisthos*, a species of rock-rose, and the Latin *colere*, "to dwell." The common name is the same as the genus name. The name was first given to the Zitting Cisticola, but it was then known colloquially as the "Fan-tailed Warbler." Kaup, who erected the genus in 1829, no doubt did so in homage to the bird's preferred habitat as Latham recalled that the bird[218]

> inhabits all the shrubby parts of the district . . . , ever darting with vast alacrity among the bushes; when disturbed, takes long flights, chirping all the way, with a remarkably loud and shrill note; at other times makes no noise whatever.

# Citril

These three species of yellow canaries are all found in East Africa. In 1663 Ray and Willughby came upon a bird known by the name "Citril" in Vienna.[16] Speaking of the "Canary-bird" from "Canaria . . . an island of the Atlantic Sea," they wrote:

> So that it differs little from those small birds, which our Countrymen call *Citrils* (Citrinella or Citrine), or those they call *Zisels*, and the Italians, *Ligurini*, save that it is a little bigger than either of those, liker in shew or outward appearance to this, something greener than that.

The name is a diminutive form of the Italian *citrinella*, which was probably used for any yellow-colored finch-like bird. The common name was probably any number of different species of canary that were sold as pets in Europe. Ray and Willughby also observed:[16]

> Now adays there be many of them brought over; nor are they sold so dear that even mean persons can afford to buy and keep them.

# Cliff-Chat

The cliff-chats are two species in the genus *Thamnolaea* that, as the name suggests, inhabit areas with cliffs—*kopjes inselbergs* (small, isolated, rocky hills), gorges, and boulder-strewn slopes—in sub-Saharan Africa. (See Chat.)

# Cochoa

Two species of cochoas in the *Turdidae* occur in South and Southeast Asia. The Nepali name कचोवा *kachowa* for this species has been adopted as the common English name. It has been transliterated from the Nepali in the past variously as *Cocho* and *Kochowa*. Hodgson, who erected the genus name in 1836, wrote:[97]

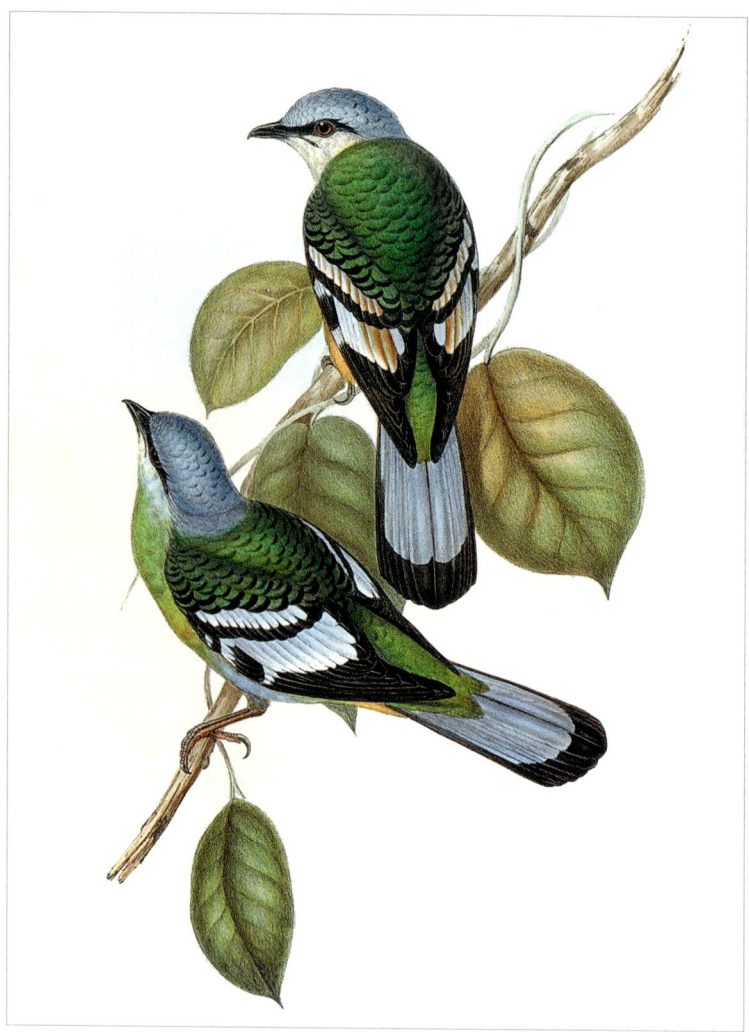

*Green Cochoa in Gould's* Birds of Asia *(1850)*

These birds are not generally or familiarly known to the Nipalese, but the foresters, whom I have met with, denominate them *Cocho*: and by that name, latinised into Cochoa, I have designated them generically in my note book.

# Cockatiel

This unique member of the Cacatuidae is the smallest member of the family, as reflected in the genus name *Nymphicus,* a little nymph, as well as the common name. It's thought that cockatiels were first imported into Europe in the mid-1800s and that an importer of exotic animals, Charles Jamrach (a renowned dealer of wildlife in nineteenth-century London) gave it the name cockatiel, from a Dutch word *kakatielje,* a compound of the word *kaketoe* ("cockatoo") and a diminutive suffix, borrowed from the Portuguese *cacatilho,* meaning "little cockatoo." Newton, who called the bird the "Cockateel," said that it was "a bird-fancier's name lately invented by Mr. Jamrach."[7]

# Cockatoo

Of the twenty-one species in the Cacatuidae, fifteen are known colloquially as cockatoos. An illustration of a cockatoo has recently been found in a book on falconry once owned by the Holy Roman Emperor Frederick II, meaning that cockatoos were known to Europeans since at least the 1200s.[588] The Dutch and Portuguese would have encountered the bird and its name during their early expeditions to the Indonesian archipelago in pursuit of the spice trade in the 1500s. Mandelslo wrote in *The Voyages and Travels of the Ambassadors from the Duke of Holstein . . . Whereto are Added the Travels of J. Albert de Mandelslo . . . into the East-Indies* about cockatoos that he saw in 1638:[406]

> L'on les appelle kakatou, à cause de ce mot qu'ils prononcent en leur chant assez distinctement. [They are called kakatou, because of this word which they pronounce in their song quite distinctly.]

An early account of a "Cacatua" appeared in *Voyage d'Orient du R. P. Philippe de la Très-Saincte-Trinité* from 1652:[258]

> J'ay vû en la ville de Goa un autre oyseau que l'on nomme *Cacatua*, parce qu'il prononce ordinairement cette parole. Il a le bec fort long & les plumes de diverses couleurs, il n'est pas neanmoins fort agreable à la veuë. Il est presque esgal en grandeur au Paon. [I saw another bird in the city of Goa called Cacatua, because it usually speaks this word. It has a very long beak & feathers of different colours, it is nevertheless not very pleasant when seen. It is almost equal in size to the Peacock.]

Given that no species of cockatoo has ever occurred in the vicinity of Goa, it may perhaps be assumed that this was a cage bird. A more accurate account appeared in 1676 in Schouten's *De Oost-Indische Voyagie* when he wrote in Dutch:[333]

> At this point, the Moluccas Islands win over all the other countries. There are Louris whose colours are admirable; Parkietes, or Parkites, whose plumage is charming; Cacataües that have strong-white plumage on the belly.

The word came into the English language from the Portuguese *kaketoe* when they brought their specimens back to Europe, and it was attested from 1630 according to *Oriente Português*.[478] But the name is borrowed from the Malay *kakatūwa*. Yule, somewhat confusingly, noted in 1886:[250]

> This word is taken from the Malay *kākātūwa*. According to Crawfurd the word means properly "a vice," or "gripe," but is applied to the bird. It seems probable, however, that the name, which is asserted to be the natural cry of the bird, may have come with the latter from some remoter region of the Archipelago, and the name of the tool may have been taken from the bird. This would be more in accordance with usual analogy. (Mr. Skeat writes: "There is no doubt that Sir H. Yule is right here and Crawfurd wrong. *Kakak tuwa* (or *tua*) means in Malay, if the words are thus separated, 'old sister,' or 'old lady.' I think it is possible that it may be a familiar Malay name for the bird, like our 'Polly.' The final *k* in *kakak* is a mere click, which would easily drop out.")

Much later in 1939, König asserted that the name is probably composed of *gagak* for "crow," heard by the Europeans as *kaka* and transposed to the parrot, combined with *tūwa* for "old," because of the advanced age that these birds can reach. He wrote in his "Overseas Words in French" from 1939:[259]

*Kaka*, crow and *tuwa*, old is combined. The second component was probably added because this parrot species usually reaches a very old age. You could say it is "onomatopoeic" as one can only understand *kaka*, but not the word in its entirety.

And, indeed, the Indonesian word *gagak* is certainly onomatopoeic. The word *cock*, used widely in English at the time for a number of species of birds, might have influenced the anglicization of the Dutch *kaketoe*.

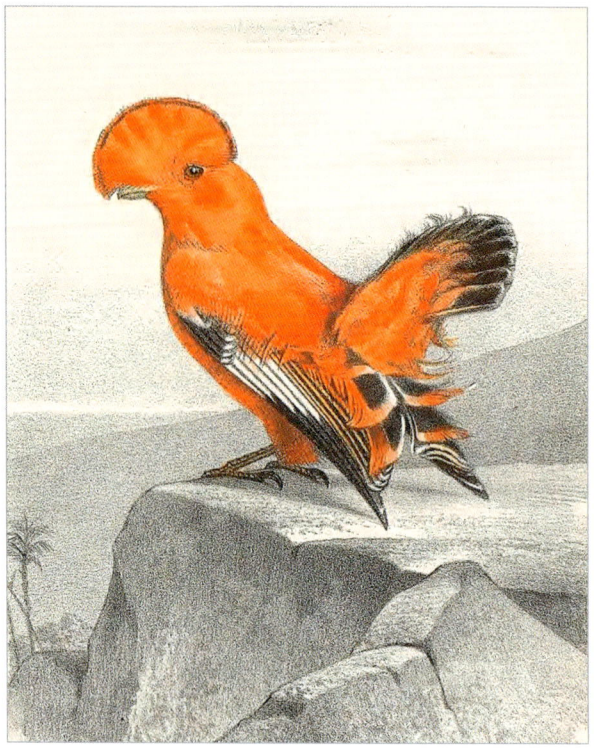

Guianan Cock-of-the-rock from Conty and Travies's Types du règne animal. Buffon en estampes *(1864)*

# Cock-of-the-rock

The two species of cocks-of-the-rock are spectacular cotingas found in the Neotropics. The genus name *Rupicola* means "rock dweller," and both this and the common name comes from the birds' habit of nesting among rocky outcrops. Given the male, the cock, doesn't assist at the nest, the name is a little incongruous. Maybe hen-of-the-rock would be more appropriate?

# Coleto

This unusual species of starling in the Sturnidae is endemic to the Philippines. In McGregor's *A Manual of Philippine Birds* from 1909, the author listed local names for the birds as[210]

*Co-ling*, Mindoro; *i-ling*, Ticao; *sa-ling*, Masbate; *co-le-to*, Manila.

The etymology of the name appears to be unrecorded, but another local name from Quezon province is said to be *kuling panot*. The influence of the Spanish colonizers on the Philippines' languages is strong, but the only possible derivation from Spanish would be from the word *coleto*, meaning "jerkin" or "doublet," a leather vest used as protection under a breastplate. It could be conjectured that the leathery bare skin on the bird's face reminded people of a *coleto* in the clothing sense, but this seems like a stretch. More likely, it is simply a local name for a bird so singular in its appearance that it was given a unique moniker.

# Comet

Three species of hummingbirds revel in the common name comet. Although the three are closely related, they are all in separate genera. The etymology of the word comet is from the Ancient Greek word *komētēs*, meaning "long-haired." Two of the three comet hummingbirds have notably long tails, thus explaining the name. Gould, who coined the common name, wrote in his introduction to his *Monograph of the Trochilidæ*:[6]

> Peru and Bolivia are the cradles of the splendid comet-tailed species of the genus *Cometes*.

# Condor

These two large New World vultures get their names from the Quechua (an Inca language) word *kuntur*.[589] The Spanish adopted the name in the seventeenth century when it morphed into the word *cóndor*. La Condamine saw the Andean Condor on his expedition to the "interior of South America" in 1778:[129]

> The famous bird called in Peru, *Contur* & by corruption, *Condor*, which I saw in several places in the mountains of the Province of Quito.

# Conebill

Eleven species of conebills in the genus *Conirostrum* are found in the Thraupidae (tanager) family. All have thin, conical bills accounting for the common name as well as the genus name. The name was coined in the early 1900s when they were known as "cone-bills." It is simply a translation of the genus name, which is from the Latin *conus* for "cone" and *rostrum* for "bill."

# Coot

Eleven species of coots belong to the genus *Fulica* in the Rallidae, the rails, gallinules, and coots. The origin of the name is debated, but it was originally used for a number of different birds; "moor coot" and "kitty coot" was the moorhen, a female Smew was a "wezel coot," the water rail was a "skitty coot," the Long-eared Owl was the "Horn-coot," and phalaropes were even called "coot-footed tringas."[121] The Eurasian Coot had many other local names, including Bellcoot, Bellkite, and Bellpoot, as well as Bald Coot and Bald Pout. Some claim it is onomatopoeic "for the harsh staccato calls"[130] but Swann believed[121]

> the derivation seems to be from the Welsh name *Cwta-iar*, lit. "short-tailed hen," from its very short tail.

Another theory is that the name is derived from the Dutch names for the common murre, *Zeekoet* (sea coot) or the Eurasian coot, *Meerkoet* (lake coot), although the Dutch *koet* is probably just a cognate of the English word, and both came from Germanic origins. The word in one form or another dates back to at least the 1500s, when Turner wrote about "those black fowls which Englishmen call Couts."[26] Newton believed that the name might be cognate with scout or scoter, especially given that the French name *macreuse* was used there for both the coots and the scoters, maybe due to a "popular estimation [of] some connexion between the birds."[7] The name, which was first given to the Eurasian Coot, was over time generically extended to all the species of *Fulica*.

## Coquette

The definition of the word coquette is a "woman who endeavors to gain the admiration of men, a flirt." It came into English from the French in the 1660s. It was, however, used originally for both sexes and came from the word *coquet*, which literally meant "little cock." No doubt, this is the explanation of the designation of the name to the eleven species of very spectacular hummingbirds with their incredible head plumes. Wood wrote in *The Illustrated Natural History* in 1862:[85]

> All the Coquettes possess a well-defined crest upon the head, and a series of projecting feathers from the neck, some being especially notable for the one ornament, the others for the other.

## Cordonbleu

The word means "blue ribbon" in French and refers to those worn by members of the knighthood. The three species of *Uraeginthus* finches with this appellation in the Estrildidae (waxbill) family all display similar light turquoise plumage. In 1890 the *Catalogue of the Birds in the British Museum* listed "The Blue-bellied Finch"; it wasn't until 1905 that Shelley adopted the name "Cordon-bleu," but he listed them as "Cordon-blues" in the index.[132] He noted:[139]

> The type species is well known in England as the Cordon-bleu, so I have adopted that name for all the members of the genus.

## Corella

There are four species of corellas in the Cacatuidae (cockatoo) family; three are endemic to Australia, and one is endemic to the island of Tanimbar in eastern Indonesia. This name is derived from the Australian Aboriginal Wiradjuri word *garila,* as listed in the *Wiradjuri Heritage Study*.[132] It was first used in the English language in 1859. However, Johnston wrote with great confidence in 1943:[191]

> Amongst the Pittapitta, Karanya and some others it was kolloora or kollora, but the Goa people called it koo-rella, whence the common European name, corella, was derived.

## Cormorant

This name came into English in the twelfth century via the Old French *cormarenc*, which in turn come from a contraction of the Late Latin *corvus marinus*, "sea raven," in the belief that they were corvids of some type. There are twenty-five species, with a worldwide distribution excluding the poles, but the English name was first applied to the species found in Europe, Great Cormorant. Another association was with Satan, possibly due to the erroneous belief that they were related to ravens. The association with ravens is demonstrated in the French explorer André Thévet's comment in 1557:[479]

> The beak [is] similar to that of a cormorant or other corvid.

And with Satan in John Milton's *Paradise Lost* written in 1667:

> Thence up he [Satan] flew, and on the Tree of Life,
>
> The middle Tree and highest there that grew,
>
> Sat like a Cormorant.

Early on, the name was used to denote gluttony as in Shakespeare's *Richard II*, written in 1595:

> *With eager feeding food doth choke the feeder:*
> *Light vanity, insatiate cormorant,*
> *Consuming means, soon preys upon itself.*
> *[The person who eats too fast will choke on his food.*
> *The hungry bird that can't get enough to eat*
> *will soon eat itself.]*

# Corncrake

Often rendered as Corn Crake, this member of the Rallidae is the "original" crake. The first element of the name, Corn, was designated to denote their preferred habitat away from water as opposed to the closely related water rails. The word in English has quite a long history, with a reference to it in the comic allegory *The Buke of the Howlat* by Holland written circa 1450:[366]

> *He gart the Empriour trowe, and trewly behald,*
> *That the Corne Crake, the pundar at hand*
> *Had pyndit all his prys hors in a pundfald*
> *For raus thaj ete of the corne in the kirkland*
> *[The Emperor's (Eagle's) horses are led off to the pound by the Corncrake,*
> *because they had been eating "of the corne in the kirkland."]*

Swainson wrote of the "Corn Crake" that it is[75]

> so called from its harsh cry; whence, and from grass and corn being its favourite haunts, are derived.

He listed many regional English names for the bird, including Creck, Cracker, Craker, Bean crake, Bean cracke, Corn drake, and Corn scrack. Many of these names, as well as the scientific name *Crex crex* (from the Ancient Greek κρέξ *kréx*), attest to the call. (See Crake.)

# Coronet

There are three hummingbirds with the common English name coronet. The word means "small crown" and came into English from the Old French *coronete*, which in turn comes from the Latin *corona*, "crown." Although Gould, who doesn't use the common name, described the crowns of these birds as being variously "a shining violet-blue" and a "luminous yellowish green," the feature isn't particularly dominant.[6] In good light, the crowns do indeed shine, but why these three should have been designated with this appellation is a bit of a mystery given none has any crown-like plumage. All three are known for their habit of holding their wings up in a V while feeding or after landing—could this be the explanation? From the front, the shape of some coronets may well resemble the shape of the bird in that posture.

# Cotinga

The cotingas are a family, the Cotingidae, of diverse and, in some cases, bizarre, arboreal, frugivorous birds of Central and South America. In many ways, they are the New World equivalent of the birds-of-paradise. Of the sixty-five species, twenty-four carry the name *cotinga* in their common English names. The word came into English via French from the ornithologist Brisson, who erected the genus in 1760, noting:[76]

> Cotinga, nom qu'on donne en Amérique à quelques especes de ce genre. [Cotinga, a name given in America to a few species of this genus.]

Many authorities state that it comes from the Old Tupi word *cutinga* or *catingá*, said to have been given by the Indigenous people to any "bright forest (bird)." According to Tupi dictionaries, the word for "forest" or "leaf" is *caá*, while *tinga* is generally translated as "white" rather than "bright." Other dictionaries state that it is akin to the Tupi *coting*, "to wash," and *tinga*, "white." The word *caátinga* is given in Portuguese as *mato-branco* [white bush] and the *Dicionario Ilustrado Tupi Guarani* asserts:[260]

> Caatinga in the language of the Indians means "white forest," due to the lack of water the plants of the caatinga become almost white.

It is possible that Brisson's observation was slightly off course. Curiously, Garcia never mentions the word cotinga in his *Names of Birds in the Tupi Language*, and his only entry for anything resembling the word is for *Ajurú-catinga*, in which he stated that it is a macaw mentioned in Marcgrave and[133]

> In my view the name . . . is derived [from] âca = tip, beak,—tinga = white, according to Marcgrav's diagnosis (loc. cit.): *rostro alvo*.

Whether this relates to the present-day word *cotinga* is difficult to ascertain.

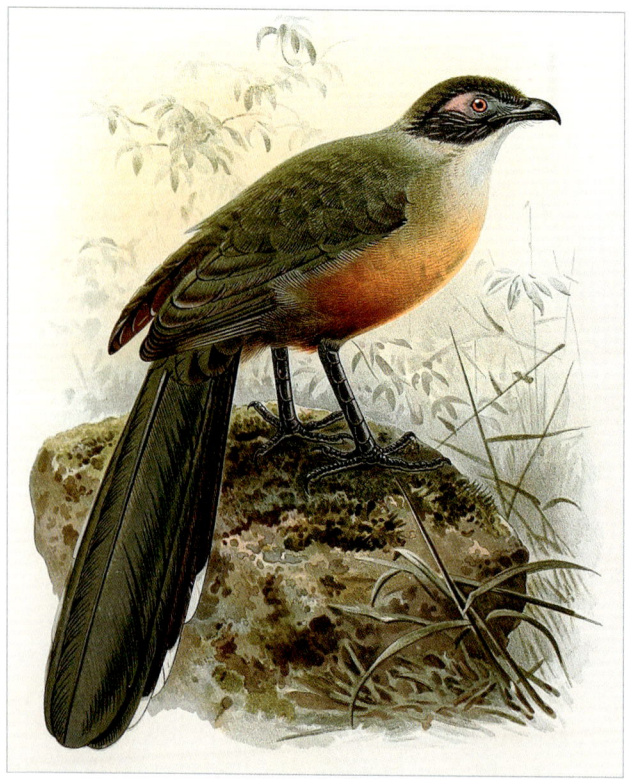

*Giant Coua in Grandidier's* Histoire Physique, Naturelle, et Politique de Madagascar *(1876)*

# Coua

The couas of Madagascar got their English names directly from the Malagasy word for the birds, *koa*, which in turn is onomatopoeic. There are ten species of cuculids (cuckoo family) in the synonymous genus *Coua*, all with the common English name *coua*, and all endemic to the island of Madagascar. Brisson was the first to use the name in the European literature in 1760 making the observation:[76]

> Les Habitans de Madagascar l'appellent COUA. [The inhabitants of Madagascar call it COUA.]

The genus name was not erected until 1821, and in 1889, Sibree commented in an article in the *Antananarivo Annual*:[134]

> These are the Couas (from a native name *Kòa*, pronounced *kooa*), which are large handsomely coloured birds.

# Coucal

The twenty-seven species in the Cuculidae (cuckoo) family, all in the genus *Centropus*, are known as coucals. The name is possibly derived from the French *coucou*, "cuckoo," and *alouette*, "lark," the latter a reference to the bird's long hallux claw, a feature that is also referenced in the genus name—Ancient Greek *kentron*, "spike," *pus*, "foot." Cuvier noted in 1836:[590]

> *Coucal, mot composé de coucou et d'alouette*; Centropus, *pied aiguillonné*. [Coucal, word made up of cuckoo and lark; *Centropus*, spurred foot.]

And that the

> species [is] from Africa and India, which have long, straight and pointed thumb nails like larks.

# Courser

The coursers are upright, long-legged shorebirds usually encountered running rapidly on the ground in open, dry habitats. They are found in Africa, the Middle East, and the Indian subcontinent, and there are nine species in three genera in the Glareolidae. Buffon, who was one of the first to describe the bird in the literature, called it "Le Coure-Vîte," justifying the choice of the name thus:[8]

> La rapidité avec laquelle il couroit sur le rivage, le fit appeler *coure-vîte*. [The rapidity with which he ran on the shore, made him call it *coure-vîte*—the fast runner.]

Influenced by Buffon, Latham erected the genus name *Cursorius* for the Cream-colored Courser, designated from the Latin *cursus* meaning "running."[17] This chain of events then led to Lewin adopting the name courser for the "European Courser" in 1800, presumably, as Newton noted, as a "rendering of Latham's word *Cursorius*."[135]

# Cowbird

Four species of cowbirds in the Icteridae (troupial) family occur in the Americas. All four are brood parasites, but the common name alludes to their habit of following cattle in search of flushed insect prey. As Catesby noted in 1729:[70]

> They delight much to feed in the pens of cattle, which has given them their name.

# Crab-Hawk

The Rufous Crab-Hawk is a raptor found in South America in the genus *Buteogallus*. The name is clearly a reference to the bird's diet, which is mostly, or possibly even exclusively, crabs.[3] As was often the case, this species was first described from a specimen, and nothing was known of its habits. Giving the bird's name as Equinoctial Eagle, Latham wrote in the 1820s:[17]

> By a label tied to the leg of one of these, we find it to be known by the name of Le Pagani roux, ou L'Aigle a plumage gris-roux. [The Red Pagani, or The Eagle with gray-red plumage.]

By the 1890s, the habits, as well as the taxonomy, of this bird were better understood, explaining the name change. Loat used the name in a paper in *The Ibis* in 1898:[345]

> The Barred Crab-Hawk (*Buteogallus aequinoctialis*), the food of which consists chiefly of lizards and crabs, is fairly abundant. The mud outside the sea-dam is simply honeycombed by crabs, affording an unlimited food-supply for this species.

# Crab-plover

The Crab-plover is an extraordinary bird, the sole member of the Dromadidae. While it is now known not to be a plover, it superficially resembles the members of that family. The first element of the common name clearly references the bird's feeding habits, as Jerdon noted in 1864:[136]

> It lives in small flocks on the banks of the rivers or sea shore, feeding, especially on the parts that have been left bare by the tide, on small crabs and other Crustacea, and perhaps also on shell fish.

# Crake

There are thirty-nine species of crakes in the Rallidae (rails, gallinules, and coots) family. They belong to a number of different genera but are all basically similar in form, being small rallids with relatively short bills. The first to receive the name was the Corncrake, historically known as the "Corne Crake." This onomatopoeic name derives from a Middle English word for the bird, *crak*, which in turn seems to have come to English from Old Norse *kráka*,

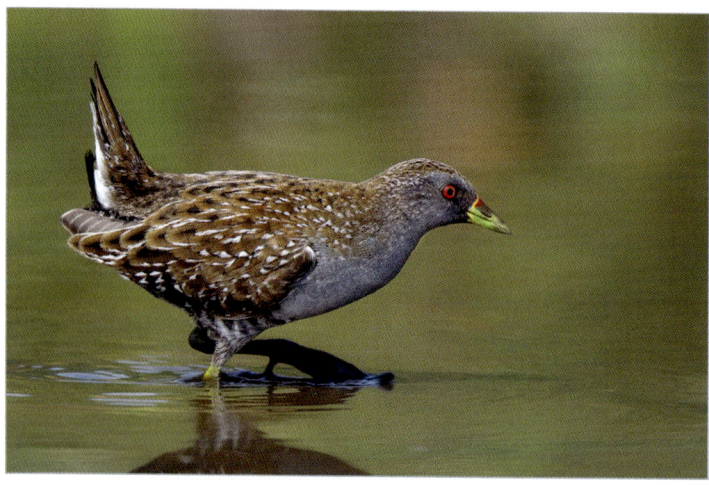

*Australian Spotted Crake (Victoria, Australia)*

meaning "crow." This came from the Proto-Germanic *krak-* or *kra-* "to croak or caw," and prior to that from Proto-Indo-European *gerh-*, an onomatopoeic term to describe the call of a crow. However, given Corncrake is the type, the name is just as likely to come from its own call as much as from a crow call. Almost like an ancient game of Rumors or Telephone, wherein the message, in the beginning, is transformed by the end of the game.

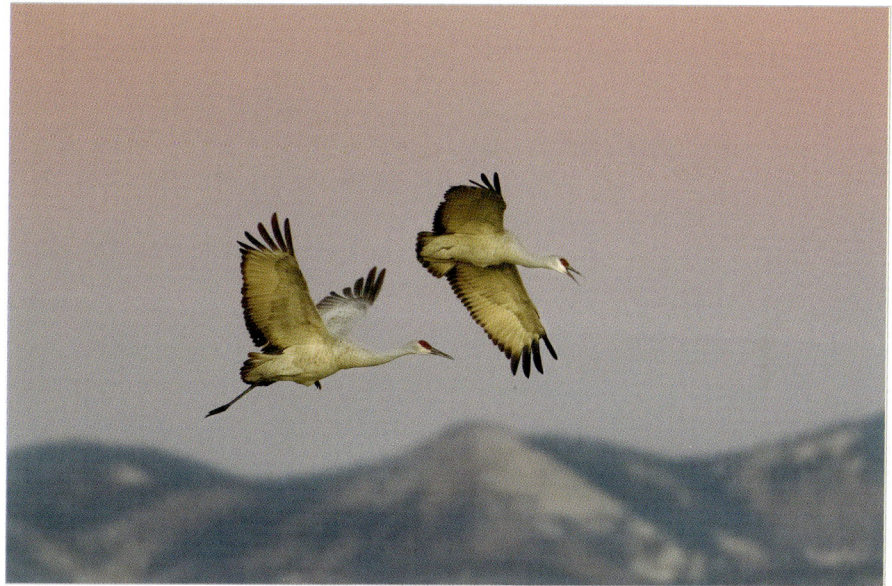

*Sandhill Crane (New Mexico, USA)*

# Crane

There are fifteen species in the Gruidae (crane) family, all but one of which bear the crane name (the Brolga is the exception—see entry). The name is rooted in the Proto-Indo-European *gerh-gerh*, meaning "to cry hoarsely," so it is probably onomatopoeic for the call of Common Crane. The Old English *cran* came via the Dutch *kraan* and Old German *kraen*. A poem from the mid-1200s in Anglo-Norman, a dialect of Old Norman French, "The Treatise of Walter de Bibbesworth," referred to a crane as "Grwe" (Wright, 1857):[137]

Le bouf mugist, la grwe growle. [The cattle low, the cranes growl.]

The present-day name first appeared in Turner (1544), who wrote in Latin of "De Grue," adding a footnote that the bird is also named[119]

[*Greek*] γερανός [*geranos*], grus, Anglice a crane, Germanice ein krän / ober ein kränich.

# Creeper

The three creepers (and the treecreepers) in the Certhiidae (treecreepers) family, as well as the seven species of Australasian treecreepers (Climacteridae), are named for their mode of locomotion on tree trunks, which looks like a "creeping" action as they search for invertebrate prey in and on the bark. One other bird, the Hawaii Creeper, is a member of the Frigillidae, and, in an example of convergent evolution, behaves in a similar manner to the other creepers in different families. (See Treecreeper.) Newton explained the name and its history best:[7]

Creeper . . . a term employed by ornithologists in a very vague sense . . . for it was customary to thrust therein almost every outlandish Passerine bird which could not be conveniently assigned to any other of the then recognized genera, provided only that it had a somewhat attenuated and decurved bill.

# Crescentchest

The Melanopareiidae, the family of crescentchests, is made of four species of passerines found in South America. All four birds have distinctive crescent-shaped black pectoral bands that contrast with the buffy throat and rufous underparts. Up until the early 1900s, these birds were named "Pied Antbirds" in the literature and placed in the Formicariidae, the antthrushes (a family that is much more restricted now).

# Crested-Flycatcher

Two members of the Monarchidae, the monarch flycatchers, and three members of the Stenostiridae, the fairy flycatchers, go by this name. As the name suggests, all have short crests and resemble the true flycatchers in the Muscicapidae. (See Flycatcher.)

# Crimsonwing

This is the choice of common name for the four species of *Cryptospiza*, crimsonwings, in the Estrlididae family, all found in Africa. All four have crimson upperparts, including the wings and in some cases the head and face. While the genus was erected in 1884, it wasn't until the early 1900s that a common name was coined, appearing as the "Crimson-wings" in Shelley's *The Birds of Africa*.[139]

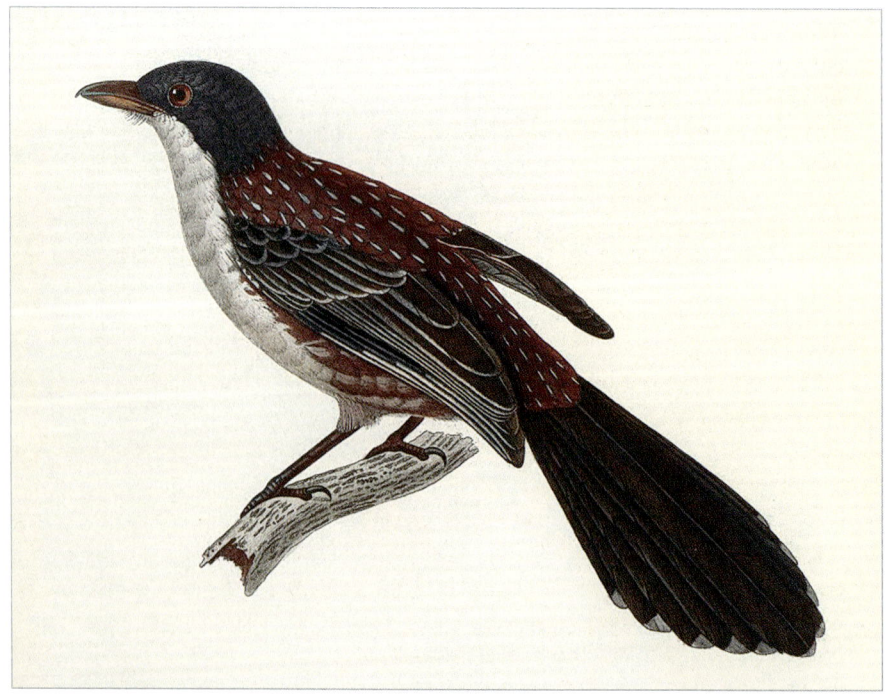

*Spotted Crocias in Temminck's* Nouveau recueil de planches coloriées d'oiseaux *(1838)*

# Crocias

There are two species of crociases in the Leiothrichidae (laughingthrush) family, both with tiny distributions—one in South Vietnam, the other in southwestern Java. The synonymous genus name was given to Spotted Crocias originally, by Temminck in 1836. The name comes from the Ancient Greek κροκίας *krokias*, meaning a "saffron-colored stone"; both species have chestnut-brown upperparts. The bird name is clearly related to the name of the flower, crocus, from which the spice saffron is derived, as illustrated in an 1875 translation of Erasmus's Colloquies from 1518:[591]

> Crocias gives the colour of crocus . . .

# Crombec

The genus *Sylvietta* consists of nine species named crombecs in the Macrosphenidae, the African warblers. The common name given to the Cape Crombec by Le Vaillant in 1802 is based on "Crombec," coined in imitation of *Krome-bec*, a term given to the bird by his Khoikhoi guides.[93] He wrote in 1802:

> J'ai donné à cet oiseau le nom de Crombec, imité de celui *Krome-bec* que mes Hottentots lui avoient donné pendant mon séjour sur la Rivière-des-Eléphans, où nous trouvâmes cette espèce en grande abondance. . . . Ce nom qui est hollandais, signifie bec courbé, et lui convient; ainsi nous le lui laisserons en le dénaturant un peu, afin d'en faire un nom spécifique, puisqu'autrement la dénomination pure et simple de bec courbé, qui convient à beaucoup d'autres oiseaux, pourroit être appliquée également à plusieurs autres espèces.

> [I gave this bird the name of Crombec, imitated from Krome-bec that my Hottentots had given it during my stay on the Rivière-des-Eléphans, where we found this species in great abundance. . . . This name, which is Dutch, means curved beak, and suits him; so we will leave it to him, changing it a bit, in order to make it a specific name, since otherwise the pure and simple denomination of curved beak, which is appropriate for many other birds, could be applied to several other species as well.]

*Krom bek* translates as "crooked beak" in both Afrikaans and Dutch. In interpreting Le Vaillant's words, it appears the Khoikhoi guides were using an Afrikaans or Dutch word, and, therefore, the name is essentially of Dutch origin.[140]

# Crossbill

The *Loxia* crossbills are a small genus of six (or seven) species nested in the Fringillidae, the finches. The origin of the common name is straightforward—it's a descriptor of the remarkable structure of the bill: the tips of the mandibles are crossed, an adaptation allowing them to open the scales of conifers on which they feed. The common name first appeared in the English literature as "Cross-bill" when Ray and Willughby wrote about it in 1678:[16]

> Its Bill is thick, hard, strong, black, and contrary to the manner of all other birds, crooked both ways, the Mandibles near their tips crossing one another: For the lower, being drawn out into a sharp point, turns upward, the upper bends downward. Neither do they always observe the same side; for in some birds the upper Chap hangs down on the right side, the nether rises up on the left; in others contrariwise, the lower takes the right side, the upper the left.

## Crow or Raven?

If you are a North American or European birder, the difference between a crow and a raven is clear cut—ravens are larger with a bigger bill, a different flight pattern, and a wedge-shaped tail, while the smaller crows have fan-shaped tails. In other parts of the world, Australia, for instance, the names are far from meaningful. There are six members of the Corvidae in Australia: three are called crows and three ravens, although there is really little difference. These Australian species are similar in size and coloration, and they can be difficult to tell apart, but they are best differentiated by range and subtle differences in plumage, habits, and calls. The bases of the feathers of the crows are white, while those of the ravens are gray, although this is useful only if birds are held in the hand or if discarded feathers are found. As for Africa, some authorities say that the Pied Crow is better thought of as a small, crow-sized raven. So, as is often the case with common English names, there are no hard and fast rules!

# Crow

The first bird to be called by this English name was the Carrion Crow, which is common in England. This very old English word is derived from the Proto-Germanic *krāwō* or *krēǫ*, which evolved into the Old English word *crauua* or *crauuae*. It is clearly imitative of the bird's call. The name is now applied to thirty-three species in the genus *Corvus*, all of which are fairly similar in size, shape, and appearance. The name has a long history in English literature, due to the folklore and mythology surrounding crows and ravens, which are often associated with death or ill omens, thanks to their black plumage and behavior. One of the first records of the word in print dates from 700 CE in the *The Épinal-Erfurt Glossary*, a Latin-Old English glossary that was probably composed in Canterbury in the late seventh century.[480] There is an entry for *crauuae*, a "cornicula." A book of Psalms from circa 1000 has the entry:[592]

> Se selþ nýtenum mete heora, and briddum cráwan cígendum hine. [Who giveth to beasts their food: and to the young ravens [crows] that call upon him.]

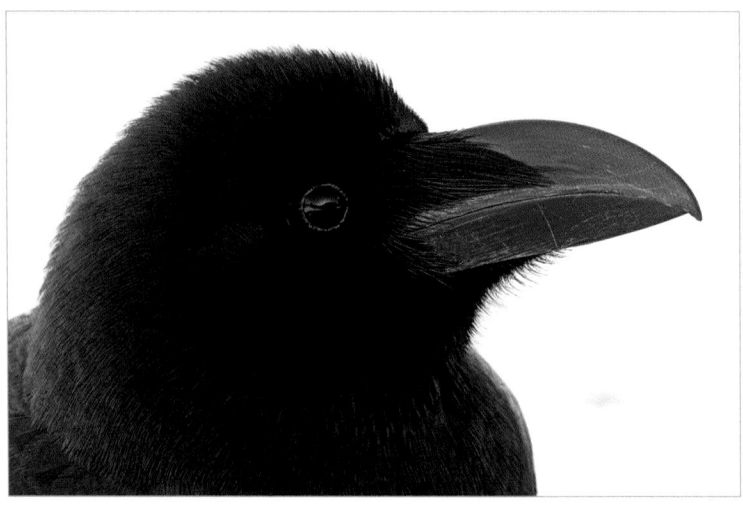

*Large-billed Crow (Japan)*

By the 1200s the name had evolved in Middle English, for example in the poem "The Owl and the Nightingale" from the twelfth or thirteenth century:[416]

> Par ich aſchevle pie & crowe.
>
> From þan þat þer is iſowe.
>
> [so I can scare off magpies and crows
>
> from what is sown there.]

# Cuckoo

Fifty-six species of several genera in the Cuculidae are known as cuckoos, but see also Ground-Cuckoo, Hawk-Cuckoo, Drongo-Cuckoo, and Lizard-Cuckoo. The derivation is onomatopoeic from the call of the Eurasian Cuckoo; it probably originated from the Latin *cuculus* and came in the mid-thirteenth century to Middle English (*cokkou*) via the Old French word *cucu* and thence became cuckoo. According to Swann:[121]

> In Old and Middle English it occurs as *coccou, cuccu, cukkow, cocow*; later it occurs as Cuckow. Chaucer spells it "Cuckowe."

Turner wrote about "De Cuckoo" in Latin in 1544 with the annotation that the names in other languages were[119]

> Κόκκυξ [*Kókkyx*], cuculus, Anglice a cukkouu, & a gouke, Germanice ein kukkuck.

By the 1600s the name was variously rendered as "Cuckoe or Guckoe" as well as the spelling "Cuckoo" as in an illustration by the artist Francis Place from 1655.[121] But even as late 1896, Newton was using the word "cuckow":[7]

> CUCKOW, or Cuckoo, as the word is now generally spelt though without any apparent warrant for the change except that accorded by custom, while some of the more scholarly English ornithologists . . . have kept the older form.

# Cuckoo-dove

The cuckoo-doves are twenty-one species of columbids (pigeons and doves) in three genera. With their slender body shape and long tails, they all superficially resemble cuckoos, especially in flight. As Blyth wrote in 1847:[313]

> The species of this division [*Macropygia*] are remarkable for their very broad, long and much graduated tail, and general Cuckoo-like figure.

# Cuckoo-Hawk

There are two species of cuckoo-hawks in the genus *Aviceda*. They are so named for their superficial resemblance to cuckoos, but, in fact, it's the other way around. It's the cuckoos that "mimic" the plumage of birds of prey, which enables them to frighten birds into briefly fleeing their nest in order to lay their parasitic eggs, an example of Batesian mimicry.[141]

# Cuckoo-roller

This unusual bird, a member of the monotypic Leptosomidae, is neither a cuckoo nor a roller, with no close relationship to either. The name is a relatively new one; it was known as the *Courol* in the English literature even up to the 1980s. In 1822, Latham wrote about the Cuckoo-roller as the "African Cuckoo":[17]

This species inhabits Madagascar, where the male is called Vouroug-driou, and the female Cromb . . . The Vouroug-driou in manners approaches to the Jay and Roller, but in feet to the Cuckow; and these being long and strong, more so than in the true Cuckow, it comes nearer to the Coucal, Coua, and Touraco. . . . M. Levaillant would have this kind called Courol.

# Cuckooshrike

Although the cuckooshrikes (or cuckoo-shrikes) are not related to cuckoos or shrikes, which are in different orders, they do bear a passing resemblance to both groups. As noted in *Australian Bird Names*:[130]

"Cuckoo-shrike" is something of a mystery, not least because it is another awful combination of names of birds of entirely different orders, which tells us very little about the group in question.

The name seems to have first been used in Australia in the mid-1800s; presumably the early European settlers, when confronted with an unfamiliar bird, noted its resemblance in color and flight to cuckoos as well as the strong, shrike-like bill and applied this compound word. There are multiple references to "cuckoo-shrikes" in Oates's *Handbook to the Birds of British Burmah* from 1833, however, suggesting a possible origin in colonial India.[112] There are thirty-five species of cuckooshrikes in five genera. Most are gray to black, except for a handful with yellow-gold plumage.

# Cupwing

The five species of cupwings in the Pnoepygidae are found in Asia, from the Himalayas through China and Taiwan, and south to Timor. These members of the genus *Pnoepyga* were classified in the babbler family and called "wren-babblers" until DNA sequencing showed them to be on a different phylogenetic branch in 2009. These small inhabitants of dense undergrowth all have cuplike, short wings with strongly curved remiges, and this is where they take their names from. This very new name was coined in 2009.[142]

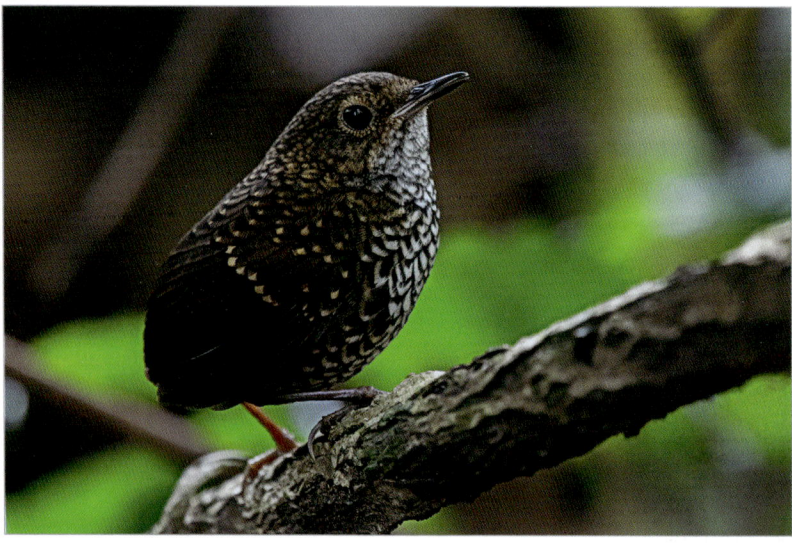

*Pygmy Cupwing (Vietnam)*

# Curassow

There are fifteen species of curassows in the Cracidae, all found in Latin America. It is said to be an anglicized version of the word Curaçao, the Caribbean island. That in turn was the name the Taino Indigenous people called themselves. Early Spanish accounts support this theory with references to the Indigenous peoples as *Indios Curaçaos*. One of the earliest references to the bird is in a book entitled *The Civil and Natural History of Jamaica* from 1756 with an entry for "The *Curaçoa* Bird" [*sic*].[147] Although curassows are not found on Curaçao, the birds may have been traded to Europe via the island, and hence acquired the name by (incorrect) association.

# Curlew

Eight species of curlews in the Scolopidae are found worldwide. The name is imitative of the loud, ringing call of the Eurasian Curlew, often rendered as *COURli . . . COURli . . . COURli*.[3] But it probably came into English influenced by the Old French *courlieu*, meaning "messenger," from *courir*, "to run," from the Latin *currere*. In Middle English there were numerous different spellings, including *curleu, curlue, curlow, curliour, corleu,* and *corlue*. One of the earliest mentions of the bird was in *Extracts of the Account Rolls of the Abbey of Durham* from 1312 when it occurred as a kitchen expense:[368]

> In 4 gallis [grouse] et pull. columpbarum [pigeons], 2s. 3½d. 4th week, curleus [curlews].

As Ticehurst noted in the journal *British Birds* in 1923:[367]

> Curlew, spelt in a variety of ways, also figure frequently in these accounts and were evidently exceedingly numerous, though less easily procured than Plover. They were bought in small lots up to about half a dozen and their price varied.

# Currawong

Three species of currawongs occur in Australia. The name is based on the Yugara-language name for the bird, *gurrawang*. It is onomatopoeic for the call of the Pied Currawong. The Yugara is an Australian Aboriginal group of the Brisbane region. Early on, as in Gould, this genus *Strepera* of artamids, the woodswallows, bellmagpies, and allies was called the "crow-shrikes."[40] A short passage in *Opals and Agates* from 1892 gives a clue to the adoption of the Aboriginal word into English with a story about the Aboriginal "Folk Lore of Australia":[145]

> Similarly the Black Magpie, Crow-Shrike, or the Butcher-bird* is the sign for the blackfish; if no "churwung," then no "dimgala"**—if plentiful the one, then plentiful the other. If the tailor fish is to be in full supply, then the wattle tree must be in extra full bloom beforehand; if the blossoms be scanty, this fish will be conspicuous by its absence for that season. The Crow-Shrike in May heralds the bream in June.

*all early names for the Pied Currawong
** presumably transliterations of the Aboriginal names for the currawong and the blackfish.

The name was adopted officially in the 1926 edition of the *Official Checklist of the Birds of Australia*, with the listing of a number of species of currawongs, the secondary name given as "bell-magpie."[128]

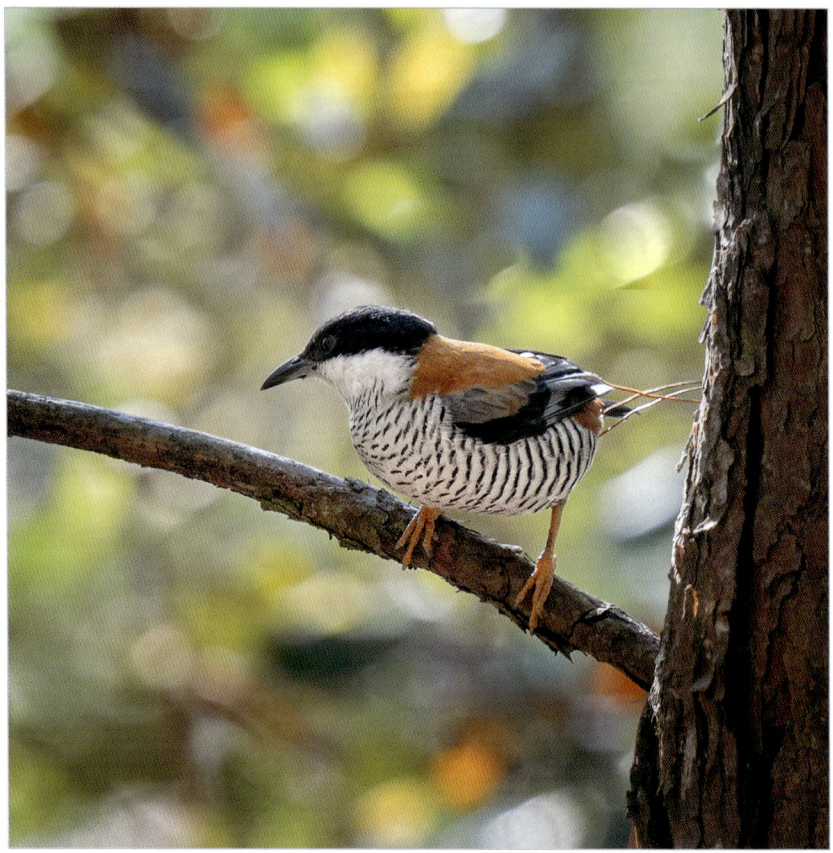

*Vietnamese Cutia (Vietnam)*

# Cutia

Two very similar species of cutias, Himalayan and Vietnamese, in the Leiothrichidae were previously considered conspecific. The synonymous genus and common names were coined in the scientific literature in 1836 by Hodgson, who pointed out that the name is "Khutya . . . of the Nipalese."[96] This is confirmed by Pittie, who transliterates the Nepali word कुटिया as *khatya* or *khutya*.[95] The Nepali word is also the word for a "small hut." According to Nepali birders (pers. comm.) the nest of the cutia is like a small hut, which is called a *kuti* in Sanskrit and Nepali; therefore, the bird got its name from this analogy.

# Cut-throat

There is only one species of Cut-throat, a small finch in the Estrildidae found in semiarid regions of sub-Saharan Africa. The etymology of the evocative name is self-evident when the male is observed—he has a bright red band across the throat from ear to ear. The female lacks this feature. An early mention of the bird can be found in Latham's *A General History of Birds* in 1822.[17] He wrote about the bird under the name of "Fasciated Grosbeak":

> In the collection of Lord Stanley is a Variety of the male, in which the band across the throat is orange-coloured, bounded above with white: in some the red band on the throat is blood-colour, and from this circumstance it has been called by some, the Cut-throat Sparrow.

# D IS FOR DACNIS

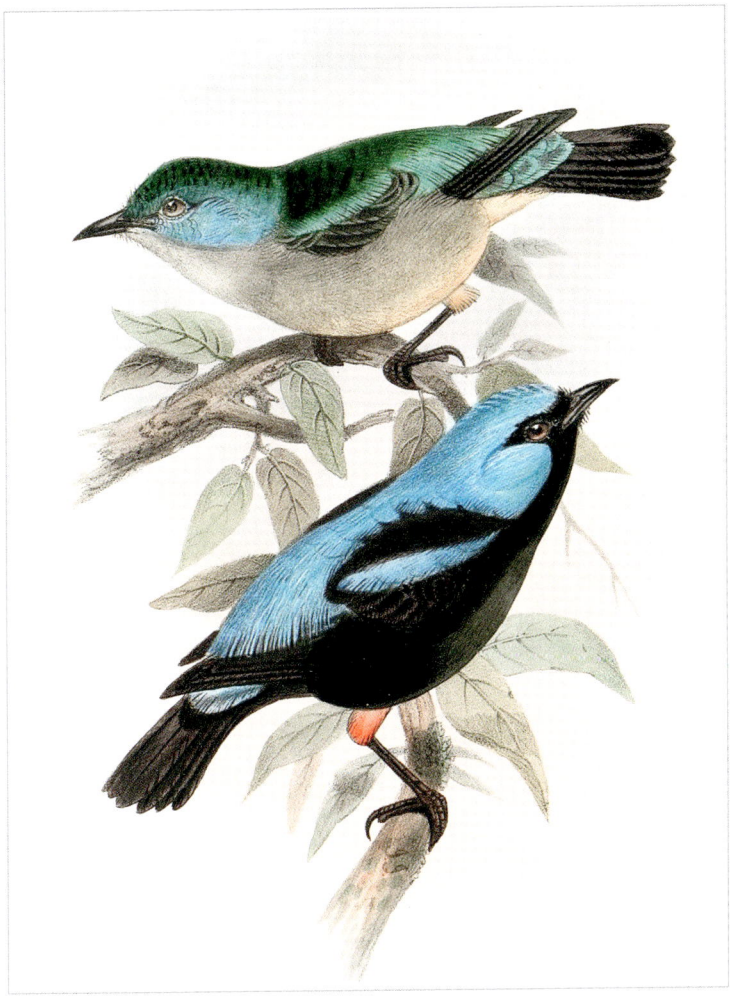

*Scarlet-thighed Dacnis in* The Ibis *Ser. 1, Vol. 5 (1863)*

## Dacnis

There are nine species of dacnises in the Thraupidae; all are confined to the Neotropics. Their common name is synonymous with the genus name, which comes from the Ancient Greek δακνίς *daknis*, an unidentified bird mentioned in Hesychius's lexicon of Greek words from the fifth century with the entry "δακνίς: ὀρνέου εἶδος [daknis: bird form/shape]." Cuvier erected the genus in 1816 without explanation, with the only comment that the birds Buffon called "Les Pit-pits" are[408]

> representing blackbirds in small size by their conical and sharp beak. They are associated with fig trees.

# Daggerbill

Two species of hummingbirds in the genus *Schistes* are known as daggerbills. Unusually for members of the Trochilidae, the birds' bills are straight and sharp—in other words, daggerlike. This name was only adopted in 2019 after *Schistes geoffroyi* was split into two species; prior to this the bird was known as the "Wedge-bill." In the discussion around the name in the South American Classification Committee, the comment was made:[409]

> Wedgebill retains the link to Wedge-billed Hummingbird but is not an accurate name in terms of shape or function . . . "Daggerbill" captures the shape and function of the bill: the distal several mm of the bill are abruptly laterally compressed to such a severe degree that it resembles the blade of a dagger. A dagger is used for stabbing, and so this match of form and function has an appeal.

# Dapple-throat

The single species of Dapple-throat in the Modulatricidae is a very unassuming and little-known bird found on only three mountaintops in East Africa. The name is self-explanatory, referring to the grayish mottling on the pale yellow throat. Earlier names were the "Dappled Mountain Robin" and "Dappled Mountain Greenbul." The name "Dapple-throat" was adopted in the 1970s.

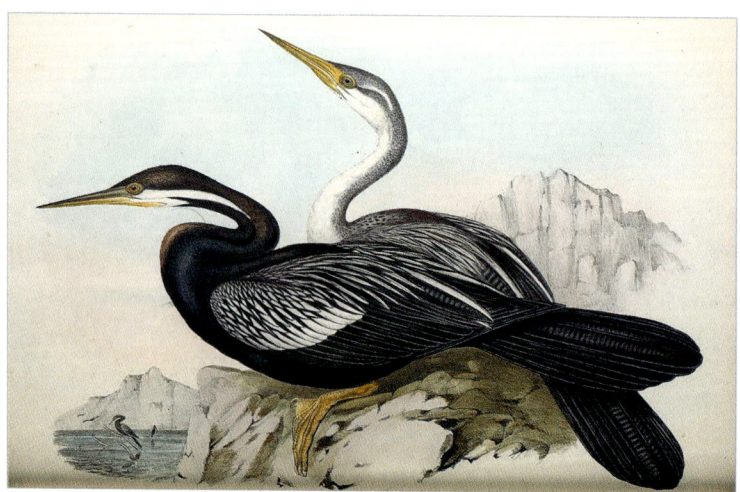

*Australian Darter from Gould's* Birds of Australia *(1840)*

# Darter

Three of the anhingas are known by the generic common name darter. The name dates back at least to 1781 when Pennant wrote under the heading "Darter" that the bird[66]

> inhabits Guinea, Ceylon, and South America. Darts out its head either at its food, or at passengers that go by; whence the name.

# Dickcissel

This very well-known bird, in the Americas at least, is abundant and very vocal on the prairies of North America in the breeding season. The name by which this member of the Cardinalidae is known in the northern part of its range is said to be onomatopoeic of the bird's vocalizations, the song of which can be transcribed as *dick-dick-see-see-see*. An amusing account of the name was coined by the writer Ernest Seton-Thompson in 1900 in a poem entitled "The Origin of Dick Cissel":[410]

> Sir Richard Cecil was a knight of very high degree,
>
> He came to preach some English fad in North Amerikey;
>
> But a clever Indian medicine man transformed him to a bird.
>
> With the funniest, drollest, dryest note that ever yet was heard:
>
> And now he sings the livelong day from mullein top or thistle,
>
> The first of his Intended speech, "Oh I am Dick. Dick Cissel."

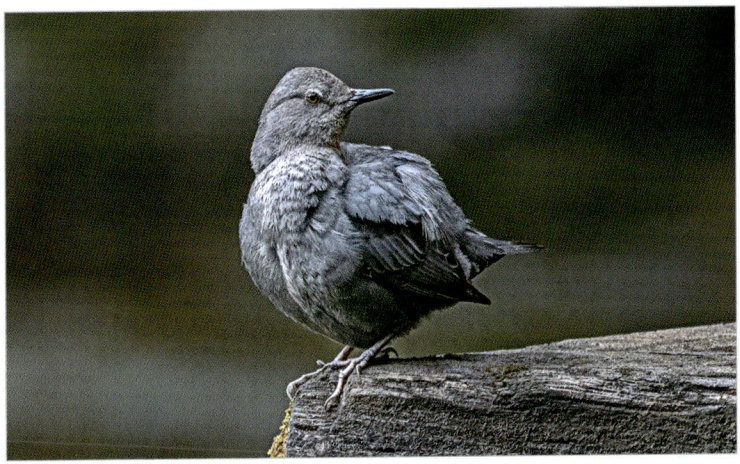

*American Dipper (Washington State, USA)*

# Dipper

The Cinclidae consists of five species of dippers with a wide distribution over four continents. They are always found near water, where they dive underwater and walk along the streambed in search of their invertebrate food. They are the only aquatic songbirds, and the name is a clear reference to their unique behavior but possibly refers to their habit of bobbing rather than their habit of dipping into the water. Ray and Willughby used the name in 1678 but they were referring to the Little Grebe.[16] Tunstall was the first to use "dipper" in what was essentially a list of British birds in 1771 for what was until then referred to as the Water-Ouzel.[369] Despite this, Newton claimed that Bewick "invented" the name in 1804 for the bird then usually called the "Water-Ousel."[7] Under the heading Water Ouzel, subtitled "Water Crow, Dipper, or Water Piot," Bewick wrote that the bird:[149]

> May be seen perched on the top of a stone in the middle of the torrent, in a continual dipping motion, or short courtesy [curtsy] often repeated, whilst it is watching for its food, which consists of small fishes and insects.

# Diuca-finch

Two species of birds in the Thraupidae in different genera are known as diuca-finches. One is found in Argentina and Chile, the other in the high Andes of Peru and adjacent areas. The word *diuca* comes from the name the Indigenous Mapuche inhabitants of present-day Chile and Argentina gave to the birds in the Mapuzugun language. This was introduced into the European literature by the Chilean naturalist Juan Ignacio Molina in 1782 without an explanation other than that:[303]

> La Diuca, Fringilla Dìuca . . . [is] of a blue color in the upper parts of the body and white in the lower parts. Its song is very pleasant. He lives in the vicinity of dwellings like the sparrow, of which he has several properties.

There are various versions of the Mapuche name, including Diuka, Fiuca, Shiwka, and Viuca. Diuca is said to be onomatopoeic for the bird's vocalizations, as Rozzi wrote in 2010:[151]

> Considered by many birdwatchers to be the loveliest bird-song of Chile, the Diuca Finch beautifies the natural environments and countryside of southern Chile and Argentina with its melodic successions of early-morning whistles: "diu-diu-diu-diu." This song inspires the onomatopoeic Mapungun name, diwka, and initiates the morning concert of birds in springtime.

# Diucon

The Fire-eyed Diucon is a member of the Tyrannidae, a tyrant flycatcher, found in Chile and Argentina. Gay described the bird as [304]

> vulgarly *Diucon* or *Tiucon* . . . This bird is found in almost all of Chile: its cry is like the sound of a bell, pronouncing the syllables *tot, tot, tot,* and so slowly that they mimic the toad.

Several Spanish dictionaries claim that the Mapuche name *Diucón* is the augmentative of Diuca, as in "from *diuca* and the suffix -on . . . bird bigger than the *diuca* and very similar to it" despite the fact the Diucon is only marginally larger on average from the diuca-finches and not likely to be confused.[481] Johnson wrote in *The Birds of Chile* in 1967:[316]

> To the unpractised eye this Tyrant-Bird with its dark grey upperparts and lighter grey under surfaces might appear similar to and be mistaken for the abundant "Diuca" Finch—hence the vernacular name "Diucón" which means "large diuca"— but to anyone at all familiar with bird families the stream-lined shape, dark head, and flycatcher-like stance and bill preclude any possibility of confusion.

Other Indigenous names include *Püdku, Huelko,* and *Wëdco.* According to a paper in the *Journal of Ethnobiology,* the name *Püdko* comes "from *püd:* to separate two clouds, *ko:* laden with water," while *Wëdco* is onomatopoeic.[150] This accounts for some local beliefs that the song of the Diucon resembles that of raindrops. They associate this song with the coming of sudden and local showers, and it is said:[580]

> cuando el Diucón se eleva desde una rama y se deja caer sobre la misma rama, es porque está anunciando el mal tiempo [when the Diucón rises from a branch and falls on the same branch, it is because it is announcing bad weather.]

# Diver

Although known as loons in the Western Hemisphere, the self-explanatory name diver is more widely used in the Old World (see Loon) for the five members of the Gaviidae. Ray and Willughby referred to them in 1678, using both names:[16]

> Greatest Diver, or *Loon, Colymbus maximus* [Great Northern Loon] . . . All these birds are also called Loons and Arsfeet, from the situation of their legs, just behind.*

*Lockwood[84] points out Ray's entry refers to a grebe, but many early publications, such as Pennant's *British Zoology,*[371] refer to *Colymbus* or *Mergus maximus* as the "Northern Diver."

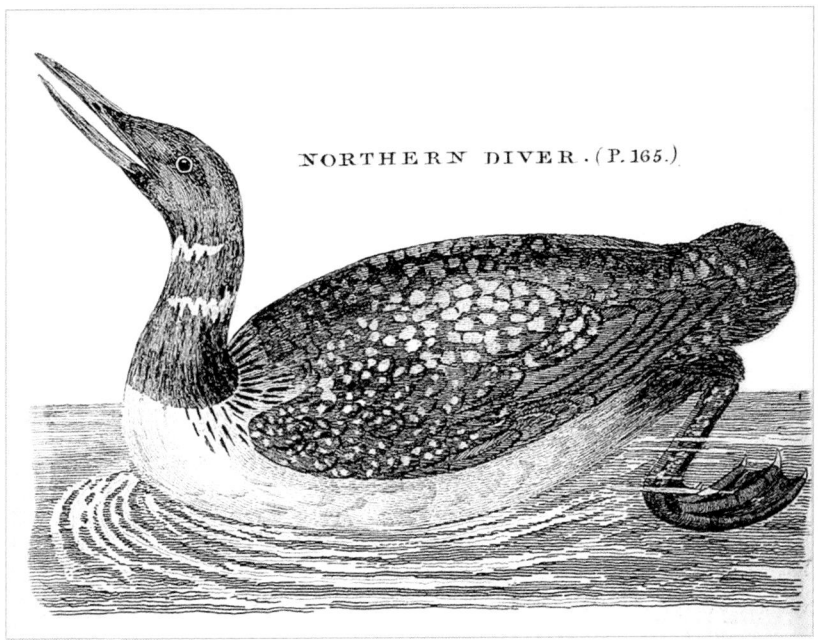

*Common Loon or Great Northern Diver from Pennant's* British Zoology (1812)

# Dodo

Although the Dodo went extinct in the early 1600s, its name is well known not just to birders, but to the public in general as a symbol of obsolescence or of stupidity. This large, flightless pigeon of Mauritius was wiped out thanks to hunting, as well as invasive species and habitat disturbance, within a century of first being described. The origin of the name is somewhat unclear. There's little doubt it comes from a Dutch word, either *dodoor*, meaning "sluggard," in honor of the ease with which the Dutch sailors clubbed the birds for the pot, or from *dodaerse*, meaning "fat-arse," a reference to the bird's bustle of tail feathers. Staub claims that the first use of the name *Dodaerse* appeared in Captain Wan Westzanen's journal *Voyage to the East Indies Amsterdam* in 1602:[285]

> Brought back to the Bruinis [the ship] very fat Dodos which tasted good to the crew. What remained was salted. On 4 August, 50 big Dodos were captured. Twenty five birds were so big and fat that two only were sufficient to feed the crew. The rest was salted. On another 3-day hunt inland, other crews of the fleet captured 20 more Dodos.

Emmanuel Altham used the word in a letter written in English in 1628, in which he claimed its origin was Portuguese. As did the English writer Sir Thomas Herbert, who used the word *dodo* in his 1634 travelogue:[581]

> Here, and in Dygarrois [Rodrigues] and no where else (that ever I could see or heare of) is generated the Dodo (a Portuguize name it is, and has reference to her simplenes,) a Bird which for shape and rarenesse might be called a Phoenix (wer't in Arabia:) her body is round and extreame fat, her slow pace begets that corpulencie; few of them weigh lesse than fifty pound: better to the eye than stomack: greasie appetites may perhaps commend them, but to the indifferently curious, nourishment, but prove offensive.

There is no direct evidence that the Portuguese used the name, although some sources still state that the word *dodo* derives from the Portuguese word *doudo*, meaning "fool" or "crazy." And to further muddy the waters, some suggest that the name is onomatopoeic of the bird's call, a two-note pigeon-like sound resembling "doo-doo."[286]

# Dollarbird

One species of roller in the Coraciidae is known as the Dollarbird, in reference to the large white spots on the underwings that appear to some to resemble silver coins. The name was coined (pardon the pun) by the early European settlers of Australia, although some ornithologists of the time preferred the name Australian Roller, despite its wide distribution through New Guinea and Asia as well. One of the earliest mentions of the name was in Lesson's *Manuel d'Ornithologie* in 1828:[152]

> Habite Java, le sud de la Nouvelle-Hollande, et toutes les îles de la Polynésie. Les naturels des environs de Sydney nomment cet oiseau *natay-kin*, et les colons *dollar bird*, à cause de la tache argentée du milieu de l'aile. [Inhabits Java, southern New Holland, and all the islands of Polynesia. The natives around Sydney call this bird *natay-kin*, and the settlers *dollar bird*, because of the silver patch in the middle of the wing.]

# Donacobius

The one species of Donacobius in a monotypic family in South America gets its common name from the genus name, which comes from the Ancient Greek δόναξ *donax*, for "reeds" and *bios*, "dwelling." And, in fact, it does dwell mostly in brushy vegetation on river edges. Swainson called it the "Babbling Thrush" in 1831, describing its preferred habitat but giving its song a very uncharitable review:[305]

> It is seldom that the notes of the feathered race are absolutely disagreeable, but we never remember to have heard a bird with a voice of such astounding discord, as that now before us. Its particular note, if note it could be called, we do not now recollect; but it was so shrill, grating, and monotinous, that we have frequently rushed out of the house, to drive away the babbling disturbers. [The house was] close to a small swamp, overgrown with reeds, among which these birds delight to dwell; and which in fact, they never quit. Clinging to the smooth stems by their strong feet and acute claws, they were incessantly uttering discord with the most provoking perseverance: all the time moving their body from one side to the other, spreading out their tail, and straining their throats, in the most grotesque way imaginable.

# Doradito

In the Tyrannidae (tyrant flycatchers) there are five species of very similar looking yellow and olive-brown doraditos in the genus *Pseudocolopteryx*. The name is borrowed from the Spanish *doradito*, which is the diminutive of *dorado*, meaning "golden," an appropriate appellation for these small flycatchers with their bright yellow plumage.

# Dotterel

The dotterels are seven species of plovers in various genera. It's not clear why some enjoy the name dotterel, while other similar-looking congeners are referred to as plovers. But that aside, the name is a little unfortunate as it was originally coined in the fifteenth century from the Middle English word *doten*, meaning "foolish," combined with the pejorative suffix *-rel*, apparently because the bird was easy to catch (and presumably eat). It first appeared in the early Latin dictionary *The Promptorium Parvulorum*, from 1440, with the entries "Dotrel, byrd" and "Dotrel, ffole" [fool].[261] The name originally referred to the Eurasian Dotterel and was applied more widely to other species later. There are other theories as to the etymology of the name, but the number of early references that mention the so-called foolishness of the birds tends to support the theory given above. Ray and Willughby wrote:[16]

> Because it is a foolish bird, even to a Proverb, we calling a foolish dull person a Dotterel.

Lockwood cited the Norfolk name *Dot Plover*, determining that the *dot* part of the name is echoic of the call with the disparaging suffix *-erel* and that the derogatory meaning came later due to the coincidental similarity to the word *dotard*.[84] This theory does not seem to have received wide acceptance and doesn't take into account other names, such as the Gaelic *Amadan mointich*, "fool of the peat bog," or the Welsh name *Hutan (hurtyn)*, literally "stupid."[121, 372]

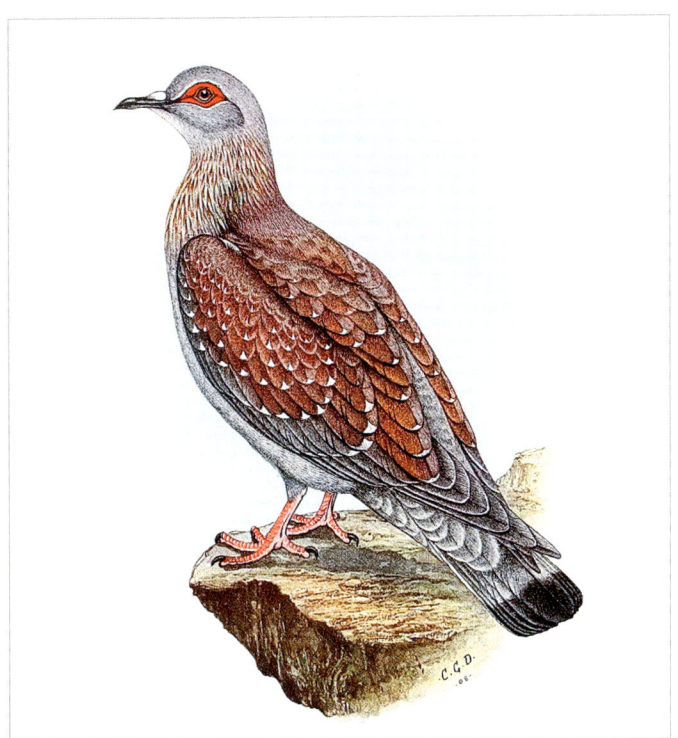

*Speckled Pigeon or (Bush-Dove) in Horsbrugh's* The Game-birds & Water-fowl of South Africa *(1912)*

# Dove

Roughly half of the members of the Columbidae are known as doves, as opposed to pigeons. This very old English bird name came into Early Middle English (twelfth century) with various spellings, such as *douve*, *dowe*, *doufe*, and *duue* from the Old English *dufe*, in turn from Proto-Germanic *dubon* or *dufa*, with the meaning of "dove" or "pigeon." One early reference appeared in 1200 in the *Old English Homilies*:[365]

> Buð admode alse duue [but humble also the dove].

Due to its antiquity, the original meaning of the word is lost, but it is perhaps related to the word "dive" and might reference the way the birds fly. Skeat wrote:[220]

> The sense is "diver," the form *dúfa* being from the verb *dúfan*, to dive, with the suffix *a* denoting the agent, as usual, for a similar formation.

The *Oxford English Dictionary* suggested that the noun is related to Old English *dūfan*, "to dive, plunge (into a liquid)," but linguists have noted that the difficulty with this hypothesis is that the noun in Germanic languages uniformly means "dove, pigeon," not an aquatic bird. The *Oxford Dictionary of English Etymology*, the etymological successor to the *Oxford English Dictionary*, abandons the *dive* connection, taking a more parsimonious approach, stating the word is simply "presumed to be imitative of the bird's note."

The name occurs in a number of hyphenated forms: Turtle-Dove, Cuckoo-Dove, Quail-Dove (see entries) and Wood-Dove, Ground-Dove, Collared-Dove as either a morphonym, bionym, or a taxonym.

## Dove or Pigeon?

There is no scientific difference between a dove and a pigeon. These common names are arbitrary and, in fact, are often used interchangeably. Most people think of doves as being a little smaller than a pigeon, but even that is not a firm rule.

# Dovekie

The Dovekie is a small auk in the Alcidae found in the North Atlantic Ocean. The name comes from its superficial resemblance to a small dove with the addition of the Scottish suffix *-kie*, a diminutive. Another early name for the Dovekie was "Sea Dove," allegedly due to the supposed dove-like bond between nesting pairs. An early account of the bird occurs in *A Journal of a Voyage of Discovery to the Arctic Regions* from 1821:[153]

> Another species of diver was seen to-day for the first time this voyage, which, like the preceding, is seldom seen except in the vicinity of ice; it is called by seamen, *Dovekey*.

# Dowitcher

Three species of waders in the Scolopidae bear the name dowitcher. The name, which first came into English in America in the mid-1800s, might come from the Iroquoian language, belonging to the Indigenous peoples of America's Northeast, as many present-day authorities suggest. It is possibly an English corruption of the word *tawístawis*, meaning "snipe." The true etymology of the name remains unclear, however. A paper from 1988 regarding the oral tradition of Cherokee (an Iroquois language) names for animals gives the name for sandpipers as *ganvsdawa* (*ganvsge*, "leg," *adowo*, "swim") with the comment that the name is "probably a general term referring to a number of sandpipers [including the dowitcher] and may allude to their long legs wading in the water."[155] But Ingersoll claimed that the name "is bad German, indicating belief in its German nativity."[9] Presumably, he is referring to one of the many local vernacular names for the bird, which was Deutscher, echoing another name, German Snipe. McAtee wrote in 1923:[154]

Dowitcher . . . also spelled dowitchee, dowiches, and dowits; all these terms are traceable to *duitsch* or *deutscher*, meaning that this is the Dutch or German snipe, to distinguish it from . . . the English snipe.

To confuse matters further, an opaque reference to the name appears in an entry for "Red-breasted Snipe," now known as Short-billed Dowitcher, in *The Birds of Long Island* in 1844:[412]

Our gunners, as if fearful that nothing would be left to connect the past with the present generation, cling to the old provincial names for birds, recognising this species by the singular and unmeaning name of "*Dowitcher.*"

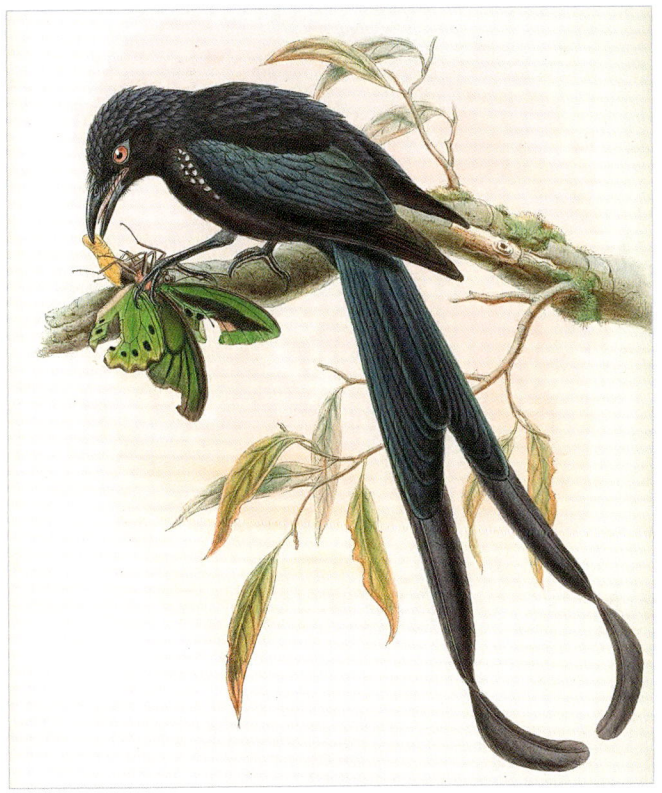

*Ribbon-tailed Drongo from Gould's* The Birds of New Guinea and the Adjacent Papuan Islands *(1875)*

# Drongo

The drongos are a family of twenty-nine species of similar-looking passerines in the Dicruridae, all sporting black plumage. The name was first used in the European literature in 1779 by Le Vaillant, who received a specimen from Madagascar sent to him by M. Poivre:[93]

Quoique les Nomenclateurs aient place cet oiseau à la fuite des gobe-mouches, il paroît en différer par de si grands Caractères . . . que nous avons cru devoir totalement l'en séparer, & lui conserver le nom de drongo qu'il porte à Madagascar [Although the Nomenclators have placed this bird after the flycatchers, it appears to differ from them by such great characters . . . that we have thought it necessary to completely separate it from it, and keep the name for it of drongo that it carries in Madagascar.]

In a later, 1805, volume, he wrote about "Le Drongear" from South Africa. In the same publication he has entries for "Le Drongri," "Le Drongo," "Le Drongup," and "Le Drongolon," all illustrated and clearly species of *Dricrurus*, despite no scientific names being given. In Madagascar there are numerous names for the Crested Drongo, as there are many dialects spoken by the different ethnic groups. The Tsimihety, one of the largest Malagasy ethnic groups from the north-central region, call it the *Lehadronga*. Richardson lists *Le'hidro'nga* in his Malagasy-English dictionary from 1885 as "Drongo, a kind of bird," but if it derives from anything other than the local name for the bird, it's unclear.[231]

## Drongo-Cuckoo

The four remarkable species of parasitic drongo-cuckoos are mimics of the drongos. Whether this has evolved in order for the cuckoo to protect itself from harassment by other birds or to allow easier access to the nests of the birds that they parasitize is still open to question. But there is evidence that host defenses can drive aggressive mimicry in a number of avian brood parasites.[157] In either case, the derivation of the name is self-explanatory.

## Duck

Some of the most familiar birds, the ducks need no introduction. All ducks belong to the same family, the Anatidae (ducks, geese and waterfowl). The theoretical Old English word *dūce* is thought to come from the Old English verb *dūcan*, "to duck, to dive," because of the way many species feed by upending or diving. Middle English spellings included *doke, ducke, dukke, dokke,* or *douke*. In 1905 Shaw wrote in *Wild-fowl*:[158]

> The Anglo-Saxon name for duck was *enid*; *enid rake*, the ruling duck, is almost undoubtedly the origin of our word drake. The word duck comes from the Danish "*duke*" or the Dutch "*duiken*," to dive or stoop.

*Ende* or *ened* was a common name for duck in Old English. As for *duck*, the Danish and Dutch words, like the Old English word, derived from the Proto-Germanic *dūkaną*, "to dive, bend down."

## Dunlin

There is only one Dunlin, a small shorebird with a very wide distribution. The English name is a dialect form of *dunling*, with various spellings occurring in the account book *The Durham Household Book* from 1531 under several entries:[413]

- *DUNLYNGGS. The bird so called.*
- *1 dd. dunlynggs, 4d., et 3 dd. styntts 6d. (1 dozen dunlins 4d, and 3 dozen tints 3d)*
- *STYNTTS A Durham name for the birds called Dunlings*
- *1 dd dunlyngs, 4d.*

The first element of the name derives from the Middle English *dŏn*, defined as "of hair or feathers, brownish-grey in colour." The second element, the suffix *-ling*, could be interpreted to mean a person or thing with the given quality as in Sanderling, or equally, it may be a diminutive, like gosling, or duckling.

Linnaeus and many contemporary authorities believed that they were dealing with two different species when discussing the birds in their breeding and nonbreeding plumage, referring to them respectively as the *Dunlin* and the *Purre*. The latter name continued to be given as an alternative in the literature right up until the end of the nineteenth century.[373]

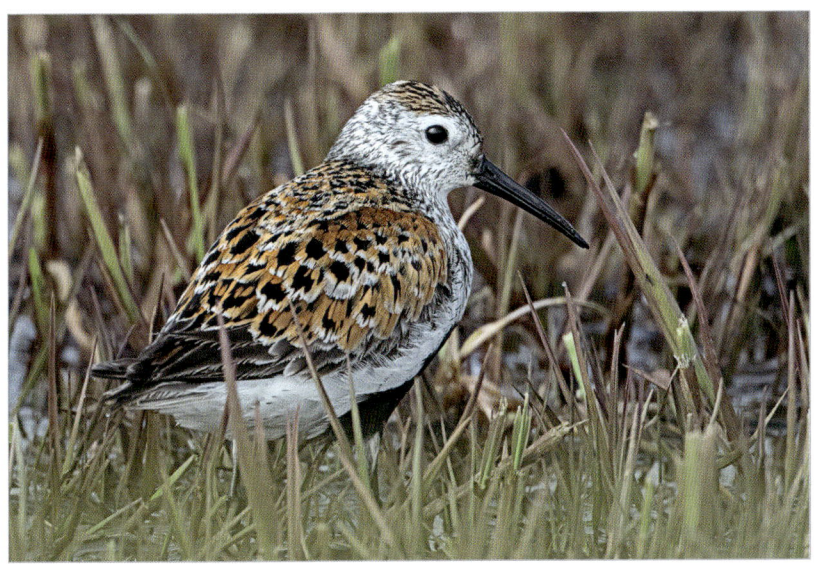

*Dunlin, in breeding plumage (Alaska, USA)*

# Dunnock

Another bird with a unique name despite having a number of closely related and similar-looking congeners, the Dunnock is a species of accentor in the Prunellidae (accentors) found throughout Britain as well as Europe and parts of North Africa. As with the Dunlin, the name comes from the Old English *dŏn*, "brownish-gray," plus the Anglo-Saxon diminutive *-ock*. The original little brown bird! As with other birds with very old English names, the Dunnock was a familiar bird in Britain from early times. Earlier spellings in Middle English include *donek* and *dunok*. There is an entry in the *Catholicon Anglicum*, an English-to-Latin bilingual dictionary from 1483:[306]

> *Dunoke*: curuca, Avis que ducit cuculum, linosa idem secundum quosdam. [Dunnock: curuca, the bird that hatches and feeds the cuckoos eggs, the linnet is the same, according to some.]

Herrtage, who provided an introduction and notes in an 1881 reprinting of *Catholicon Anglicum* wrote:[306]

> Cotgrave gives . . . '*Mari cocu* [Cuckoo husband]. An hedge-sparrow, Dike-smowler, Dunnecker: called so because she hatches and feeds the cuckoes young ones, esteeming them her own.' Cooper explains *Currucca* as "the birde that hatcheth the cuckowes egges; a titlyng." *Dunnock*, from *dun*, the colour, as ruddock = *redbreast*, from *red*. Harrison, Descript. of Eng . . . , mentions amongst the birds of England the "*dunock* or redstart." Withals gives Pinnocke, or Hedge-sparrow, which bringeth up the Cuckoe's birdes in steade of her owne.

# E IS FOR EAGLE

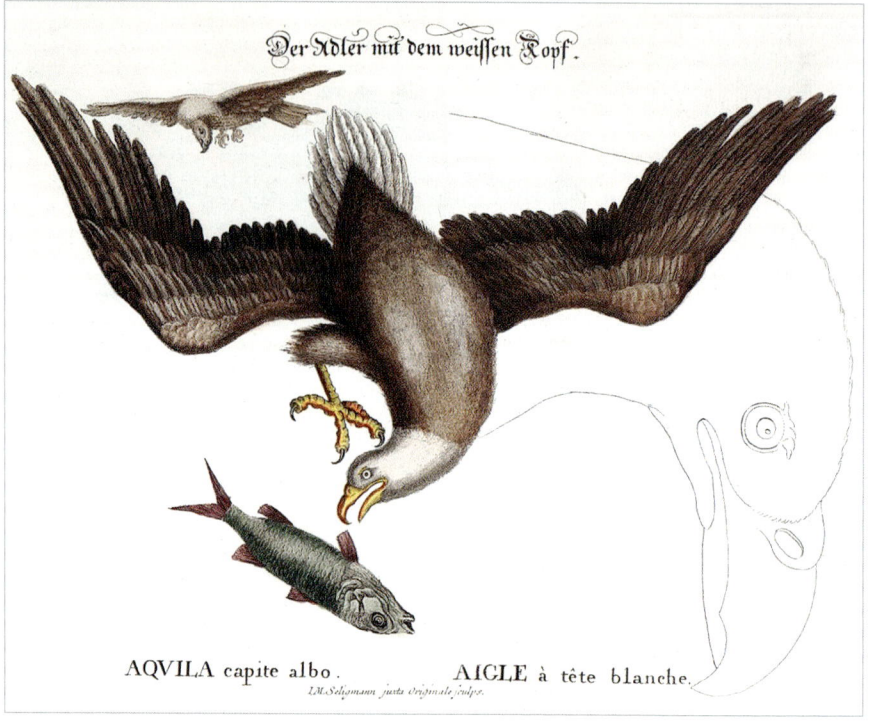

Der Adler mit dem weißen Kopf.

AQVILA capite albo.     AICLE à tête blanche.

*I.M.Seligmann juxta Originale sculps.*

*Bald Eagle in Catesby's* The natural history of Carolina, Florida, and the Bahama Islands, Vol. 1. *(1729)*

## Eagle

There are sixty-nine members of the Accipitridae with *eagle* in the name. Various hyphenated eagle names reference the morphology (Hawk-Eagle, Buzzard-Eagle), diet (Serpent-Eagle, Snake-Eagle, Fish-Eagle), or preferred habitat (Sea-Eagle) of the species in question. The word came into Middle English as *ēgle, egel,* or *agle.* One of the earliest uses of the name appeared in one of the Stowe Manuscripts, a glossary called the *De Natura Jumentorum, Bestiarum, et cunctorum Animalium,* including a list of names of animals with the Anglo-Saxon or Norman designations sometimes added in another hand from circa 1175.[374] There is an entry for "Aquila: ærn i. eigle." The glossary suggested that "eagle" is derived from the Latin word *aquila.* It is thought to have arrived in English by way of the Old French *aigle* or *egle.* The Old English word for an eagle was *earn* or *erne,* which had a Germanic origin. The origin of *aquila* is uncertain, but it probably derives from the Latin *aquilus,* meaning "dark-colored, swarthy, or blackish" as a reference to the plumage of eagles. As De Vaan wrote in 2008:[307]

> It is possible that "eagle" was derived from *aquilus* "dark" when this had received its colour meaning. It may not be the only dark bird, but it is certainly one of the biggest and most majestic of them.

## Eagle or Hawk or Buzzard?

The assignation of many of these names has historically been quite arbitrary. In North America, a number of accipiters are known as hawks, where elsewhere they would be called goshawks or sparrowhawks. The buteos are termed buzzards everywhere except North America, where they are called hawks, and sometimes the New World vultures are even referred to as buzzards. Likewise, the term "eagle" refers to several birds of prey, some of which do not have a close genetic relationship. Generally speaking, there are significant differences between eagles and hawks in terms of their size, eagles being significantly larger than most other birds of prey.

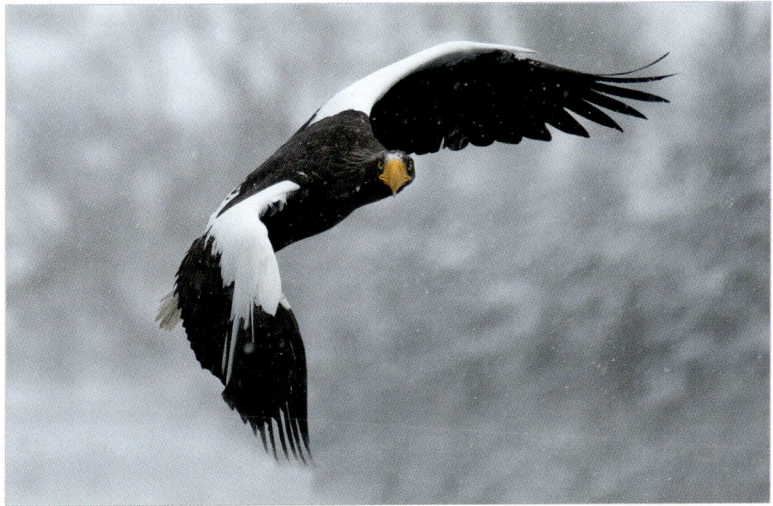

*Steller's Sea Eagle (Japan)*

# Eared-Nightjar

The two species of eared-nightjars in the genus *Lyncornis* are distinctive nightjars with prominent and distinctive "ear" tufts. (See Nightjar.) As Blanford noted in 1888:[315]

> This genus is distinguished from *Caprimulgus* by the want of rictal bristles and the presence of ear-tufts or aigrettes, consisting of a few elongate feathers just above and behind the ear-coverts.

# Eared-Pheasant

There are four species of *Crossoptilon* eared-pheasants found in East Asia. All have prominent ear-coverts that could euphemistically be called "ears." Gould poetically described the head plumage of Blue Eared-Pheasant:[33]

> The male has the short, velvety, and partially curled feathers clothing the head deep glossy black; sides of the head devoid of feathers, and of a deep blood-red, below which is a conspicuous lengthened tuft of silvery white feathers directed backwards and upwards.

(See Pheasant.)

# Earthcreeper

There are ten species of earthcreepers in different genera in the Neotropical family, the Furnariidae. As the self-explanatory name suggests, all are somewhat terrestrial, although not exclusively so. When on the ground, they cock their tails in a characteristic style. A description of their habits was given in *Argentine Ornithology* in 1888:[159]

> Their legs are short, but on the ground their movements are very rapid, and . . . they fly reluctantly, preferring to run rapidly from a person walking or riding, and at such times they look curiously like a very small Curlew with an extravagantly long beak.

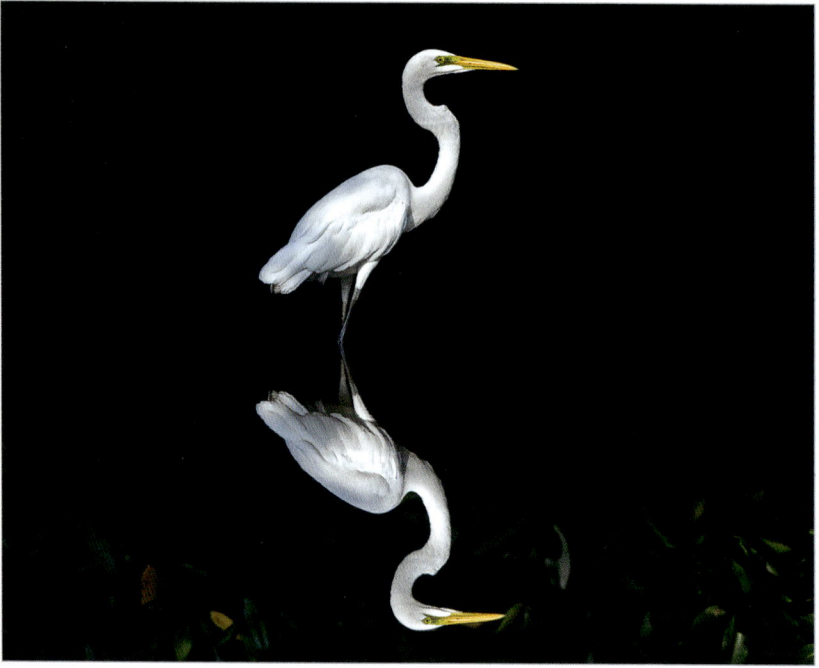

*Great Egret (Malaysia)*

# Egret

There are fifteen species of egrets, wading birds of the genera *Egretta* or *Ardea* closely related to the herons. The name came into English in the mid-fourteenth century from the Old French *aigrette*, a diminutive of *aigron*, "heron," perhaps of Germanic origin from the word *heigir*. In Middle English, there were various spellings, including *ēgret*, *egrete*, and *egrette*. The word in English dates back at least to the mid-1400s. One of the *Forest Laws* from 1450 stated:[262]

> If ther be ony man that cometh in to the forest and taketh away ony facouns, Jerfacouns, egrettes..her eiron or her berdes, ye shul do vs to wite. [If any man comes into the forest and takes any falcons, gyrfalcons, egrets . . . her eggs or her chicks, they shall be punished.]

# Eider

Four species of ducks known as eiders are found in the freezing waters of the Northern Hemisphere. The name came into English in the mid-eighteenth century from the Icelandic *ædar*, or from the German or Dutch *eider*. In turn, both derive from the Old Norse *æþar*, meaning "duck." Ray and Willughby referenced the origin of the name in 1678:[16]

> Here hath been brought me . . . from the *Ferroyer* [Faroe] Islands a certain sort of *Duck* they call there *Eider*.

In Iceland *æðardúnn*, or eider down, has been harvested for centuries, for use in duvets and pillows and as insulation in cold clothing.

# Elachura

The Spotted Elachura is the sole member of the Elachuridae. Until molecular studies were undertaken in 2014, the Spotted Elachura was treated as a member of the babbler family and called the Spotted Wren-Babbler. It turns out it is not even remotely related, so the name wren-babbler was replaced with the genus name *Elachura* from the Ancient Greek ἐλαχύς *elakhus*, for "short, little" and οὐρά *ura*, for "tail." Oates established the genus name *Elachura* in 1889 with the justification:[308]

> The Wren which forms the type of this new genus differs conspicuously from [Troglodytes] in having a much stouter bill and a more graduated tail. The plumage is moreover spotted, not barred.

# Elaenia

Twenty-eight species of elaenias are found in the Tyrannidae of the Neotropics. They belong to two genera, one of which is the genus *Elaenia*. The name comes from the Latin word *elaia*, the color olive, which derives from the Ancient Greek ἐλαινεος *elaineos*, "of olive-oil" or "oleaginous." Although the word oleaginous means "of or belonging to the olive tree," the name refers to the olive-green color sported by the majority of these New World flycatchers. When Sundevall erected the genus in 1836, he noted:[160]

> Olivaceae-cinevascentes; sæpe subtus, interdum vertice flavae.—Ελαινιος, oleagineus. [Olive-cinereous; often below and sometimes the top yellow.- Ελαινιος, of olives.]

# Elepaio

The elepaios are three species of monarchs (family Monarchidae) endemic to Hawaii. The Hawaiian language name for these birds is onomatopoeic for the birds' distinctive songs, which are pleasant and loud warbles. The birds are considered to bring good luck and were used in Hawaiian augury to select the best koa tree to make canoes, as detailed by Kanahele in 1995:[161]

> Canoe makers revered its presence. If the bird pecked at a fallen tree, it was a sign that the wood was insect ridden and thus not suitable for a canoe. Hence the saying *Ua 'elepaio 'ia ka wa'a* "The canoe is marked by the 'elepaio."

# Emerald

There are twenty-seven species of emerald hummingbirds in three genera. As one would expect, given the name, all of them sport beautiful emerald-green plumage. Gould was the first to use the common name, but Buffon made the comparison to the gemstone in 1808:[8]

> The emerald, the ruby, the topaz, all are to be seen in the brilliant colours of [the humminbird's] plumage.

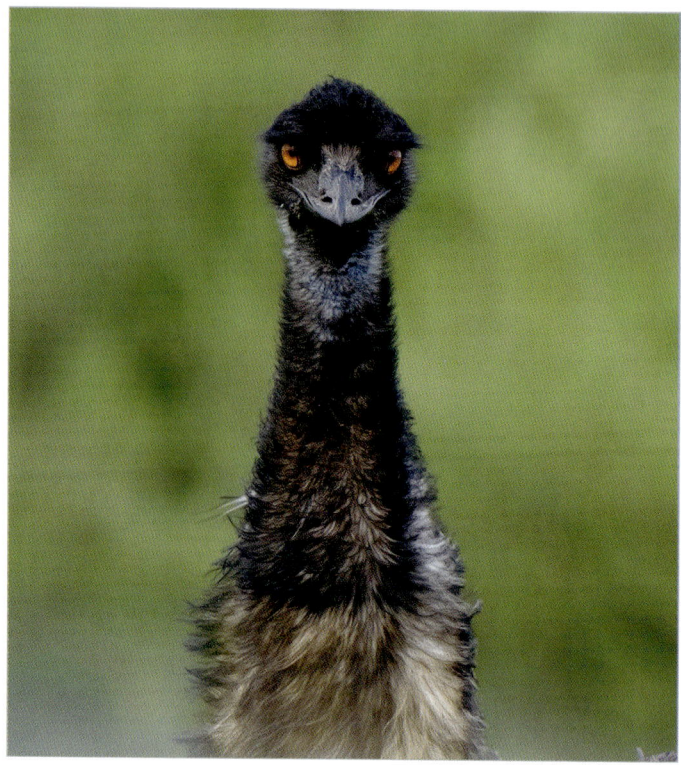

*Emu (Victoria, Australia)*

# Emu

The unmistakable Emu is a large, flightless bird endemic to mainland Australia. The name is often mistakenly thought to be derived from an Aboriginal language, but in fact comes from the Portuguese *ema*, the name originally used for the cassowaries, rheas, and ostrich. There are early uses of the name, but they refer to the cassowary, such as in the chapter "Some Kindes of Birds Their Egges, Beaks, Feathers, Clawes, and Spurres," in the *Musæum Tradescantianum* written in 1656:[263]

> A legge and claw of the Cassawary or Emeu that dyed at *S. James's, Westminster.*

And an earlier publication from 1576 by the Portuguese historian Gandavo gives an account of a rhea:[414]

> They're called *Hemas*, they'll have as much meat as a big ram and their legs are so big they're almost as big as a man's height.

Boles claimed that the name has its roots in the Arabic name for Ostrich, which is نعامة *naeama*, pronounced "nahma," believing that the word came to Portugal during the Muslim occupation from the seventh to ninth centuries but was used as a general term for any type of large bird.[118] The Portuguese were possibly the first Europeans to encounter rheas and cassowaries during their expansions in trade to Brazil and the Indonesian archipelago in the 1400s. In Indonesia, the word was replaced over time by the Papuan words, and, eventually, cassowary became the accepted common name. (See Cassowary.) The Emu of Australia was not encountered until much later, when the members of the First Fleet initially referred to it as the "New Holland Cassowary" and the "Emu." One of the first uses of the name for the bird we now know as the Emu was in John Hunter's journal of a voyage to New South Wales on the *Sirius* in 1788:[415]

> There have been several large Birds seen, since our Arrival in this Port; They were suppos'd by those who first saw them, to be the Ostrich, as they cou'd not fly when pursued, but ran exceedingly fast, so much so, that a very strong & fleet Greyhound cou'd not come near them . . . Some were of Opinion that it was the Emew . . . others imagined it to be the Cassaware, but it far exceeded that Bird in its size, it was when Standing, 7 feet 2 inches from the feet to the upper part of its head, the only difference which I cou'd perceive between this Bird & the Ostrich was in the Bill, which appear'd to one to be Narrower at the point it had three toes which I am told is not the case with the Ostrich.

## Emuwren

There are three species of emuwrens in the *Stipiturus* genus, all endemic to Australia. All of them possess long tails with filamentous rectrices. These loosely webbed feathers account for their name, as Gould noted:[40]

> The decomposed or loose structure of these [tail] feathers, much resembling those of the emu, has suggested the colonial name of Emu Wren for this species, an appellation singularly appropriate, inasmuch as it at once indicates the kind of plumage with which the bird is clothed, and the Wren-like nature of its habits.

## Eremomela

The eremomelas are eleven species of warbler-like birds in the Cisticolidae, all found in sub-Saharan Africa. The common name derives from the genus name, which references the bird's habitat and pleasant song. The genus was erected by Sundevall in 1850 who wrote in Latin:[162]

> Radices nominis: Ερημος, desertum; μέλος, carmen, hinc Eremomela, deserti cantor. [The roots of the name *Erimos*, desert; *mélos*, a song, here Eremomela, desert singer.]

## Erpornis

There is just one species of erpornis, the White-bellied Erpornis, a vireo found through large swathes of Asia. The genus and common name are the same and have their origins in the Ancient Greek ερπω *herpo*, "to creep about," and ορνις *ornis*, for "bird." The genus was erected by Blyth in 1844, who said that it is "shy; creeps among foliage, buds and flowers."[163]

## Euphonia

The *Euphonias* are a big genus of twenty-nine species with a synonymous common name, in the Fringillidae, the finches. The name was coined by the French zoologist Desmarest in 1806 from the Ancient Greek εὖ *eu*, meaning "good," and φωνή *phōnē*, meaning "sound" or "voice," for their complex and pleasing songs, with the comment:[164]

> La voix de Teités ressemble beaucoup à celle du Bouvreuil, ce qui détermine les colons à les élever en cage, où ils se plaisent assez, pourvu qu'ils soient plusieurs ensemble. [The voice of Euphonias closely resembles that of the Bullfinch, which determines the colonists to raise them in cages, where they are quite happy, provided they care for several together.]

# F IS FOR FAIRY

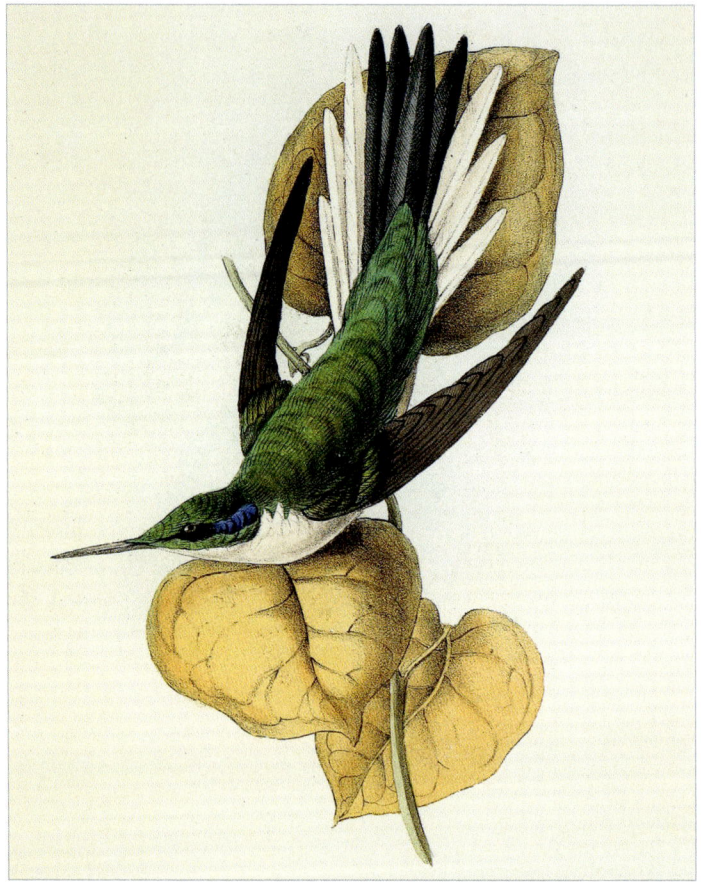

*Black-eared Fairy in Swainson's* A Selection of the birds of Brazil and Mexico *(1841)*

## Fairy

Two species of hummingbirds in the genus *Heliothryx* are known as fairies, presumably for their delicate and pretty appearance. In fact, Gould, who named the bird, stated:[6]

> The trivial name of Fairy will be very appropriate, since the elegance of their form and the peculiar chasteness of their colouring readily recall to memory the ideas of grace and beauty connected with those imaginary beings.

## Fairy-bluebird

The Irenidae contains just two species of fairy-bluebirds, one endemic to the Philippines, the other more widespread in Southeast Asia and India. They are characterized by their shining blue and contrasting black plumage in the males. According to Sir Stamford Raffles, the founder of Singapore:[482]

Nothing can surpass the richness of the colours which distinguish the male of this species; they far exceed what any painting can convey.

Latham was probably the first to use the common name in the literature in 1787, stating that it

inhabits India, where it is known by the name of the Blue Fairy Bird.[166]

## Fairy-Fantail

The Yellow-bellied Fairy-Fantail is one of only a handful of members of the Stenostiridae, the fairy flycatchers. Until this newly formed family was established, the species was classified with the *Rhipidura* fantails. The new English name was coined in 2011 in order to differentiate this spritely forest-dweller from the more robust fantails.

## Fairywren

There are fifteen species of fairywrens in the Maluridae found in Australia and New Guinea. They are described as "perky, light-footed jewels" by the Cornell Lab of Ornithology.[1] They were initially called "wrens" by the homesick European settlers of Australia for their superficial resemblance to the Winter Wren (now Eurasian Wren) of their homeland. The adoption of the fairy prefix came later, in the early 1900s (as fairy-wren), presumably to avoid confusion with the true wrens and to denote the birds' delicate and pretty appearance. The ornithologist Tom Iredale wrote floridly about the "fairy-wrens" in 1924:[168]

The poverty of imagination seen in these nominations [the former names] is being redressed slowly, but surely some more appropriate names can be invented for these whiffs of beauty. Whiff is descriptive of their flight across a path, as with their long tail they fly undulating and zigzagly as if wafted by a gnome-driven wind, disappearing into the thicket, leaving an impression of dainty fairiness. Fairy Wren would be an appropriate name for the group, then Purple Fairy Wren could be used for the Banded Wren, and so on.

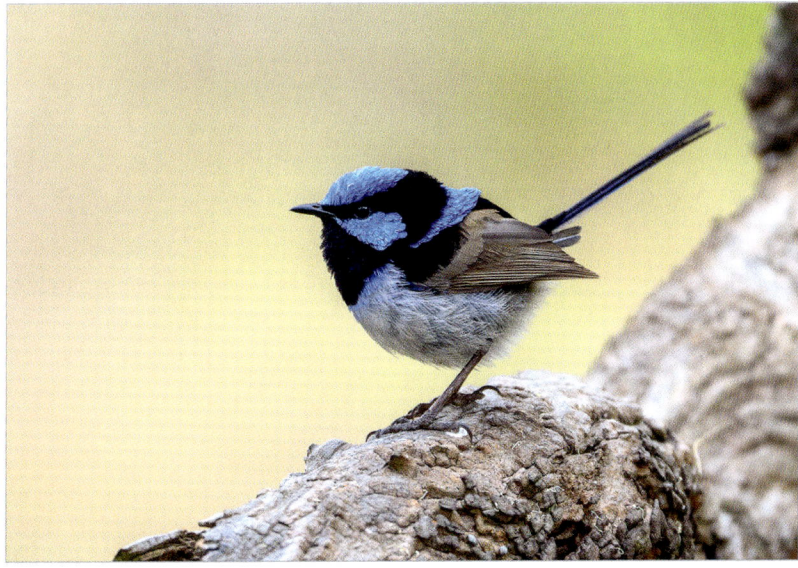

*Superb Fairywren (Victoria, Australia)*

# Falcon

Among the sixty-six members of the Falconidae, twenty-nine are known by the common generic name falcon. This very old English bird name arrived into the language in the mid-thirteenth century in a number of forms, including *faucŏun, fauken, falcon,* and *faucon,* from the Old French name *faucon* for various species of *Falco.* This in turn came from the Latin *falconem* or *falco,* and probably derives from the Latin *falx,* meaning "curved blade, pruning hook, sickle, war-scythe," said to refer to the shape of the bird's talons, legs, or beak, but also possibly from the shape of its wings in flight. The word appeared as early as circa 1250 in "The Owl and the Nightingale," a Middle English poem:[416]

> Þe faukun wes wroþ wiþ his bridde,
>
> *[the falcon was angry with her chicks,]*
>
> & lude yal and sturne chidde;
>
> *[and yelled noisily and harshly chided them;]*

Another theory that the Latin bird name *falx* is of Germanic origin and means "gray bird," from the PIE word *-pel,* meaning "pale," is supported by the antiquity of the word in Germanic. But falconry was brought to Germany by the Romans, suggesting the word flowed from Latin into the Germanic languages rather than the other way around.

# Falconet

The name of these six species of small falcons means just that, *falcon* + *-et,* a diminutive suffix derived from Old French.

# Fantail

There are fifty-four species of fantails in the Rhipiduridae, and, as one might expect, they all have fan-shaped tails. These insectivores flit around dense foliage or on the ground, actively fanning their tails in order to flush their prey. Although the name "fan-tail" was used for certain types of domestic pigeons in English, it was in Australia that the name was given to the *Rhipidura* fantails. The name was probably used locally in the "Colonies" before it first appeared in print in a paper presented in the "Transactions of the Linnean Society of London" in 1827:[169]

> Mr. Caley [the English botanist] thus observes on the manners of this bird. "Fan-tail .—There is something singular in the habits of this bird. It frequents the small trees and bushes, from whence it suddenly darts at its prey, spreading out its tail like a fan, and to appearance turning over like a tumbler Pigeon, and then immediately returning to the same twig or bough from whence it sprang. These actions it continues constantly to repeat."

# Fernbird

The fernbirds are two species (one extinct) of Locustellidae endemic to New Zealand. They were first described by the French zoologists Jean Quoy and Joseph Gaimard in 1832, but the English name wasn't seen in print, at least, until 1882 in the *Manual of the Birds of New Zealand,* in which the author noted:[268]

> It frequents the dense fern (*Pteris aquilina*) of the open country, and the beds of the raupo (*Typha angustifolia*) and other tall vegetation that cover our swamps and low-lying flats.

# Fernwren

The single species of Fernwren from northeastern Australia is named for its passing resemblance to the true wrens and its favored habitat of shady gullies, where it hops around on or near the ground in thick vegetation consisting of ferns, vines, and thickets. A note appeared in *The Emu* in 1905, giving us the origin of both the common and genus names.[182] The latter was until then *Sericornis*, which explains its former common name, "Collared Scrub-Wren."

> Mr. A. J. North, Ornithologist Australian Museum, has assigned the species a new genera—namely, *Oreoscopus*—and the very distinctive vernacular name "Fern-Wren."

# Fieldfare

The Fieldfare would have been a familiar bird to English speakers since antiquity, as evidenced by the Old English name *felde-ware* or *feldefare*, said to be pronounced with four syllables by Chaucer. The name entered into Middle English as *ffeldefare*, *feldfar*, or *feldifare*. Most sources state that the name means "field dweller," but Liberman says:[593]

> Despite the support of some of the most authoritative dictionaries, *–fare* in this word has probably nothing to do with "faring."

Lockwood proposed an alternative theory that the name is a corruption of the Old English *fealu fearh*, meaning "gray piglet," with the reasoning that other regional names include the Welsh *socen lwyd*, "gray pig," and the West Frisian *feale lister*, "gray thrush," the pig reference relating to what he called the bird's "grunt or other animal-like sound."[84] Relating the bird name back to other Old English names for animals, however, Liberman, citing Lockwood, suggested:

> William B. Lockwood, perhaps the greatest modern specialist in the history and origin of bird names, noted that the concept "goer" or "dweller" is alien to the popular ornithological nomenclature. The fieldfare was, consequently, not a field-farer or a field-dweller but a bird whose search for food is restricted to a field, which makes much better sense.

This thrush, a member of the genus *Turdus*, featured on the menu in the *The Boke of Keruynge* in the 1500s for the second course in the chapter "Dinner Courses from Michaelmas to Christmas":[483]

> Large byrdes, snytes, feldefayres, thrusshes, fruyters, chewettes, befe with sauce gelopere. [Great birds, snipes, fieldfares, thrushes, fruit, meat pies, beef with clove sauce.]

The migratory Fieldfares overwinter in Britain, during which time they favor lowland areas in open habitats, such as fields, pastures, and moorlands.

# Fieldwren

The fieldwrens are three species of Acanthizidae (thornbills and allies) endemic to Australia. As with fairywrens, the early colonists from Britain were reminded of the Winter Wrens of Europe, due no doubt to their size and habit of cocking their tails. The *field-* prefix was added later to denote the birds' preferred habitat of heathlands, shrublands, and saltmarsh. Hall appears to have been the first to use the common name "Field-Wren," although in the same entry he refers to one of the congeners as a "Desert-Wren."[170]

# Figbird

The three taxa of figbirds in the Oriolidae (Old World orioles) have been frequently lumped and split in the past. At present, three species are recognized from Australia, New Guinea, and Eastern Indonesia. The common English name is a clear reference to the preferred diet of fruits, and especially figs. In 1889, North wrote about the *Sphecotheres* figbirds with the comment:[124]

> It is locally known as the "Mulberry-bird," from the decided preference it evinces for that species of fruit amongst many others attacked by this bird.

Hall was calling them "Fig-Birds" by 1899, and he appears to be the first to have used the name in print, although he probably took it from a local name.

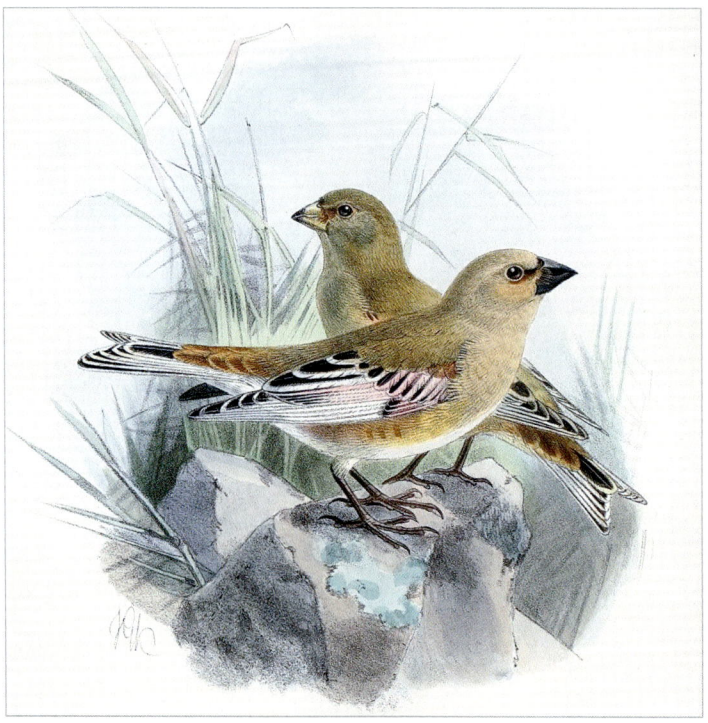

*Desert Finch in Dresser's* A History of the Birds of Europe *(1895)*

# Finch

The name "finch" might be one of the most widely applied bird appellations in the English language. Forty-one species are referred to simply with the generic group name "finch," but there are many, many others with prefixes or suffixes attached to the word. These include cactus-finch, diuca-finch, ground-finch, inca-finch, koa-finch, mountain-finch, rosy-finch, sierra-finch, tree-finch, warbling-finch, parrot-finch, yellow-finch, and so on. The Old English word *finc* came from the Proto-Germanic *finkiz*, "finch," and changed over time to *fynche* or *fynch* in Middle English, and thence to the modern-day finch. The Germanic word was thought to have descended from the echoic PIE word *(s)pink*, echoic of the single-note *pink* call of the Chaffinch, and we can see related influences in the Greek σπίγγος *spingos*, for "finch," and even the Sanskrit फिङ्क *phingaka* (drongo). An entry in the *Museum of Natural History* in 1887 explained the origin of the name:[65]

It usually consists of a sharp repetition of a sound resembling the syllable *Fink* or *Pink* from the former of these words the word finch is derived.

Then in 1896, Newton expanded on the etymology:[7]

This call-note [of the Chaffinch], which to many ears sounds like "pink" or "spink," not only gives the bird a name in many parts of Britain, but is also obviously the origin of the German Fink and our Finch.

## Finchbill

The two species of bulbul (Pycnonotidae) known as finchbills both possess short, conical bills that are somewhat reminiscent of the bills of finches. A common name for the *Spizixos* finchbills was coined in the 1890s as the "Finch Billed Bulbul," as in an article entitled "The Bulbuls of North Cachar" in the *Journal of the Bombay Natural History Society*.[171] The English name "finchbill" wasn't used until the mid-1900s.

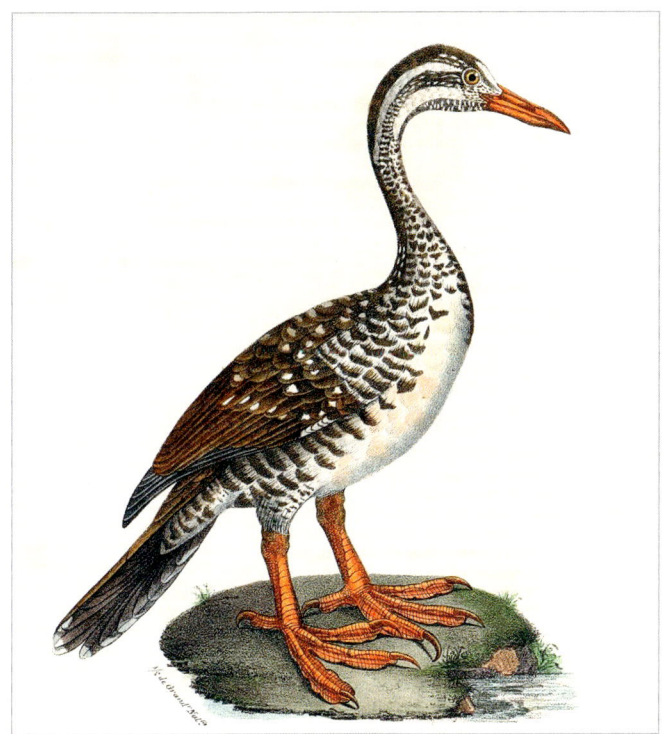

*African Finfoot in Vieillot's* La Galerie des Oiseaux *(1825)*

## Finfoot

Two species of finfoots in two different genera belong to the Heliornithidae and are closely related to the only other member of the family, the Sungrebe. The English name was coined by Latham in 1824 when he described the African Finfoot and the "American Finfoot," now known as the Sungrebe.[17] The members of the family are all characterized by their webbed lobed toes, which, as Latham noted, are

all disunited, but furnished on each side with a triple, scolloped membrane, in the manner of the Phalarope, or Coot.

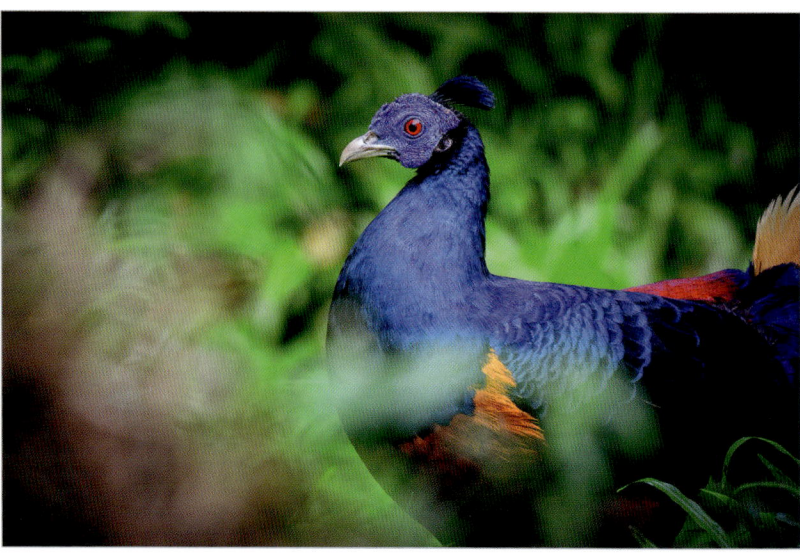

*Crested Fireback (Sabah, Malaysia)*

# Fireback

Three species of *Lophura* fireback pheasants are found in Asia. One of their most prominent features is the red and golden plumage on the lower back to rump. Beebe wrote about the Siamese Fireback in 1921:[309]

> Sportsmen in Siam who sit up on the look-out for tigers, sometimes see this magnificent bird step out from the jungle and walk slowly past, its fiery golden back flashing even in moonlight. It lives in dense bamboo thickets, and comes into more open jungle, often near the banks of a river, to feed and drink. It is as easy to trap as it is difficult to observe, and the Siamese bring many to the Bangkok market.

# Firecrest

The *Regulus* firecrests are two species of kinglets, one widespread and found in Europe, the other on the tiny island of Madeira. They were formerly treated as conspecific. The common name recognizes the deep orange-colored crown stripe, but earlier literature referred to the European bird as the "Fire-crested Wren" or "Fire-crested Regulus."

# Firecrown

The males of the two *Sephanoides* firecrown hummingbirds sport glittering fiery green-yellow to fiery flame-orange crown patches. Gould, who called them the "Fire-crowns" said:[6]

> The male has the crown of the head rich, deep, glittering, fiery orange-red.

# Fire-eye

The five species of *Pyriglena* fire-eyes in the Thamnophilidae (typical antbirds) live up to their names with their bright fiery red irises. The first use of the name appears to have been in *Cassell's Book of Birds* in 1875, in which they were called the "Fire Eyes" with the observation:[106]

> The eye, as the name of the bird indicates, is of a brilliant fiery red.

# Firefinch

Eleven species of Estrildidae finches, all in the genus *Lagonosticta* from Africa and known as firefinches, sport variations on a theme of bright red plumage. Latham described a "Fire Finch" as a bird with "the general colour of the plumage a glossy brownish red" in 1783, giving "Fire Bird" as an alternative name.[17]

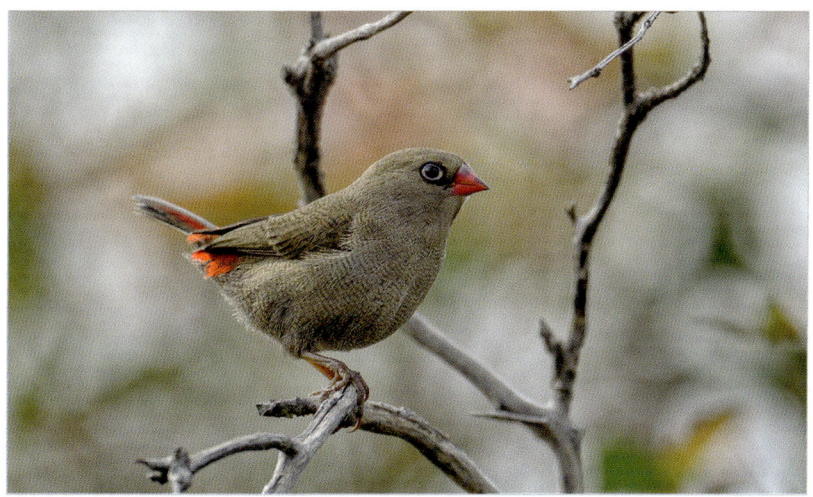

*Beautiful Firetail (Victoria, Australia)*

# Firetail

The firetails are four species of finches in the genus *Stagonopleura*, endemic to Australia and New Guinea, all with bright red rump and undertail coverts. The name first appeared in print when Gould labeled a plate of Beautiful Firetail as the "Fire-tailed Finch" with a side note that the name "Fire-tail" is used by the "Colonists of Van Diemen's Land."[40] By the early 1900s, the name had been modified to the present day "firetail."

# Firethroat

The Firethroat is a congener of the rubythroats and the Blackthroat in the genus *Calliope*. Although Oustalet didn't use a common name for the bird, in 1901 he described the namesake feature beautifully:[172]

> En limitant une zone d'un rouge ardent qui couvre le menton, la gorge et la région thoracique . . . un rouge légèrement orangé et extrêmement intense (rouge de Saturne). [. . . a fiery red area that covers the chin, throat and thoracic region . . . a slightly orange and extremely intense red (Saturn red).]

# Firewood-gatherer

This single species of furnarid (ovenbirds and woodcreepers) owes its unique name to its habit of constructing huge nests of sticks, and thus it spends quite a lot of time conspicuously collecting sticks and twigs that look, to the human eye, like firewood. Sclater explained the origin of the English name in 1888:[22]

> This is a common and very well-known species throughout the Argentine country and Patagonia, also in Uruguay and Paraguay, and is variously called *Espinero*

(Thorn-bird), *Tiru-riru*, in imitation of its note, and *Añumbi* (the Guarani name); but its best known name is *Leñatero*, or "Firewood-Gatherer," from the quantity of sticks which it collects for building-purposes.

# Fiscal

The fiscals are seven species of shrikes found in Africa. The word *fiscal* is an old one for public officials, usually a taxman. The fiscal birds all display handsome black and white plumage that is reminiscent of the suit-and-tie attire of the taxman ("fiscal"). Schlegel wrote in 1872:[173]

> De voorouders der Hollandsehe inwoners van de Kaap de Goede Hoop noemden dezen vogel, die tot in de tuinen der Kaapstad aangetroffen wordt, *Fiskaal*, omdat hij met zijne patienten even kort proces maakt als hun fiskaal deed, ofschoon deze, zoo als hun vogel-fiskaal, niet tevens de functiën van beul waarnam. [The ancestors of the Dutch inhabitants of the Cape of Good Hope called this bird, which can be found in the gardens of Cape Town, *Fiskaal*, because he makes just as short a process with his patients as they did their tax, although they did not also act as executioner, like their bird-fiscal.]

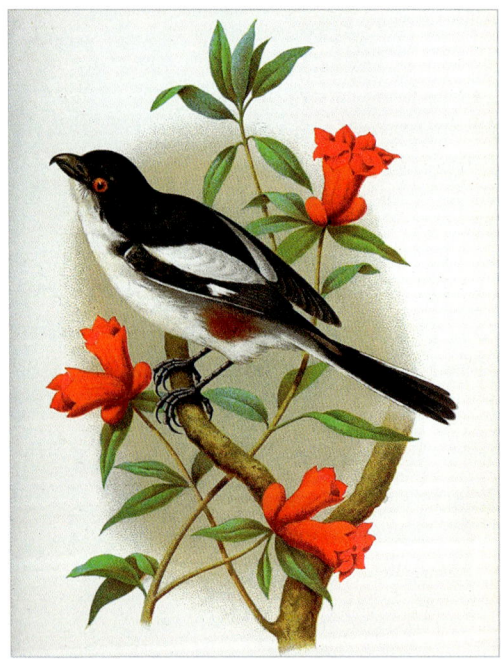

*Southern Fiscal from Holub and Pelzeln's* Beiträge zur Ornithologie Südafrikas *(1882)*

# Flameback

Twelve species of large woodpeckers in two genera are known as flamebacks. All of them have a large patch of bright red or flame-colored plumage on the lower back and rump. Prior to the mid-1900s, the English name for the known species of flamebacks was "golden-backed woodpecker"; the name flameback appears to have first been used in the literature in 1947.[484] Even in the recent Indian-English ornithological literature, the common names "Lesser Golden-backed Woodpecker" and "Greater Golden-backed Woodpecker" [*sic*], as well as simply Goldenback, are used.

# Flamecrest

This kinglet, a member of the Regulidae, is endemic to Taiwan. Ogilvie-Grant was the first to describe the bird in the ornithological literature in 1906 when he wrote that *Regulus goodfellowi* (now known as Flamecrest) is[174]

> most nearly allied to the male of *R. ignicapillus* [Common Firecrest]. . . from which, however, it differs greatly in the more brilliant fiery orange-red of the crown, the more distinct black-and-white markings on the sides of the head, the canary-yellow lower back and rump, and the somewhat paler yellow sides of the breast, flanks, and under tail-coverts. . . . This brilliantly-coloured Fire-crested Wren differs widely from every known form.

Despite Ogilvie-Grant's assertion, the Flamecrest was often treated as a subspecies of the Firecrest and invariably referred to as the Taiwan Firecrest.

# Flamingo

The flamingos are a family (Phoenicopteridae) of six species, all of which sport strikingly pink plumage. The name comes from the Portuguese or Spanish *flamengo*, which came into that language from the Provençal *flamenc*, from *flama*, "flame" and the Germanic suffix *-enc* "belonging to." Ultimately, the name derives from the Latin *flamma*, meaning "flame," but originally the name may have come from the Spanish word for the people of Flanders, known as the Flemings, who were singled out by their neighbors to the south for their ruddy complexions. This was arguably the impetus for the coining of the Spanish word "flamingo," a metonym of "a Fleming," taking on the meaning of "a bird of a hue reminiscent of the Flemings." The word *flemengo* was first used in English in a 1589 account of discoveries by the English nation, which included an entry written by the naval commander John Hawkins of a voyage "begun in An. Dom. 1564":[175]

> Fowles also there be many, both upon land and upon sea: but concerning them on the land I am not able to name them, because my abode was there so short. But for the fowle of the fresh rivers, these . . . I noted to be the chiefe, whereof the Flemengo is one, having all red feathers, and long red legs like a herne, a necke according to the bill, red, whereof the upper neb hangeth an inch over the nether.

# Flatbill

There are seven species of flatbills in two genera, the *Rhynchocyclus* and *Ramphotrigon* in the Tyrannidae. These unassuming birds all have large, wide bills. Up until the early 1900s, the name "flat-bill" was used for various other birds of the Neotropics, some of them not always closely related, including some species of *Conopophaga* gnateater, *Todus* todies, *Tolmomyias* flycatchers, and *Platyrinchus* Spadebills. In 1921, Chubb wrote that a member of the *Rhynchocyclus* genus is

> rather a small bird with a depressed bill which is hooked at the tip of the upper mandible, the width at the base is about equal to the length of the exposed culmen.[176]

# Flicker

There are seven species of woodpeckers known as flickers. The word "flicker" came into Modern English from the Old English word *flicorian*, meaning "to flutter, flap, or move wings" and originally pertained to birds. It was believed to be an onomatopoeic word, and

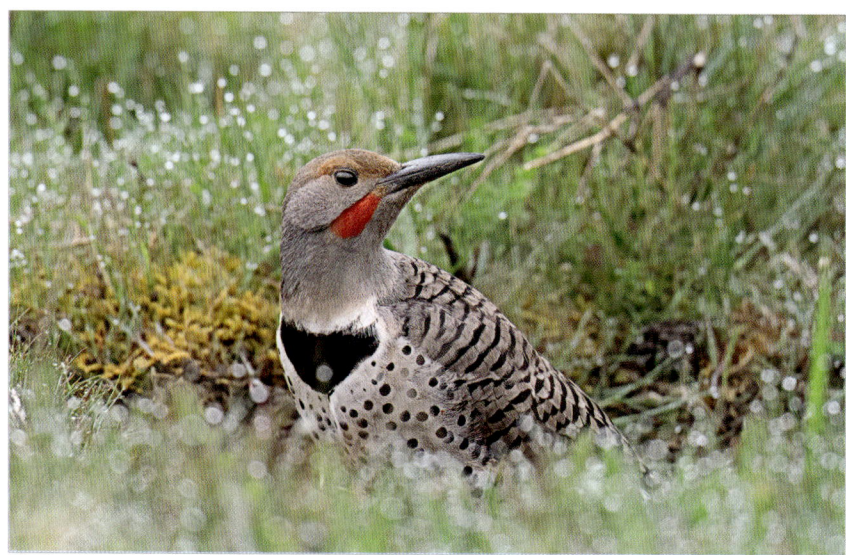

*Northern Flicker (Washington State, USA)*

then it evolved into the sense of a flickering light in the 1600s. There is some thought that the bird's name arose from the bird's call, but it seems far more likely that a reference to the spots on the bird's wings that appear to flicker in flight is the true origin of the name. That said, Ingersoll claimed that the name is all about the call, although many species of woodpeckers have a rhythmic *flika flika* call:[9]

> Who first coined the nickname "flicker" for the ubiquitous golden-winged woodpecker is unknown, but it fits fairly well the repeated shout of this fine species as it flies from place to place, and has generally replaced a large number of insignificant local names, such as pigeon-woodpecker, highhole and yellow-hammer.

## Florican

The floricans are two species of small bustards in the Otididae found in Asia. The origin of this unusual name is uncertain. Jerdon wrote in 1862:[136]

> I have not been able to trace the origin of the Anglo-Indian word "Florikin," but was once informed that the Little Bustard in Europe was sometimes called Flanderkin.

While Yule wrote:[250]

> Of some very familiar words the origin remains either dubious, or matter only for conjecture. Examples are hackery (which arose apparently in Bombay), florican, topaz. . . . The origin of the word Florican is exceedingly obscure. It looks like Dutch. (The N.E.D. suggests a connection with Flanderkin, a native of Flanders).

Possibly the earliest use of the word in the literature was written by Munro, whose mastery of ornithology was no match for his military expertise, when he wrote in 1780's *A Narrative of Military Operations on the Coromandel Coast*:[334]

> As the army moves along, the officers oftentimes enjoy upon the road a fine chase after antelopes and hares, with which this country abounds. . . . They also meet with covies of partridge, wild-duck, and the floriken, a most delicious bird of the buzzard kind.

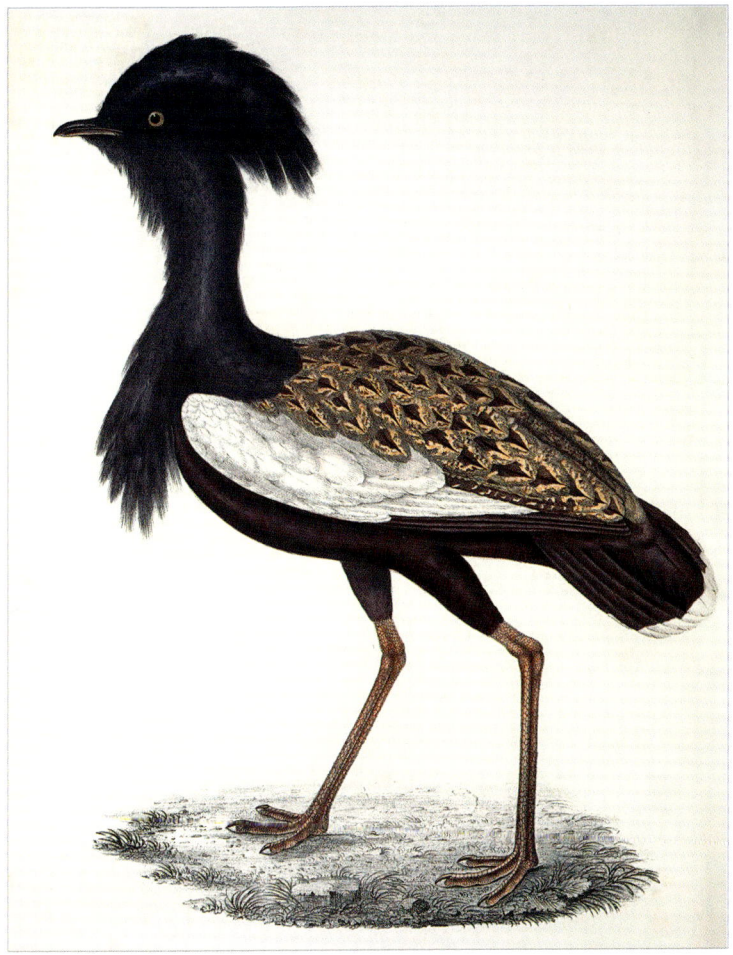

*Bengal Florican in Gould's* A Century of Birds from the Himalaya Mountains *(1831)*

And Newton summed up the name as[7]

> the Anglo-Indian name for the smaller Bustards, the origin of which neither Jerdon nor Yule can trace. The latter shews that it was used in 1780 (Munro, Narrative, p. 199), and says "it looks like Dutch" but from analogy a Portuguese derivation would seem more likely.

# Flowerpecker

The Dicaeidae, the flowerpeckers, consists of forty-seven species of flowerpeckers, all of which are restricted to Australasia and Asia. These tiny birds feed primarily on small fruits and nectar, as well as small invertebrates, and despite the name, aren't known to peck at flowers. A number of flowerpeckers have a mutualistic relationship with mistletoe plants—they preferably feed on the berries and the plant, in turn, relies on the flowerpeckers for seed dispersal. Gould noted:[33]

> The little birds may be seen in flocks of from ten to fifteen in one tree, twisting and turning themselves about these flowers and clinging to them back downwards with the active movements peculiar to the family.

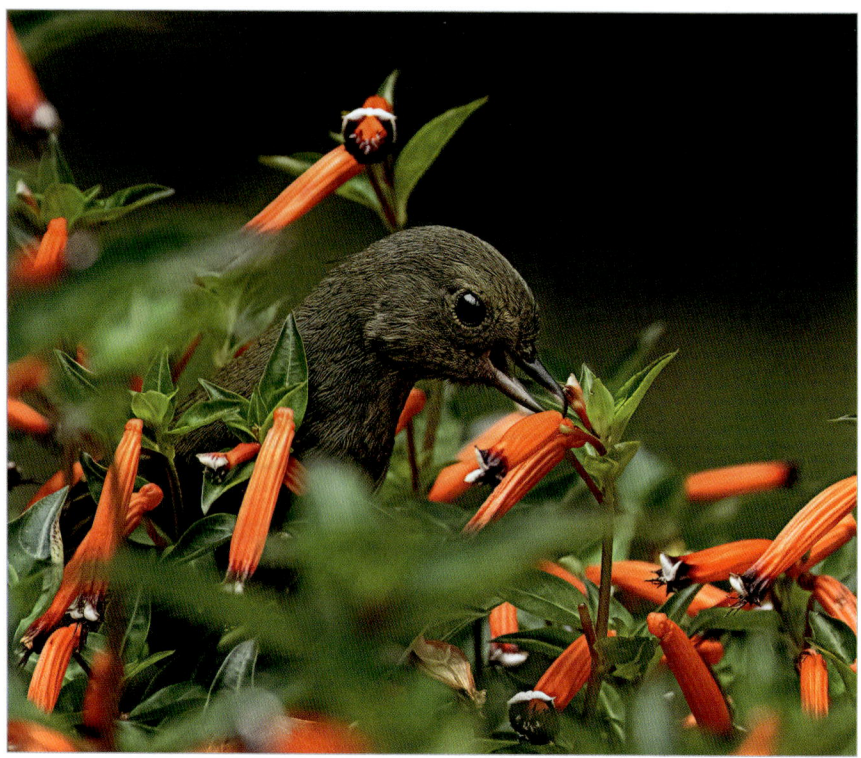

*Slaty Flowerpiercer (Costa Rica)*

# Flowerpiercer

The eighteen species of flowerpiercers belong to the Thraupidae (tanagers and allies). All belong to the *Diglossa* genus and possess remarkable hooked bills with which they pierce the bases of flowers to consume the nectar. The genus was first named in 1832 by the German naturalist Johann Georg Wagler, who wrote:[310]

> Three crooked wrinkles behind the hook of the upper mandible, as well as the complete absence of a chin angle give the bill a unique/peculiar look.

# Flufftail

The Sarothruridae houses eleven species of small flufftails found in Africa and Madagascar. Although they closely resemble crakes, and in the past were referred to as such, they are unrelated. One of their unique features is their soft and decomposed tail feathers—in other words, fluffy tails.

# Flycatcher

The name flycatcher is one of the most widely used in English-language birding, with at least 246 species in five families, as well as many other compound names possessing the appellation. The origin of the name is very straightforward, referring to the feeding habits of these insectivorous birds, which take invertebrate prey on the wing. The first known use of the word, as pertaining to birds, was in the late 1600s when Ray and Willughby wrote:[16]

> Aldrovandi describes another bird by the name of his first Muscicapa, or Flie-catcher.

Prior to this, in the 1500s, the word *muscicapa* had been widely used in the ornithological literature, which was usually written in Latin.[178] The word literally means "fly catcher," from the Latin *musca*, for fly, and *căpĭo*, "to seize or to take hold of," and was used for a large number of unrelated birds, including the *Merops* bee-eaters. Brisson erected the genus *Muscicapa* in 1760 in the French-language publication *Ornithologie*, giving the colloquial name as Le Gobe-Mouche, the fly-catcher.[76] And later in the late 1700s, Pennant would use the name for Spotted Flycatcher and European Pied Flycatcher.[177] Subsequent to that, as Newton wrote, the moniker (which is now often used in hyphenated forms as well *viz* forest-flycatcher, slaty-flycatcher, blue-flycatcher, jungle-flycatcher)[7]

> has since been used in a general and very vague way for a great many small birds from all parts of the world, which have the habit of catching flies on the wing.

# Flycatcher-shrike

The charming flycatcher-shrikes are two species in the genus *Hemipus* found in Asia. In the literature of the early to mid-1800s, the birds were variously labeled "pied shrikes" and "Shrike-like Flycatcher." This is another case of the early European observers, on encountering unfamiliar birds, assigning the names of superficially similar and familiar, yet unrelated birds. Oates alluded to this in 1883 when wrote:[112]

> Sykes's Pied Shrike . . . takes its post on a branch of a tree and darts on passing insects, catching them on the wing like a Flycatcher . . . this bird also searches the leaves like a Wood-Shrike.

# Flycatcher-Thrush

Two African members of the Turdidae in the *Neocossyphus* genus are unusually flycatcher-like thrushes that forage from perches, sallying out to take insects on the wing or from foliage.

# Flyrobin

The flyrobins from New Guinea and Eastern Indonesia are two species of *Microeca* in the Petroicidae, the Australasian robins. The name "fly-robin" was used widely up until the early 1900s in Australia for another member of the family, the Ashy Robin. Given that most members of the family are called either flycatcher or robin, it's not surprising that this portmanteau was coined by early naturalists who couldn't decide which European bird they most closely resembled.

# Fody

The fodies are eight species of little red or yellow weavers in the Ploceidae found in Madagascar and the surrounding Indian Ocean islands. The English word "fody" and the name of the genus *Foudia* are from the Malagasy name *foudi* or *fodi* for the Red Fody that is widespread and common on the island. The genus was erected by Reichenbach in 1850;[72] however, Finsch and Hartlaub disagreed in no uncertain terms, to no avail:[179]

> The name "Foudi," with which Sakalaven and Malagasy designate some of the birds belonging to this area and which is also written Soudi (Verr.), Fouli (Grandidier) and Fody (Newton), has no right to scientific application.

Richardson related a Malagasy saying that references the Fody:[231]

> *Manaò fodilàhy mitsindroka àlina*—to steal at night, literally to act as the cardinal bird [a species of weaver-finch] picking up food at night.

## Foliage-gleaner

The foliage-gleaners are thirty-two species of Furnariidae (ovenbirds and woodcreepers) from the Neotropics. As the name suggests, all of them glean foliage for their invertebrate prey as they forage in thick vegetation. Ridgely et al. described their feeding behavior thus:[180]

> [It] forages actively, sometimes even acrobatically, at middle and upper tree levels, clambering along branches, often hanging upside down or moving out onto terminal twigs, pausing to inspect epiphytes and dead leaves.

## Forktail

The forktails are a beautiful group of seven species of Muscicapidae (Old World flycatchers) from Asia. All of them are strongly associated with fast-flowing waterways and their long, forked tails (except for Little Forktail), which are constantly flicked up and down and may be used to signal conspecifics—but that is still an open question. In 1867, Adams charmingly described the Spotted Forktail in the Himalayas:[181]

> The rich white and black colourings are particularly attractive, and its habits so eccentric as to arrest the attention of even the most indifferent observer—now running sprightly along the margin of the torrent, with its forked tail expanded like a beautiful black and white fan; anon with extended neck and wings it turns its well-marked body from side to side as if on a pivot, until, gathering up its snow-white legs, with an austere screech it shoots rapidly along the windings of the stream.

## Francolin

There are forty species of francolins in a number of different genera in the Phasianidae (pheasant, partridges and allies). They are widespread in Africa, with a handful of species in Asia. These chicken-like ground dwellers derive their names from the Italian word *francolino*, possibly meaning "little hen." The name appeared in *Icones Avium Omnium* when Gessner wrote:[266]

> Italice Francolino, ut circa Brixiam uocant: Pernis alpedica, Perdice alpestre, Fasanella. Sed Francolini nomen alteri potius avi debetur, nempe Attageni. [Italian Francolino, and around Brescia called mountain hawk, alpine partridge, Fasanella. However, the Francolini, another name given to the bird, is assuredly the grouse.]

## Friarbird

There are seventeen species of, some would say, unattractive honeyeaters from Australasia. Versions of the name, including "fryar" and "frier," are recorded from the late 1700s. The name was probably first applied to the Noisy Friarbird in the 1700s by the first European settlers of southeastern Australia. Most of the friarbirds have balding pates (another old name for them was leatherhead) and thus they take their name from a supposed resemblance to the Franciscan monks known as friars with their shaved heads and "the semblance of a hood about its shoulders, formed by a ruff of soft recurved feathers, and the sad hue of its plumage."[7] Latham called the Noisy Friarbird the "Knob-fronted Honey-Eater" but noted:[17]

> It is called by the English in New-Holland, The Friar-Bird . . . the hindhead projecting, and being of a black, downy texture, gives some resemblance to a cowl or hood, and has occasioned it, as well as the last, to be called the Frier.

# Frigatebird

The five species of frigatebirds have their own family, the Fregatidae. Also known colloquially as man-o'-wars, the name frigatebird originated with French sailors who likened them to the fast warships known as frigates. The origin of the word frigate is uncertain. Du Tertre wrote of the birds in 1667:[183]

> L'oyseau que les habitans des Indes appellent *Fregate* (à cause de la vistesse de son vol) n'a pas le corp plus gros qu'une poule. [The bird that the inhabitants of the Indies call "frigate" (because of the speed of its flight) has a body no larger than a chicken's.]

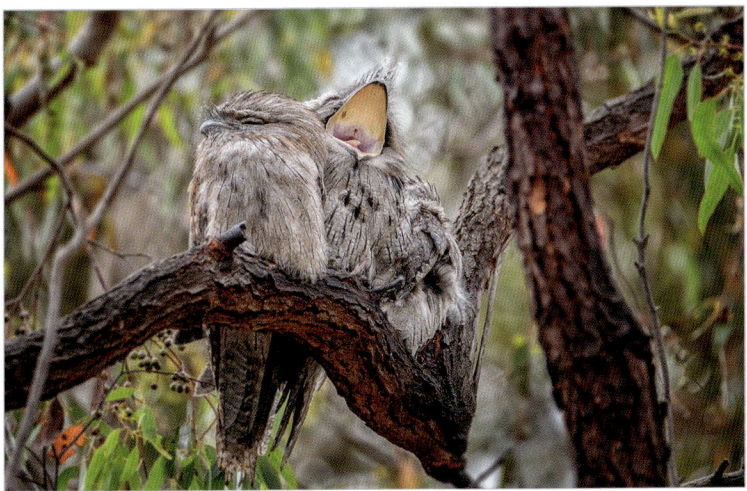

*Tawny Frogmouths (Victoria, Australia)*

# Frogmouth

The frogmouth family, the Podargidae, consists of sixteen species of remarkable nocturnal and cryptic birds from Asia and Australasia. One of their most prominent features is their enormously wide bills, with which they'll consume anything from insects to lizards and even small mammals. It gives their faces a frog-like visage and hence, the name. In 1838, Gould established the genus *Batrachostomus*, from the Ancient Greek βάτραχος *batrakhos*, "frog," and στόμα *stoma*, "mouth," which Jerdon rendered as "Frog-Mouth" in 1864.[485]

# Fruitcrow

The four species of cotingas known as fruitcrows belong to four different monotypic genera. They all have a somewhat crow-like appearance and were thought to possibly be related to the corvids in early ornithology. As well, they are frugivores, although all will feed on invertebrates too. Wood wrote in 1862:[85]

> The Fruit Crows are placed by some systematic authors among the chatterers, while others, as in the catalogue which we follow, have considered them to be nearly related to the true Crows. They are all natives of Southern America, and are distinguished by their straight flattened beak, with its upper mandible round, and a notch at its extremity. The nostrils are placed in two membranous groves [*sic*] at each side of the bill. Most of the Fruit Crows are of considerable dimensions, some species equalling the Crows of Europe, while others are a little less.

# Fruiteater

The fruiteaters are twelve species of cotingas in the genus *Pipreola*, the exception being one species in the monotypic genus *Ampelioides*. Most of them have very similar green, yellow, and black plumage. All are strict frugivores, except for the *Ampelioides*, which will also consume insects and snails. One of the first uses of the name was in Swainson's *Zoological Illustrations* in 1820 when he labeled an illustration of the closely related Black-headed Berryeater as the "Black-headed Fruiteater."[50]

# Fruit-hunter

The single species of Fruit-hunter in the monotypic genus *Chlamydochaera* is a member of the Turdidae, the thrushes and allies. The taxonomic position of this unique Bornean forest dweller has been argued over since it was first described in 1887 when Sharpe classified it as a member of the Campephagidae, the cuckooshrikes,[486] and the vernacular name "Black-breasted Triller"[487] into the mid 1900s. Subsequently, it was moved to the Turdidae and then was often known as the "Black-collared Thrush." Whitehead, the English explorer and naturalist, wrote in 1893 that he[184]

> found [the birds] on my first expedition at about 3000 feet, and they were, so far as we observed, always feeding on berries . . . it is strictly a forest bird, and decidedly local.

The current name, which came into use in the 1990s as Black-breasted Fruithunter (the descriptive part of the name now considered redundant), tells the tale of its dietary preference for all sorts of fruits, but it is also known to consume invertebrates, including snails.

# Fulmar

Two species of *Fulmarus* in the Procellariidae, the shearwaters and petrels, derive their name from the Old Norse name for the bird *fúlmár*, a compound of the words *fúll*, meaning "foul," and *már* for "gull" (see Gull), in reference to the foul-smelling vomit it ejects to deter predators. Although the English name has been in use since the 1700s, at least, it was in 1826 that Stephens named the genus *Fulmarus* with a colorful description of their habits:[185]

> Its food consists principally of fish, but it will devour indiscriminately any floating putrid substances, such as the filth of ships, which it fearlessly follows. These birds also follow the tracks of the wounded whales, and, when they are exhausted, alight on the carcases by hundreds, and ravenously pluck off and devour lumps of the blubber until they are satiated. This gross food causes them to become excessively fat, and their stomachs are always charged with oil, which they have the power of ejecting with force from the beak; and when attacked squirt it into the face of their enemy.

# Fulvetta

There are twenty-six species of fulvettas in three different families, although they were previously included in one family, the Timaliidae. (See Babbler.) Many of these small passerines, found throughout Asia, look like the classic "little brown bird," and the origin of the name reflects this. It derives from the Latin *fulvus*, meaning "fawn" or "tawny brown." The *-etta* suffix is a diminutive, so the name literally means "little brown thing." David and Oustalet coined the genus name in 1877 with the justification:[92]

> Verreux had classified this species, as well as the following ones, first in the genus *Siva*, then in the genus *Proparus*; but it seemed necessary to us to form a distinct group, whose name alludes to the fawn hues dominant in the plumage of these birds.

# G IS FOR GADWALL

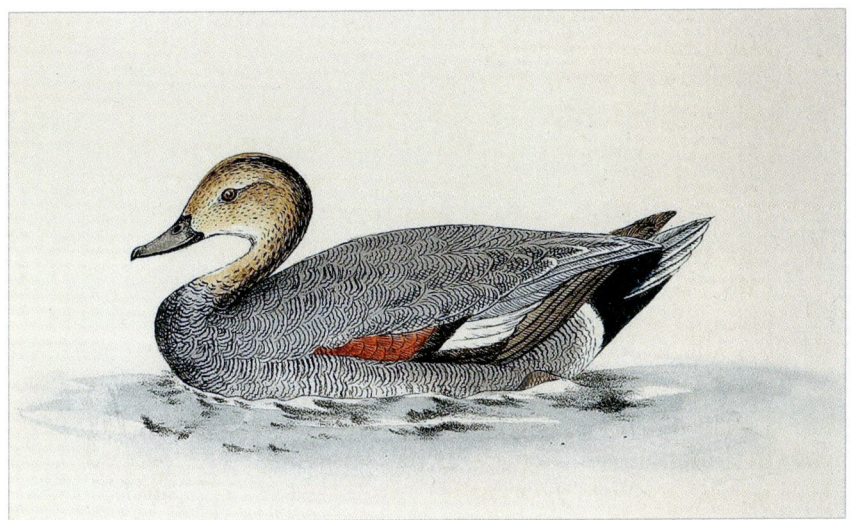

*Gadwall in Meyer's* Coloured illustrations of British birds, and their eggs *(1849)*

## Gadwall

There is just one Gadwall, an understated yet handsome duck. The name has been in use since the mid-1600s, but nobody seems to really be sure of the origin. There are suggestions that it may come from the words the "gad well" meaning "get about," but Newton asserted that this "is nonsense."[7] A version of the name appeared in 1667 in Merrett,[122] who wrote that the bird-dealers called it the "Gaddel," a name that Lockwood[84] emphatically asserts is onomatopoeic. "Gadwall" was first mentioned by Ray and Willughby in 1678 but with no hint as to its origin.[16] The English philologist Robert Gordon Latham suggested a solution in his 1866 volume *A Dictionary of the English Language*.[186] In one entry he conjectures that the names Garganey and Gadwall may both be

> a transformation of *Querquedula* [Latin for teal]: gadwall also probably the same, i.e. taken from the last two syllables, *-quedul-*, the present word being from the fuller form.

## Galah

The Galah is one of Australia's most famous birds. The name of this pink and gray cockatoo comes from the Yuwaalaraay[104] and Wiradjuri[132] (Aboriginal languages from New South Wales) name for the bird, *gilaa*. Johnston recorded that Roth (who was both a champion of Indigenous rights and an exploiter), the first Northern Protector of Aboriginals in 1898:[586]

> mentioned many names for the galah amongst north-west-central Queensland tribes, some of them being ga-la- ga-la (Walookera tribe), ge-la-ro (Woonamurra), boombabaro (Karanya), kelun-ji (Pittapitta). Our name galah is obviously derived from ga-la just mentioned.

It's likely that similar-sounding names were used in a number of different languages. Interestingly, it has come into Australian slang to mean "a fool or an idiot," presumably from a correlation with the bird's playful antics. The name was first recorded in 1862 in John McKinlay's *Journal of Exploration in the Interior of Australia*, in which he wrote that there was[187]

> a vast, number of gulahs, curellas, macaws, cockatoo parrots, hawks, and crows here [near the Leichhardt River].

Later in his journey, he found, as is often the case in Australia:

> Some few days ago not a bird was to be seen scarcely, but a few kite, crows, and gulahs; now the whole country seems to be alive with ducks of various kinds, macaws, currellas, cockatoo parrots, and innumerable small birds.

## Gallinule

The seven species of gallinules are all types of large Rallid. Brisson coined the name for the genus *Gallinula* in 1760, noting that "La Poule-d'Eau" or the Eurasian Moorhen[76]

> est à peu près de la grosseur d'une petite Poule [is about the size of a little hen].

But the name preceded that date when Gessner used it to refer to a moorhen in 1555.[178] The name derives from their chicken-like appearance—the Latin for "cock" is *gallus*, and *gallina*, "a hen." In Latin a *gallinula*, the diminutive of *gallina*, "hen," is "a little hen, a chicken." Later the name was anglicized to *gallinule* and more widely applied to other members of the Rallidae.

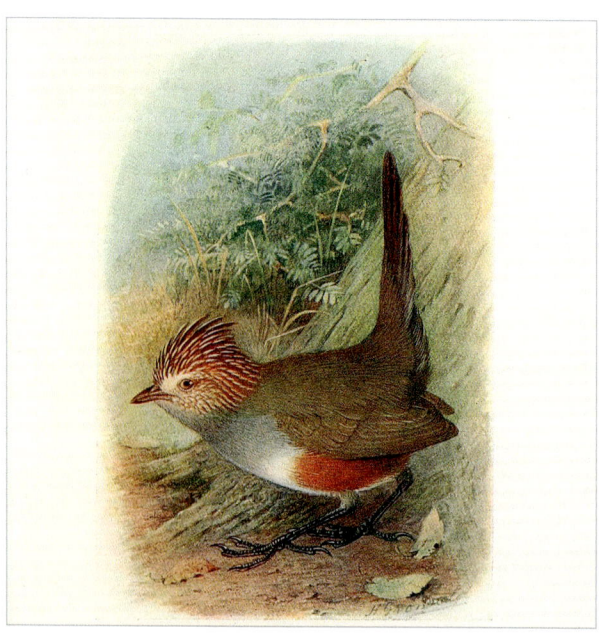

*Crested Gallito in Hudson's* Birds of La Plata *(1920)*

## Gallito

Two species of gallitos are found in the Southern Cone of South America. They are members of the Rhinocryptidae, the tapaculos, whose name comes from the Spanish diminutive of *gallo*, "cockerel or little cock." The name was first used by de Azara in 1805, who justified his designation with this delightful account of the bird:[188]

Asi le llamo, por la figura de la cola . . . El macho se eleva algunas veces espaciosamente casi á plomo, batiendo con mucha priesa las alas, elevando mucho la cola, pareciendo en esta disposicion mariposa mas que páxaro y quando llega á diez ó doce varas de altura, se dexa caer obliqüamente, posándose en alguna paja. No es arisco ni bullicioso; y aunque rara vez se hallan dos machos á doscientas varas, es freqüente encontrar dos y hasta seis hembras bastante inmediatas, porque al parecer son á lo menos quadruplicadas en número que losmachos. Estos en mi juicio no hacen los referidos ascensos, ni levantan la cola como el gallo sino en tiempo de amor. [That's what I call it, because of the shape of the tail . . . The male sometimes rises airily almost vertically, beating his wings very quickly, raising his tail very much, in this arrangement looking like a butterfly more than a bird, and when he reaches ten or twelve varies (8–10 meters) in height, he lets himself fall obliquely, perching on some straw. It is neither surly nor boisterous; and although two males are seldom found at two hundred yards, it is common to find two and even six fairly close females, because they seem to be at least four times in number than the males. In my judgment, these do not make the aforementioned ascents, nor do they lift their tails like a rooster except in times of love.]

# Gannet

The three species of gannets are closely related to the boobies in the Sulidae, but they have escaped that rather derogatory epithet, maybe because they are so much more graceful looking. The name came into Old English as *ganot* or *ganet*, from the Proto-Germanic *ganzô*, meaning "goose," and morphed over time to gannet with the general sense in English as well of "goose" up until the late 1400s. The word "gander" for a male goose has the same root word. A Middle English text of "Medical Books" from 1450 has gannet, probably referring to some type of goose, on the menu in a remarkable recipe for an ointment that heals scars:[488]

Tak cattes grece, ganates grece, bausones grece, bores grece, Mery of an hors, & grece of a dogge, & alle þese tempre to geder. [Take cats grease, gannets grease, badger grease, bores grease, oesophagus of a horse, grease of a dog, and mix them all together.]

# Garganey

The single species of Garganey is a duck whose name was first mentioned in the scientific literature by Gessner in 1554 when he wrote (in Latin):[178]

The inhabitants of Bellinzona use the diminutive form Garganey from the Italian garganello.

The name Gargenello was subsequently used in Aldrovandi's *Ornithologiae* in 1637[101] when discussing the "Querquedula" [Latin for teal] and later as Garganey in Ray and Willughby.[16] Its root word, the Old Italian *garganello*, means "throat" or "gargle," and the Bellinzona word in the Lombard language was probably a reference either to the duck's skimming method of feeding or to the male's unusual throaty rattling call (or both). As discussed under Gadwall, the philologist Robert Gordon Latham suggested that the name might be "a transformation of *Querquedula* [Latin for teal]" but this doesn't seem to have gained wide acceptance.[186]

# Geomalia

Although it's now classified as a thrush in the genus *Zoothera*, this enigmatic species found on the island of Sulawesi was previously assigned to the monotypic genus *Geomalia*. The name was coined in the late 1800s from the Ancient Greek *geo-*, meaning "ground" and

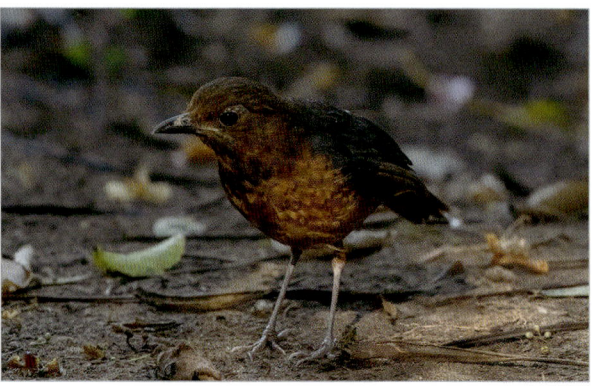

*Geomalia (Sulawesi, Indonesia)*

from the genus name *Malia*, which in turn is a nod to the *Malia*'s and the *Geomalia*'s perceived relationship to the so-called Timalia-group, the babblers. The name was suggested to Stresemann in 1931 by Heinrich, who sent him the specimen, and presumably thought it was a babbler in the Timaliidae.[189] Heinrich wrote to Streseman:

> *Geomalia* only lives in the thickest bushes of the rainforest on the western slope of the Latimodjong Mountains between 1800 and 2800 m. It does not seem to leave the ground at all, but it runs at great speed on earth. I finally caught the mysterious animal, which lives much more hidden than *Androphilus* [Australian whipbirds] and is even more difficult to kill than this, in mistnets, after which I had tried in vain to capture it for a long time.

# Gerygone

The nineteen species in the genus Gerygone (pronounced juh-rig-uh-nee) are found from Southeast Asia through Indonesia and Australia to New Zealand. The most outstanding feature of these small, not terribly colorful birds is their delightful song, and the name reflects this. It comes from the Ancient Greek word *gerygone*, the feminine form of γηρυγόνος *gērygonos*, meaning "born of sound" or "producing sound," from γῆρυς *gērys*, "sound" or "voice," and γόνος *gonos*, "that which is born."[190] Gould erected the genus name without elaboration in 1841 when he mentioned it in Grey's *Journals of Two Expeditions of Discovery in North-West and Western Australia*.[489]

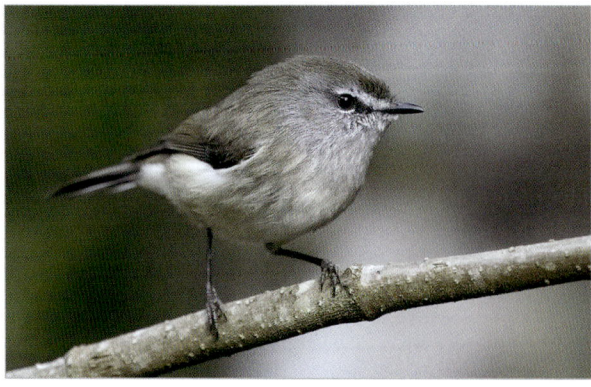

*Brown Gerygone (Queensland, Australia)*

# Gibberbird

This unique Australian chat in the monotypic genus *Ashbyia* is named for its preferred habitat—the gibber plains of the Australian outback. "Gibber" is an Australian word for desert pavement where the ground surface is covered with closely packed, rounded pebble-sized rock fragments; the term is also used to denote some ecological communities. The name is derived from the Sydney language or Dharug word *giba* for "stone" or "rock." Early on, the bird was called "Desert Chat," while the Inland Dotterel, which also inhabits the gibber plains, was often referred to as the "gibber bird" in the 1930s. An entry in a journal from 1943 listed the Gibberbird as:[191]

> *Ashbyia lovensis* Desert chat—Wee-icka (yellow and brown gibber bird or paper-bag bird resembling the orange desert chat). Onomatopoeic name.

A local legend of the Mitaka people of Queensland was related by Duncan-Kemp in 1933:[192]

> In ancestral times the huge dragon-lizards, Printhee, Kwoolcudee and Boolah-dee fought for possession of a woman, Wee-icka (white flower) and fought so hard and dug the ground so much that the mountains fell and covered the plain with stones (gibbers). White Flower died, so Printhee, the victor, was without a wife. He lived in a cavern, from which he emerged regularly to raid the tribes and capture the most desirable women for wives and tore to pieces those refusing to wed. Wee-icka became transformed into the little gibber bird or desert chat which has to make its nest in a hole scraped out in the stony soil, and its penetrating call, "*wee-icka wee-icka*" tells the tribesmen that White Flower still lives amongst the gibbers.

# Gnatcatcher

The gnatcatchers are a small family, the Polioptilidae, of twenty-one species of small insectivorous birds in the Americas. The sixteen species of gnatcatchers use their slender bills to pluck and glean small invertebrates from foliage. They are literally gnat catchers. The word "gnat" comes from the Old English *gnæt*, "a small flying insect," from the Proto-Germanic *gnattaz*.

# Gnatwren

In the same family as gnatcatchers (Polioptilidae), the four species of gnatwrens have similar behavior and much longer bills. Their cocked tails give them a superficial resemblance to the typical wrens.

# Go-away-bird

The go-away-birds are three species of turacos with a rather intimidating name. They are noisy and gregarious and take their names from their loud alarm calls that encourage any would-be intruder to go away. In 1925, Sydney Porter wrote about the bird in *Notes from Rhodesia*:[193]

> They are extremely noisy birds and their chatter is incessant, but the most remarkable thing is the call-note, which is "Go-away," repeated several times, and from which it derives its African name the "Go-away Bird." The call is wonderfully distinct, clear, and resonant, and can be heard from quite a long distance, and when once heard can never be forgotten.

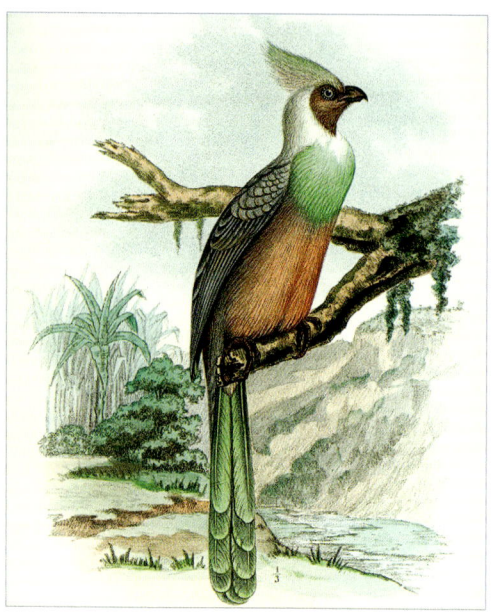

*Bare-faced Go-away-bird in* Heuglin's Ornithologie Nordost-Afrika's *(1871)*

# Godwit

The four species of godwits are large sandpipers in the Scolopacidae. According to Swann, the unusual name was first attested as the Anglo-Saxon name *god-wihta* and literally meant "good eating."[121] The name first appeared in the European literature in 1544 when Turner mentioned as an aside that "*Anglorum goduuittam* [the English call it godwittam]" stating that it is[119]

> much like the Woodcock, that, if it were not a little larger, and did not the breast verge upon ash-colour, the one of them could hardly be distinguished from the other. It is found in marshy places and on river banks.

One theory of the etymology is that the name derives from the Anglo-Saxon words *gōd*, "good," and *with* or *wight*, meaning "creature." Given that the bird was hunted as a delicacy, and still is in France, this explanation certainly makes sense. It appeared as an item on the menu in the *Poly-Olbion*, a topographical poem from 1612 as "The *Puet, Godwit, Stint,* the pallat that allure."[490] In 1611, the French philologist Isaac Casaubon Latinized the word as *Dei ingenium*, which might translate as "God's wit or nature":

> Vidimus etiam aves aliquas exquisitas quae saginabantur vendendae. Erat inter caeteras avis quae dicitur Godwie . . . Dei ingenium, quae mirifice commendatur. [Also we saw some excellent birds being fattened for sale. Among others there was a bird which is called Godwie . . . God's character, which is greatly recommended.]

However, the philologist Lockwood favors the theory that the name is onomatopoeic and points to other local names such as *shrieker, barker, yarwhelp,* and *yarwhip* that are echoic of the bird's varied vocalizations.[100] As Lockwood noted in 1981, "although godwit is fundamentally echoic in origin, the name as we have it is nevertheless 'meaningful,' the Latinization *Dei ingenium* . . . it will therefore be correct to recognize the distinct possibility of folk-etymological interference" [in other words, a popularly held but false belief about the origin of the word.]

# Goldcrest

The single species of Goldcrest is a kinglet found from Britain all the way across to Japan. The obvious origin of the name is a reference to its bright yellow crown stripe. Ray referred to the Goldcrest as the "Golden-crowned Wren," while Pennant called it the "Golden-crested Wren." By the mid-1800s, ornithologists were unhappy with the bird being referred to as a wren and changed the name to kinglet, a direct translation of the genus name *Regulus*, as Yarrell did in 1845:[344]

> The little Golden Crested Regulus, or Kinglet . . . has a soft and pleasing song, somewhat like that of the common Wren.

But people were resistant to this change and continued to refer to the Golden-crested Wren, eventually dropping the "wren" and shortening the first part of the name to create a noun.

The legend of the "King of Birds" is often said to refer to the Eurasian Wren, but, in fact, the true king in the legend is almost certainly the Goldcrest:

> All the birds gathered to choose one amongst them to be king. The crown was to be awarded to the one who could soar to the highest heights in a competition. The eagle rose higher than all the others, but when it was about to descend again joyfully, a little bird that had sneaked unnoticed on the eagle's back soared even higher into the air. Despite the eagle's wrath, the kingdom was given to this bird—the smallest of them all—and he was adorned with a crown.[194]

As Hugo Suolahti discussed in his 1909 book *Die Deutschen Vogelnamen*:[194]

> The fact that the Goldcrest was the real king is supported by the crown with which nature has endowed it—the vividly colored head feathers which, when raised, assume a crown-like shape. There is little doubt that it was precisely this peculiar headdress of the bird that gave rise to the legend of the choice of a king.

The confusion that existed between the Goldcrest and the wren probably stems from the use of the same Ancient Greek word βασιλίσκος *basiliskos* for both the wren (in Aesop) and the kinglet or Goldcrest (in Philagrius), a name that also meant "chieftain or princeling" as well as the similar βασιλεύς *basileus* for the wren (in Aristotle), also meaning "king."[195]

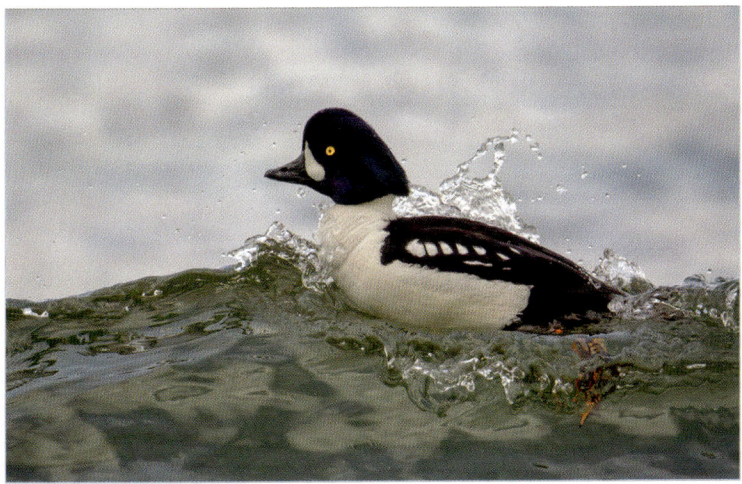

*Barrow's Goldeneye (Washington State, USA)*

# Goldeneye

The goldeneyes are two species of sea ducks in the genus *Bucephala* from North America and Eurasia. The derivation of the common name is straightforward—both species possess golden-colored irides. The name for a duck was first mentioned in Ray and Willughby in 1678:[16]

> The Golden-Eye, Clangula . . . This was sent us from Cambridge by the title of Shelden, I suppose so denominated from its being particoloured of black and white, that is Sheld, so other pied birds are called Sheld-fowl . . . The Irides of the Eyes are of a lovely yellow or gold-colour.

Early on they were referred to as "Golden Eye," and in the United States War Department's *Reports of Explorations and Surveys* in 1858, the common name "Whistle Wing" is also listed for Common Goldeneye, a reference to the sound made when the birds take flight.[196]

# Goldenface

A species in the Acanthizidae (thornbills and allies) found in New Guinea, the Goldenface is a montane dweller with a bright yellow face and underparts. Gould labeled his illustration of the bird as the "Yellow-and-grey Thickhead," alluding to its early placement in the Pachycephalidae, the whistlers, although he did question the taxonomy, stating, "this is by no means a typical *Pachycephala*."[40] By the 1970s, the name "Golden-faced Pachycare" was being used. Later, the bird became the "Dwarf Whistler," but, following molecular studies in 2009,[594] the Clements Checklist[491] proposed that "As this species is not a whistler, change the English name to Goldenface."

# Goldenthroat

The name of the goldenthroats, three members of the genus *Polytmus*, belies the fact that all three of these hummingbirds are green. The throat appears golden only when seen in certain lights. A version of the name, "Golden-throat," appears to have been first used in the 1918 issue of the *Catalogue of Birds of the Americas*.[197]

# Goldfinch

The European Goldfinch *Carduelis carduelis* was known in Old English as the *goldfinc*, literally "a gold + finch," so-called for its yellow wing markings. The three goldfinches of the Americas, in a different genus, *Spinus*, all feature much more yellow in their plumages than the *Carduelis*. One of the first records of the bird name in the literature is in one of Chaucer's *Canterbury Tales*, "The Cook's Tale" from the late 1300s, in Middle English:

> A prentys whilom dwelled in oure citee,
>
> [An apprentice once dwelt in our city,]
>
> And of a craft of vitailliers was hee.
>
> [And of a craft of food merchants was he.]
>
> Gaillard he was as goldfynch in the shawe,
>
> [Gaily dressed he was as is a goldfinch in the woods,]
>
> Broun as a berye, a propre short felawe,
>
> [Brown as a berry, a good-looking short fellow,]
>
> With lokkes blake, ykembd ful fetisly.
>
> [With locks black, combed full elegantly.]

# Gonolek

The gonoleks are four species of strikingly red and black bushshrikes in the Malaconotidae. A number of authorities state that the name is imitative of the birds' calls, but this is almost certainly coincidental or incorrect. Buffon wrote in 1770:[417]

> L'oiseau qui nous été envoye du Sénégal par M. Adanson fous le nom de Pie-grieche rouge du Sénégal, & que les Nègres, dit-il, appellent gonolek, c'est-à-dire, mangeur d'insectes. C'est un oiseau remarquable par les couleurs vives dont il est peint. [The bird which was sent to us from Senegal by M. Adanson is called the Red Shrike of Senegal, and which the Negroes, he says, call gonolek, that is to say, eater of insects. It is a bird remarkable for the vivid colors with which it is painted.]

And, indeed, in Wolof, the main language of Senegal, "insect" is *gunóor* and *lekk* is "food" or "to eat," making *gunóor-lekk*.

# Goosander

The Common Merganser, or Goosander, as it is referred to in Britain, is a species of duck in the genus *Mergus*. The name is a portmanteau of *goose* and either *gander*, from the Proto-Germanic *ganzô*, a "male goose," or from the Old Norse *andar*, for "duck." The oldest known use of the word is by Drayton in the *Poly-Olbion* poem from 1612:[490]

> As they above the rest were Lords of Earth and Ayre.
> The Gossander with them, my goodly Fennes doe show
> His head as Ebon blacke, the rest as white as Snow,

# Goose

With the exception of the Magpie Goose (the single member of the Australasian family, the Anseranatidae), all the many species of geese are members of the Anatidae. The English group name is a very old one, coming from the Proto-Germanic name for "goose," *gans*. Cognates, words having similar forms, exist in many other languages, leading linguists to conclude that the root word was the three thousand-year-old Proto-Indo-European *ghans*, probably imitative of the goose's honking, and that "goose" is possibly the oldest bird name in the English language. A very early example of the word in the English language can be found in the *The Exeter Book*,493 recognized by UNESCO as "the foundation volume of English literature, one of the world's principal cultural artefacts,"[492] believed to have been written between 960 and 990, in Riddles 25: line 3:

> Hwīlum iċ grǣde swā gōs.
> [Sometimes I cry like a goose]

# Goshawk

The goshawks are large members of the *Accipiter* genus in the Accipitridae. There are twenty-four species of goshawks. The Old English name, *gōshafoc*, literally means "goose-hawk," from *gōs*, "goose," and *hafoc*, "hawk" (see those entries). In other words, these are birds that prey on geese, if not literally, then large birds of a similar size. As with many Old English names, the word was originally applied to the species found in Britain, Northern Goshawk, and extrapolated out to related birds worldwide over time. In 1686 an English publication entitled *The Gentleman's Recreation . . . hunting, hawking, fowling, fishing* told its readers:[198]

The *Goshawk* preyeth on the Pheasant, Mallard, Wild-goose, Hare, and Coney; nay, she will venture to seize on a Kid or Goat; which declareth the inestimable courage and valour of this *Hawk*.

But earlier in the entry the author asserts that "there are several sorts of Goshawks, and they are different in goodness, force, and hardiness" so, although Northern Goshawks are capable of taking large prey items, it may be that the name originally was used for a number of different species of "long-winged Hawks, or true Falcons" as Newton suggested.[7]

## Grackle

The grackles are eleven species of New World birds in the Icteridae. The name is derived from the Latin word *grăcŭlus*, supposedly from its *gra gra* note and referring to the Eurasian Jackdaw, *Corvus monedula*.[199] However, this may be pure conjecture on the part of the English-speaking translator of the Latin tome in 1920, and the true identity of the bird is lost in the mists of time. Historically, a number of members of the Sturnidae, the starlings, have been called grackles, and the name survives in the scientific nomenclature for a number of relatively large, black birds, such as the *Gracula* hill mynas and the now-defunct genus *Graculus* for cormorants. In the early history of American ornithology, the *Quiscalus* grackles were thought to be members of the Sturnidae, thus accounting for the common name. Ingersoll, in *An Adventure in Etymology*, stated:[9]

> Early European ornithologists transferred the generic term, as *gracula*, to the starlings of India, particularly the sacred myna of the Hindoos, and early American writers borrowed the term for use here when our blackbirds were mistakenly classified as starlings.

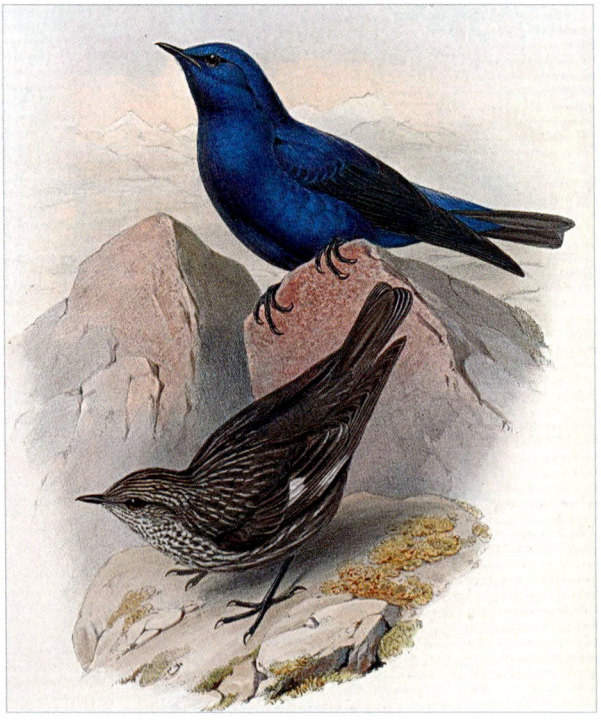

*Grandala in Gould's* Birds of Asia *(1850)*

# Grandala

The Grandala is a member of the Turdidae in the synonymous genus *Grandala*. Hodgson erected the genus in 1843, derived from the Latin *granda*, for "big," and *ala*, for "wing." He commented that it is[253]

> a singular bird, having the general structure of a Thrush, but with the wings vastly augmented in size and the bill of a Sylvian.

Gould called it the "Celestial Grandala" and noted its singular character:[33]

> That all the operations of nature are governed by certain fixed laws is so evident, that few persons, I believe, will venture to gainsay it. Thus the birds of the sandy deserts assume the colouring of the soil, the brilliant-plumaged birds and gaily coloured insects of the tropics are surrounded with plants and flowers equally gorgeous in their hues, and the spotless Ptarmigan is a denizen of the snow-clad hills. At the same time we do not fail to observe that these laws, like all others, have their exceptions; and thus we occasionally find brilliantly coloured birds inhabiting regions so elevated that they are almost perpetually clothed with snow . . . Who for a moment would suppose that this beautiful bird is an inhabitant of the lofty snow-clad ranges of the Himalayas, and that it never leaves these icy regions, neither the temperate valleys nor the hot terri [terai] having any temptation for it?

# Grassbird

Not surprisingly, this somewhat unimaginative name has been used especially at a local level for many birds in the past. For example, Latham noted in 1823 that the Vesper Sparrow "inhabits New York, where it stays all the winter, and [is] known by the name of Grass Bird."[17] In the present-day formal checklists, one species of Macrosphenidae, the African warblers; two species of Pellorneidae, the ground babblers and allies; and ten species of Locustellidae, the grassbirds and allies, are known as grassbirds. Of the Striated Grassbird, Newton wrote:[7]

> The plumage above is striated, and they skulk in thick grass so as to be seldom seen, flying but a short way when forced to take wing.

Clearly, all the birds with this appellation are associated with grassy habitats, whether that be marshlands, rice paddies, reed-beds, meadows, or even scrublands.

# Grasshopper-Warbler

The grasshopper-warblers are six species of Locustellidae in the *Locustella* genus. Their names invoke, to lesser or greater degrees, their insectlike, reeling vocalizations. They are secretive and more often heard than seen, so it's not surprising that their songs inspired the name. In fact, the grasshopper-warbler in England had many analogous local names, including cricket-bird, birrl-bird, reeler, reel-bird (for the "resemblance of its song to the noise of the reel used by the hand spinners of wool"), and the remarkable "Sibelous Bush-Hopper."[121] Ray and Willughby were the first to write about the Common Grasshopper-Warbler in 1678 under the heading:[16]

> Tit-lark that sings like a grasshopper, *Locustella*.

Pennant wrote about the "Grasshopper Lark" in his 1766 edition of *British Zoology*, saying that "it is the same with that Mr. Ray describes as having the note of the grasshopper, but louder and shriller."[342] But in later editions, he called it the "Grasshopper Lark Warbler." Latham was the first to call it the Grasshopper-Warbler, noting that "the hind claw sufficiently crooked to prove it does not belong to the Lark genus" adding that "its note [is] so like that of a Grasshopper, as to be mistaken for it."[17]

# Grassquit

There are seven species of grassquits in the New World tanager family, the Thraupidae. The word *quit* is used in several other bird group names and derives from a Caribbean word applied to small birds. It is probably onomatopoeic, from the call note described as *guit-guit*. All grassquits favor grassy and scrubby habitats, hence the first element of the name. (See Bananaquit and Orangequit.)

# Grass-Warbler

The Moustached Grass-Warbler is a Macrosphenid African warbler in the monotypic *Melocichla* genus. This unique bird's contrastingly pedestrian name is a relatively recent attribution. An early mention of the species is from 1874 when Brehm used a name that ties the bird to the Fante Confederacy of Ghana in the 1800s:[201]

> Gordon is enchanted with the song of a cousin of the above, the Fantee Warbler (*Melocichla mentalis*), an inhabitant of Western Africa. The "splendid" melody of this bird reminded him more than anything of the woods and copses of his English home and their bright minstrels.

By the early 1900s, the name used in the *Annals of the Transvaal Museum* was Rufous-fronted Thrush-Warbler, while the cisticolas were referred to by the colloquial name "grass-warbler."[494] Van Someren called it the Chyulu Great Moustached Warbler in 1939 with an explanation of its habits:[200]

> Here and there along the forest edges where the tall grass merged into the fringing woody herbage, and in the ravines of the moorland where the grass was often shoulder high, this species was noted. It was difficult to procure except in the early mornings or late afternoon; at these times one might see the birds sitting on the grass stems enjoying the sunrise or the dwindling rays of the setting sun. During the day they were hidden, or if seen it was just for a moment and they disappeared into the grass.

By the 1950s, contributors to the *Journal of the East Africa Natural History Society* were referring to the bird as the "Moustache Warbler," saying that it is "an abundant resident of the sage-bush."[395] The present-day name appears not to have been used until the 1990s.

# Grasswren

The grasswrens are thirteen members of the genus *Amytornis*, endemic to Australia. With their cocked tails giving them a superficial resemblance to the true wrens and their preference for grassy or scrubby habitat, the group name "grasswren" was an obvious choice made by the influential Australian ornithologist Robert Hall back in 1899.[170]

# Graveteiro

A small and unassuming bird, the one and only Pink-legged Graveteiro may well be said to have a name bigger than itself. The origin of the name lies in the Portuguese word *gravetos*, meaning "sticks," and refers to this unique furnariid's nests. An oval collection of twigs and sticks, the nest is situated in the fork of a tree and constructed with the aid of many family members. This bird is a good example of a common and conspicuous bird overlooked by ornithologists until quite recently. In the 1996 paper in which the species was first described, the authors noted that they chose the common name Graveteiro in order[202] to call attention to . . . the fact that it gathers twigs and sticks (*gravetos* in Portuguese) to construct its nest, as do several other groups of furnariids, such as the canasteros ("basket-makers" in Spanish).

# Grebe

The grebes are a family of twenty-two species of diving birds. The group name came into English in the 1700s from the French *grèbe*, which is of unknown origin. The name was first used in the ornithological literature by the Swiss naturalist Gessner in 1555, but his description and illustration of a bird he tells us is "Anglis secob, seegell" (in English secob, seagull) are that of a gull, albeit a rather strange-looking one, rather than a grebe:[178]

> Gavia cinerea, quæ ad flumina & lacus ascendit . . . Hæc Italice circa Comum galedor uocatur, circa Verbanum lacum et alibi galetra, Gallis gauian uel mouette, uel glaumet, Sabaudis grebe, uel griaibe, uel beque, uel heyron, quamuis ardeæ potius id nomen conueniat. [An ash-gray seabird, which goes up to the rivers and lakes . . . This (bird) is called in Italian *galedor* around Como, *galetra* around Lake Maggiore and elsewhere; (it is called) by the French *gavian* or *mouette* or *glaumet,* and by the Savoyards *grebe* or *griaibe* or *beque* or *heyron,* this (last) name applying rather to the heron.]

It is thought that Gessner recorded the name accurately but assigned it to the wrong bird. Later in 1765, an entry for Grebe in Diderot and d'Alembert's *Encyclopédie* illustrated "Le Grébe Hupe," the Great Crested Grebe.[317] *The Merriam-Webster Dictionary* says:

> Efforts to etymologize the Franco-Provençal word are speculative. The *Glossaire* [*des patois de la Suisse Romande*] suggests a possible relationship with dialectal grébo "of more than one color, variegated," itself of obscure origin.

But another theory exists that it possibly comes from the Breton (a Celtic language from the Brittany region of France) *krib*, "a comb," maybe in reference to the shaggy, comb-like crest of the Great Crested Grebe, which is widespread in Europe. Prior to the first use of the word in the English literature by Pennant in 1768, the grebes were known by a number of apropos names such as dabchick, doucker or ducker, diver, and even dipper-duck. Sadly, the colorful name *Arsfoot*, for the "backward position of the legs" is no longer in use.[121]

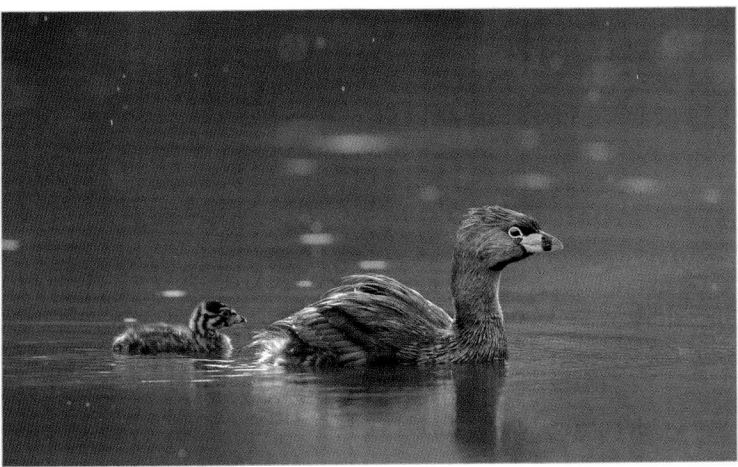

*Pied-billed Grebe (Washington State, USA)*

# Greenbul

(See Bulbul.) The forty-six species of greenbuls are bulbuls in twelve different genera with green to greenish plumage. All have distributions confined to the continent of Africa.

# Greenfinch

The five species of greenfinches in the genus *Chloris* are quite literally green finches in the Fringillidae. They have varying amounts of green in their plumage; some appear to be more yellow than green, however. The first attestation of the name was in *The Sherborne Missal*, a late-medieval English service book dating from 1399 to 1407, in which depictions of a number of British birds are illustrated in the margins.[377] An illustration of what is clearly a European Goldfinch is labeled "Grene fynch." Ray and Willughby wrote about the "Greenfinch" in 1678, noting "it is called by some the Green Linnet."[16]

# Greenlet

The greenlets are fourteen species of small vireos all with varying shades of green and the *-let* suffix being a diminutive, so these are "little green" birds. The name may have originated with Swainson in an 1831 publication, but it's possible he coined it from a local name.[305] Prior to this, the vireos were often called "Red-eyed Flycatchers," as in *The Museum of Natural History* from 1877 when the author noted that the birds[65]

> belong to a small group of Flycatchers peculiar to America, and to which the name of Greenlets has been given, from the prevalence of green or olive tints in their plumage.

# Greenshank

There are two species of greenshanks in the genus *Tringa*. The name was coined by Thomas Pennant in 1785, more than 200 years after the name Redshank came into being, probably by analogy with its similar congener.[19] Bewick wrote about the Common Greenshank in 1804 under the heading "Greenshank," giving alternative names "Green-shanked Godwit, or Green-legged Horseman" and noting its namesake feature with the comment:[149]

> The legs are long, bare about two inches above the knees, and of a dark green colour.

# Grenadier

The Purple Grenadier in the Estrildidae is the proud owner of a unique name. (Common Grenadier has been renamed Violet-eared Waxbill). French *grenadiers* of the 1670s were soldiers who were apparently "dexterous in flinging hand-grenades," so their name came from the French *grenade*. They also wore smart uniforms of blue and red, as did the grenadiers of Portugal. And, not coincidentally, our Purple Grenadier is decked out in blue and orange-red plumage with a bright red bill. Edwards told us in 1751 the "Grenadier" was[78]

> brought from *Lisbon*, and is call'd by the *Portugueze, Grenidiero*, but for what Reason I know not, except for its being a bold Bird, that will fight through the Wires of its Cage; or perhaps the Uniform of the Grenadiers of *Portugal* may be of an Orange-Colour.

# Graytail

There are two species of graytails in the Furnariidae, the ovenbirds. They are small and they have gray tails. One species was first described in the *Proceedings of the Zoological Society* in 1885,[495] and the second was described in *The Ibis* in 1886,[203] but no common English name was used until 1912, when it was dubbed a Spine-tail in Brabourne's *Birds of South America*,[45] perhaps, as Wetmore stated in 1953, "because of [its] small size and lack of striking markings it is often overlooked."[204] The name graytail was adopted in the 1980s.

# Griffon

Four species of *Gyps* vultures bear the group name griffon (although some maintain that all the *Gyps* should be known by the appellation).[254] The name has an illustrious history, but the true origins are uncertain. It comes to us from the Old French *grifon* via the Latin *gryphus*, in turn from the Ancient Greek Κύφός *griffos*, meaning "bent or curved," probably referring to the talons or the beak. Possibly the earliest reference to the name occurs in *La Chanson de Roland*, the famous four-thousand-line epic poem from around 1090, describing the battle of Roncevaux Pass in the year 778 in line 2,544:

> Grifuns I ad, plus de trente millers. [Griffins were there, thirty thousand, no less.]

But the *grifuns* were included with "adversaries, dragons, wyverns, [and] serpents," so it's hard to know if this was a reference to *Gyps* vultures or to the mythical *griffins*—powerful and majestic creatures with the head of an eagle and the body of a lion. But it's probable the myth arose from fanciful tales of real Griffons. In 1734, Perrault wrote a chapter called "Le Grifon" in *Histoire Naturelle des Animaux* referring to the Eurasian Griffon:[255]

> The Description that the old Authors give of the Grifon does not suit any animal that is known: besides the monstrous figure they give it, making it have the head & wings of an Eagle, & the rest of the body of a Lion, they still attribute to it an absolutely incredible strength. This is why the name of Grifon is sometimes given to the Unknown Birds when they had a size & a force or some other peculiarity which had to do with what is said about the Grifon. We have known for a hundred years of an African bird to which we gave the name of Grifon because of its strength and its size which is prodigious.

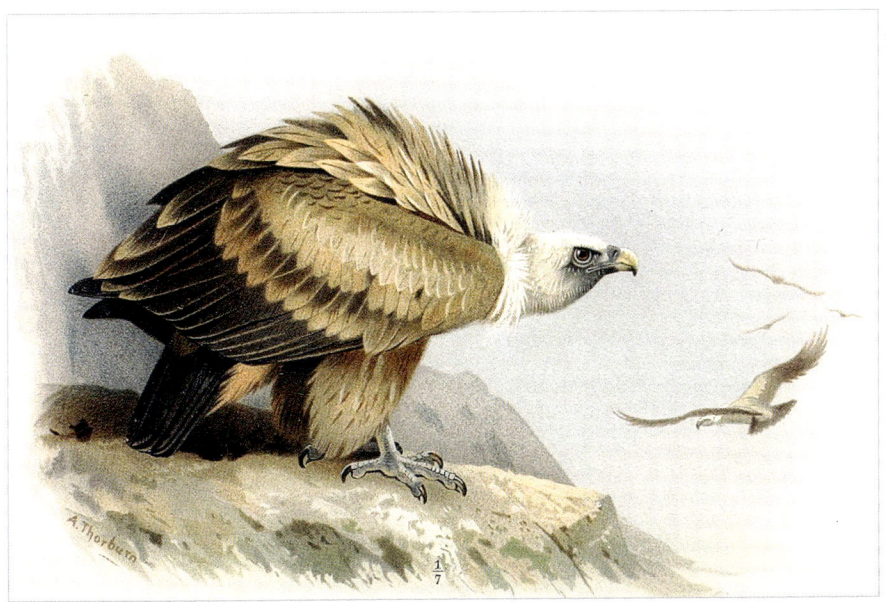

*Eurasian Griffon in Lilford's* Coloured figures of the birds of the British Islands *(1885)*

Perrault also suggested some sort of Hebrew origin of the word, but there is no evidence for this. In 1760 Brisson used *Grifon* for one of the "Le Vautour fauve," the Eurasian Griffon, in his *Ornithologie*.[76] And then in Latham's *A General Synopsis of Birds* from 1781 under the chapter heading "Vulture," there is a subheading "Le Griffon," but this is a reference to the vultures from a French journal.[17] Then, at last, in *An Introduction to the Ornithology of Cuvier* by Bowdich in 1821, we see what appears to be the first use of the name Griffon in English, although this actually refers to *Gypaetos*, the Bearded Vulture (Lammergeier).[257] The first use of the group name for a *Gyps* vulture appears to be in Martin's 1835 volume *An Introduction to the Study of Birds*.[256] There is an entry labeled "Griffon Vulture *Vultur fulvus*" [sic] with the descriptor:

> This magnificent species is spread from the south of Europe throughout Greece, Turkey, Persia, and Africa, everywhere taking up its abode among lofty mountain scenery. . . . It is not by any means improbable that this species is the Vulture so often alluded to in the scriptures. At all events it is common in Judea, and agrees with the descriptions to be met with in the holy volume.

And in *A Familiar History of Birds: their nature, habits, and instincts* by Edward Stanley, also from 1835, an illustration appears labeled "The Griffon, or Fulvous Vulture," but that is the only time the name is used in a long chapter called "The Vulture."[418] In 1913, Swann believed that the Griffon[121]

> is the Grype or Gryffon of Aldrovandus, from which the name seems to be derived.

## Grosbeak

Fifteen species of grosbeaks of various genera are contained in the Fringillidae, the Cardinalidae and the Thraupidae; these days the grosbeaks are a polyphyletic assemblage of distantly related birds, all with notably large, conical bills. An early reference to "Du Grosbec," *gros*, for "large" or "thick," and *bec*, for "beak," occurs in Belon's *Histoire de la Nature des Oyseaux* in 1555, with the observation:[378]

> Encor n'avons trouvé autre propre nom Françoys mieux à propos pour nômer cest oiseau, que de l'appeller Grosbec: Car il à le bec moult gros pour fa corpulence. Il est bien vray qu'es autres contrées on luy donne quelques autres noms: car les Manceaux le nomment Pinson royal. [Yet we have not found any other proper French name better aptly to name this bird, than to call it Grosbec: Because it has a very large beak for its corpulence. It is quite true that other countries have given it some other names: for the Manceaux call it Royal Finch.]

In 1678, Ray and Willughby captioned a chapter "Of the Gros-beak or Haw-finch, called by Gefner, Coccothrauftes."[16] Discussing the "*The common Gros-beak*: Coccothraustes vulgaris," which is actually the Hawfinch, they wrote:

> This Bird for the bigness of its body, but especially of its Bill, in which it exceeds all others of this kind, doth justly challenge the first and chief place among thick-billed birds. The *French* from the bigness of its Bill do fitly call it *Grosbec*.

## Groundcreeper

The Canebrake Groundcreeper is a species of furnariid ovenbird of the Atlantic forest of the Neotropics. It derives its name from its behavior of creeping around in its habitat of canebrake—dense growths of cane grass—often in pairs.

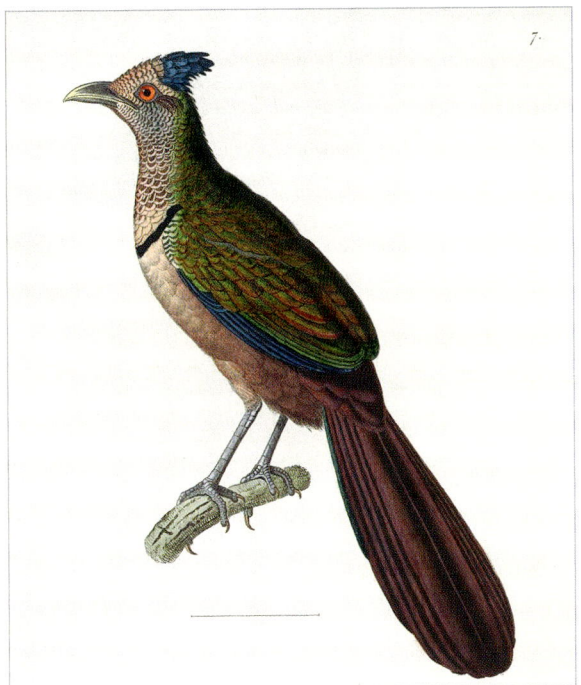

*Scaled Ground-Cuckoo in Temminck's* Nouveau recueil de planches coloriées d'oiseaux *(1838)*

# Ground-Cuckoo

Nine species of terrestrial cuckoos in three different genera go by the group name ground-cuckoo. In the 1800s, the name was often used for the coucals, (such as in *The Ibis* in 1898 when Woodward wrote that "The large Red Ground-Cuckoo [Senegal Coucal] was common everywhere"),[496] as well as for the couas of Madagascar. Wallace mentioned the "pheasant ground-cuckoo (*Carpococcyx radiatus*)," now called the Bornean Ground-Cuckoo in *The Malay Archipelago* in 1869.[205] (See Cuckoo.)

# Ground-Jay

The four species of *Podoces* ground-jays from Central Asia are weak fliers, preferring to run along the ground, but they will also readily perch on trees and bushes. Hume dubbed them the "Chough-Thrushes" in the literature from the 1870s.[206] Later, due to confusion about the taxonomic position of the genus, they were known as "thrush-choughs," "desert-choughs," and "ground-choughs." Dresser wrote in 1902:[207]

> It is essentially a ground bird and but seldom perches on a bush, but runs with extreme swiftness and is not often seen on the wing. Its flight is not unlike that of a Jay, and is usually but a few feet above the ground, and is but short.

# Ground-Roller

The gorgeous Brachypteraciidae ground-rollers are a family of five species endemic to Madagascar. They are members of the Coraciiformes, the order that includes the kingfishers, bee-eaters, and the rollers, which they most closely resemble. R. B. Sharpe wrote in an 1871 article in *The Ibis*:[208]

We come to the Ground-Rollers, which I propose to raise to the rank of a subfamily. They are distinguished from the true Rollers by their long legs. . . . [Crossley] tells me that the only specimens he obtained were procured by the natives, who informed him that they only came abroad in the night-time, and sought their food on the ground, this latter fact being confirmed by the earth which always remained on their bills. Madagascar is the home of these curious birds, which, though Coraciine in their affinities, are among the most aberrant forms of which that wonderful island gives us so many examples.

# Grouse

Plump ground dwellers, the grouse are fourteen species of Phasianid in a number of different genera. The name was perhaps first used in English in the 1530s as *grows*, a plural form. In a manuscript dated "*mense* Jan. 22 Hen. VIII" (1531), the bird is mentioned as a pantry item:[209]

Among fowl for the tables are crocards, winders, runners, grows, and peions, but neither Turky or Guiney-fowl.

The origin of the name is unknown, but it possibly came from Latin *gruta*, meaning "crane," or it may derive from the German word *graus*, meaning "gray." There are also theories that it may simply be onomatopoeic. Newton asserted:[7]

The most likely derivation seems to be from the old French word *Griesche*, *Greoche*, or *Griais* (meaning speckled, and cognate with griseus, grisly or grey), which was applied to some kind of Partridge.

In 1678 the name occurred in *The Ornithology of Francis Willughby of Middleton in the county of Warwick, esq.* with a reference headed "*Grous*, the Black Grouse":[16]

The Black Cock, his Hen the *Grey hen*, his Brood the *Grey Game*. This Bird is called also the *Heath-Cock*, and *Grous*, *Tetrao*, sive *Urogallus minor*. In great Heaths in many places of England.

Since the 1800s the name "grouse" in Britain is usually taken to refer to the Red Grouse, known elsewhere as the Willow Ptarmigan. As has often been the case, "grouse" was originally the name used for the birds of the British Isles, but later it was extended to similar birds in other places.

# Guaiabero

The Guaiabero is a single species of cute, chunky, and small parrot of the Philippines. The Spanish name for the bird, also *Guaiabero*, means "guava-harvester" in homage to its penchant for fruit. It particularly is known to feed on orchard trees, including guava plantations, but will take all types of ripe fruits. The Spanish word for "guava" is *guayaba*, which comes from the Latin American languages Taino and Arawak *guayabo* for "guava tree." Although an ocean divides the Philippines from South America and the Caribbean, this little parrot provides a fascinating linguistic link between two continents, thanks to the Spanish colonialist expansion. McGregor noted in the 1909 publication *A Manual of Philippine Birds* that the bird was called *bo-bó-toc* or *gua-ya-bé-ro* on Luzon.[210]

# Guan

These twenty-five species of large, terrestrial, and pheasant-like birds in the Cracidae, the guans, chachalacas, and curassows, are restricted to the Neotropics. Their group name came into English from the Spanish name, which in turn came from the Indigenous Kuna

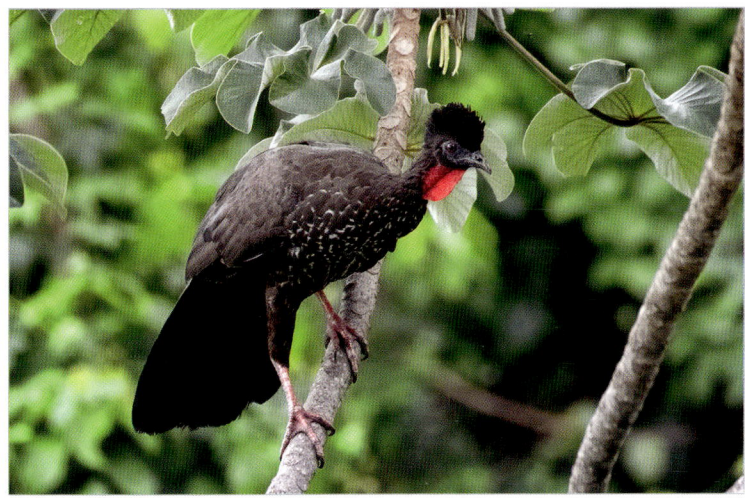

*Crested Guan (Costa Rica)*

word *kwama*. Kuna is a language of Panama. It seems most likely that the Kuna name is onomatopoeic for one of the vocalizations of Crested Guan, which has been described as "a soft, nasal, low-pitched *whúaan*."[318] In the 1743 book *A Natural History of Uncommon Birds*, George Edwards appears to be the first to have used the name in the European literature when he wrote, without elaboration:[78]

> The Quan or Guan, [is] so-called in the West Indies.

# Guillemot

Guillemot is a name for three species of Alcidae that came into English directly from French. In 1555, Belon wrote in French that the name was used for the juvenile "Le Pluvier," a name then in use for the guillemot because [378]

> pour ce qu'on le prend mieux en temps pluvieux qu'en nulle autre saison. [for we take it more easily in rainy weather than in any other season.]

Ray and Willughby were the first to use the name in the English literature when they titled a chapter:[16]

> The Bird called by the Welsh and Manks-men, a Guillem; by those of Northumberland and Durham, a Guillemot, or Sea-hen; in Yorkshire about Scarburgh, a Skout; by the Cornish, a Kiddaw: Lomwia Hoieri in Epift.ad Clusium.

Newton believed that the name [7]

> seems to be cognate with or derived from the Welsh and Manx *Guillem*, or *Gwilym* as Pennant spells it. The association may have no real meaning, but one cannot help comparing the resemblance between the French *Guillemot*—though that appears to have been originally applied to the young of the Golden Plover—and *Guillaume* with that between the English *Willock*, another name for the bird, and *William*.

Most authorities now believe the word is of French origin, derived from *Guillemot*, a pet form of *Guillaume*, the French form of William. Other British names, such as the Scottish *Wilkie* and the Sussex *Willy*, reflect this, too. The name is onomatopoeic based on the call of the juvenile. Gurney explained the conflation of a forename and a bird name:[324]

It is easy to understand how [a name] would be selected as a common Christian name by seafaring men, who are never slow in applying a nick-name; indeed there are several such instances in Ornithology. . . . *Guillemot* is said to be derived from the French *Guillaume*, and "Willock," an appellation for the same bird, from *William*.

# Guineafowl

The guineafowls are a family of six species, the Numididae. They are chicken-like ground dwellers found in sub-Saharan Africa. This accounts for the fowl part of the name, but the origins of the word *guinea* are more complex. With relation to the bird name, it is clearly a reference to the geographic area it was first known from. Although the guineafowl is thought to have been recorded in human history from as early as 3600 BCE, with depictions of them in Egyptian artifacts, the domesticated guineafowl in Europe originated from one of several wild species on what used to be called the Guinea Coast of West Africa. Portuguese sailors gave the bird its present name in the sixteenth century. The word "guinea" is originally a Portuguese word but perhaps came from the Tuareg word *aginaw*, meaning "black people." Pennant wrote in 1785 that Prosper Alpinus, the Venetian physician and botanist of the late 1500s,[19]

describes the *Meleagrides* of the antients; and only proves that the *Guinea*-hens were brought out of *Nubia*, and sold at a great price at *Cairo* . . . It was introduced into *Italy* from *Africa*, and from *Rome* into our country.

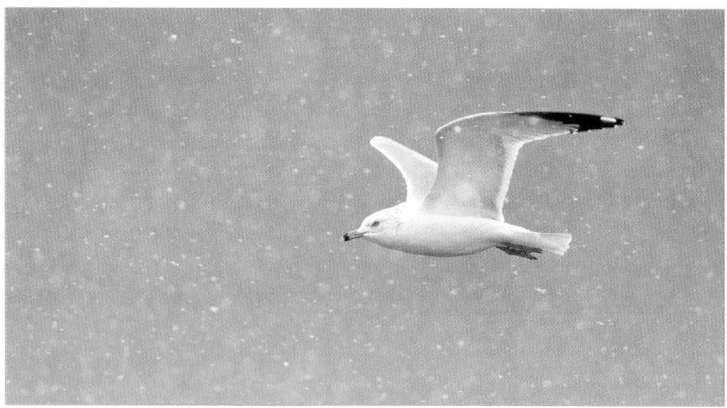

*Ring-billed Gull (Washington State, USA)*

# Gull

The ubiquitous gulls, all fifty species of them, have their group name in Celtic language roots. In Old English, the gulls were called *mæw*, which was derived from the Old Norse *mári*, imitative of the bird's mewing calls. This name survives in the Modern English name Mew Gull. Swainson wrote in 1886:[75]

The name of gull appears to have been given to this family of birds from their wailing cry. (Corn. *Gullan*, Welsh, *Gwylan*; Breton, *Gwelan*; Italian, *Golano*; French, *Goeland* ).

The name dates from the early 1400s; it appeared in a cookbook from 1450 as the plural *gullys* with a comment in the nineteenth-century translation that "even the fishy Gull was eaten."[211] Most authorities now agree that it came into Middle English as *gulle* from the Proto-Celtic word *wēlannā* for all types of gull.

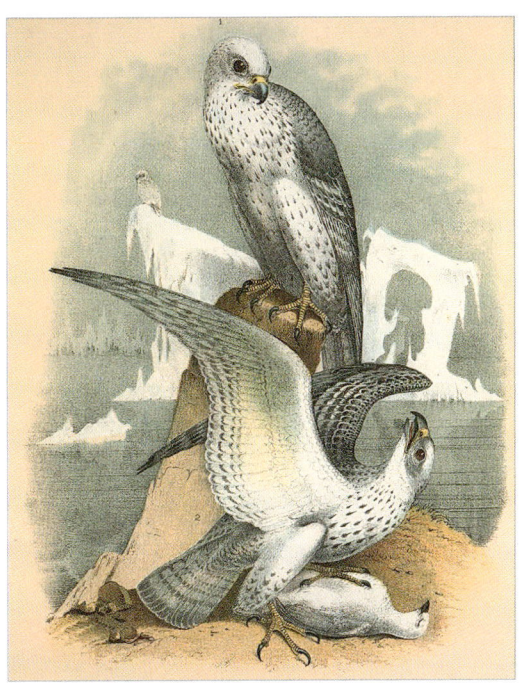

*Gyrfalcon, labelled as "Jer-falcon, or Gyr-falcon" in Studer's* The Birds of North America *(1903)*

# Gyrfalcon

A single species of *Falco* was, and still is, highly prized in the world of falconry. *Geier* is the root word from which, via the Old French *Gerfaucon*, the present-day name Gyrfalcon emerged in English. The first part of the name, *gir*, is "vulture" in German, from the Proto-Germanic adjective *giri*, for "greed." Folk etymology since the Middle Ages has connected it with the Latin *gyrus* in reference to "circling" in the air, but this is not the true etymology. Coues explained the name in 1903:[375]

> *Gyrfalcon* is the worst, *gerfalcon* the better, *jerfalcon* the best, spelling of the name, if we regard the etymology of the word, which was formerly in English also *gerfaulcon, gerfaucon, gerfawcon, jerfaucon, gierfalcon, girefaucon, gyrfalcon, gerfauk*, etc., with many identical or similar forms in other European languages . . . the first element of the word is not connected with Lat. *gyrus*, a gyration, circle, but with German *geier*, greedy.

The name was first used in a Middle English manuscript from around 1330, in which the *gerfauk* was mentioned frequently:[376]

> He schal bring to þe turment þat day . . . A gerfauk þat is milke white. [He shall bring to the tournement that day . . . a gyrfalcon that is milk white.]

And later, one of the *Forest Laws* from 1450 stated:[262]

> If ther be ony man that cometh in to the forest and taketh away ony facouns, Jerfacouns, egrettes . . . her eiron or her berdes, ye shul do vs to wite. [If any man comes into the forest and takes any falcons, gyrfalcons, egrets . . . her eggs or her chicks, they shall be punished.]

(See also Falcon.)

# H IS FOR HAMERKOP

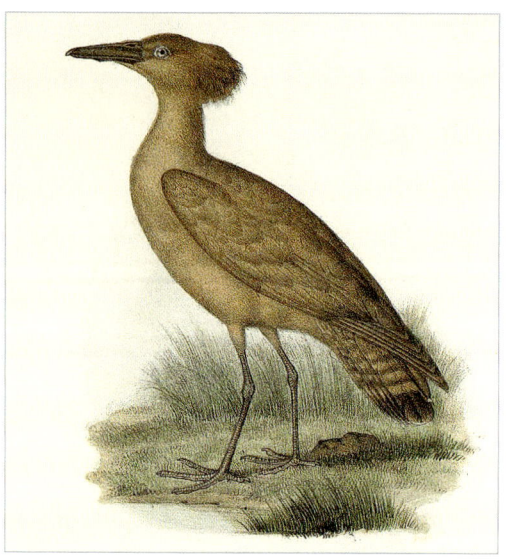

*Hamerkop in Fitzinger's* Bilder-atlas zur Wissenschaftlich-populären Naturgeschichte der Vögel in ihren sämmtlichen Hauptformen *(1864)*

## Hamerkop

The Hamerkop is a unique African, somewhat heron-like bird placed in a monotypic family, the Scopidae. The name is a clue to its appearance. It derives from two Afrikaans words—*hamer*, meaning "hammer," and *kop*, meaning "head." An early account of the bird is found in an 1857 article in *Konglika Swedish Academy of Sciences Documents* in which Victorin, the Swedish zoologist and explorer, told the author that the specimen sent to him[212]

> is a full-grown female in almost pure plumage; iris nut brown. A single pair of this species has been visible here for a few days. It's not shy at all. The scream can be expressed with *qui* or *kwi* or *kowi*; this tone is very short, as of a Kingsfisher (*Alcedo*). It feeds on frog larvae, among other things. The colonial name Hamerkop (Hammarskalle), is quite apt, for the bird's head is quite similar to a hammer, if you let the comb be the blunt and the beak the pointed end.

## Hanging-Parrot

The hanging-parrots are fourteen species of tiny green psittaculids in the *Loriculus* genus. Jerdon wrote about the *Loriculus* genus, and although he didn't call them hanging-parrots, he mentioned that they were called *Latkan* in Bengal, which means "the pendent" [*sic*] and that "they are occasionally caged, and become very tame, sleeping with their heads downward."[136] A study in 1968 found:[213]

> Group upside-down pseudo-sleeping as a response to the appearance of potential predators is frequent in both species: it is not certain if this is a displacement reaction or a normal escape reaction to the safety of an inaccessible refuge.

# Harrier

The harriers are sixteen species of birds of prey in the genus *Circus*. An Old English word, *hergian* means "to make war, the plunder." From there it evolved into the Middle English *herien*, and then into Modern English as the words "harrass," "harried," or "harrying," and the verb "to harry." The harriers, therefore, got their names from their hunting strategies and are "ones who harry." In 1544, Turner called the Hen Harrier the "hen harroer," writing in Latin:[119]

> Rubetarium esse credo accipitrem ilium, quern Angli hen harroer nominant. [The Rubetarius I think to be that Hawk which English people name Hen-Harrier.]

He goes on to describe the female Hen Harrier, calling it the Ringtail, apparently believing it to be a separate species. This name is still often used colloquially for females and immatures. Hogg described the name in 1845:[214]

> The name *harrier* was applied to this species, either from its usually flying low, and carefully skimming over the fields, like a hound in search of game, or it is a corruption of *harrower*, from the verb *harrow*, to pillage, strip, or tear up; or from the Scotch word *harry*, which is derived from the old French *harer* to rob.

# Harrier-Hawk

The harrier-hawks are two species of raptors in the *Polyboroides* genus, one found over much of the African continent and one from Madagascar. They combine features of their two namesakes. (See Harrier and Hawk.)

# Hawaiian Thrushes

The names of the Hawaiian members of the Turdidae in the *Myadestes* genus have a confused history, perhaps largely due to a lack of understanding on the part of the early European naturalists. The 1800s naturalists Bloxam, Andrews, and Dole noted that the original name of the thrushes on all the islands was Amaui and that the different island names were merely corruptions of this—so that the names of the individual species are all phonetically quite similar—we have Kamao on Kauai; Omao on Hawaii; Olomao on Maui, Molokai, and Lanai; and Amaui on Oahu. These names are all corruptions of *manu-a-Maui*, meaning the "bird of the demigod Maui." It's likely that the four very similar-looking thrushes were not differentiated by the local inhabitants, the names differing only according to island dialects.

Maui is a Polynesian ancient chief and folk hero who was able to transform himself into a bird. In an issue of a magazine in 1930, the unknown author gives us some insight into the Māori (Polynesian) beliefs regarding augury, although, of course, different birds to those on Hawaii:[48]

> Here it has an honorific title, Te Manu a Maui, or "Maui's Bird." Maui is the traditional tutelary deity of the food gardens. Its call when it first arrives on these shores in spring sounds like "Kwee, kwee, tio-o." The Maoris interpret the "Kwee, kwee" as "Koia, koia" ("Dig away"), and they proceed to obey its musical injunction. Later on its clear whistling cry sounds to the Māori ear a paean of rejoicing at the warmth of midsummer, "Kui, Kui, whiti-whiti ora, tio-o!" This high call, with its melodious whistle at the end, is peculiarly the pipi-wharauroa's cry; it can never be mistaken for that of any other bird.

# Hawfinch

The English name Hawfinch, for the single species of fringillid that is widespread across Europe and Northern Asia, was first used, in print at least, by the ornithologists Ray and Willughby in their seminal tome written in 1676.[215] This large finch is named in honor of its favorite food—namely, berries. According to Ray and Willughby:

> It breaks the stones of Cherries, and even of Olives with expedition, the Kernels whereof it is very greedy of. The Stomach of one we dissected in the Month of December was full of the stones of Holly-berries. It feeds also upon Hemp-seed, Panic, &c. and moreover upon the buds of trees, like the Bulfinch.

Haws are the red berries of the common hawthorn, being one of a number of types of berries, seeds, and buds favored by the Hawfinch. The word *haw* comes from Middle English *hawe*, and from the Old English *haga*, a berry of the *hagaporn*, the hawthorn hedge. (See also Finch.)

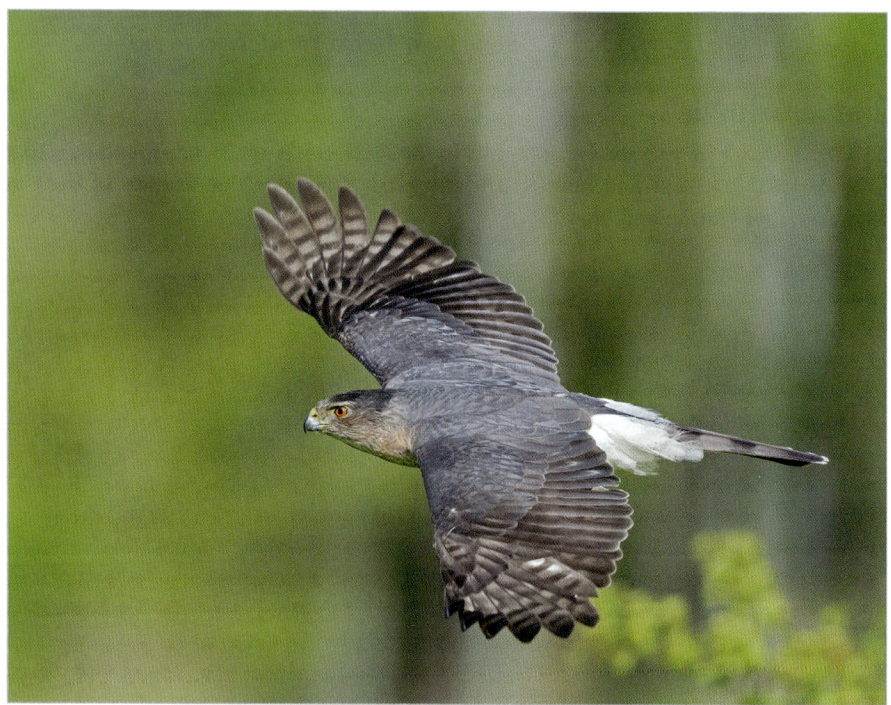

*Cooper's Hawk (Washington State, USA)*

# Hawk

A very familiar word in the English language, the name hawk is not only applied to thirty-eight species of Accipitridae, but it appears in many other compound bird names in the same family as well as in the Cuculidae. (See Hawk-Cuckoo.) The word is thought to have originated from the Proto-Germanic *habukaz* in the 900s, *haf* meaning "to take or seize" and the diminutive suffix *-ukaz*. This Old English word *hafoc* evolved into the Middle English *havek*, *haefuc*, or *hauk* by the 1200s. The *hauke* featured prominently in *Layamons Brut, or Chronicle of Britain*, a Middle English poem written sometime between 1190 and 1215:[379]

*Summe heo gunnen wondrien; swa doð þe wilde cron.*

*i þan mor-uenne; þenne his floc is awemmed.*

*& him halde[ð] after; hauekes swifte.*

*hundes in þan reode; mid reouðe hine imeteð.*

*þenne nis him neouðer god. no þat lond no þat flod.*

*hauekes hine smiteð; hundes hine biteð.*

*þenne bið þe kinewurðe foȝel; fæie on his siðe.*

*[Some they start to wander, as the wild crane doth*

*in the marshlands, when his flight is impaired,*

*and swift hawks pursue after him,*

*and hounds with mischief meet him in the reeds;*

*then is neither good to him, nor the land nor the flood;*

*the hawks him smite, the hounds him bite,*

*then is the royal fowl'at his death-time; adread on each side!]*

# Hawk-Cuckoo

There are eight species of hawk-cuckoos in the genus *Hierococcyx*, which also means hawk-cuckoo from the Ancient Greek ἱέραξ *hierax*, for "hawk," and κόκκυ *kokku*, for "cuckoo." They are visual mimics of the *Accipiter* hawks, and Davies and Welbergen noted in their paper entitled "Cuckoo–hawk mimicry? An experimental test" in 2008:[314]

Cuckoo–hawk resemblance may reflect convergent evolution of cryptic plumage that reduces detection by hosts and prey, or evolved mimicry of hawks by parasitic cuckoos, either for protection against hawk attacks or to facilitate brood parasitism by influencing host behaviour.

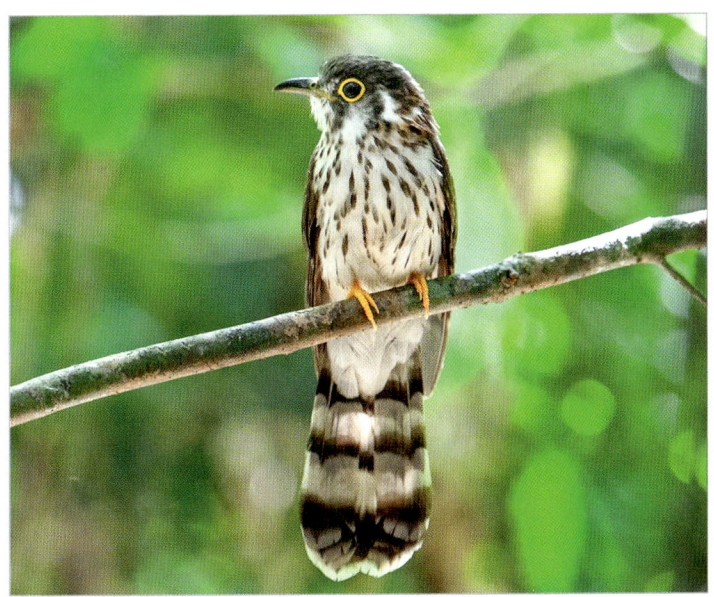

*Moustached Hawk-Cuckoo (Sabah, Malaysia)*

The authors came to the conclusion that the mimicry serves the latter function. Either way, the similarity was noted by ornithologists since the early 1800s, one example being an entry that Gould wrote for Large Hawk-Cuckoo:[227]

> The apparently inexhaustible stores of Ornithology which Nature displays in the Himalayan Mountains have furnished several species of this rather limited genus . . . The present species is one of the largest of the group, and differs from the common Cookoo in the marking of its plumage, the tail and wings being varied with several broad bars of brown, and the breast blotched with patches of the same tint. This peculiarity of colouring gives it a great resemblance to some of the *Falconidae*, particularly the *Falco sparverius*, whence it derives its specific name.

# Heathwren

The heathwrens are two species of Acanthizidae endemic to southern Australia. This is another example of the early European naturalists naming an unrelated bird on the opposite side of the world after a more familiar bird from home. Almost any small bird with a cocked tail was dubbed a "wren." In early accounts of the heathwrens, such as in Gould, they were simply referred to as "Red-rumped Wren" and "Cautious Wren."[40] By the 1890s, a prefix had been attached and the birds were being called ground-wrens.[170] Cayley was probably the first to formally use the name "heath-wren," but he gave alternative names for both species as ground-wren, scrub-warbler, reed-lark, and even mock quail.[216] Most of these names reference the habitats favored by the birds, and Cayley noted that they are

> usually in pairs or small parties, frequenting heath-lands and low stunted scrub-lands. A shy bird, spending most of its time on the ground, over which it hops or moves among the undergrowth with great celerity.

# Helmetcrest

Four species of hummingbirds are known as helmetcrests. As the name suggests, they all boast long, narrow helmet-like crests. As Wood says in *The Illustrated Natural History* in 1862:[85]

> The Helmet-crests are very curious birds, and are at once known by the singular pointed plume which crowns the top of the head, and the long beard-like appendage to the chin.

# Helmetshrike

The helmetshrikes are not shrikes but eight species in the Vangidae that sport an array of striking, helmet-like head adornments consisting of crests, forecrown tufts, and hindcrown plumes. Brehm wrote that the Grey-crested Helmetshrike is[106]

> easily recognisable by the remarkable plume, composed of stiff, hairy feathers, with which the head is decorated. Some of these hairy feathers cover the nostrils and base of the beak, and incline forwards, whilst the rest rise directly from the top of the head, and combining, form a crest that in shape resembles the upper part of a helmet.

Their upright stance, hooked bills, and possibly their dietary preferences, led to some comparisons with the true shrikes, explaining the suffix.

# Hemispingus

*Hemispingus* is a now-defunct genus name for twelve species that retain the word in their common names. They are now known to be members of the Thraupidae, the tanagers and allies, but when Cabanis erected the genus in 1851, he believed them to be members of the subfamily Pitylinae, no longer recognized, that included species from the Passerellidae, Thraupidae, and Cardinalidae.[63] He introduced the name with the German subheading Halb-Ruderfink, "half rowfinch." The *hemi* part of the name, of course, comes from the Ancient Greek ἥμι *hemi*, meaning "half" but also "small," while *spingus* comes from the Ancient Greek σπίνος (σπίγγος) *spínos*, meaning "finch."

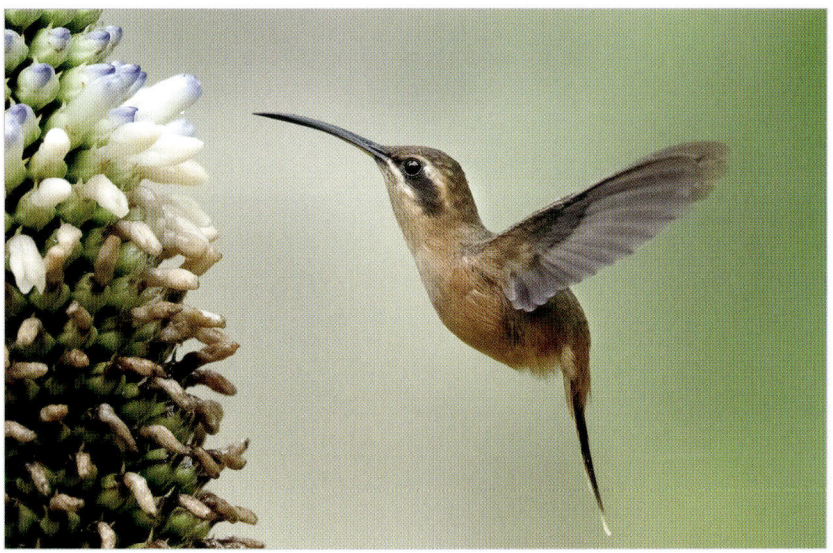

*Stripe-throated Hermit (Costa Rica)*

# Hermit

The hermits reside in the Trochilidae, the hummingbird family. The thirty-two species in various genera are all somewhat drab when compared to many of their brightly adorned congeners. The word hermit, meaning "solitary," comes from the Ancient Greek ερημικός *erimikós*, meaning "of the desert" or "a recluse," coming into English from the Old French *hermite*. Like most hummingbirds, they are generally solitary when not lekking. According to Gould:[8]

> Of the members of the genus *Phaethornis*, a group of Humming-Birds, popularly known by the name of Hermits, from their frequenting the darkest and most retired parts of the forest, three-fourths are natives of Brazil.

# Heron

Herons are well known to all, comprising a group of long-necked and long-legged waterbirds of forty-two species in the Ardeidae. The Old English name for the Grey Heron was *hrágra* and has the same roots as the Old French *hairon*, which replaced it. The name came into Middle English as *heron, heroun, heiron* in the 1300s from the Old French *hairon*, which derived from the Proto-Germanic *hraigran*. Ultimately, the common Proto-Indo-European

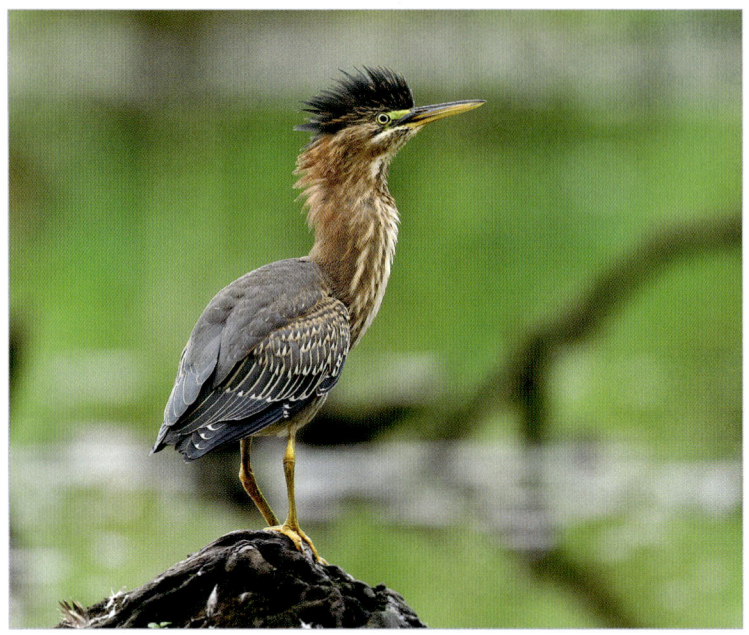

*Green Heron (Washington State, USA)*

root words *(s)kreik- or (s)kreig-* "to screech, creak" are imitative of the harsh, croaking vocalizations of the Grey Heron. "Sir Orfeo," an anonymous narrative written in 1330, had the entry:[419]

> Of game þai founde wel gode haunt: Maulardes, hayroun, & cormeraunt. [Of game they found many dwelling: mallards, herons and cormorants.]

# Hillstar

Nine species of hummingbirds, seven in the genus *Oreotrochilus* and two in the genus *Urochroa*, are known as hillstars. All are found in high-altitude montane habitats, and their bright, shiny plumage accounts for the star epithet. The sisters Mary and Elizabeth Kirby wrote eloquently about the Ecuadorian Hillstar in 1874:[217]

> A star of beauty indeed he is! His costume is magnificent. The head and throat are violet, and there is a ring of shining green round the neck. The white of the under part of his body contrasts with the brilliant colours that adorn him . . . This star-like creature haunts the grand mountain of Chimborazo, and ascends nearly to the line where snow commences.

# Hoatzin

This strange bird belongs to the monotypic Opisthocomidae. The name came into English from the American Spanish word *hoazín*, which in turn came from the Nahuatl (an Aztec language) *Huāctzin*. Under the heading of Crested Pheasant, Latham in *A General Synopsis of Birds* from 1783 listed the alternative names *L'Hoazin* and *Hoactzin*, with the comment:[218]

> This inhabits Mexico, and parts adjacent, where it feeds on snakes: makes an howling kind of noise, and is found on trees near rivers; is accounted an unlucky bird. Met with chiefly in autumn, and is said to pronounce a sound not unlike the

word *Hoactzin*. We learn from others that it may be domesticated, and is seen in that state among the natives; and further, that it feeds on ants, worms, and other insects, as well as snakes.

The entry for *Huāctzin* in the authoritative *Dictionary of Nahuatl* from 1992 stated that it is[267]

> a large bird with a distinctive call known in Spanish as pájaro vaquero [cowboy bird] (*Herpetotheres cochinans*) [sic] / cierta ave canora, grande como la gallina; es también nombre de otra ave de la misma magnitud, pero no canora. [a certain songbird, big as the hen; It is also the name of another bird of the same magnitude, but with no song.]

This complies with the view by many that the name was actually originally for a different bird, as *Herpetotheres cachinnans* is the Laughing Falcon. Furthermore, there are no Hoatzins even on the same continent as the Aztecs, so it can't be a Nahuatl name for *Opisthocomus hoazin*. Buffon wrote about "L'Hoazin" in 1772, suggesting that the name is onomatopoeic:[417]

> His voice is very loud, and it is less a cry than a howl: it is said that he pronounces his name apparently in a dismal and frightening tone, he does not falter.

But he goes on to discuss a conflicting account of another bird given the same name in the earlier European literature, the description of which fits that of the Laughing Falcon:

> Fernandez speaks of another bird to which he gives the name of *hoazin*, although by his story it is very different from the one we have just spoken of, for besides being smaller, its song is very pleasant, & sometimes resembles a man's burst of laughter & even a mocking laugh & if one eats his flesh it is neither tender nor in good taste: besides, it is a bird that cannot be tamed.

In an examination of the phonograms of the Nahuatl writing system, Lacadena confirms that the word *huāctzin,* a "bird from whose song omens were derived," is the Laughing Falcon.[269] The author discussed three elements of a compound glyph—the leaves sign, a sign representing a bird, and the buttocks sign TZIN—stating that it depicts the wa-WAK-TZIN, *Wāktzin* breaking it down to *wāk·tli,* "falcon," and *-tzin,* "a diminutive or honorific," and asserting:

> There is no reason to suppose that the bird <*huactli*> designated by the gloss is anything other than the one depicted . . . the circle or collar under the bird's head should serve as an additional diagnostic.

# Hobby

There are four species of hobbies in the genus *Falco*, small, fast-moving falcons also with common names Merlin, kestrels, and falcons. The name came into Middle English as the words *hobi* or *hoby* via the Old French *hobet*. It is first attested in *The Promptorium Parvulorum* in 1440 with an entry for a "*Hoby, hawke.*"[261] There appear to be two theories for the origin of the name. One is that the Old French *hobet* comes from the Latin word *harpe*. This word has its roots in the Ancient Greek ἄρπη *hárpē*, meaning "bird of prey, falcon, scimitar." If we go way back to the Proto-Indo-European, it's thought the word might relate to *serp-*, meaning "to reap, harvest; sickle." This reference to a sickle parallels the origin of the word "falcon" and invokes the bird's wings, bill, or talons. However, most authorities attribute the etymology to an Old French word meaning to "jump about," referring to the bird's renowned agility. In 1888 Skeat wrote in *An Etymological Dictionary of the English Language*:[220]

Like other terms of falconry, it is of French origin; being merely the corruption of the Old French *hobreau* . . . So named from its movement.—O. F. *hober*, "to stirre, move, remove from place" . . . This etymology is confirmed by noting that the O. F. verb *hober* was sometimes spelt of an *auber* . . . ; corresponding to which latter form, the hobby was also called *aubereau*.

The present-day French name for Eurasian Hobby is *Faucon hobereau*. The French *hoberel*, "little bird of prey," is attested from around 1195 and *hobereau* from 1377, a form enlarged from the Old French *hobel*, "kind of little bird of prey," and the suffix *-reeau* or *-eau*. This probably relates to the obsolete French verb *hobeler*, "skirmish, harass the enemy, plunder," itself from the Middle Dutch *hobelen* "to turn, or roll."[380] Historically, falconers used the name "Hobby" for the female, while the supposedly inferior male was called "Jack" or "Robin." Swann wrote:[121]

In addition to being a favourite species for hawking, this bird was formerly employed in what was called the "Daring of Larks," an ancient usage in fowling, in which a Hobby was let off to prevent the larks from rising while they were being netted.

# Honey-buzzard

Six species of honey-buzzards in two genera are members of the Accipitridae. These Old World raptors inherited their name from the European Honey-buzzard, which is known to seek out Hymenoptera nests in order to prey upon the larvae, pupae, and adults. As Newton wrote:[7]

The name Honey-Buzzard is admittedly misleading, for honey forms no part of its food, though the immature stages of Wasps and Humble-bees have a particular attraction for it; and it may be seen on the ground, where it runs swiftly . . . , scratching out their nests, and feeding on the living contents of the combs, regardless of the stings of the infuriated owners, against which the short, rounded and closely-adpressed feathers covering its face are said to form a protective vizor.

# Honeycreeper

The Laysan Honeycreeper is a now-extinct Hawaiian member of the Fringillidae, while the other honeycreepers, of which there are six, are all members of the Thraupidae (tanagers and allies). Although the honeycreeper members of the Thraupidae have very different habits from the Certhiidae treecreepers, they were initially classified as members of the genus *Certhia* based on specimens sent to Europe from the Neotropics. They don't feed in the same manner as the *Certhia* creepers; instead, most are specialized nectar feeders with curved bills for probing flowers, but they are also known to forage by foliage gleaning and sallying and are typically found in mixed feeding flocks. In the 1700s, the *Certhia* genus was something of a catch-all taxon; Latham illustrated this in a comparison with the hummingbirds, which in present-day ornithology would seem to be quite remarkable:[16]

This genus has, by many naturalists, been confounded with the [hummingbirds]; but a little consideration will point out the difference. In the *first* place, *Creepers* are not confined to any climate, being found in all quarters of the globe; while *Humming-Birds* are met with only in the warmer parts of *America*. *Secondly*, The genus here treated of, has the bill pointed at the end, be the shape of it however different; but that of the *Humming-Bird* is bluntish. *Thirdly*, The *Creeper* genus feeds principally on insects; while the *Humming-Bird's* food consists only of the juice extracted from the *nectaria* of *flowers*.

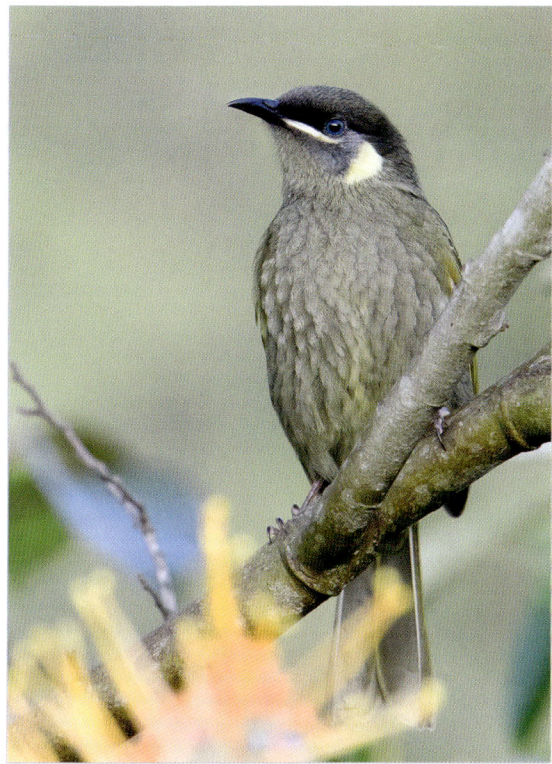

*Yellow-cheeked Honeyeater (Queensland, Australia)*

# Honeyeater

This very large Australasian family, the Meliphagidae, is known as the honeyeaters. Of the 187 species, 106 of the birds are assigned the group name honeyeater. The name for the family is derived from the Ancient Greek μέλι *méli*, for "honey" and φάγω *phágo*," to eat." These very familiar birds of the Aussie bush have highly adapted tongues and bills for feeding on nectar, but most will readily consume fruit and invertebrates. Early names for the birds included "honey-sucker" and "honey-creeper." The family name was taken from the genus name *Meliphaga*, erected by Lewin in 1808 with the justification:[497]

> The birds of this genus feed on the nectarine juice, concentrated in all the flowers of this country, which they extract with the bunch of hair at the end of their long tongues.

He gave the names of the birds as "honeysuckers," however. Latham used honey-eater indiscriminately in 1822 in *A General History of Birds*, calling it a "genus" and noting two types—those "With Thrush-like Bills" and those "With Creeper-like Bills."[17] In his honeyeater "genus" he included bee-eaters, whistlers, Australian robins, and even whipbirds—called the Coach-whip Honey-eater.

The name honeyeater was earlier used for what was probably the African honeyguides as seen in a publication from 1768:[613]

> The honeyeaters, or gnatsnappers, live intirely on honey, bees, gnats, and flies . . .
> These birds are a sort of guides to the Hottentots in their searching for honey, which the bees deposit in the cliffs of the rocks.

Then in 1793 Buffon mentioned the name when he explained the Ancient Greek origin of the genus name for the "Bee-eater," not the Meliphagid honeyeaters:[8]

> Aristotle calls the Bee-eater Μεροψ, which Pliny writes in Roman characters Merops: it was also termed Αεροψ [Aerops], Φλωρος [Floros], Μελισσοφάος [Melissofáos], contracted for Μελισσοφάγος [Melissofágos] (honey-eater).

# Honeyguide

The honeyguides are seventeen species of predominantly African birds in their own family, the Indicatoridae. While all honeyguides feed on beeswax, and to a lesser extent honey, only one species, *Indicator indicator*, is known to "guide" large mammals to beehives, where it feeds on the wax, honey, and larvae after the mammals have dismantled the hives and left the scene. Sparrman was the first to formally describe the bird that he called the "Honey-guide, or *Cuculus indicator*" in 1777 when he wrote:[221]

> The Dutch settlers thereabouts have given this bird the name of *Honig-wyzer*, or Honey-guide, from its quality of discovering wild-honey for travellers . . . Not only the Dutch and the Hottentots, but likewise a species of quadruped, which the Dutch name a *Ratel*, are frequently conducted to wild bee-hives by this bird, which as it were, pilots them to the very spot. The honey being its favourite food, its now interest prompts it to be instrumental in robbing the hive, as some scraps are commonly left for its support.

# Hookbill

There are two unrelated hookbills, one in the Furnariidae and a now-extinct Hawaiian fringillid. As the name suggests, both have strongly hooked bills. The heavy, hooked bill, the furnariid's namesake feature, is also referenced in the genus name *Ancistrops* from the Ancient Greek ανκίστρων *ankistron*, for "hook" or "fishing hook," and όψ *öps* for "face."

# Hoopoe

The three species of hoopoes in the Upupidae, with their distinctive fanning crests, also have memorable vocalizations, and it is these that give it the name. The mellow song is a far-carrying *hoo-poo-poo*. The name, in similar form, probably dates back to Ancient Greece as έποψ *epops*, and the genus name *Upupa* was given to it by Linnaeus. There's another theory that the name derives from the Old French *huppé*, which means "crest," but in fact, it is the other way around. Littré noted in 1883, the etymology of *huppé*, "Qui a une huppe sur la tète, en parlant des oiseaux. [That which has a crest on its head, speaking of birds]" is *huppe*, the Hoopoe or the "Oiseau de la grosseur d'un merle, qui a une petite touffe de plumes sur la tète, *upupa epops*, Linné, ordre des passereaux. [Bird the size of a blackbird, with a small tuft of feathers on the head, *upupa epops*, Linnaeus, order of passerines.]"[222]

# Hornbill

The hornbills are a family of fifty-nine species, the Bucerotidae, possessed of remarkable, large bills usually topped with a casque. The family name itself refers to the bill—from the Ancient Greek *boukerōs*, meaning "cow horn" (βους *bous* + κέρας *kéras*). The common English name is an obvious reference to the birds' most striking feature—their amazing bills. An early account can be found in 1773 when Pennant described the bill of the hornbill as a[66]

> great bending Bill, oft-times a large protuberance resembling another bill on the upper mandible.

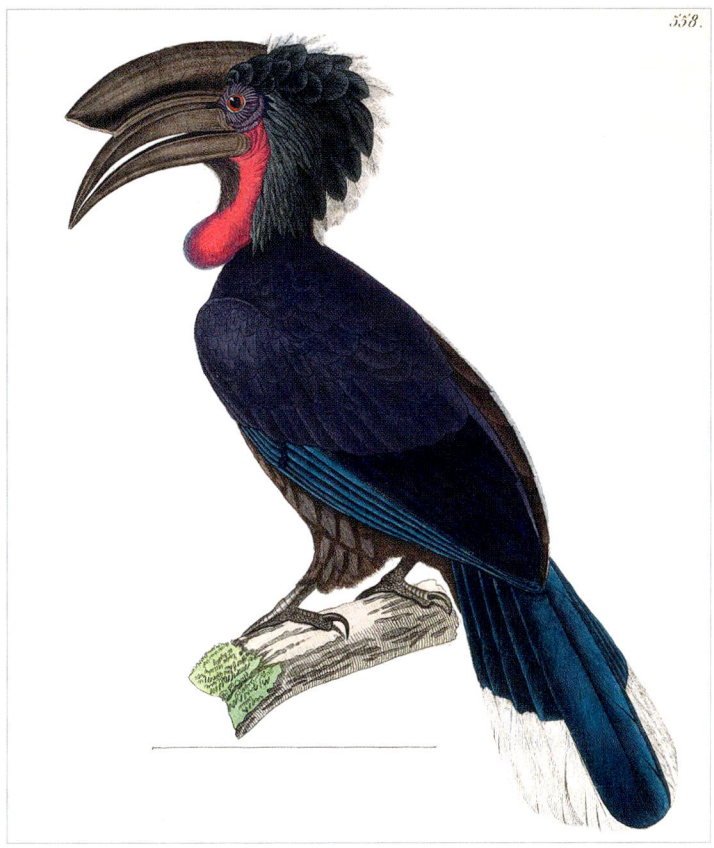

*Black-casqued Hornbill in* Tableau Méthodique *(1838)*

A few years later, in 1793, presumably Gmelin rather than Carl von Linné wrote about the hornbills, giving an excellent description of their namesake feature:[223]

> The bill is large, convex, bent downwards, sharp edged, mostly jagged or serrated outwardly, and having a large horny protuberance on the upper mandible at its base, or on the forehead: The nostrils are placed behind the base of the bill . . . This genus, with the Toucan, Motmot, and Scythrops, have all singularly disproportioned bills, the use of which is not apparent.

## Hornero

The *Furnarius* horneros are known for building mud nests that resemble old-fashioned wood-fired clay ovens. The six species of horneros in the Furnariidae all construct or use so-called oven nests, unique multichambered constructions made of mud usually situated on tree limbs. The Rufous Hornero is the national bird of Argentina, and its nests can be seen on every phone pole and fence post, while *El Hornero* is the peer-reviewed scientific journal of Aves Argentinas-Asociación Ornitológica del Plata. The Spanish word *horner* means "baker," from the word *horno* for "oven." Azara expanded on the etymology in 1809 in his account of his travels in South America as a Spanish military officer:[188]

> L'oiseau de cet article porte, à la rivière de la Plata, le nom de *hornero* (fournier),
> et au Tucuman, celui de *casero* (ménagère); ils font tous deux allusion à la forme

extérieure du nid, qui ressemble à celle d'un four. On l'appelé au Paraguay, je ne sais pourquoi, *alonzo garcia*. [The bird in this article bears the name of *hornero* (fournier) in La Plata River, and in Tucuman that of *casero* (housewife); they both refer to the exterior shape of the nest, which resembles that of an oven. They called him in Paraguay, I don't know why, *alonzo garcia*.]

# Huet-huet

The two species of huet-huets are found in a small area of southern South America. The name is pronounced "wet wet" and is imitative of the Black-throated Huet-huet's call. The other species, the Chestnut-throated Huet-huet, has a different call, described as "*put put put*" but retains the name huet-huet due to its close affinities with the aforementioned species. The name is derived from onomatopoeic names used by the Indigenous Mapuche inhabitants of present-day south-central Chile and southwestern Argentina. These have been transliterated in the past as Huahueta, Huedhued, Waweta, Wed-wed, and Wüzwüz.[151] But another version of a Mapuche name, *Wed-wed*, is said to

> generally mean "crazy" and, more specifically, a person with a joke flowering on his lips who knows what he says is funny and dares to say, and show it. From far away in the ravines, the Black-throated Huet-Huet sounds like the talking of a crazy person who accompanies travelers through the forests, telling curious stories that amuse them as they go.

# Huia

It's so sad to think that this remarkable bird from New Zealand with its sexually dimorphic bill structure is but a memory. The name is a Māori word, *huia*, for the bird in imitation of its vocalizations. Wakefield wrote about the bird in the field in 1845:[224]

> [They] engaged some native guides one day to go and look for some birds called *huia*, which were said to abound in this part of the country. The *huia* is a black bird about as large as a thrush, with long thin legs, and a slender semicircular beak, which he uses for seeking in holes of trees for the insects on which he feeds. In the tail are four long black feathers tipt with white. These feathers are much valued by the natives as ornaments for the hair on great occasions; and are highly esteemed as presents from the inhabitants of this neighbourhood to those of the north, where the bird is never found. Near the insertion of the beak, a fleshy yellow wattle is placed on either side. . . . Among the forests on the top they remained ensconced in the foliage, while the natives attracted the birds by imitating the peculiar whistle, from which it takes the name of *huia*.

The Huia was a *tapu*, or sacred bird, in Māori culture, with many legends surrounding it. One such legend was that[225]

> a chief . . . placed a magic spell on his favourite huia, so that it would appear before him on command. On one occasion when the chief summoned the bird, it had been nesting and its tail feathers were ruffled. The chief was angry and asked the bird why it had appeared before him in such a dishevelled state. On hearing the reason, the chief said: "I will provide you with a means that will enable you to keep your tail feathers in good order when next I call you." He picked up the bird, which was a female, and bent its beak into a circular shape, telling it that whenever it sat on its eggs it was to pick up its tail feathers with its beak and lift them clear of the nest.

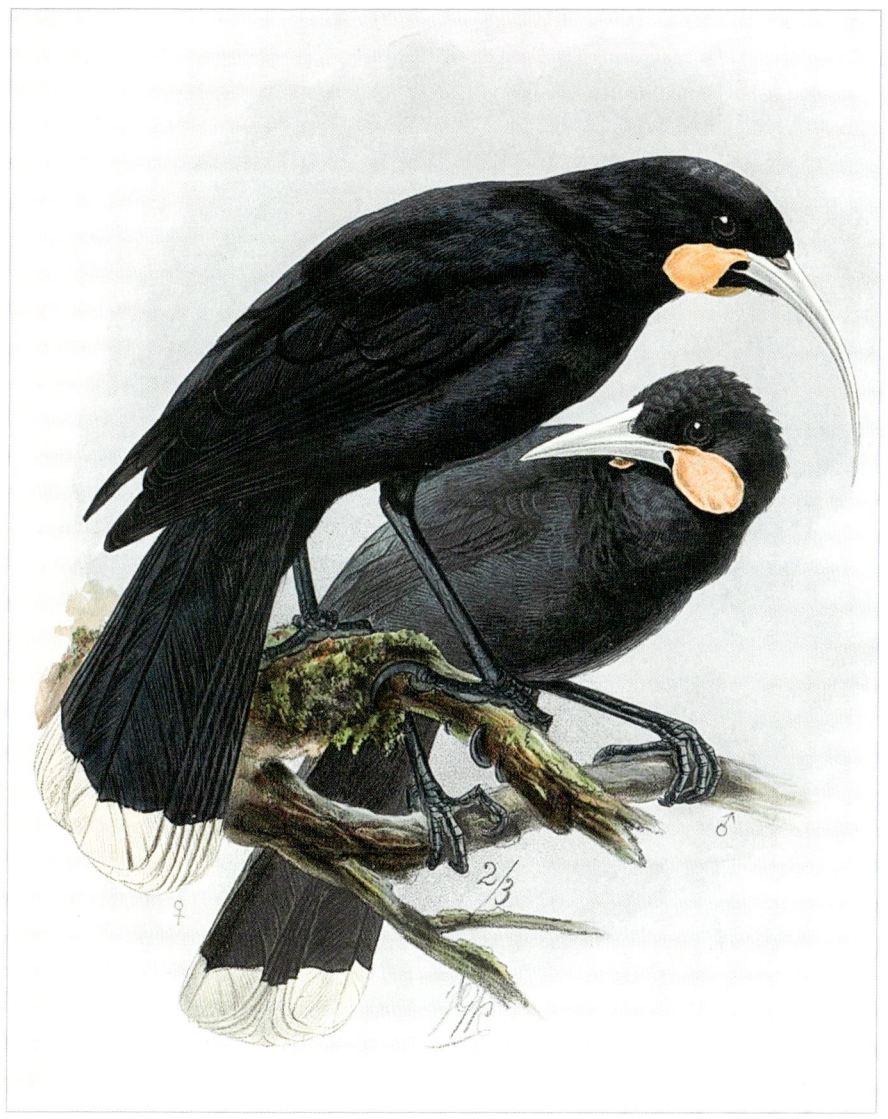

*Female and male Huia in Buller's* A History of the Birds of New Zealand *(1873)*

# Hummingbird

At least eighty-one species of trochilids are known by the familiar name hummingbird. The British ornithologist John Gould stated in the introduction to his first volume of *A Monograph of the Trochilidæ, or family of humming-birds*:[6]

> The questions have often been asked, whence is the term Humming-Bird derived, and why is the bird so called. I may state in reply that, owing to the rapid movement of the wings to most of the members of this group, but especially of the smaller species, a vibratory or humming sound is produced while the bird is in the air, which may be heard at the distance of several yards, and that it is from this circumstance that the trivial name by which these birds are known in England has arisen.

# Hummingbird Names and Their History

Many of the common English names for hummingbirds still in use today first appeared in John Gould's multivolume series of treatises, *A Monograph of the Trochilidæ, or family of humming-birds*, as well as *An introduction to the Trochilidæ, or family of humming-birds*.[6, 80] Some of the still-extant hummingbird names that appear in these volumes are Barbed-throat, Blossom-crown, Brilliant, Carib, Comet, Coquette, Emerald, Fairy, Fire-crown, Golden-tail, Helmet-crest, Hermit, Hill-Star, Inca, Jacobin, Lance-Bill, Mango, Mountaineer, Pied-tail, Plover-crest, Puff-leg, Ruby, Sabre-wing, Sapphire, Sapphire-wing, Shear-tail, Snow-cap, Star-frontlet, Sun-Angel, Sun-beam, Sylph, Thorn-bill, Thorn-tail, Topaz, Train-bearer, Velvet-breast, Violet-Ear, Vizor-bearer, White-tip, and Wood Star. Of course, in modern usage, the hyphens have been dropped in many cases. Sadly, others such as Brassy Tail, Cazique, Flame-bearer, Flutterer, Garnet, and Panoplites are no longer in use. As for their provenance, many, such as "The Ruby," "The Sapphire," and "The Amethyst," came from the pen of the French ornithologist Buffon, who clearly had a penchant for gemstones. One name coined by Buffon, "The Carbuncle," seems unfortunate in modern light until one realizes that a now-obsolete use of the word meant "a rounded red gemstone." Many of the other "trivial names," as they were referred to at the time, were coined by Gould.

Buffon wrote in 1808:[607]

> Of all animated beings the humming bird is the most elegant in its form, and the most brilliant in its colours. Stones and metals polished by art are not to be compared to this jewel in nature . . . she has bestowed upon it all those gifts which she has only distributed in part to other birds; celerity, rapidity, grace, rich ornament, all are to be found in this little favourite. The emerald, the ruby, the topaz, all are to be seen in the brilliant colours of its plumage; it never sullies these colours by the dust of the earth, and, in its existence, purely aerial, it is hardly seen to touch the flowery turf, even for a moment; it is always in the air flying from flower to flower; it possesses their freshness and their splendour.

In 1923's *The Birds of California*, William Leon Dawson succinctly described the task of naming the hummingbirds:[270]

> Ornithologists have been hard put to it to provide names for these most exquisite of birds, the Hummers. In default of an authoritative dictionary of the Colibrian dialect by a native Hummingbird, we have been obliged to call poetic imagination to the aid of our own poor stammering tongues. The realm of callilithology, chromatics, esthetics, astronomy, history, classical mythology, and a score beside, have been laid under tribute to secure such fanciful and high sounding titles as Fiery Topaz, Ruby-and-Topaz, Allied Emerald, Sapphire-breasted Emerald, Green-crowned Brilliant, Equadorian Rainbow, Equatorian Sunbeam, Parzudaki's Sun-Angel, Gould's Heavenly Sylph, Fanny's Wood-Star, Compte de Paris's Star-frontlet, Mrs. Stewart's Star-throat, Isaacson's Puff-leg, Baroness de Le-fresnaye's Plumeleteer, Blossom-crown, Little Violet-ear, Pallid Hermit, Bearded Mountaineer, Green Mango, Darker-Green Carib, Sparkling tail, Tyrian-tail, White-booted Rocket-tail, Black-billed Streamer-tail, Curve-winged Saber-wing, Julia's Train-bearer, the Sappho Comet, the Circe, Rivoli, and Lucifer Hummers, the Frilled, Spangled, Festive, and Adorable Coquettes, the Charming, Beautiful, and Lovely Hummingbirds, and, last but not least, the truly Marvelous Hummingbird (*Loddigesia mirabilis*).

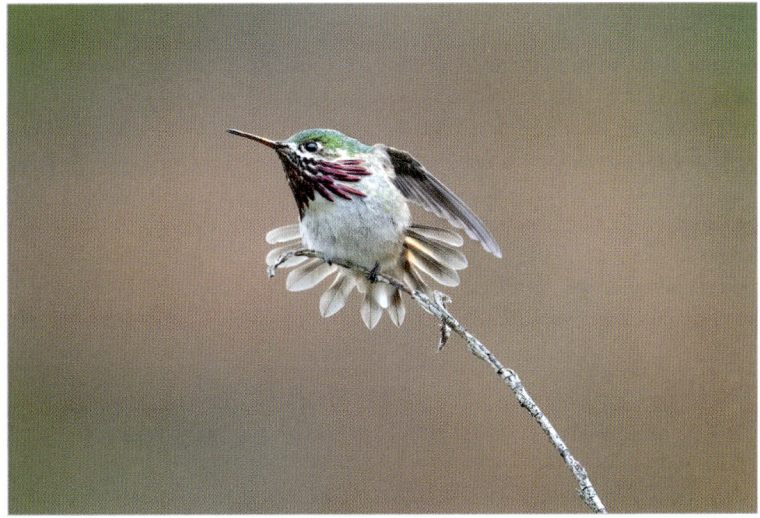

*Calliope Hummingbird (Oregon, USA)*

# Hwamei

The two species of hwameis, which were formerly treated as conspecific, are members of the *Garrulax* genus in the Leiothrichidae, the laughingthrushes and allies. One of the few members of the family to retain a name from its country of origin, the etymology is from the Chinese language name for the bird 画眉 Huàméi, the characters of which stand for "painted eyebrow," a reference to the bold white postocular stripe of the Chinese Hwamei. Swinhoe may have misunderstood the nature of the name, confusing the homonyms 花 Huā for flower and 画 Huà for paint, when he wrote in 1863:[226]

> The *Hwa-mei* (Flowered-Eyebrow) or Song-Thrush of the Chinese is so universally met with as a cage-bird in China, that every European possessed of ordinary observation that has visited the Celestial realm must be acquainted with it.

# Hylia

The Green Hylia and the Tit-hylia are closely related species in the Macrosphenidae, the African warblers. The Green Hylia is in the synonymous genus *Hylia*, erected in 1859 by Cassin, who wrote rather unconvincingly:[311]

> This bird is by no manner of means a *Stiphrornis*. . . , nor is it a *Chloropeta*. . . , nor moreover do we know any other genus to which it belongs, and so set up for ourselves as best we may.

These small forest dwellers derive their name from the Ancient Greek ὕλη *ýli* or *hulē*, meaning "woodland" or "forest."

# Hyliota

The hyliotas are members of the small family the Hyliotidae, consisting of only four species. The common and genus names are synonymous. These small insectivores of Africa are usually found in mixed feeding flocks in the canopy of broad-leaf forests. For this reason, in 1837, Swainson designated a genus name derived from the Ancient Greek ὕλη *ýli* or *hulē*, meaning "woodland" or "forest."[312]

## Hylocitrea

The Hylocitrea of Sulawesi belongs to a monotypic family, the Hylocitreidae. Furthermore, the common name is synonymous with the genus name. The name *Hylocitrea* derives from the union of the Greek word ύλη *ýli* or *hulē*, meaning "woodland" or "forest," with the second and third syllables of *Muscitrea*, an obsolete synonym of *Pachycephala*, which means "forest whistler," in reference to its supposed connection to these birds. In turn, the name *Muscitrea* was derived from *Tchitrea*, from the Malagasy name *Tchitrec* for a paradise flycatcher. This species previously had been placed in Pachycephalidae, but recent molecular phylogenetic studies have shown it to be more closely related, albeit weakly, to the *Bombycilla* waxwings, and to the Hypocolius in particular.

## Hypocolius

Another bird of a monotypic family, the Hypocoliidae, the Hypocolius has long been something of a taxonomic conundrum. This synonymous common and genus name is derived from the union of the Ancient Greek prefix ύπο *hypo*, meaning "under or beneath (placed within)" with *colius*, indicating a supposed affinity with mousebirds. When Bonaparte described the genus in 1850, he considered it to be closely related to the mousebirds and cotingas.[52] The suffix *-colius* was previously used for two genera of starlings, the *Coccycolius* and the *Lamprocolius*, and the Hypocolius has also been placed in the Sturnidae in the past.

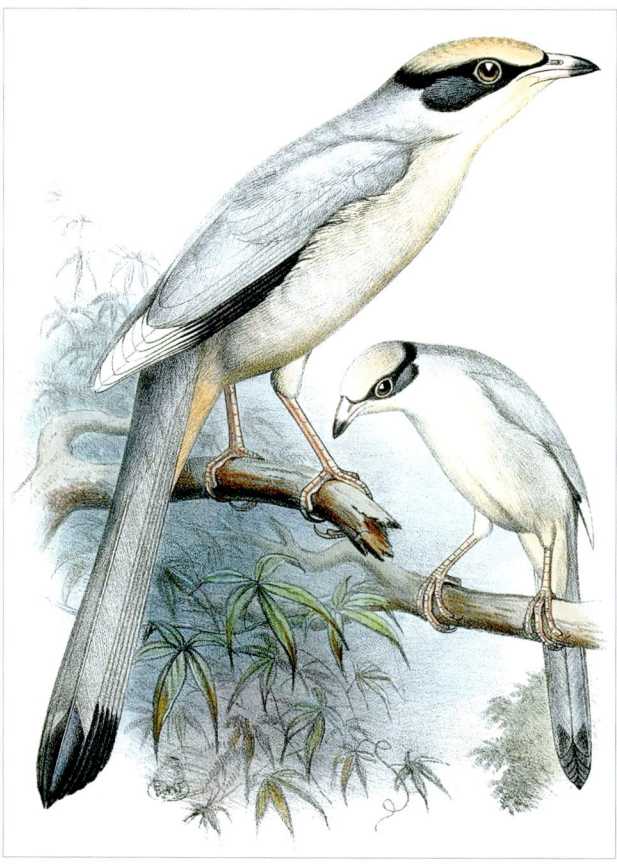

*Hypocolius by F. W. Frohawk in* Proceedings of the Zoological Society of London *(1890)*

# I IS FOR IBIS

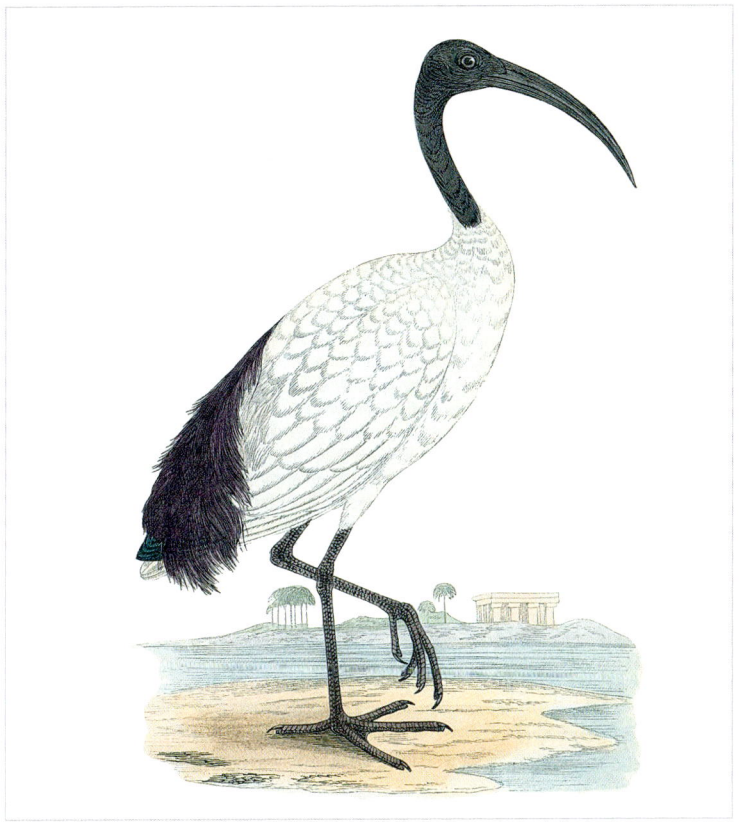

*Sacred Ibis in* A history of the birds of Europe, not observed in the British Isles *by Charles Bree (1863)*

## Ibis

Forty species of Threskiornidae are called by the name ibis. These long-billed birds have a long and storied history in human culture, and the origins of the name reflect that. The name came into English in the 1300s from the Latin *ibis* via the Ancient Greek ἶβις *ibis*. But the origins go back even further to the Egyptian *hbj* (usually anglicized as *hebi*), the sacred bird of Egypt. The Egyptian hieroglyph consists of three characters—"H" for the reed shelter, "B" for the foot, and "J" for the numeral two. These merely have phonetic values, although it could be interpreted as a "two-footed animal that lives in the reeds or grasslands"—when written, only consonants were used, and the vowels were implied. The Egyptian language exerted some influence on Classical Greek, so that a number of Egyptian loanwords, including this bird name, survive into modern usage (other words include ebony, ivory, lily, and oasis). Thoth, with the head of an ibis, was the God of wisdom and writing. In fact, the ibis has three Egyptian hieroglyphs for three different species, among only twenty-seven (with variations) featuring birds. First mentioned in the Middle English in Wycliffe's

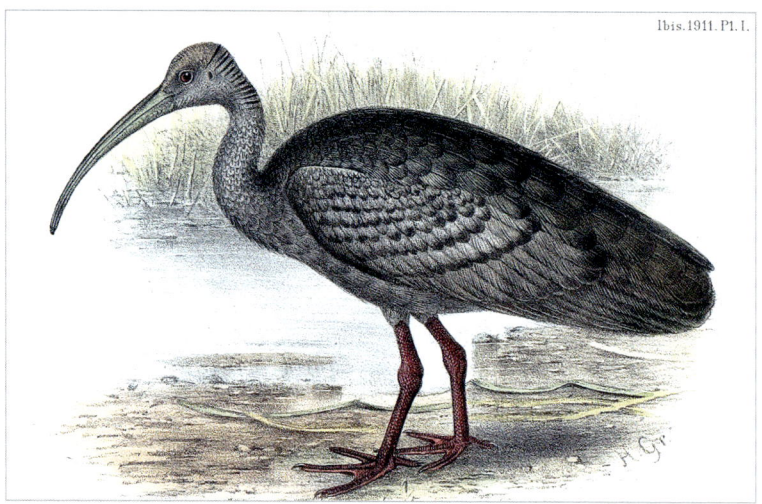

*Giant Ibis in* The Ibis Vol. 5 *(1911)*

Bible as *ybyn* or *ibin*, then as *ibys* from the sixteenth century, and *ibis* shortly after. The name appeared in the very popular *The Travels of Sir John Mandeville*, a memoir that first circulated between 1357 and 1371:[498]

> Aboute this ryuere [Nile] ben manye briddes & foules, As Sikonyes þat þei clepen Ibes. [Around the river there were many birds and fowls, like storks that they call ibises.]

# Ibisbill

The wonderful Ibisbill is the sole member of the Ibidorhynchidae. As both of these names suggest, the bird's long, downcurved bill is similar to those of the ibises. Soon after Vigors assigned the genus name *Ibidorhyncha* in 1832,[499] Gould wrote in *A Century of Birds from the Himalaya Mountains*:[227]

> Throughout the whole of our new discoveries in the vast district which has furnished the subject of the present work, it would be difficult to point out a more interesting species than that before us, or one which has supplied ornithological science with characters more striking and peculiar. It may be observed to form a union between two groups generally considered as widely separated from each other; the body, the general form, and the legs of the *Ibidorhyncha Struthersii* being similar to those of the *Haematopus*, while the bill is strictly that of the *Ibis*.

# Ibon

The single species of ibon, the Cinnamon Ibon, is a small passerid sparrow in the monotypic genus *Hypocryptadius* found only on the island of Mindanao in the Philippines. The name derives simply from the Filipino- or Tagalog-language word for "bird," *ibon*.

# Ifrita

The Blue-capped Ifrita is the sole member of the Ifritidae and is endemic to the island of New Guinea. The name derives from the Arabic word *'ifrit* (عفريت), meaning "spirit or djinn." The ifrit djinns are always malevolent, so the name could be due to the fact that handling or eating these birds was known by the Papuans to cause upper respiratory tract irritation, sneezing, and other unwanted symptoms. Although it took until the year 2000 for Western science to realize that the ifrita is actually a poisonous bird (they sequester batrachotoxins from beetles they consume to their skin and feathers), the local people have known this since antiquity, and the fear of the bird may have been conveyed to European naturalists who first collected and named the species.[500]

# Iiwi

Linguists derive the Hawaiian-language word *ʻiʻiwi* from the Proto-Polynesian (from which all modern Polynesian languages descend) word *kiwi*, which is thought to originally refer to the Bristle-thighed Curlew (*Numenius tahitiensis*).[239] The long, decurved bill of the curlew could be said to resemble that of the single species of *ʻiʻiwi* in the Fringillidae. The origin of the Polynesian root word is undoubtedly onomatopoeic.

## Proto-Polynesia KIWI

The reconstruction of the Nuclear Polynesian language reveals the word "kiwi," thought to be the name for a shorebird, probably onomatopoeic, the Bristle-thighed Curlew, a migrant to the region. Variations of the name for various birds exist not only in English, but in a number of other Polynesian languages. In English we have "Kiwi" and "Iʻiwi," but these are loanwords from Māori and Hawaiian respectively. A panoply of other birds in other Polynesian languages have variations on the name—and all of them have long, curved bills or have *keewee* calls. In Mangareva, the curlew is known as the "Kivi," and on the Cook Islands the same bird is the "Kihi"; on Tupuaki the Wandering Tattler is called the "Iivi"; in the Marquesas, Nukumanu, Nuguria, and Takuu, a number of shorebirds are called "Kivi"; the Māori also call the Snares Island Snipe the "Kiwi"; on the atoll Nukuoro the Ruddy Turnstone is called "Givi"; while on Luangiua the Bar-tailed Godwit is the "Ivi."[239]

# Illadopsis

The common English name is synonymous with the genus name *Illadopsis*, eight species of rather nondescript African babblers, members of the Pellorneidae. Heine, the German ornithologist and collector who erected the genus name in 1860, wrote the etymology as being[228]

von ἰλλάς, Drossel und ὄψις, Aussehen. [from *illas*, thrush and *opsis*, appearance.]

# Imperial-Pigeon

The word "imperial" comes from the Latin *imperium*, meaning "authority, command," and the imperial-pigeons are some of the largest and certainly most regal looking of the columbids. In 1836, Hodgson erected the name *Ducula*, the genus to which all thirty-nine species of imperial-pigeons belong, with the note that it is "*Dukul* of the Nipalese," and, indeed, the Nepali for pigeon is डुकुर *dhukura*.[96] Hodgson called them the "tree pigeons" at that time, but by the late 1800s the name imperial-pigeon was being widely used. Although *Ducula* is derived from the Nepali word, the common name seems to have arisen either from a misunderstanding or perhaps from wordplay. As, for example, in Gotch (1981), who

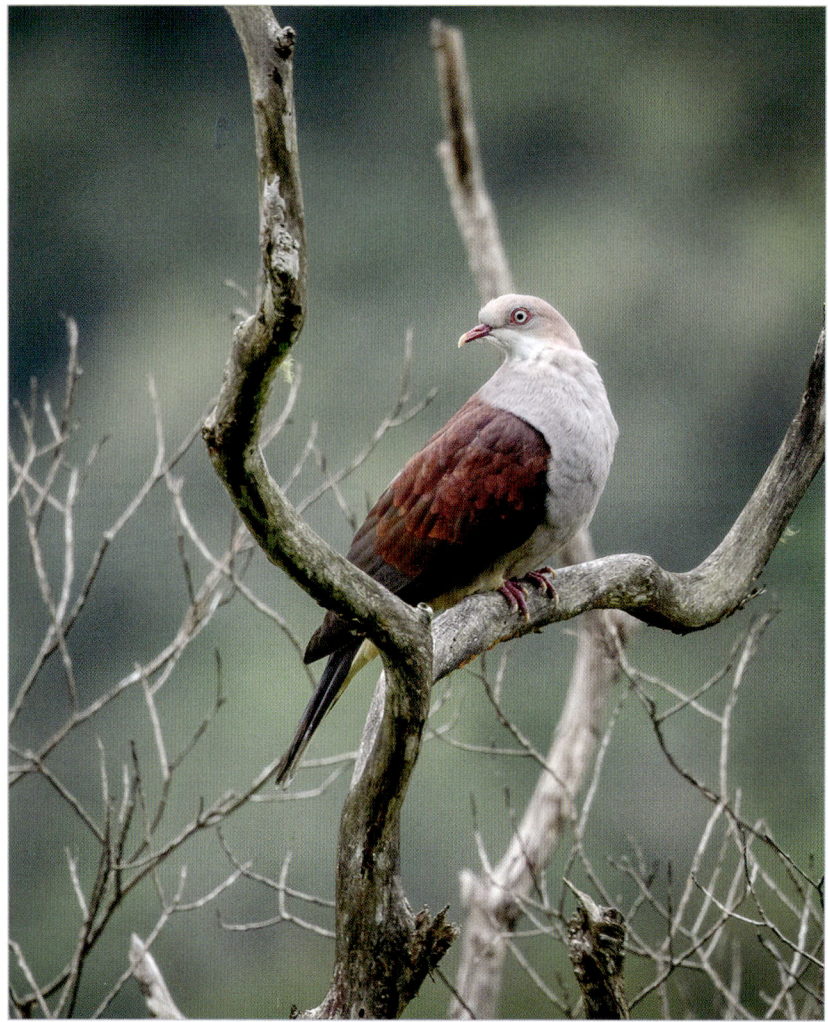

*Mountain Imperial-Pigeon (Sabah, Malaysia)*

maintained that it was derived from the genitive *ducis* of the Latin *dux*, "a leader, a duke."[398] In other words, an "imperial" pigeon. Jerdon provided some hints of the provenance of the name when he wrote "*Imperial Pigeon* of Europeans in the South of India."[136] In the same entry he noted the Hindi names as "*Dunkul* or *Doomkul*."

# Inca

The incas are four hummingbirds in the genus *Coeligena*. The word *inka* is a Quechua word for the ruling class or the ruling family. The Inca Empire, as it was known by the Europeans, was the largest empire in pre-Columbian America. In Quechua it was known as *Tawantinsuyu*, meaning "four parts together." Gould was the first to use "inca" as a bird name in his *Monograph of the Trochilidæ* in 1849, but he gave no explanation for his choice.[6] It's possible that it relates to the purple in the plumage of most of the members of the genus, which is considered the first color and associated with Mama Ocllo, the founding mother of the Inca race.

## Inca-Finch

The five species of inca-finches all belong to the genus *Incaspiza*, which was established by Ridgway in 1898. The common name is simply an anglicization of the genus name, σπίζα *spiza* being the Ancient Greek for "chaffinch" (from σπίζω "to chirp," related to *spingos*). They are all found in the region of the former Inca Empire.

## Indigobird

The indigobirds are ten species of *Vidua*, in the same genus as the whydahs, very similarly plumaged birds in the Viduidae, the indigobirds. It should come as no surprise that the males all have different shades of indigo-colored plumage.

## Iora

The ioras are a family of four species, the Aegithinidae. The Ancient Greek word φρουρός *frourós*, which may be transliterated as *ioros*, means "watchman" or "town-crier," an apt description for these vocal and feisty birds. Horsfield, who coined the now disused genus name *Iora*, described its habits in 1824:[167]

> The *Iöra scapularis* is a bird of social habits, and resorts to the vicinity of human dwellings: indeed it appears to have retired from the forests, and established itself in the trees and hedges which surround the villages and plantations. . . . It is most lively in the middle of the day, when, under a burning sun, the inhabitants uniformly retire to rest: early in the morning, and towards the approach of night, it is rarely perceived; but during the silence prevailing at noon, it enlivens the villages with the song of *Cheetoo, Cheetoo*, which it repeats at short intervals, during its sportive sallies between the branches.

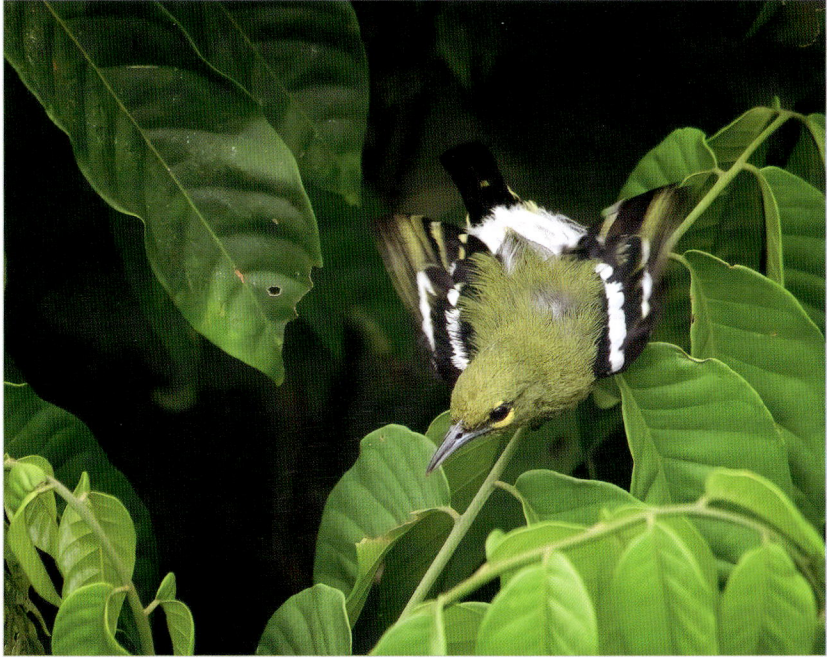

*Green Iora (Sabah, Malaysia)*

# J IS FOR JABIRU

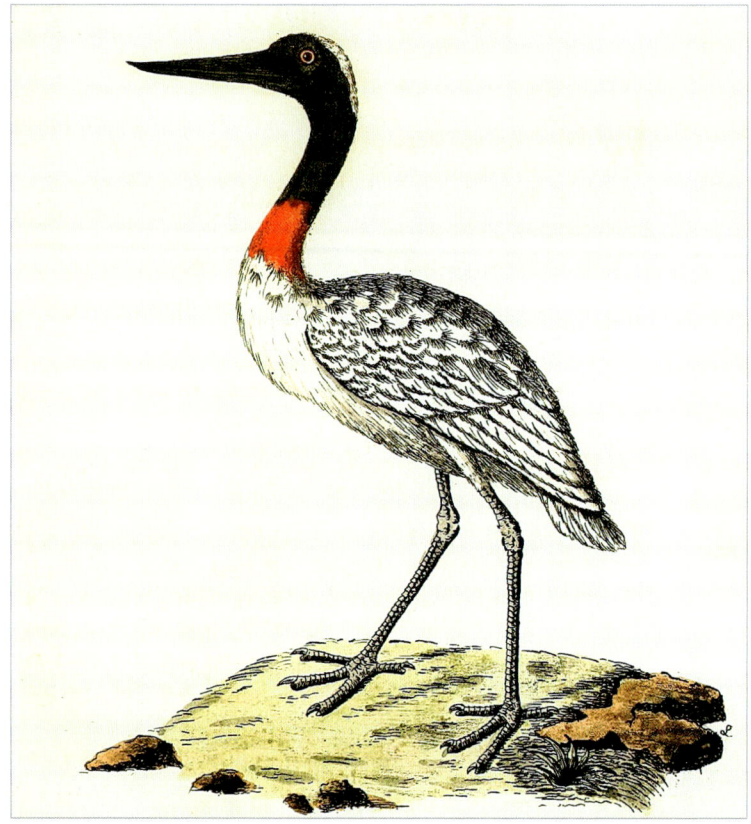

*Jabiru in Latham's* A General Synopsis of Birds *(1781)*

## Jabiru

One of the most distinctive features of the unique stork, the Jabiru, is the bird's namesake feature, its swollen bare black neck with a large red patch at the base. This large member of the Ciconiidae is found in South America. Jabiru came into English in the late 1700s from the name the Portuguese gave the bird in Brazil. Ultimately, that word came from the Indigenous Tupi-Guarani language name for the bird *jabirú*, which according to Garcia is from[133]

> de *y*, demonstrative (= what, that who, what has, or is) + *abirú* = full, replete, swollen: what is full, or replete—allusion to the great crop of the bird.

The common English name and the genus name are synonymous. According to the *Guaraní Concise Dictionary*, *i* is "to have" and *haviru* is "swollen."[229] The name is very often incorrectly applied in Australia to the Black-necked Stork in the mistaken belief that it is an Aboriginal word. This may have come about after Latham's use of the common name for all the (then) members of the genus *Mycteria*, including the "New Holland Jabiru."[17]

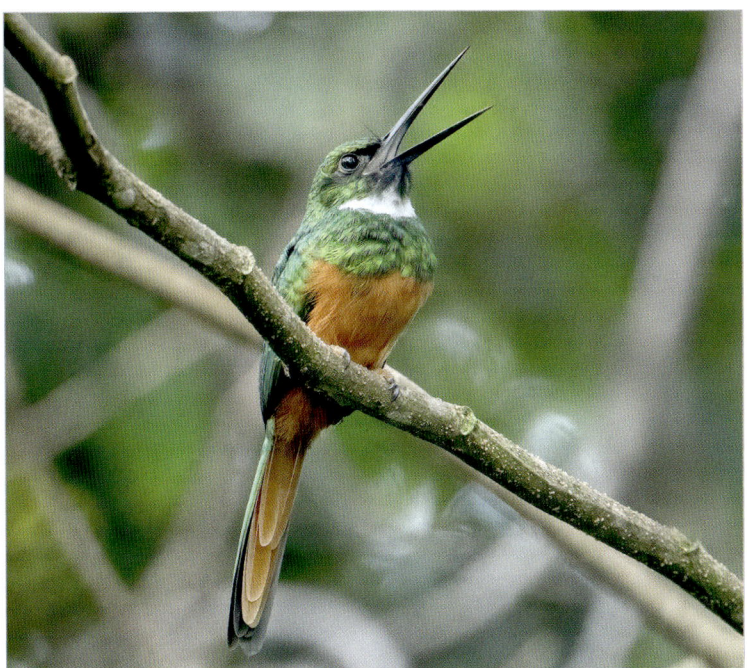

*Rufous-tailed Jacamar, Guatemala*

# Jacamar

The jacamars are eighteen species in the jacamar family, the Galbulidae. These birds of the Neotropics have notably long, pointed bills. The name came into English from the French in the early 1800s by way of the French naturalist François Le Vaillant, but it was first mentioned in the literature by Marcgrave in 1648 as "Jacamaciri Brasiliensibus."[90] He dubbed the bird *jacamaralcion* using a portmanteau of jacamar and halcyon, meaning "kingfisher." Obviously, he was struck by the similarity the birds bore to kingfishers in their feeding habits. This name lives on in the genus name of just one species of jacamar, the Three-toed Jacamar *Jacamaralcyon tridactyla*. All that aside, the present-day name is a modification of the Portuguese *jacamacira* from the Tupi-Guarani name *jacamá-ciri*. It's possible that it relates to the Guarani word *jagarra*, meaning "to catch," and *siri* for "crab," although they don't eat crabs and the word may refer to arthropods in general. Brisson wrote in 1760:[76]

> *Jacamar*, nom formé de *Jacamaciri*, qui est celui que les Brasiliens donnent à la premiere espece de ce genre. [*Jacamar*, name formed from *Jacamaciri*, which is that which the Brazilians give to the first species of this kind.]

# Jacana

The jacanas are a family of eight species, the Jacanidae. Their nicknames, "Jesus-bird" and "lily trotter," hint at their habits—their extraordinarily long toes allow them to walk easily on floating vegetation, giving the impression that they can walk on water. Despite having a pantropical distribution, the common English name was originally applied to the Wattled Jacana *Jacana jacana* of South America. Marcgrave was possibly the first to introduce the name to European ornithology with the entry:[90]

IACANA Brasiliensibus, gallon aquatica, *Waterhun*.

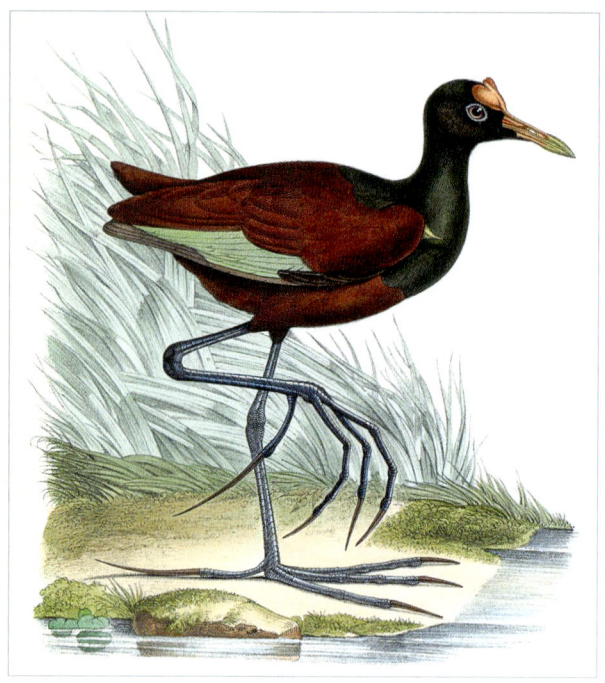

*Northern Jacana in Des Murs's* Iconographie Ornithologique *(1849)*

The Wattled Jacana was first named *Parra jacana* by Linnaeus, in 1766. The specific epithet came from the Portuguese *jaçanã*, which in turn came from the Tupi-Guarani name for the bird, *jasanã* or *ñaha'nã*. Garcia found that the derivation of the name in Tupi was[133]

> from *y*, demonstrative (= what, who has or is), *eçá* = eye + *enã* = alert, attentive, vigilant: what is with an alert eye? . . . ; or, better, *y-açã-nã*: the one who screams loudly, the one who has the intense scream.

## How Do You Say Jacana?

The pronunciation of the word "jacana" is debated. The Portuguese word is pronounced approximately zha-sah-NAH. But the Spanish pronunciation, which is, of course, the predominant language of South America, is ha-KA-nah. Elsewhere, the anglicized ja-KAH-na is used and considered correct.

## Jackdaw

Two species of corvids bear the name jackdaw. The name is a portmanteau of *jack* and *daw*, which first appeared in English in the 1500s as an entry in *The Promptorium Parvulorum*:[261]

> Cadaw or keo or chowghe: Monedula [= Eurasian Jackdaw *Corvus monedula*]

The nickname "Jack" was often used in Old English to signify a small form of an animal group, for example, as in the name "Jack Snipe." But, equally, in the case of the Jackdaw, it was imitative of one of the bird's characteristic calls. The second part of the name, *daw*, came from the Old English *dāwe*, Old Low German *daha*. Originally, the bird was called simply "the Daw," but the nickname Jack was commonly affixed and jackdaw came into favor over time as the "vulgar and redundant name," Newton presumably believing that two onomatopoeic names were superfluous.[7] This is another of Shakespeare's birds, appearing in *Othello* in Act 1, Scene 1 from 1603:

> I will wear my heart upon my sleeve, For daws to peck at.

# Jacky Winter

This little Australasian robin in the Petroicidae revels in a unique name with murky origins. Some say the Jacky Winter acquired its name from its rapid and strong vocalizations, which sound like *jacky-jacky winter-winter*. Another old name for the Jacky Winter is "Peter Peter," also imitative of the song. The mnemonic could easily also be transliterated as *winter winter* or even *jacky jacky*. In Edward Morris's *Dictionary of Australasian Words*, he stated that the name was[20]

> ascribed to the fact that it is a resident species, very common, and that it sings all through the winter when nearly every other species is silent.

Other local names for the bird include Stumpbird, Post Sitter, and Postboy, in honor of its habit of perching on stumps and fence posts.

# Jacobin

The two species of little jacobin hummingbirds in the genus *Florisuga* both sport smart contrasting dark and white plumage. The epithet "jacobin" was applied to certain species of *Lonchura* munia by Buffon in the late 1700s because their pied plumage resembles the black and white clothing of the Jacobin friars of Europe.[8] Buffon also called the hummingbirds "L'Oiseau-Mouche a Collier; dit La Jacobine [The Necklaced Fly-Bird; called the Jacobin]," noting:

> This bird-fly['s] . . . head, throat & neck are a beautiful dark blue changing to green; along the back of the neck, near the back, he wears a white half-cap; the back is golden-green; the tail white at the tip, edged with black, with the two middle pinnae & the green-golden coverts; the chest & the flank do the same; the belly is white: it is apparently from this distribution of white in its plumage that the idea came to call it Jacobine.

Note that this is the feminine form of the noun and therefore refers not to Dominican monks, but to their counterpart nuns, who wore variations on the theme of white dresses with a black veil. Prior to Buffon's use of the name, the bird was called simply the "White-collared Hummingbird," but by the mid-1800s, the name had been adopted, albeit in its masculine form, and this was soon popularized by Gould in his *Monograph of the Trochilidæ*.[6]

# Jaeger

There are three species of jaegers, seabirds that are closely related to the skuas in the Stercorariidae. The name, as a foreign appellation, first appeared in English in the late 1700s in Buffon's *Natural History of Birds*, when he wrote:[8]

> The people of the north report that its object is to obtain the excrements from those unhappy little mews [gulls]; and they have, for that reason, called it *Strund-jager* [Dung-hunter], to which *Stercorius* is synonymous.

The present-day English name comes from the German word *jäger*, the noun for "hunter or fighter" (from *jagen*, "to hunt") in homage to their aggressive kleptoparasitic habits. As with many birds, there is an interesting relationship between the Old English names, the scientific name, and the present-day common name. Brisson[76] erected the genus *Stercorarius* in 1760 from the Latin *stercus*, meaning "dung," relating to their older English names, Dung-bird and Dung-hunter, from the belief, as Buffon mentions again, that the Black-legged Kittiwake[8]

> is perpetually pursued and harrassed by the bird *Strundt-jager* [Parasitic Jaeger], and constrained to void its excrements, which the latter greedily swallows.

The German word *strunt* or *strundt* means "dirt," "crap," or "garbage," so that Strundt-jager translates into English, to put it politely, as "excrement hunter." Another widespread English name at the time was "the Labbe" from the Swedish name for the bird derived from *labba*, "to walk heavily and clumsily." Buffon described the Jaegers thus:[8]

> Considering its size and figure, this bird might be ranged with the mews [gulls]. But if it be of that family, it has lost all fraternal affection; for it is the avowed and eternal persecutor of its kindred, and particularly of the kittiwake. It keeps a steady eye on them, and when it perceives them betake to flight, it pursues without intermission. The people of the north report that its object is to obtain the excrements from those unhappy little mews; and they have, for that reason, called it Strund-jager, to which Stercorarius is synonymous. Most probably, however, this bird does not devour the dung, but only the fish which the kittiwakes drop from their bill or disgorge.

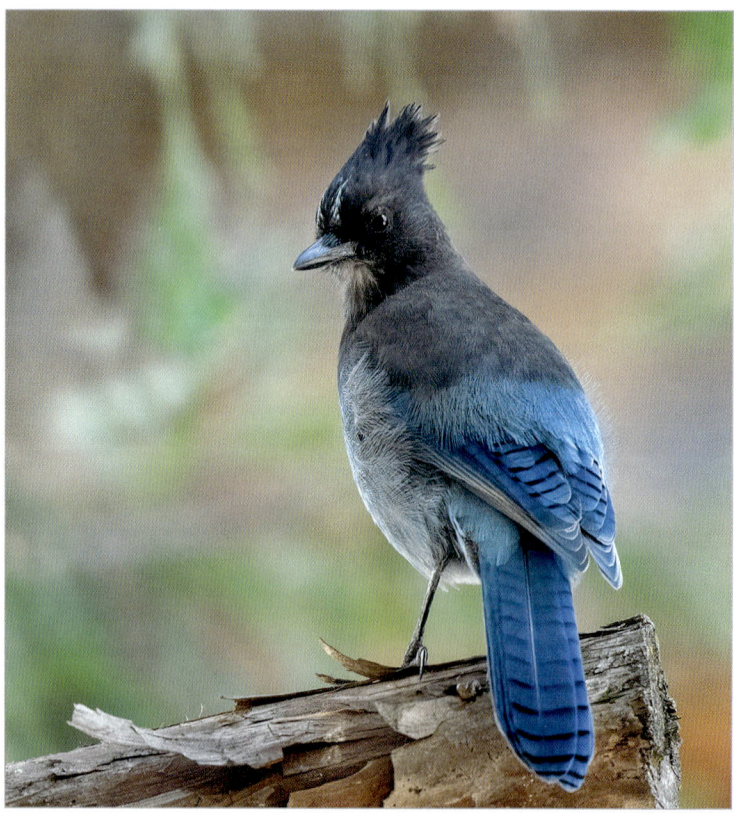

*Steller's Jay (California, USA)*

# Jay

The jays are very familiar birds to people throughout the Americas and in parts of the Old World, too. They are members of the Corvidae and known for their strikingly good looks and high intelligence. The name came into English for the Eurasian Jay *Garrulus glandarius* as the Middle English *jai*, derived from the Old French *géai*. Ultimately, the names come from the fifth-century Latin *gaius*, "a jay," echoic of the bird's harsh warning cries. The earliest attestation of the English name occurs in the early 1300s when it was written in *The Harley Lyrics*:[381]

> Ientil iolyf so þe Iay [gentle, joyful, so the jay.]

It was also mentioned in Turner in 1544 as a bird he encountered in Italy that was "a certain bird like a Pie, in English called a Jay."[119] And Shakespeare mentioned it in his play from 1623, *The Winter's Tale*:

> *The lark, that tirra-lyra chants,*
>
> *With heigh! with heigh! the thrush and the jay,*
>
> *Are summer songs for me and my aunts,*
>
> *While we lie tumbling in the hay.*

The belief that the Latin name for the bird derived from the popular Roman name *Gaius* is almost certainly erroneous as it's believed the use of Christian names for birds didn't come into fashion until the Middle Ages.

# Jery

The four species of jeries are drab little birds in two families—the Cisticolidae and the Bernieridae—that are all endemic to Madagascar. Richardson's *Malagasy-English Dictionary* from 1885 lists:[231]

> *Je'ry* "the canary. [Fr. serin]"
>
> *Je'rika* "the name of a little bird"

There are no canaries or serins on Madagascar, so no doubt Richardson, who was not an ornithologist, used the word simply to refer to generic small passerines. Other local Malagasy names for the birds include *jijy*, *kimitsy*, *zeazea*, and *joijoika*. The names are probably all imitative of the birds' varied vocalizations.

# Jewel-babbler

These four members of the *Ptilorrhoa* belong to the Cinclosomatidae, the quail-thrushes and jewel-babblers. While these birds are closely related to the similarly misnamed quail-thrushes, their colorful plumage that brightens the dark forest floor when spied inspired the first part of their name, in much the same way that pittas were known as "jewel-thrushes." And although they are now known not to be babblers, their initial placement in the Timaliidae accounts for the second part of the name. Gould [550] gave them a common name synonymous with their old genus name, *Eupetes*. In a description of "Blue-bodied Eupetes" (Blue Jewel-babbler), he claimed:

> The genus *Eupetes* is no doubt one of the great group of strong-legged Thrushes, or *Timaliidae*, so plentifully distributed throughout the tropical portions of the Old World.

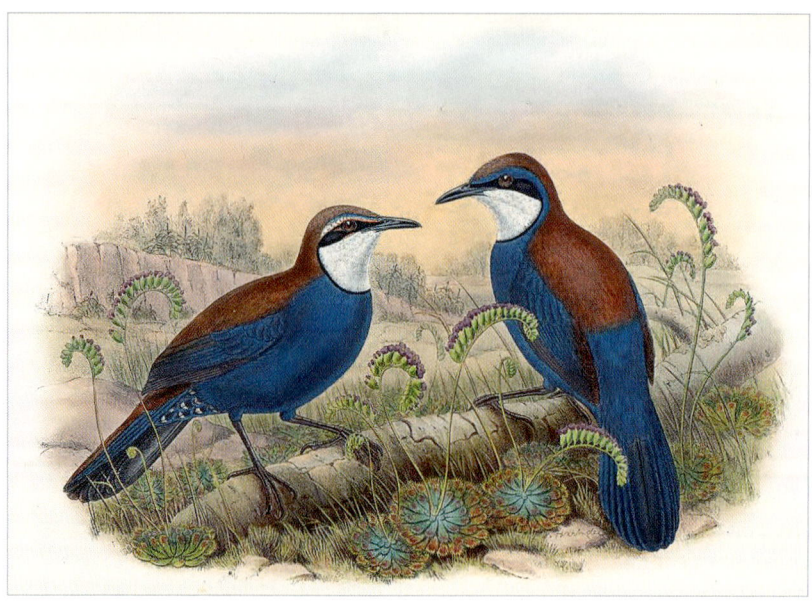

*Chestnut-backed Jewel-babbler in Gould's* The Birds of New Guinea and the Adjacent Papuan Islands *(1875)*

# Jewelfront

The jewelfront is a single species of hummingbird, Gould's Jewelfront, found in the Amazon. Needless to say, given the name, this is a little forest gem with glittering green plumage topped with a sparkling purple forecrown and a bright rufous chest band. Gould was the first to describe this species, in 1846. He then went on to give it the English name "Banded Ruby"; by the early 1900s it was known as the "Gould's Ruby."[232] The current name appears to have been adopted in the 1970s.[233]

# Junco

The juncos are five species of New World sparrows, the Passerelidae. The provenance of the name is complex. In 1544, Turner reported that Aristotle wrote in his *Historia Animalium* from the fourth century BCE:[26]

> Lacus & fluuios petunt iunco, cinclus, & trynga, quae inter haec minora, maiuscula est, turdo enim aequiparatur: omnibus his cauda motitat. [The Junco and the Cinclus live on lakes and streams, as does the *Trynga*, which among these little birds is somewhat largest, for it equals in its size a *Turdus*; all these wag their tails.]

According to Turner's calculations, Aristotle was writing of the "junco, in English a rede sparrow . . . since I know no other little bird which sits upon the rushes and the reeds, save the Reed Sparrow of the English, I believe that kind to be the Junco."[119]

The root of the word in English is from the Latin *iunco*, meaning "reeds or rushes." In 1599, Aldrovandi wrote about the *Iunco palustri*, mentioned later in Ray and Willughby as the "Greater Reed-Sparrow."[16] By 1789 Cornelius Nozeman and Maarten Houttuyn had dubbed the Eurasian Reed Warbler (*Acrocephalus scirpaceus*) *Turdus junco* in their *Nederlandsche vogelen* (*Dutch Birds*).[420] By the mid-1800s the name was being used to refer to the juncos of the Americas when Wagler appropriated it for this new genus of New World sparrows. Turner's "rede-sparrows" are believed to be the bird we now call Reed Bunting, and with its black head and paler body, it bears a passing resemblance to the birds of the genus *Junco*. So, despite the fact that the juncos are not associated with reeds, they are saddled with the moniker. Wagler, who died at the age of thirty-two, never traveled to the Americas and so was probably not familiar with the behavior of the juncos. Somewhat amusingly, Ingersoll wrote in 1937:[9]

> With the first sign of winter storms there appear in the United States from the North small, plump "ground-birds," dark slate-color above sharply contrasted with the white of the breast. Every one used to call them simply snow-birds; but the somewhat elevated tribe of "*bird lovers*" have taught us to say junco instead—that is the generic name of the species.

# Junglefowl

The junglefowl are four species of ground-dwelling game birds of the genus *Gallus* in the Phasianidae. The Red Junglefowl is the ancestor of arguably the most famous bird in the world, the domestic chicken. The name itself is a compound—two words combined to make one. The origins of the word *fowl* are discussed in the Introduction. The "jungle" part of the word combo is also interesting. It clearly refers to the bird's preferred habitat, but where did the word come from? It came into English in the 1700s, borrowed from the Hindi जंगल and Urdu جنگل (*jaṅgal*), and ultimately from the Sanskrit जङ्गल (*jaṅgala*), meaning "arid, sterile, desert."[421] Sanskrit is the classical language of South Asia and could be said to be the equivalent of Latin and Ancient Greek to many Asian languages. The true meaning of the word morphed when it came into English and came to refer more to dense, tangled vegetation. The English word is even often used to refer to tropical evergreen forests, which is very far from the original meaning of the word.

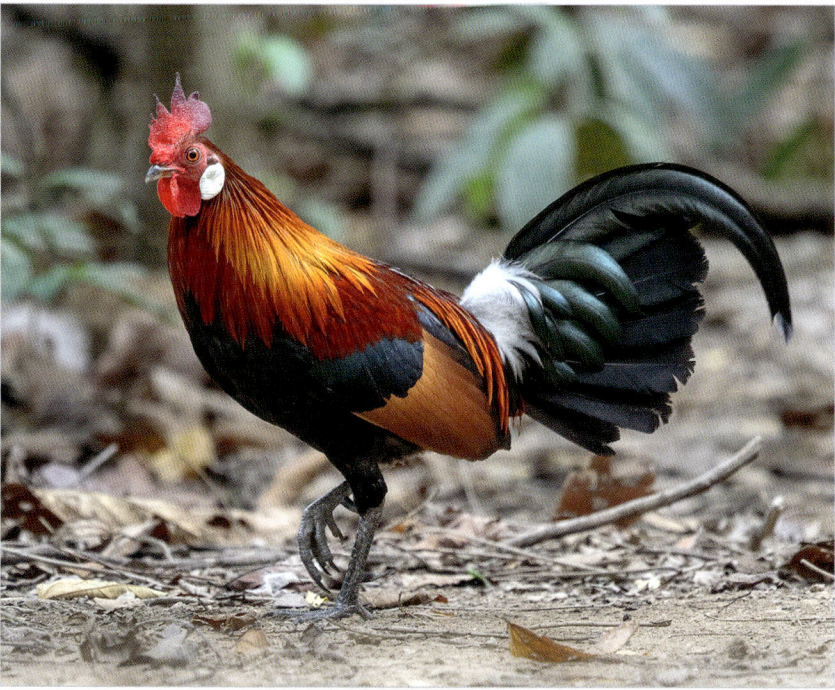

*Red Junglefowl (Vietnam)*

# K IS FOR KAGU

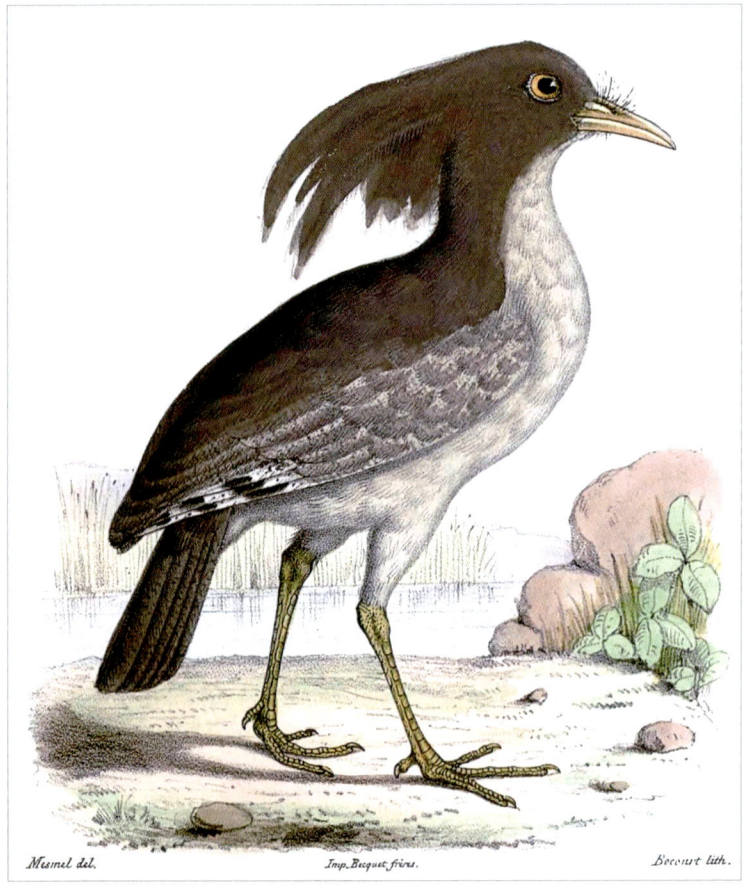

*Kagu in* Revue et Magasin de Zoologie *(1860)*

## Kagu

The Kagu is a strange, flightless bird found only on the island of New Caledonia and belonging to the monotypic family the Rhynochetidae. *Kagou* or *kavu* is a Kanak (a collective name for the New Caledonian languages) name for the bird and is said to mean "ghost of the forest," certainly in reference to its pale gray plumage. The genus was erected by Verreux in 1860 with the comment that the bird comes from[422]

> Nouvelle-Calédonie, où les indigènes le nomment Kagu. [New Caledonia, where the natives call it Kagu.]

## Kaka

The raucous parrot known as a New Zealand Kaka and its now extinct congener, the Norfolk Island Kaka, get their names from the Māori-language name for the two species of *kākā*, which means "parrot" and is an emphatic repetition of the word *kā*, "to screech," and an accurate representation of the bird's rowdy vocalizations.[165] The Kaka plays an important role in Māori tradition and is said to possess strong *mauri*, a life force through which *mana*, spiritual power, flows between objects and living things.[165] Buller wrote in 1873:[74]

> With the earliest streaks of dawn, and while the underwoods are still wrapped in darkness, the wild cry of this bird breaks upon the ear with a strange effect. It is the sound that wakes the weary traveller encamped in the bush; and the announcement of his ever active Māori attendant "*Kua tangi te Kaka*, [the Kākā has cried]" is an intimation that it is time to be astir.

# Kakapo

The amazing Kakapo is a nocturnal parrot found only on New Zealand. The Māori word for the bird, *kākāpō*, literally translates as "night parrot."[165] (See Kaka.) Buller seemed to be determined to link the birds, formerly known as the owl-parrots, phylogenetically with the Strigidae:[74]

> This is one of the very remarkable forms peculiar to New Zealand, and has been appropriately termed an Owl Parrot. As its name *Strigops* indicates, its face bears a resemblance to that of an Owl; and our knowledge of the structure and habits of the bird would seem to prove that it supplies in the grand scheme of nature the connecting link between the Owls and Parrots . . . Although exclusively a vegetable-eater, its habit of hiding during the day in holes of trees and dark burrows exhibits a further point of resemblance to the nocturnal birds of prey. As these latter are in reality night Hawks, so is this bird, what the native name, Kakapo, implies, a night connecting link between the *Accipitres* and *Psittaci*.

# Kakawahie

This extinct Hawaiian fringillid honeycreeper got its English name from the Hawaiian-language name for the bird, Kākāwahie. In his 1890 book *Aves Hawaiienses: the Birds of the Sandwich Islands*, Scott B. Wilson wrote:[47]

> The name applied to this bird in the Hawaiian language means fire-wood; but whether this is given to it from the note, which, as remarked above, resembles the sound of chopping wood, or from the brilliant flame-colour of its plumage, I am unable to say.

Later, authors Pukui and Elbert (1986), experts on the Hawaiian language, translated the name as "wood chopping."[234]

# Kamao

One of five Hawaiian thrushes in the genus *Myadestes*, this is another extinct Hawaiian bird, which probably vanished as recently as 1989. The English name comes from the Hawaiian name for the bird, *kāmaʻo*. (See Hawaiian Thrushes.)

# Kea

The Kea is a parrot found in the mountains of New Zealand, and the name comes from the Māori name, which is imitative of its call: *keee aaa*. The Waitaha tribe believed that *Kea*, along with the *Kāhu* (harrier) and *Ruru* (morepork), were *kaitiaki* (guardians) of their people.[235]

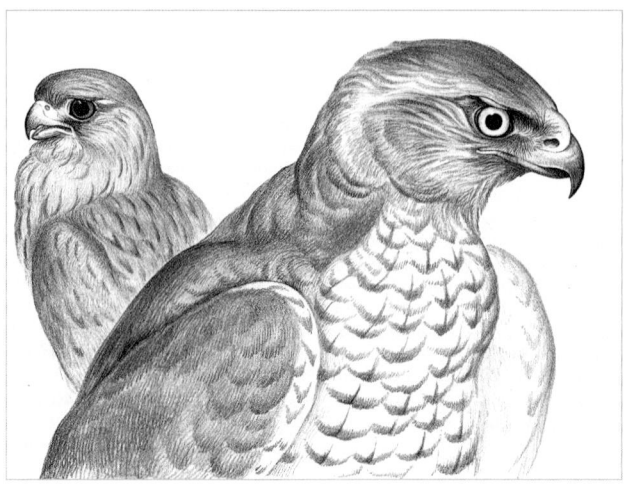

*Kestrel and Goshawk from* Birds drawn from nature *by Blackburn (1868)*

# Kestrel

There are fifteen species of kestrels, a familiar little bird of prey in many parts of the world. But the name was first applied to the Common Kestrel *Falco tinnunculus*, which was certainly known to the inhabitants of England from early times. The name came into Middle English as *castrel* or *kesterell*, from the Middle French *cresserelle* or *crecerelle*, derived from *crecelle*, "a rattle or wooden reel," referring to the vocalizations. The origin of the word *crecelle* is unclear, though. It may be from the Latin *crepitāre*, "to crackle," or perhaps from a root word *krek-*, "to crack, rattle, creak, emit a bird cry." The name in English dates back at least to the first half of the 1400s in an entry in the *Harley Manuscript* entitled "Treatises of Hawking":[381]

> A gosehawke for a pore man. A tersell for a zeman. A spawke for a prest. A muskett for an haly watyr clarke. A kesterell for a knafe. [A goshawk for a poor man, a tercel for a yeoman, a sparrowhawk for a priest, a musket for a holy water clerk, a kestrel for a knave.]

In 1871, Harting wrote in his *Ornithology of Shakespeare* that the bird name was thought[423]

> by some [to be] derived from "coystril," a knave or peasant, from being the hawk formerly used by persons of inferior rank, as we learn [in *Treatises of Hawking*]. This opinion is strengthened by the reading "coystril," in Twelfth Night . . . and "coistrel," in Pericles.

However, this association of the words was dismissed by later academics, such as Dyer:[236]

> The name kestrel, says Singer, for an inferior kind of hawk, was evidently a corruption of the French quercelle or quercerelle, and originally had no connection with coystril, though in later times they may have been confounded.

# Killdeer

The Killdeer is a widespread and much-loved little plover of the Americas. The strange name does not refer to any preternatural ability to murder deer but instead comes from a human interpretation of their shrill calls as "*kill deer.*" If that seems like a stretch, we could revert to some older names given by early naturalists who variously dubbed them Chattering Plover and Noisy Plover. Around 1730, Catesby appears to have been the first to use the name in print:[70]

These Birds are very frequent both in *Virginia* and *Carolina*; and are a great hinderance to Fowlers, by alarming the game with their screaming noise. In *Virginia* they are called *Killdeers*, from some resemblance of their noise to the sound of that word.

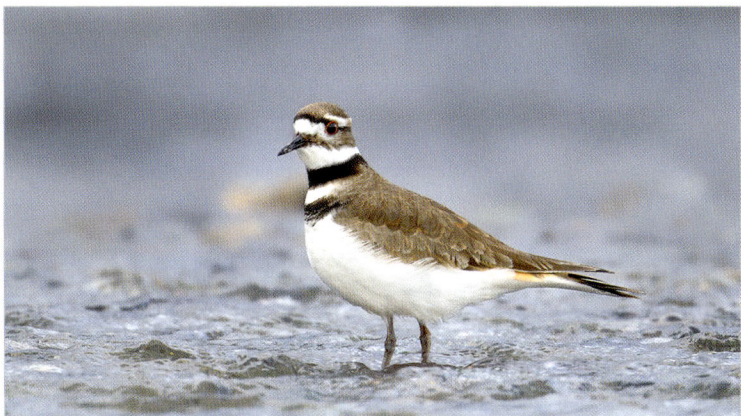

*Killdeer (Washington State, USA)*

# Kingbird

The kingbirds are some of the largest members of the tyrant flycatcher family, the Tyrannidae. They are all members of the genus *Tyrannus*, and the roots of this name provide a clue to their common name. The name for these members of the Tyrannidae relates directly to the common name "tyrant" (see that entry) and comes from the Latin *tyrannus*, meaning "lord, master, or monarch"—in other words, a king. When threatened, the kingbirds will occasionally expose a small patch of red feathers on the crown, just like a king's crown. Secondarily, the birds' aggressive habits when defending nests may account for the moniker. Vieillot wrote in 1807:[68]

> Les Américains le nomment King Bird, Oiseau Roi, soit parce que sa tête paroit comme couronnée de jaune, lorsqu'il redresse les longues plumes qui la couvrent, soit parce qu'il domine en maître absolu dans le canton qu'il habite. [The Americans call it King Bird, either because its head appears crowned with yellow, when it straightens the long feathers which cover it, or because it dominates as absolute master in the canton where it lives.]

# Kingfisher

The kingfishers are a large family, the Alcedinidae, with a worldwide distribution. The oldest use of the name is for the Common Kingfisher *Alcedo atthis*, a bird very familiar to Europeans from way back in history. According to a French folk story, the kingfisher was originally a plain gray bird that acquired its bright colors when it flew toward the sun from Noah's ark:[577]

> The kingfisher left, well before daybreak; at that moment such a strong wind rose on the waters, that in order not to be submerged in the waves, it took off towards the sky. He flew with extraordinary speed, having not used his wings for a long time. He soon arrived in the blue of the firmament where he did not hesitate to settle. From grey as it was before, its plumage turned celestial blue. Arriving at a great height, he saw the sun rising far below him; an irresistible curiosity pushed him to go and look closely at this star; the closer he approached the sun, the greater the heat;

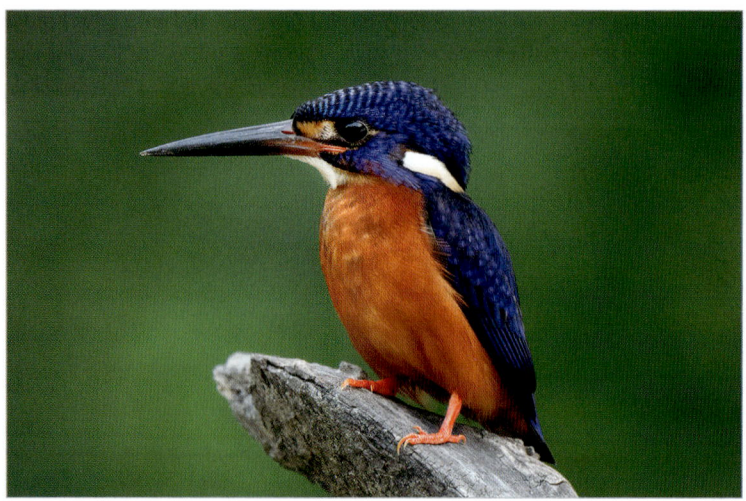

*Blue-eared Kingfisher (Sabah, Malaysia)*

soon even the feathers of his belly began to scorch and catch fire. He abandoned his task and rushed back to dive in the waters that covered the earth. After having plunged several times into the refreshing wave, he remembered his mission, but no matter how much he looked, the ark had disappeared. Indeed, during the absence of the kingfisher, the dove had returned with an oak branch, and the ark pushed by a great wind that God had sent, had touched down, and Noah, leaving the ark, had demolished it to make a house and stables. The kingfisher, no longer seeing anything on the waters, began to utter sharp cries to call Noah. Even today, we see him looking along the banks, searching for the ark. He has kept to this day on the upper part of his body the blue sky plumage that he acquired in the firmament, and his belly is still quite scorched because of his recklessness approach to the sun.

Although the second part of the name is a clear reference to the bird's feeding methods, the relationship to kings is less clear. The name appeared in 1440 in *The Promptorium Parvulorum* with the entry:[261]

Kyngis fyschare, bryde.

Then in 1658 Phillips wrote in *New World of English Words*:[424]

King's-fisher, a Bird so call'd, because it feeds upon Fish, and has Blue Feathers resembling a King's Purple Robe.

In early publications such as *Genera of Birds* by Thomas Pennant in 178[281] and *A General Synopsis of Birds* by John Latham in 1782,[218] the authors referred to the birds as the "King's Fisher," as did Carl von Linné in his *Systema Naturæ*.[223] Lockwood believed it came from a local French name for the kingfisher, *roi pêcheur*, formed from the Old French *Roi Pêcheur*, the Fisher King of Arthurian legend.[84] Joannes Richter proposed a theory for the bird's association with kings.[237] The tombs of the kings of England in the 1100s were bedecked in blue and orange, which, coincidentally or otherwise, correspond to the colors of the Common Kingfisher. As Richter noted:

The word "King" has been applied in nature for species deemed remarkably big or dominant (e.g. king crab, 1690s). The bird's colours however reveal a remarkable correlation to the tomb of the Plantagenets at the abbey Fontevraud in France.

Later, in the 1300s, English nationalism was on the rise following victories in France. Richter postulated that this may have prompted the promotion of the Norse word *kungsfiskar*, which was used in the Norman language, as a symbol of royalty having been reminded of the colors of the tombs of Henry II and Richard I. So, in a nutshell, the bird probably came to symbolize royal power, based on the colors red and blue.

# Kinglet

Among the six species of the Regulidae, two bear the common name kinglet. With their brightly colored crowns and small size, they are like little kings,—in other words, kinglets, as the suffix *-let* is a diminutive. Interestingly, the genus name *Regulus* (and by extension the family name Regulidae) also translates as "little king."

There is also a fable that could explain the name kinglet, but this is clouded by confusion between the names kinglet and wren (the Ancient Greek names for both birds were very similar, and many didn't differentiate the two). The story is known as "The Fable of the King of Birds." In brief, it goes something like this:

> All the birds of the world gathered together to decide who would be crowned the King of Birds and it was decided that whoever could fly the highest would receive the crown. The birds set out to achieve their goal and naturally the Eagle reached the highest heights but at the very last minute, the Wren jumped from the Eagle's back and claimed the crown. The other large birds were angry but the Wren retorted "The Eagle would have won through strength and brawn, why is that better than cunning and intelligence? If you have your doubts name another challenge and I will win once more."

This legend is very old, having been told even by Pliny the Elder in *Naturalis Historia, Book X583* from AD 73. Given the mention of a gold crown on the "wren," it seems that the legend stems from a reference to the Goldcrest (a kinglet) rather than to Winter Wren, as Hugo Suolahti discussed in his 1909 book *Die Deutschen Vogelnamen*.[194] (For details see Goldcrest.)

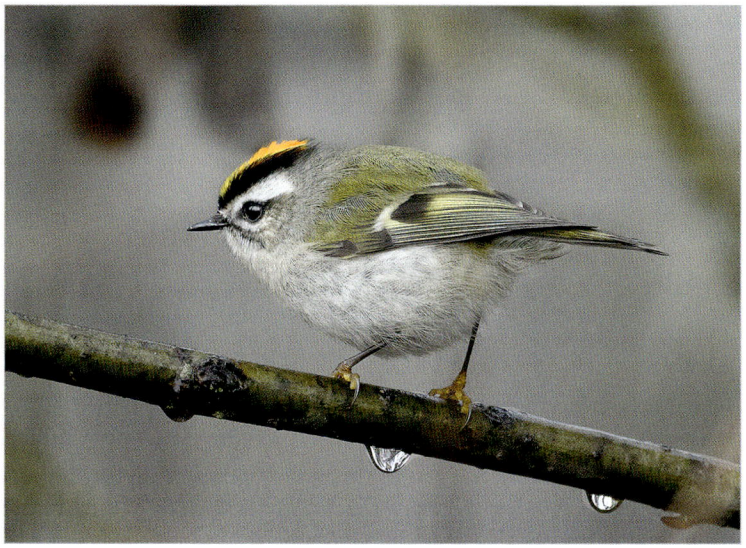

*Golden-crowned Kinglet (Washington State, USA)*

# Kioea

The name of this extinct species of Hawaiian honeyeater, the largest of its kind on the islands, is said to mean "to stand tall." Supporting this thesis, Perkins discussed the name in *Fauna Hawaiiensis* in 1913:[244]

> The native name of *Chaetoptila* appears to have been the same as that applied to the bristle-thighed curlew (*Numenius tahitiensis*) but there is no reason to doubt the correctness of this name, which is applicable enough to either species.

Earlier, he stated:

> Onomatopoeic names . . . may have an applicable meaning as well as imitating the cry of the bird, e.g. Kioea (*Numenius*), which is onomatopoeic and at the same time refers to the height at which the bird stands from the ground.

# Kiskadee

The two kiskadees are large Neotropical flycatchers of the Tyrannidae. The name is onomatopoeic, from the Great Kiskadee's call. In a paper entitled "Notes on the Raptorial Birds of British Guiana" from 1853, Dr. G. R. Bonyan wrote:[319]

> The Kiskadee, a tyrant shrike, is the little champion who thrashes the Snake-eater. Sometimes two or three of these birds–will be seen, always keeping above it, pecking the Hawk most unmercifully, and they seldom fail in bringing it to the ground, when the sight of its powerful talons I presume, reminding them that the better part of valour is discretion, causes them to fly off to some neighbouring tree and set up a glorious "*lo Pæn*" of *Kiskadee, Kis-kis-kiskadee* over their victory.

# Kite

There are two groups of kites in the Accipitridae, the small black, gray, and white *elanine* kites and the more robust *pernine* kites. Altogether, twenty-one species are known colloquially as kites. The name first appeared in English as the Anglo-Saxon word *cȳta*, which may be imitative of the bird's cry. It derives from the Proto-Germanic word *kūts* for "bird of prey." A very early reference can be found in a manuscript from circa 1335, *The Kildare Poems*, written in an Irish dialect of Middle English. The passage is part of a poem entitled "Satire" that portrays and satirizes religious, merchants, traders, and artisans in an Irish marketplace on Michaelmas Day:[238]

> Hail, seint Franceis wiþ þi mani foulis,
>
> [Hail Saint Francis with his many birds,]
>
> Kites and crowis, reuenes and oules,
>
> [Kites and crows, ravens and owls,]

# Kittiwake

Two delicate gulls of the Northern Hemisphere derive their names from their shrill *kittee-wa-aaake, kitte-wa-aaake* calls. In the 1778 tome *Histoire Naturelle des Oiseaux*, Buffon wrote:[8]

> The spotted seagull or *ze kutgeghef* [which evolved to "kittiwake" in English]: "In the old days," Martens said, "when we cut up the whales heads, many of these birds came crying thus near our vessel; they seemed to pronounce *kutgeghef*." This name indeed makes the type of sneezing, *keph, keph*, which various captive gulls made us hear.

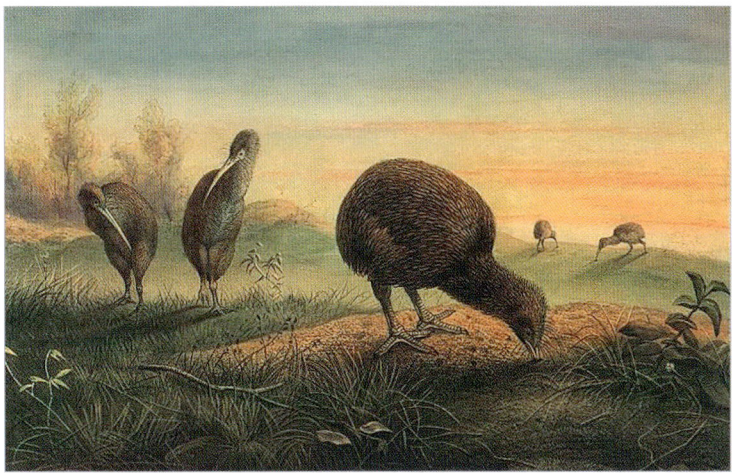

*North Island Brown Kiwi in Joseph Wolf's* Zoological Sketches *(1861)*

# Kiwi

The five species of the beloved flightless kiwis of New Zealand in the Apterygidae are often said to have derived their name from an association with their vocalizations. However, many linguists now believe that the name is derived from a Proto-Polynesian word, *kiwi*, for a shorebird that is thought to have been the Bristle-thighed Curlew *Numenius tahitiensis*. Variations of the name for the curlew occur in a number of other Polynesian languages, such as *kivi* in the Marquesas and *ivi* on Lord Howe Island. The name for the Hawaiian honeycreeper, the Iiwi, is also related.[239] Presumably the long, decurved bill of the kiwi—a case of mistaken identity by the first Polynesian arrivals on Aotearoa—is the source of this misapplication.

# Kiwikiu

The Maui Parrotbill *Pseudonestor xanthophrys* has been renamed by some authorities to the Hawaiian name Kiwikiu. The original Hawaiian name is lost in history, but this new name was created by the Hawaiian Lexicon Committee in 2010. The first element of the name, *kiwi*, means bent or curved (see Kiwi), referring to the shape of the bill. "Kiu" has a double meaning. The first, meaning "to spy, observe secretly" refers to the bird's secretive habits; the second to a strong, moderately cold northwesterly wind, referencing the bird's habitat.

# Knot

The knots are two medium-sized waders in the Scolopacidae with an unusual name. There are theories that it gets its name and specific epithet from King Cnut, the eleventh-century king of England, Norway, and Denmark. One of these theories is that the name would refer to the knot's foraging along the tide line and the story of Cnut and the tide. But in 1678's *The Ornithology of Francis Willughby of Middleton in the county of Warwick, esq.,* there is an entry for "The KNOT, that is King *Knout* or *Knute* [*Canutus*] his bird. *Cinclus Bellonii*, an *Callidrys cinerea*?"[16] And later in the 1804 *History of British Birds* under the heading "KNOT, KNUTE, OR KNOUT, (*Tringa Canutus,* Lin.—*Le Canut,* Buff.) [*sic*]" Bewick wrote:[149]

> This bird is said to have been a favourite dish with Canute, king of England; and Camden observes, that its name is derived from his—Knute, or Knout, as he was called, which, in process of time, has been changed to Knot.

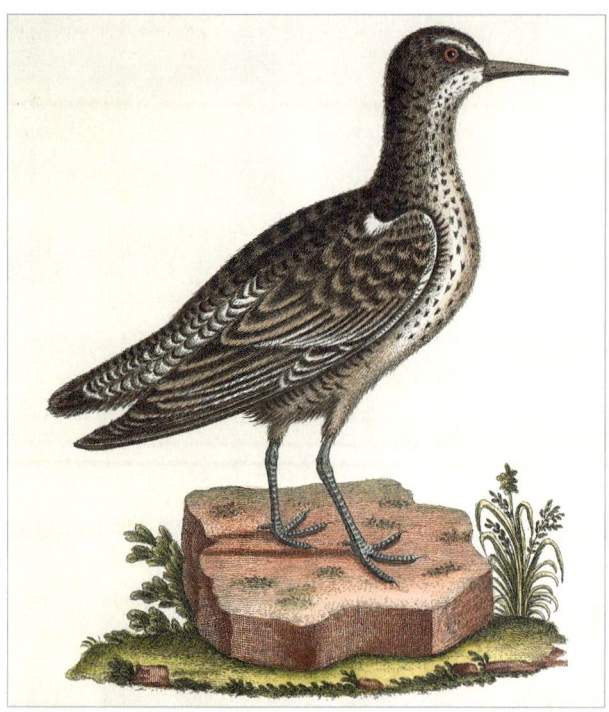

*Le Canut, the Knot, in Huth's* Recueil de divers oiseaux étrangers et peu communs *(1768)*

There appears to be no concrete historical foundation for this etymology, but this seems to be the most plausible explanation, especially in light of this verse from the seventeenth-century poem, the *Poly-Olbion*:

> The *Knot*, that called was *Canutus* Bird of old,
>
> Of that great King of *Danes*, his name that still doth hold,
>
> His apetite to please, that farre and neere was sought,
>
> For him (as some haue sayd) from *Denmarke* hither brought.

More prosaically, another etymology is that the name is onomatopoeic, based on the bird's grunting call note.

## Koa-finch

The two koa-finches are large, chunky Hawaiian endemic fringillids with robust bills. These bills are adapted for cutting through the large and hard leguminous seedpods of the endemic koa tree, *Acacia koa*. The word *koa* is a Hawaiian word for the plant and also means "warrior."

## Koel

The five species of koels are large, sexually dimorphic cuckoos found in the Australasian and Asian regions. Their characteristic *koel* calls are a feature of the soundscape of the Asian forests and woodlands. The name might seem to be obviously onomatopoeic but does actually have different roots in the languages of South Asia. The Sanskrit कोकिला *kokiláh* evolved into the Hindi words कोयल *koel* or *kuyil*. *Kokiláh* is thought to have derived from an autochthonous Dravidian or Austroasiatic language such as the Kannada *kukil*, Tamil *kuyil*, or Santali *koya*, *kuya,* meaning "black, smirched," and *kuila,* meaning "black, dark skinned, charcoal" from the Proto-Munda *ko(y)ila,* the postulated ancestral language of the Munda languages of India and Bangladesh. Hammer made a study of the etymology of the name in 2017 and noted:[335]

> This morphologically determined reference to the black glossy shining plumage of the male chanced to match the phonetics of the call rather well and the terms *Kuyil, Koel* or better *Koïl,* which are more onomatopoetically appropriate than Sanskrit *Kokiláh,* came to denote the species *Eudynamys scolopacea.*

Jerdon wrote that the name koel came from Hindi and:[136]

> The Koel is by no means a shy bird . . . when it takes wing . . . it is remarkable for its noisy cries . . . About the breeding season the Koel is very noisy, and may be then heard at all times, even during the night, frequently uttering its well-known cry of *ku-il ku-il,* increasing in vigour and intensity as it goes on. The male bird has also another note, which Blyth syllables as *Ho-whee-ho,* or *Ho-a-o,* or *Ho-y-o.*

# Kokako

The kokakos are two large wattlebirds in the endemic New Zealand family the Callaeidae. In the Māori languages, various names such as *hokako, pakara, werewere* (meaning "wattles"), and *kōkako* were used for the birds. These names are said to be mimetic of the beautiful vocalizations. Kōkako were considered sacred by many Māori so were not often eaten:

> In the Māori myth "Maui's Thirst," the superhero Maui tamed the sun so that it would move more slowly across the sky. After he had accomplished this, Maui lay exhausted and very thirsty, so he asked the birds of the forest, one by one, if they would bring him some water. Many ignored his request, but Kōkako obeyed, collecting the water in his wattles. To reward Kōkako, Maui stretched Kōkako's legs so he would have longer legs than any other bird in the forest and so be the envy of them all. Kōkako would be able to move through water and mud without getting wet. Kōkako leaps and bounds with joy to this day, in thankfulness.[501]

# Kookaburra

There are five species of kookaburras, large kingfishers in the Alcedinidae, but the best known and the namesake is the Laughing Kookaburra. The name is taken from the Wiradjuri word *guuguubarra,* of onomatopoeic origin for the bird's famous call. The Wiradjuri are the largest Aboriginal group in New South Wales. Robert Mathews, the Australian surveyor and self-taught anthropologist, was among the first to record the Aboriginal word for the "Laughing Jackass," *guguburra,* in his lengthy article "The Wiradyuri and Other Languages of New South Wales."[240] Other iterations of the name include *Goo-goor-gaga.* Initially, the English name used for the bird was Great Brown Kingfisher, and it wasn't until 1926 that the arbiters of such things, in this case, the Checklist Committee of the Royal Australasian Ornithologists Union, replaced the somewhat dull moniker with the much more interesting Laughing Kookaburra. The early settlers also called the kookaburras Laughing Jackass, and in the case of the tone-deaf Blue-winged Kookaburra, Howling Jackass.

The Aboriginal artist Michael J. Connolly related a Dreamtime story from the Kullilla of southern Queensland:[502]

> Long ago in the Dreamtime, when the animals were first on the earth which were very much bigger than they are today, there was a time when there was no sun, only a moon and stars.

One day, Dinewan the emu and Brolga the beautiful dancing bird, were out on a large plain arguing and fighting. Brolga got so angry that she ran over to Dinewan's nest and grabbed one of her large eggs, which she threw up into the sky with all her might. It landed on a heap of firewood breaking, spilling the yellow yolk, which burst into flames. This lit up the whole world below to the astonishment of all the creatures as they had only been used to the semi-darkness and were dazzled by such brightness. A good spirit who lived in the sky saw how beautiful the earth looked when it was lit up by this blaze. He thought it would be a good thing to make a fire every day; which he has done ever since. All night the good spirit and his helpers collect wood and stack it up. When the stack is nearly big enough, the good spirit sends out the morning star to let them know on earth that the fire will soon be lit.

However, the spirits found that sending out the morning star was not enough because those who slept did not see it. The spirits decided they must have a noise made at the dawn of each new day to announce the arrival of the sun, which would wake the sleepers—but what noise? Then one day the spirits heard the laughter of Goo-goor-gaga, the kookaburra ringing through the air. This was the noise the spirits were looking for. They asked Goo-goor-gaga that as the morning star faded and the day dawned, every morning would he laugh his loudest to awaken all the sleepers before sunrise. Goo-goor-gaga agreed and has done so ever since—making the air ring with his early morning laughter: goo-goor-gaga—goo-goor-gaga—goo-goor-gaga.

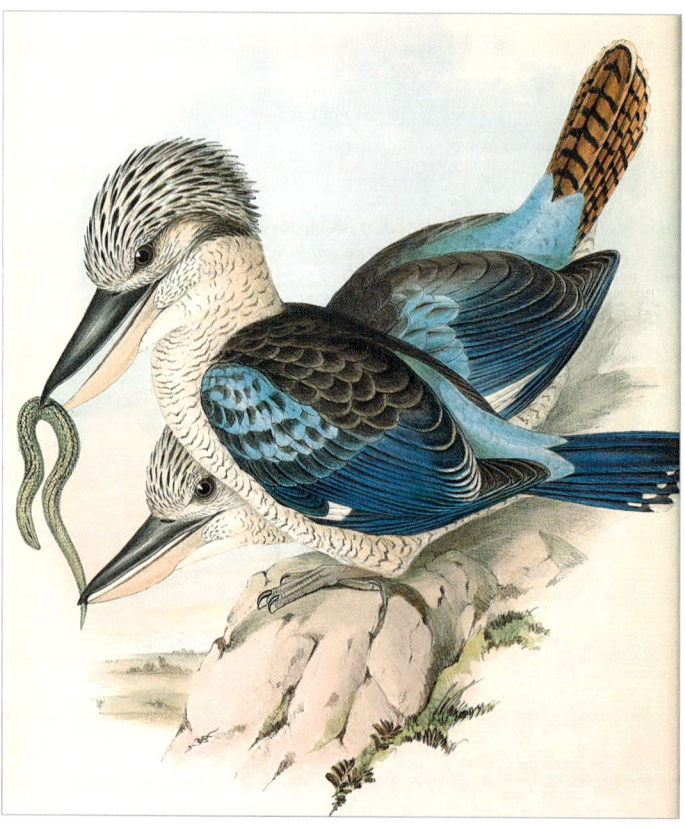

*Blue-winged Kookaburra from Gould's* Birds of Australia *(1840)*

# L IS FOR LAMMERGEIER

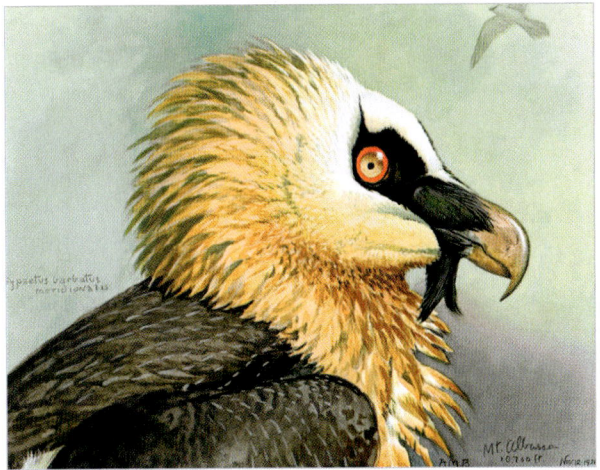

*Lammergeier in Fuertes's* Album of Abyssinian birds and mammals *(1930)*

## Lammergeier

The Lammergeier these days is usually known as the Bearded Vulture, although the older name is still widely used, so I include it here. It is a surprisingly controversial name, too. It derives from the German name *Lämmergeier*, a compound word from *Lämmer*, the plural of *lamm* ("lamb") and *geier* ("vulture"). The name stems from the belief that the bird preys on lambs. An exchange in the 2008 volume of *British Birds* illustrates the angst surrounding the name.[10] Everett related that at a meeting of the Council of Europe in the 1980s, the ornithologist Paul Geroudet voiced "strong criticism of the UK and other English-speaking countries, and the Dutch, for persisting with the use of an obsolete German name which (given the bird's parlous status and history of persecution in Europe) sent all the wrong messages." At the same meeting, "Some of them also accused the British in particular of using an exotic-sounding name they neither understood nor even thought about . . . A formal recommendation followed on the disuse of 'Lammergeier' in future publications, but nothing ever came of it." In a later issue, the Rev. Christopher Carter wrote, "It has long puzzled me that no-one seems aware of the fact that the bird has a perfectly good and accurate English name which has inexplicably been supplanted by the German one. That name is 'Ossifrage,' derived from the Latin and meaning 'bone-breaker,' accurately describing the bird's characteristic feeding habit . . . So, what about a campaign to restore the ancient English name of this splendid bird to its rightful place in the language?"

## Lancebill

Two species of lancebill hummingbirds, as the name suggests, have long, spear-like bills. The name appears to have first been used by Gould in his 1849 *A Monograph of the Trochilidæ* as the hyphenated name "lance-bill" after having been first described in 1847 by the French naturalist Bourcier, when he was consul to Ecuador.[6] Gould wrote:

The [Green-fronted Lance-Bill] is remarkable for its short and robust body, the length of its wings, and the straight, sharp, needle-like form of its bill, features in its structure which doubtless in some way modify its habits; probably the peculiar shape of the bill is especially fitted for procuring its insect food from the smaller kinds of tubular flowers, or from among the prickly spines of the Cacti, tribes of plants which abound in the country frequented by the bird.

# Lapwing

These familiar birds are twenty-four species of long-legged and noisy Charadriidae in the genus *Vanellus*. The unusual, albeit very old, name first appeared for the Northern Lapwing as the Anglo-Saxon word *hlēapewince*. One of the oldest records of the word is from the eighth century *Corpus Glossary*, which has an entry for *lepe-wince*. The name is a blendword of the Old English *hlēapan*, "to leap, run" (cf. English *leap*) and *wincan*, "to move aside, totter, or turn" (cf. English *wink*) referring to the distinctive irregular flight that lapwings employ—the eye-catching swooping and aerobatic display flight, accompanied by the *peewit* call, and perhaps also to the broken wing behavior often used to distract predators. The name evolved into the Middle English *lappewinke* in the late 1300s, and then to *lapwyngis* in the early 1400s. Skeat wrote in his *An Etymological Dictionary of the English Language* published in 1882:[220]

> The first part is *hleápe-* connected with *hleápan* to run, spring, leap. The second part literally "winker," but we assign to the verb wink its original sense. This original sense [is] to move from side to side, a sense preserved in modern German *wanken*, to totter, stagger, vacillate, reel, waver, etc. Thus the sense is "one who turns about in running or flight," which is fairly descriptive of the habit of the male bird . . . Popular etymology explains the word as "wing flapper," but *lap* does not really take the sense of flap; it means, rather, to droop, hang down loosely. This interpretation is wrong as to both parts of the A.S. form of the word, and is too general.

# Lark

The Alaudidae is a large family of ninety-nine species, all of which bear the moniker "lark" or a hyphenated form thereof. The true origin of this very old English word has been much debated. It undoubtedly descends from the Anglo-Saxon *læwerce*, with many authorities believing it stems from the Proto-Germanic *laiwarikon* or *laiwazakō*. Lockwood postulated that *laiwiz-* is the stem of an onomatopoeic word for "song" and that the suffix *-ikō* is a diminutive:[84]

> The name then literally means little song.

In Middle English, there were various forms of the name, including *lerke*, *laurche*, and *leverock*. Over time, the word *læwerce* contracted to lark. But Lockwood's premise is far from the last word on the matter. The English linguist and lexicographer John Minsheu wrote in 1617 that lark came from *leef–werck* "life work" because:[425]

> This bird flies seven sundrie times every day very high, so sings hymns and songs to the Creator, in which consists the lives worke.

This unlikely etymology never gained credence. Skeat, drawing inspiration from the Icelandic name *lævirki*, "lark" (*læ-*, "treason, deceit" and *virki*, "work"), interpreted the name as a derivation of *læw-werca*—"guile-worker":[220]

> Whence *lewjan, leiwjan*, to betray. The name points to some superstition which regarded the bird as of ill omen.

Another etymology that seems highly unlikely. The renowned etymologist Anatoly Liberman tells us that in 1846, Wilhelm Wackernagel, a German-Swiss philologist, proposed that the old forms of the word consisted of *lais*, for "furrow," and *waker*, the lark being thought to alert the ploughman that morning has arrived and work should begin.[11]

Liberman went on to state in his 2012 Oxford University Press article, "Oh, what lark!":

> Two features of the lark are especially noticeable to humans: it is an early bird (whence its association with daybreak), and its songs (trills) are loud and melodious . . . In *lai-* most researchers recognize a sound imitative complex . . . Among other things, it often refers to sound. Here we find such different words as Russian lai "barking," Engl. lullaby, Engl. ululate "howl" (from French, from Latin), Engl. hoopla (from French), and a host of others. It matters little whether the lark's call resembles la-la-la; in this situation, anything goes. Most dictionaries . . . state that, although the etymology of lark is debatable, the word is onomatopoeic.

So perhaps *lark* is simply onomatopoeic. This may be supported by the origin of the Latin for lark *alauda*, said by Pliny to come from the Celtic names *al* for "great" and *auda* for "song." Liberman concluded:

> [It has been] suggested that the hopelessly obscure word for "lark" had been borrowed from some other language. If we accept this hypothesis, the form in both Celtic [*Alauda*] and Germanic [*laiwazakon*] will emerge as an adaptation of the etymon we have no chance of finding . . . [T]he name of the lark does look like a loan from a lost source, for the etymology of Latin *alauda* is as impenetrable as that of *laiwazakon* . . . [P]erhaps we can risk the conclusion that lark is neither a Celtic nor a Germanic word . . . and that it probably contains an onomatopoeic element. This is a familiar denouement: the sought-for answer escapes us, but we seem to be closer to the truth than we were at the outset of our journey.

Lark also appears in a number of hyphenated names, including magpie-lark, torrent-lark, hoopoe-lark, and sparrow-lark.

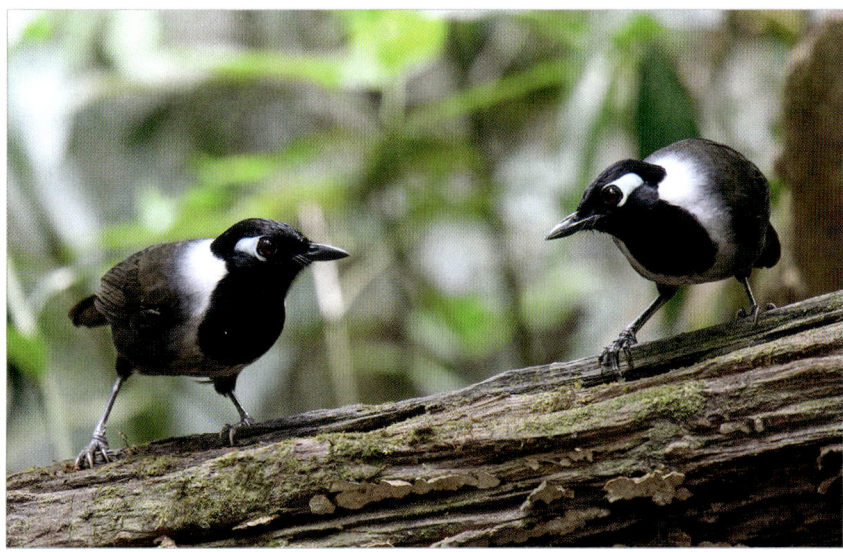

*Black-hooded Laughingthrushes (Vietnam)*

# Laughingthrush

The laughingthrushes are sixty-six species of thrush-like birds of Asia in the Leiothrichidae. This superficial resemblance to thrushes, along with the birds' lively vocalizations, explains the common names. Nearly all of the laughingthrushes were previously contained in the genus *Garrulax* (recent molecular phylogenetic studies now recognize four different genera), the derivation of which is the Latin word *garrulous*, a suitable descriptor of these animated and often noisy birds with their laughter-like utterances. As Jerdon wrote in 1844:[275]

> The cry of the "laughing thrush" is very peculiar, and once heard cannot be forgotten.

# Leafbird

A small family, the Chloropseidae are eleven species of small, leaf-green birds restricted to Asia. It's from these lovely green hues that the birds get their names.

# Leaf-love

A single species found in West-Central Africa, how did this drab bulbul come to own such an individualistic name? In 1854's *The Naturalist's Library*, Sir William Jardine discussed the bird, which he called the Climbing Leaflove, in comparison with its congener Terrestrial Brownbul.[58] The Leaf-love is endowed with the specific name *scandens*, meaning "climbing." He entered into a rather long-winded discussion of a comparison of these rasorial (terrestrial) and scansorial (climbing) congeneric birds. Clearly, at the time of writing, the Leaf-love's behavior had not been well documented, as there is no reference to any leaf loving apart from "it differs sufficiently in the structure of its feet to make us believe it is rather a scansorial than a terrestrial bird." The Terrestrial Brownbul's behavior had been described in detail by Le Vaillant, though. Jardine said, in his discussion of the morphology of the two species that the Leaf-love:

> To the scientific ornithologist, in short, this is the most interesting bird contained in our volume.

We can take it that the common name is a derivation of the word scansorial and intended to lie in contrast to the Terrestrial Brownbul, which maybe should be renamed Ground-love!

# Leaftosser

Seven species of very similar-looking Neotropical furnariids clearly like to toss leaves. These members of the genus *Sclerurus* are all secretive forest birds that forage on the ground, where they flip and toss leaves in search of their invertebrate prey. Skutch, who coined the name, wrote:[116]

> The leaftossers (*Sclerurus*) hunt on the floor of lowland forests, where with their bills they tirelessly toss the dead leaves and other litter right and left.

# Leiothrix

There are two members of the genus *Leiothrix* but only one of them, the Red-billed Leiothrix, is known by the synonymous common name. The genus name was coined by Swainson in 1832 from the Greek λεῖος *leios*, "smooth," and θρίξ *thrix*, "hair." Then in 1836 in a chapter entitled "The Family of Ampelidae, or Chatterers" Swainson wrote:[64]

> The name of *Leiothrix* will express the soft and silky texture of its plumage.

Later, in 1869's *The Museum of Natural History* it received the charming name "The Yellow Bud-hunter," but, sadly, this never gained wider usage.[65]

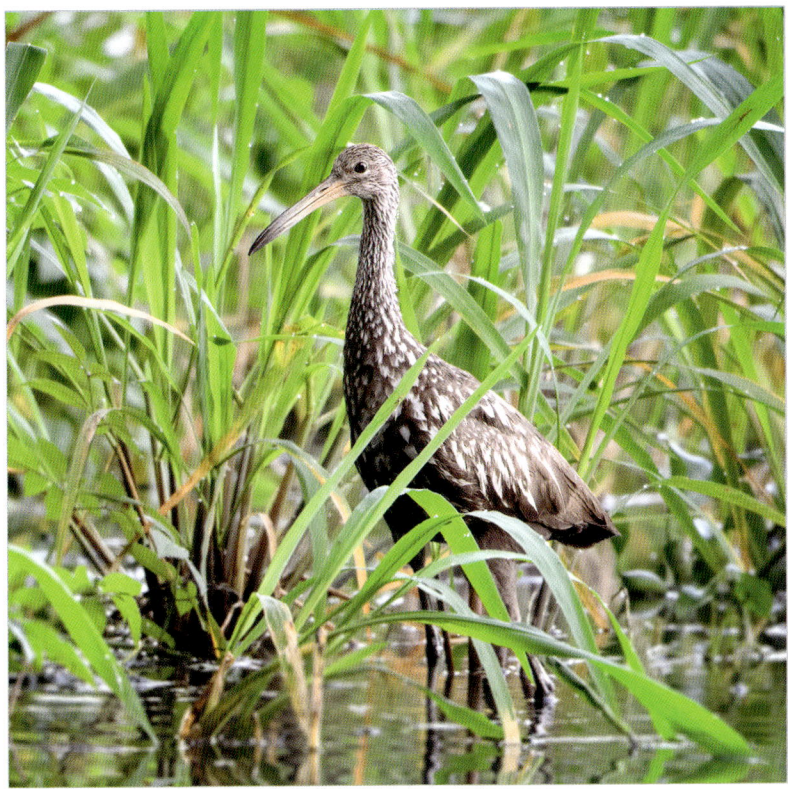

*Limpkin (Florida, USA)*

# Limpkin

The single species of Limpkin in the monotypic family the Aramidae looks like a cross between a heron and a rail with an ibis thrown in. Early on it was known by the name "Crying Bird," and according to Roosevelt, writing in 1884:[289]

> It is a bird about the size of a fish-hawk, but it roars like a lion and screeches like a wild-cat, although it occasionally whistles like a canary . . . It is conversational and talks to you in a friendly way during daytime, but at night it harrows up your soul and makes your blood run cold with the fearful noises it utters. If you hear any charming note or awful sound, any pretty song or terrifying scream, and ask a native Floridian, with pleased or trembling tongue, "What is that?" he will calmly answer, "That? that is a Limpkin."

However, the name Limpkin comes not from the vocalizations, but from its ungainly movements. As birdsoftheworld.com states, "The Limpkin often walks with what seems to be a light-footed step coordinated with movements of the neck and head, suggesting to some a limping gait."[12] Newton wrote:[7]

> A bird so called in Florida, because, though swift of foot, some of its movements resemble those of a limping man.

So, the name comes from the word *limp* with the suffix *-kin* from the Old English *cynn*, meaning "kind, sort, rank," which is often seen in affix words used to suggest a likeness or resemblance.

# Linnet

There are three species of *Linaria* finches known by the common name linnet. The Eurasian Linnet would have been a familiar sight in Britain since historic times, and the name reflects that fact. It comes from the Old French *linette*, a word consisting of the Latin *linum* for "flax" and the diminutive or associative suffix *-ette*, due to its perceived predilection for the seeds of the flax plant, although they consume the seeds of numerous plant species. The name eventually morphed into the Old English *linete*, and by the 1600s the name *linnet* was being used in publications such as *The Ornithology of Francis Willughby of Middleton in the County of Warwick, esq.* in 1678.[16]

# Liocichla

There are five species in the genus *Liocichla* belonging to the Leiothrichidae. The common and genus names are synonymous. It comes from the Ancient Greek *leios* for "smooth" and *kikhlē* for "thrush." The genus name was coined by Swinhoe in 1877 when he described Steere's Liocichla, likening it to the Leiothrix, which he spelled "Liothrix."[596]

# Lizard-Cuckoo

Four species of Caribbean cuckoos have been described rather inadequately as having "reptilian features."[1] This has nothing to do with the common name—the former genus name *Saurothera* comes from the Ancient Greek roots *sauro*, meaning "lizard," and *thera*, meaning "hunting," and, indeed, they are carnivorous birds that particularly like to eat lizards.

# Locustfinch

This is one African species of finch; the common name is reflected in the specific name *locustella*, too. Unlike the *Locustella* grasshopper-warblers, the name is not about the vocalizations, which sound nothing like the namesake insects. It almost certainly relates to their penchant for the seeds of annual grasses, as they can occur in very large flocks at certain times of the year. While this is true, the name may also relate to its similarity in appearance to the Red-billed Quelea, which is often called "Africa's feathered locust." Ironically, the birds themselves eat pest insects such as locusts.

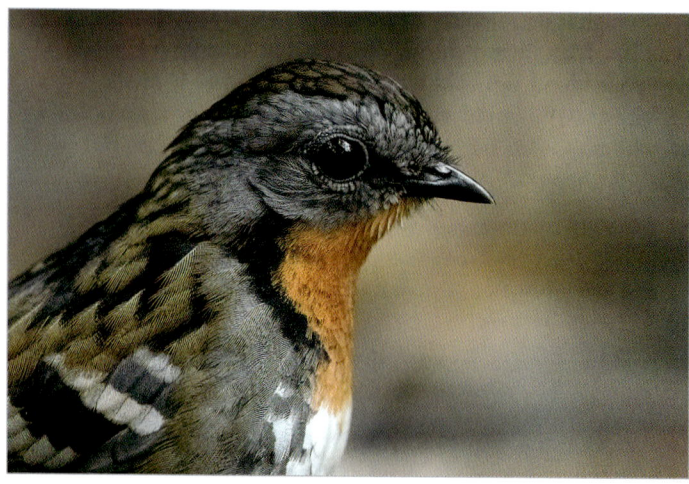

*Australian Logrunner (Queensland, Australia)*

# Logrunner

The two species of logrunners in the small Australasian family the Orthonychidae are rainforest birds that clamber around the fallen logs and low branches on the forest floor as they forage for their invertebrate prey. One of the first uses of their name is in Robert Hall's *A Key to the Birds of Australia and Tasmania with their geographical distribution in Australia* (1899), in which the bird is referred to as Spine-tailed Log-runner.[170]

# Longbill

With their long bills, the derivation of the common names of these four species in the Melanocharitidae family (berrypeckers and longbills) endemic to New Guinea is no mystery. As an aside, this is a fascinating example of convergent evolution, as the longbills bear such a striking similarity in appearance and behavior to the unrelated Asian spiderhunters in the Nectariniidae that are found just on the other side of Wallace's Line. There are also five species of longbills in the Macrosphenidae. These African warblers, in the genus *Macrosphenus,* all have long bills in comparison to other members of the family.

# Longclaw

Like all members of the Motacillidae (wagtails and pipits), the eight species of *Macronyx* longclaws all found in Africa have long claws on the hallux (the digit of the foot that faces backward). This is reflected in the genus name, with *macros* meaning "big" and *onyx* meaning "claw."

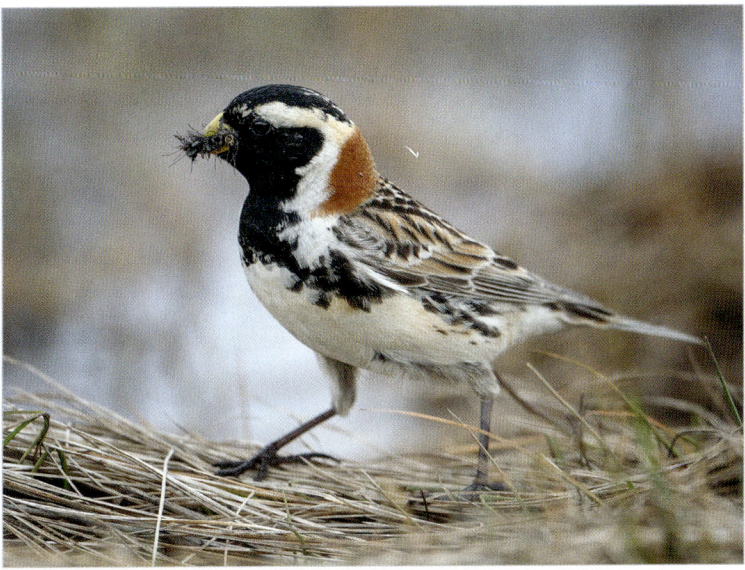

*Lapland Longspur (Alaska, USA)*

# Longspur

Like longclaws, the four species in the family Calcariidae (longspurs and Snow Bunting) also have elongated claws on the hallux. Bonaparte wrote in 1828:[597]

> From the typical Emberizae they differ remarkably by the length and straightness of their hind nail.

# Loon

Five species known as divers in Europe comprise the family Gaviidae, the loons. Most people assume the name is associated with the word "lunatic" or "loony," but for what reason? Even Newton asserted that loon is[7]

> a name applied to water-birds of three distinct Families, all remarkable for their clumsy gait on land . . . [and] is probably connected with lame. The signification of loon, a clumsy fellow, and metaphorically a simpleton, is obvious to any one who has seen the attempt of the birds to which the name is given to walk.

However, this is not a correct interpretation of the name. It actually traces its origin from the Old Norse name *lómr* for Red-throated Loon, which is not only onomatopoeic for the bird's eerie call but has the secondary meaning of "moan," according to Lockwood.[84] It came into English in the 1600s, when, as early as 1634, William Wood wrote in *New England's Prospect*:[370]

> The Loone is an ill shap'd thing like a Cormorant; but that he can neyther goe nor flye; he maketh a noise sometimes like a Sowgelders [one who spays sows] horne.

The *American Heritage Dictionary* states that the name is "Perhaps [an] alteration of dialectal *loom*, guillemot, diver, from Old Norse *lómr*." Interestingly, the Scandinavian word is also the root for the English word *lament*, which could well describe the call of the loon.

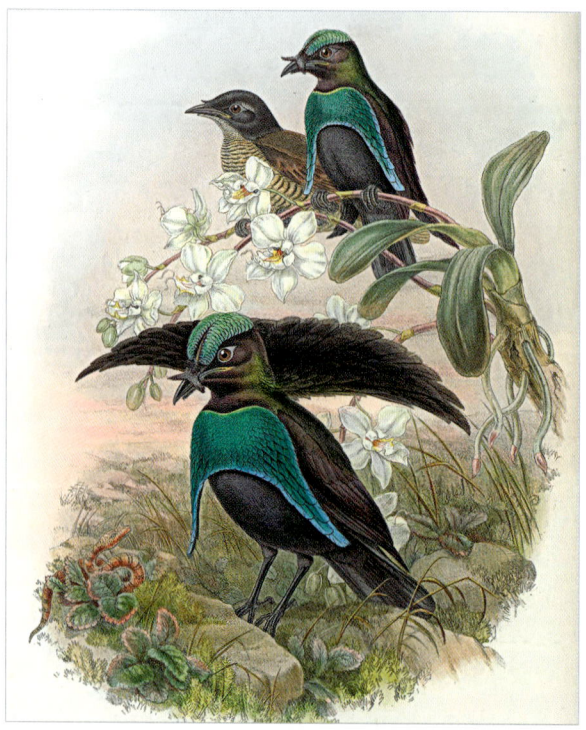

*Lophorina superba from Gould's* The Birds of New Guinea and the Adjacent Papuan Islands *(1875)*

# Lophorina

Three species of birds-of-paradise in the genus *Lophorina* go by the same common name. The derivation is the Ancient Greek *lophos*, for "crest or tuft," and *rhinos*, for "nostrils." The name points to the tuft of elongated feathers around the nostrils, especially pronounced on the male.

# Lorikeet

The parrot family, the Psittaculidae, is among the best known in the world. Of these 189 species with an Old World distribution, 41 species are known as lorikeets and are found in the Australasian region—Australia, New Guinea, and Eastern Indonesia. The lorikeets possess long, slender brush tongues adapted for collecting nectar. Lorikeet is a hybrid word, derived from two languages. The first part, *lori*, is a modification of the Indonesian word *nuri* for "parrot" in Indonesian, which in some dialects is pronounced *luri*. It came into English from the Dutch corruption of the local word. (See Lory.) The second part of the word, *-keet*, is a diminutive of French origin. (See Parakeet.)

## Lory or Lorikeet? Parrot or Parakeet?

The usage of these names is subjective and has no true scientific basis. The subtribe Loriini contains sixty-one species in nineteen genera, all with specialized brush-tipped tongues for feeding on nectar of blossoms. Generally, the lorikeets have longer, tapering tails, while the name lory is given to closely related species with blunt tails. The generic names parrot and parakeet are also used very loosely—the name parakeet is usually given to smaller parrots that have long, tapering tails.

# Lory

There are eighteen species of lories in the parrot family, the Psittaculidae. Lory is a corruption of the Malay word for "parrot," *nuri* or *luri*, which came into English via Dutch. In the fabulously titled *A Natural History of Birds: most of which have not been figur'd or describ'd, and others very little known from obscure or too brief descriptions without figures, or from figures very ill design'd* by George Edwards (1743), the following passage appears:[78]

> The first Black-capped Lory
>
> I have taken the Name *Lory* from Nieuhoff. Our Countryman, Albin, has exhibited a Bird of this Kind, which he calls a *Laurey*, and says it is from the Brasils; but I dare say he is mistaken.

Later, in 1896's *Dictionary of Birds*, Alfred Newton wrote:[7]

> The anonymous author of a *Vocabulary of the English and Malay Languages*, published at Batavia in 1879, in which the words are professedly spelt according to their pronunciation, gives it *Looree*. Button (*Hist. Nat. Ois.* vi. p 125) states that it comes from the bird's cry, which is likely enough in the case of captive examples taught to utter a sound resembling that of the name by which they are commonly called. Nieuhoff (*Voyages par mer et par terre a differents lieux des Indes*. Amsterdam: 1682–92) seems to have first made "Lory" known (*cf.* Ray, *Synops. Avium*, p. 151). Crawfurd (*Dict.Eng. and Malay Languages*, p. 127) spells it *nori* or *nuri*; and in the first of these forms it is used, says Dr. Finsch (*Die Papageien*, u. p. 732), by Pigafetta. Aldrovandus (*Ornith. lib.* xi. cap. 1) noticed a Parrot called in Java *nor*, and Clusius (*Exotica*, p. 364) has the same word. This will account for the name *noyra* or *noira* applied by the Portuguese, according to Buffon (*ut supra*, pi. 125–127); but the modern Portuguese seem to call a Parrot generally *Louro*, and in the same language that word is used as an adjective signifying bright in colour. The French write the word *Loury* (cf. Littré, *sub voce*). The Lory of colonists in South Africa is a Touracoo; and King Lory is a name applied by dealers in birds to the Australian Parrots of the genus *Aprosmictus*.

# Lovebird

The lovebirds are nine species of small parrots from Africa and Madagascar in the *Agapornis* genus in the Psittaculidae. The birds are known and loved for their strong, monogamous pair-bonding behavior, as well as their habits of perching huddled close together, hence the common name. This is reflected in the genus name, too, which combines the Ancient Greek αγάπη *agape*, meaning "love," and ὄρνις *ornis*, meaning "bird." Martin described their habits in 1835:[256]

> it is very interesting to see them dress each other's plumage, caress and fondle each other, and by various actions indicate their mutual attachment.

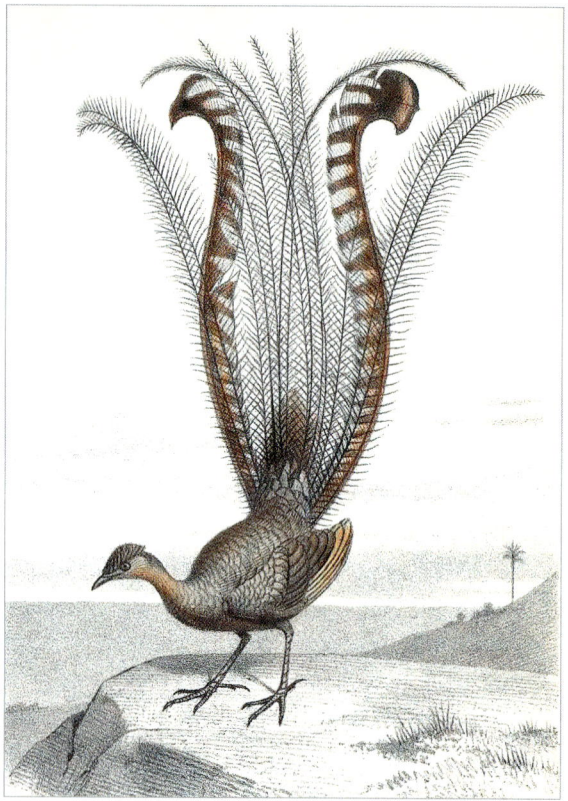

*Superb Lyrebird in Conty and Traviés's* Types du règne animal. Buffon en estampes *(1864)*

# Lyrebird

The two Australian endemic members of the Menuridae are known for their incredible vocal gymnastics; their name nevertheless derives from another remarkable feature—the shape of the spectacular tail plumes of the displaying male, which resembles the shape of the stringed instrument of Ancient Greece. An early account of the bird from 1800 is found in the Rev. Bingley's *History of Birds*:[73]

> Its chief beauty is in the plumage of its tail, which is very elegant, assuming the form of an ancient Lyre. The tail is composed of three different sorts of feathers, of which the upper side is a dark grey. . . . The tail feathers are detached entire from the bird, and are sold in the stores at quite fancy prices.

# M IS FOR MACAW

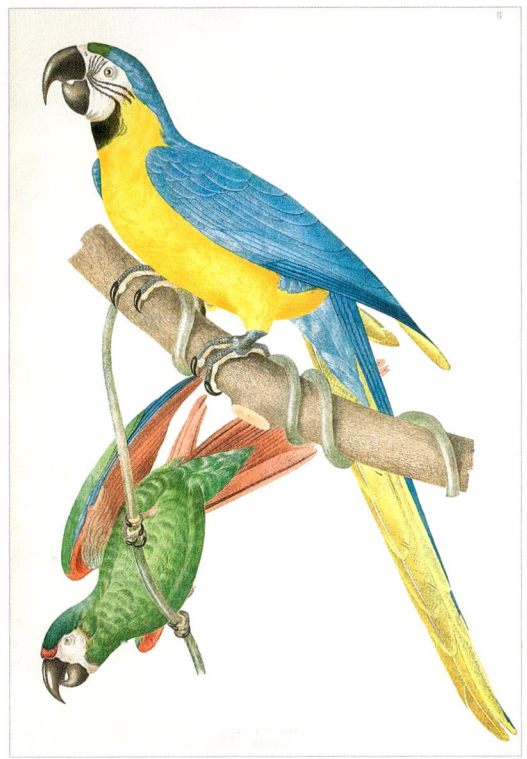

*Blue-and-yellow Macaw and Chestnut-fronted Macaw in Descourtilz's*
Ornithologie Brésilienne, ou, Histoire des oiseaux du Brésil *(1854)*

## Macaw

Of the New World and African parrots in the Psittacidae, there are eighteen species of macaws, all from the Neotropics. The original name is thought to have been a Tupian word that migrated into Portuguese as the word *macao* before its anglicization to macaw. The Tupian word may have been Tupi *macavuana*, which in turn may have been the name of a type of palm tree the fruit of which the birds eat. As early as 1678, the names appear in *The Ornithology of Francis Willughby of Middleton in the county of Warwick, esq.* as *Maccaw* or *Macao*:[16]

For so it seems the Brasilians call [them] Maccaws.

## Madanga

The Madanga is a strange little forest bird found only in the mountains of the tiny east Indonesian island of Buru. It has long been something of a mystery, but recent molecular studies have shown it to be a type of pipit in the Motacillidae, despite its nuthatch-like behavior. The genus and common names are synonymous, having been named by Rothschild and Hartert after being collected by the Pratt brothers in 1922. In *Novitates Zoologicae: a journal of zoology in connection with the Tring Museum* (1924), the type locality is given as the Madung Range on Buru with later comments:[426]

> Three specimens were shot at Wa Fehat, 2,700 feet, in April, one at the "Mada range," 5,000 feet, also in April. One at Wa Fehat, 8.iv, one at Mada Range, 9.iv.

The highest point on the island is now known as Mount Kepala Madan and also called Ghegan. So, the bird is most likely named for the location where it was first found by European naturalists. But the origin of the *-nga* part of the name is not clear. It's possible the word is an amalgamation of the words "Madan" and "Ghegan" or that it is a misunderstanding of the Indonesian suffix *-nya*. The suffix *-nya* has unique functions, but one of them is to serve as a possessive to relate to the subject, so *Madan-nya*, "belonging to Madan."

# Magpie

The true magpies are members of the Corvidae, the crows, jays, and magpies. This name is now used in many names of birds of a number of different families, especially in hyphenated forms, including green-magpie, blue-magpie, and magpie-jay (also members of the Corvidae). This is an old word in the English language, though, and the original titleholder was the familiar Eurasian Magpie, *Pica pica*. Prior to the 1600s, when Rowley wrote about being "as merry as a magge pie," the bird was known simply as *the Pie*, a derivation of the Latin name *pica* for magpie or woodpecker.[387] Although the word *pied* has come to mean black and white (from the magpie's plumage), it's thought the Proto-Indo-European root *pi-* denoted pointedness of the beak. Now, the magpie has long had a reputation in Europe for its noisy habits and curiosity, and this is how the first part of the present-day name came about. Mag is a nickname for Margaret and was often used as shorthand for a "gossipy woman." One very early reference to the "pie" occurred in "The Owl and the Nightingale," the Middle English poem from circa 1250:[416]

> *Þe faukun leuede his ibridde,*
> *& nom þat fule brid amydde,*
> *& warp hym of þan wilde bowe,*
> *Þat pie and crowe hit todrowe.*
> *[The Falcon believed her chicks,*
> *and seized that dirty chick by the middle,*
> *and threw it off that wild branch,*
> *where magpies and crows tore it to pieces.]*

# Magpie-lark

One of Australia's most familiar birds, the Magpie-lark is a member of the Monarchidae in the genus *Grallina* along with its only congener, the Torrent-lark. The first part of the name is a reference to the black and white plumage; the second part is another example of a misapplication of a name given by the colonists from a more familiar European bird. Gould called it the Pied Grallina, in his typical manner of applying the genus name to the common name, but noted that it was also known by other names, including:[40]

*Magpie Lark*, Colonists of New South Wales.
*Little Magpie*, Colonists of Swan River.
*Bÿ-yoo-göol-yee-de*, Aborigines of the lowland, and
*Dil-a-but*, Aborigines of the mountain districts of Western Australia.

# Magpie-Robin

The four species of magpie-robins in the Muscicapidae, the flycatchers, are neither magpies nor robins, although they were formerly treated as members of the Turdidae. They are both pied and robin-like. One of the earliest uses of the name in print was in Gould's *Birds of Asia* (1850–1883), but the entry was headed "Dial Bird," with an array of common names listed:[33]

> The Little Indian Pye; Dial Grackle; Magpie Robin, English in Ceylon; Dayal, Beng.; Day-yur or Deyr, Hind; Polichia, Cing.; Caravy cooroori, "Charcoal Bird," Mai.; and Chuy-kam-chay, at Amoy.

Gould continued:

> Mr. Layard informs us that, in Ceylon, "this familiar household bird is called the 'Magpie-Robin' by Europeans; and the natives regard it with as much interest as we do our own red-breasted favourite, of which it is the Eastern representative . . . Dr. Buchanan adds that at Calcutta it is commonly called Doil by the Bengalese; in Persia, Dahool or Dahale, and there kept only for its song."

(See also Magpie and Robin.)

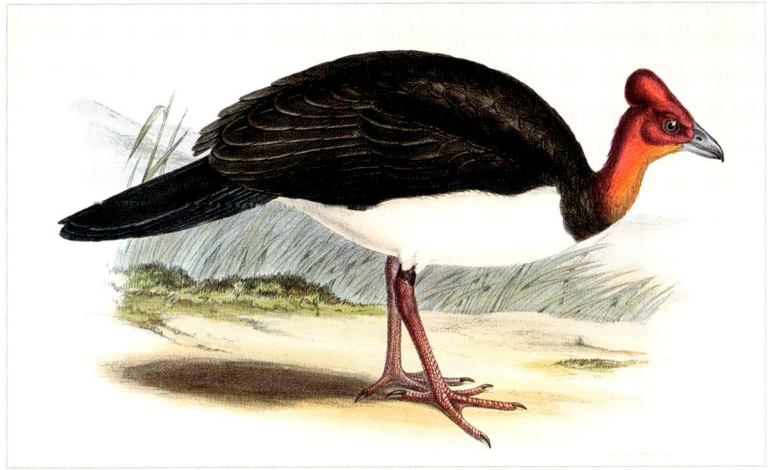

*Maleo in Gray's* The Genera of Birds *(1849)*

# Maleo

The remarkable Maleo *Macrocephalon maleo* is a chicken-like megapode found only on the Indonesian island of Sulawesi. Most authorities designate that the name was originally a Galelarese word from northern Halmahera for the Moluccan Scrubfowl but *Macrocephalon maleo* doesn't occur on Halmahera. An early written reference to maleos was in the *Verhandelingen over de natuurlijke geschiedenis der Nederlandsche overzeesche bezittingen* [*Treatises on the Natural History of Dutch Overseas Possessions*] (1839–1844):[320]

> The four known species (*Megapodius la-Perousii, Freycinetii, rufipes* and *Duperreyi*) differ little from each other in size and color; the latter involuntarily reminds, by the dull and sombre of its hues, of the forest floor, on which these birds dwell mainly, and where, among the withered leaves, they can hide themselves with the greatest security from any enemy. The Amboinese call the Megapodii *Moleoe* or *Maleo*, with the latter

name they are also called by the natives on the eastern beaches of Celebes. In Ceram they are called *Moma*, in the Philippines *Tawon*, with the Papuans, on the west coast of New Guinea, *Manoh Soewa*, near that on the north-east coast of *Manok Irio*; the inhabitants of Waigioe know them under the name of *Manok Sakkej*, and the natives of Pulau Samauw, near Timor, under the name of *Manoe Boklaka*.

This passage refers to megapodes in general, not specifically to the bird we know in English today as the Maleo, the author suggesting the word originated in the South Moluccas and the far east of Sulawesi. Müller was the first to describe *Macrocephalon maleo* from Sulawesi, then known as Celebes, in 1846, and he prefaced his description with an explanation of the word:[321]

> Characteristic of the eastern half of the archipelago are also the strange *Maleos* or big-footed fowl (*Megapodius*) . . . Their range is within the Philippine Islands, New Guinea, Timor and Celebes, i.e. in the physically distinct region in which the eastern monsoons are stormy and rainy. These chickens are distinguished from all birds by the surprising fact that they neither breed themselves nor, like the cuckoo and the cow finch, put their eggs on other birds to hatch . . . they conceal them in little mounds 3 to 5 feet high and 20 to 24 feet peripheral, consisting of sand and dry leaves, and only by means of the heat of the sun and the fermentation process, thus resembling those of the crocodiles, of the turtles and other cold-blooded animals. I myself have observed many such mounds of earth in the damp coastal forests on the west side of New Guinea, which my Amboinese companions unanimously declared to be maleo nests . . . The largest species of Maleo, the most beautiful in its feather suit and the most remarkable in its adult state due to a thick hump on the back of the head, inhabits Celebes. [Translated from German]

# Malia

This bird of the Sulawesi mountains has had a varied taxonomic life. It's been classified in the past as a babbler, as a bulbul, and now, following some recent molecular studies, it has settled in as a member of the Locustellidae (grassbirds and allies). This explains the synonymous common and genus names. It is a contraction of the family name Timaliidae, that "wastebasket taxon" which previously contained all the so-called Old World babblers in which the Malia was placed in the 1800s. Schlegl, who established the genus in 1880, wrote "The bird in question belongs evidently to the group of the Timaliae" but then went on to state:[427]

> In vain have I tried to find for the apparently unknown species a place in one of the numerous genera established in favor of the *Timalia*-group. It deviates from all of them in its general appearance, by its coloring, and by other modifications in the bill, wings and legs.

# Malimbe

The malimbes are a group of ten species of weavers in the Ploceidae, all found in Africa. The first of the birds that now bear this name was apparently collected in Malimba or Malimbe, in the country then known as Portuguese Congo. This is present-day Cabinda Province, an exclave of Angola, located within the Republic of Congo. The first reference to a bird with the name malimbe attached to it is in the *Annales du Museum National d'Histoire Naturelle* in 1802, by F. M. Daudin:[428]

> The Malimbe Tanager deserves the attention of naturalists, not only because it serves to prove that birds of this genus do not all live in America, but also because

it has a plumage whose colors, although similar in both sexes, are not arranged in the same way.

> *Tanaga malimbica*:
> T. nigra; comb, the face, throat, neck, and chest scarlet (male).
> Head not crested; above top, nape and collar scarlet (female).
> This is the bird now known as Crested Malimbe *Malimbus malimbicus*.

Apparently, Daudin believed these birds new to Western science were tanagers, presumably owing to their red and black coloration. Later in *Histoire Naturelle des plus Beaux Oiseaux Chanteurs de la Zone Torride* (1805),[503] Vieillot, the naturalist who raised the genus name *Malimbus*, wrote under the heading of "The Natural History of Malimbes. Le Malimbe, Buffon, Sonnini edition. *Malimbus cristatus*":

> This beautiful bird is found in the kingdom of Congo, located on the west coast of Africa; but he only lives in the breeding season in the country of Malimbe, whose name M. Sonnini has imposed on it. . . . We owe to the zeal of this enlightened naturalist, the only one who has been in Malimbe, the knowledge of the only three species that make up this small family, as well as of a large number of others from the same country.

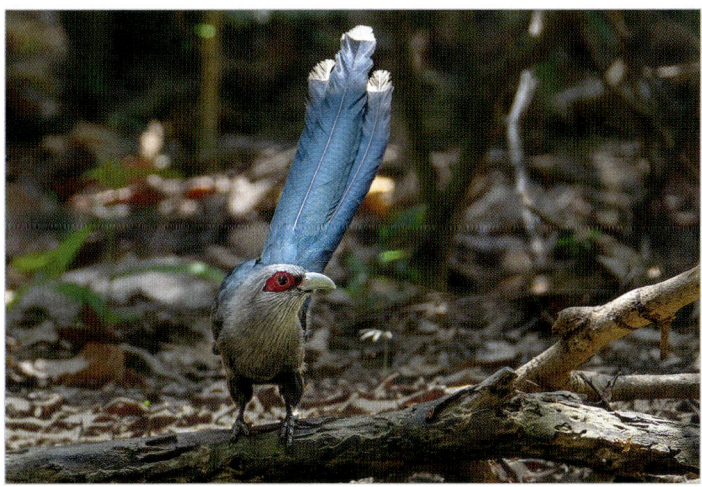

*Green-billed Malkoha (Vietnam)*

# Malkoha

The malkohas are a diverse and spectacular group of fourteen nonparasitic cuckoos (order Cuculidae) belonging to seven genera. The name is borrowed from the Sinhalese (Sri Lankan) name මල්කොහා *Mal koha* and the Nepali name is मालकौवा *mālakauvā*.[2] The Sinhala word translates as "flower cuckoo." The name probably came to English in the eighteenth century as the first printed use of the word appears in *Indian Zoology* by Thomas Pennant in 1769:[35]

CUCULUS PYRRHOCEPHALUS.
The red-headed Cuckoo.
The Cingalese [Sinhalese] give this species the name of *Malkoha*: it inhabits the woods, and lives on fruits.

This is a reference to the Sri Lankan endemic now known as Red-faced Malkoha *Phaenicophaeus pyrrhocephalus*.

# Mallard

One of the world's most familiar birds, the Mallard is also one of a minority of birds with single-word names. The name dates back to the 1300s in Britain, arriving in the islands from the Old French for a male duck, a drake—*mallart*. Originally, the name would have been applied to any wild duck, particularly drakes. The root of the French word may be the Latin for male, *masculus*, or from the Old French *malle* ("male") and the suffix *-ard* (in the sense of a specific condition). However, Salerne hinted the name is derived from a French proper name when he wrote in French in 1767's *L'Histoire Naturelle*:[382]

> It appears . . . that the name of Maillard, which is a proper name of man, was given to a domestic duck (because it is also given to the wild duck & to the tame duck), like that to Margot was given to a Magpie, and that Henry to a Donkey. If this is so, the English will have taken it from us; as they call the Wild Duck the Mallard or Mallart.

# Malleefowl

Unique among the megapodes, the Malleefowl's habitat is arid woodland and scrub. This sole member of the genus *Leipoa* inhabits a plant community of southern Australia known as "the Mallee." The word "mallee," also used for several species of multistemmed, low-growing eucalypts, is thought to originate from *mali*, in Wemba Wemba, an Aboriginal Australian language group from northwestern Victoria and southwestern New South Wales.[358] Earlier names included Mallee Pheasant, Native Pheasant, and Lowan, from the Wemba Wemba name for the bird, *Lawan*. Hall wrote about the names of the Superb Lyrebird and the Malleefowl in 1900:[359]

> While Victoria . . . has two species of so-called "Native Pheasants," one the "Mallee Pheasant," and the other the "Gippsland Pheasant." Neither belongs to the true pheasants, being known as such in the vernacular only.

This illustrated well the propensity of the early colonists of Australia to assign the names of familiar European birds to the unrelated Australian birds that seemed to the uneducated eye to be similar. Arguably, the earliest use of the name "Mallee Fowl" is in an 1893 volume of *The Victorian Naturalist* in a report entitled "A Decade in Australian Oology."[360] (See Scrubfowl.)

# Mamo

The mamos are two extinct Hawaiian honeycreepers that derive their name from the very fine cloaks of feathers worn by Hawaiian royalty. In *The Hawaiian Archipelago: Six months among the palm groves, coral reefs, & volcanoes of the Sandwich Islands* (1876), Isabella Bird related this story:[429]

> A COSTLY MANTLE.
>
> The object of our visit was to procure a *lé* of birds' feathers which they had been making for her, and for which I am sure 300 birds must have been sacrificed. It was a very beautiful as well as costly ornament,* and most ingeniously packed for travelling by being laid at full length within a slender cylinder of bamboo.
>
> * A small bird, Melithreptes Pacifica [*sic*], inhabits the mountainous regions of Hawaii, and has under each wing a single feather, one inch long, of a bright canary yellow. The birds are caught by means of a viscid substance smeared on poles. Formerly they were strictly *tabu*. It is of these feathers that the *mamo* or war-cloak of Kamehameha I., now used on state occasions by the Hawaiian kings, is composed.

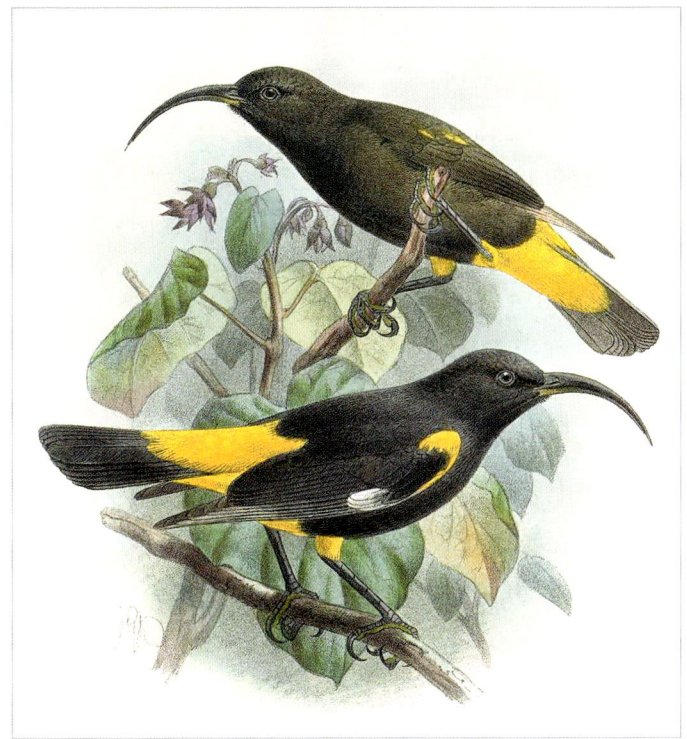

*Mamo from Rothschild's* The Avifauna of Laysan and the Neighbouring Islands *(1893)*

This priceless mantle is four feet long, eleven and a half feet wide at the bottom, and its formation occupied nine successive reigns. It is one of the costliest of royal ornaments, if the labour spent upon it is estimated, and the feathers of which it is made have been valued at a dollar and a half for five.

# Manakin

The manakins are a family of tiny frugivorous birds of the Neotropics. There are fifty-three species, all bearing the group name manakin. In *Genera of Birds* (1781) Thomas Pennant writes:[66]

Manacus (is) from the Dutch, Manakin, the name they bear in Surinam.

The Dutch word *mannekijn* means "little man" (the same word gives us the English word "mannequin"). The Brazilian ornithologist Helmut Sick suggests the word "manaquim" is derived from an Amazonian Amerindian name[14]. If this is the case, then it's possible the similarity of the words in the two different languages accounts for the derivation of the word.

# Mango

The seven species of mango humingbirds are all predominantly green, except the purple-hued Jamaican Mango. The easy assumption is that the birds are named for their colors, but this is almost certainly not the case.

A paper by Olson and Levy (2013) entitled "Eleazar Albin in Don Saltero's coffee-house in 1736: how the Jamaican mango hummingbird got its name, *Trochilus mango*" throws light on the issue:[430]

The Jamaican hummingbird that Eleazar Albin called the "Mango Bird," which was the basis for the Linnean name *Trochilus mango*, is shown likely to have been based on a specimen he saw in Don Saltero's Coffee-House in Chelsea, London, in 1736, that was probably a gift of Sir Hans Sloane. The name "mango-bird" has long been in wide use for certain south Asian orioles, especially the Indian Golden Oriole (*Oriolus kundoo*), at least one specimen and nest of which was also on display in Don Saltero's. Albin's text concerning two species of Jamaican hummingbirds contains numerous dubious or erroneous statements and his use of "Mango Bird" for the hummingbird was most likely a lapsus confounding another bird he had heard of at Don Saltero's, particularly in light of the fact that the mango tree (*Mangifera indica*) was not introduced into Jamaica until 1782. Thus, the modern use of the word "mango" in connection with an entire group of hummingbirds arose through a purely fortuitous mistake and the birds never had any association with the mango tree.

## Mannikin

The mannikins are six species of estrildid waxbill finches. The name munia has replaced the word in the case of many species, due to confusion with the manakins, which are totally unrelated. The Dutch origin in both species is the same, "little man," presumably a reference to their diminutive size. In 1883, *A List of British Birds compiled by a Committee of the British Ornithologists' Union*[504] made a reference to a Dutch name for Bearded Reedling:

the modern Dutch name "Baardmannetje = bearded mannikin."

And in 1893 a book entitled *Birds in Town and Village,* when relating an outing in the English countryside, W. H. Hudson has the following passage:[582]

I had, I imagine, a swarter [swarthy] skin and firmer flesh when I could ride all day over great summer parched plains, where there was not a bush that would have afforded shelter to a mannikin, and think that I was having a pleasant journey.

Neither the mannikins nor the munias are found anywhere near Europe, so perhaps the name mannikin was applied to any sort of small bird?

## Manucode

With their glossy black plumage, the five species of manucode birds-of-paradise are certainly crow-like in appearance. Maximilianus Transylvanus, the author of the earliest account of Magellan's circumnavigation of the world in the early 1500s, used the term *Mamuco Diata*.[431] He wrote:

They [Indonesians] call the bird *Mamuco Diata*, and they hold it in such reverence and religious esteem, that they believe that by it their kings are safe in war.

This would have been a corruption of the Indonesian *manuk dewata*, meaning "birds of the gods." The word *Manucodiatus* even appears in Latin tomes such as *Ulyssis Aldrovandi philosophi ac medici Bononiensis historiam naturalem in gymnasio Bononiensi profitentis* [*Ulysses Aldrovandi, philosophers and physicians in the Natural History Library of Bologna*] from the 1500s and refers to birds-of-paradise in general.[101] And in 1678, Chapter VII of *The Ornithology of Francis Willughby of Middleton in the county of Warwick, esq.* is headed "Of the Bird of Paradise, or Manucodiata, in general."[16] And Buffon, in 1793's *The Natural History of Birds*, wrote:[8]

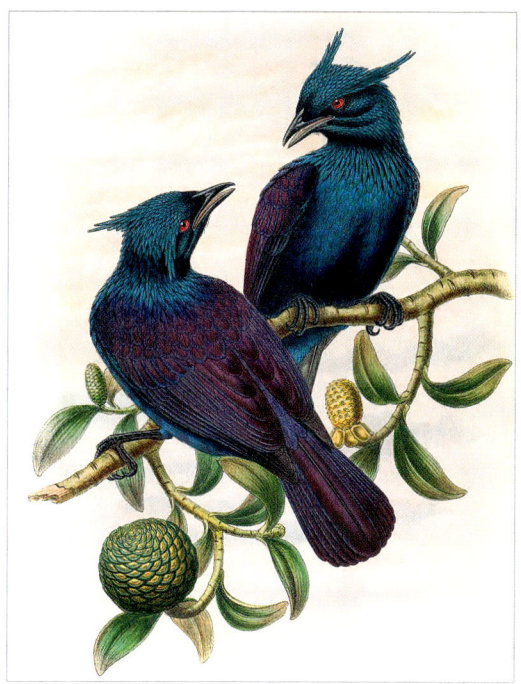

*Trumpet Manucode in Gould's* The Birds of New Guinea and the Adjacent Papuan Islands *(1875)*

The MANUCODE
I ADOPT this name from the Indian appellation *Manucodiata,* which signifies Bird of God. It is usually called the King of the Birds of Paradise; but this appellation is drawn from fabulous accounts.

It wasn't until the 1800s that the appellation bird-of-paradise was adopted and the name manucode came to be assigned to arguably the most boring representatives of the family. One of the earliest references to the manucodes as we know them today is in Gould's *The Birds of New Guinea and the adjacent Papuan Islands: including many new species recently discovered in Australia* from 1875.[505]

# Mao

The Mao is a large species of honeyeater in the genus *Gymnomyza* found on two islands in Samoa. The Samoans call it *Ma'oma'o,* the name to which it is referred in Mitchell's *Birds of Samoa* from 1909[351] and which is given in *Pratt's Grammar and Dictionary of the Samoan Language* (1911).[352] The name is almost certainly echoic of the bird's vocalizations, which are described in the Samoan government's *Recovery Plan for the Ma'oma'o or Mao*:[353]

> The bird's most remarkable feature is its range of extraordinary calls described as mechanical-sounding chips and short squeaks and its song which includes cat-like squeaky wails and cries and hoarse low notes.

# Marshbird

There are two marshbirds, brown and yellow icterids (troupials and allies) found in South America. In both species, the favored habitat is, not surprisingly, marshes and wetlands.

# Marsh-Harrier

There are four species of marsh-harriers, all in the genus *Circus*, which also contains a number of other species known simply as harriers. As the name suggests, they tend to favor marshy habitats. The name was originally adopted for the Eurasian Marsh-Harrier and later extrapolated to other similar species in the Old World. Newton noted:[7]

> One European species indeed, *C. aeruginosus*, though called in books the Marsh-Harrier, is far more commonly known in England and Ireland as the Moor-Buzzard. But Harriers are not, like Buzzards, arboreal in their habits, and always affect open country, generally, though not invariably, preferring marshy or fenny districts, for snakes and frogs form a great part of their ordinary food.

# Martin

There are twenty-six species of martins, including hyphenated forms crag-martin, river-martin, and house-martin, in the Hirundidae. The first parts of the hyphenated names are a clear reference to their preferred habitats. The legend goes that the original martin, the Common House-Martin, was named after St. Martin of Tours, who died in 397 CE. He was a "military" saint of the early Christian church who is considered one of the most important saints after he converted large swathes of Europe to Christianity. Migratory birds are used widely in Christianity as symbols of the cycle of life and rebirth and are represented in European legends by species such as the martins and swifts, the geese and swans, and shorebirds, such as the plovers. In those days it was thought that house martins and swallows hibernated during the winter in the mud at the bottom of ponds.

Martinmas, or St. Martin's Day, is celebrated in Europe on November 11, supposedly on the day he died. Some dictionaries maintain that this happens to be about the time that the migratory martins depart for more southerly locations, but this is incorrect; any martin in Europe or Britain would be a major rarity, as by then they are well south of the Sahara. However, there is another more ancient midsummer feast, Martinus aestivus, celebrated on July 4.

Given that St. Martin brought Christianity, with its promise of renewal not in this world but another one, to the Pagan cultures of north Europe, his name seems like a fitting one for a bird that could be said to represent the soul flying from the body after death to be renewed in a far-off place before returning again in a new incarnation. As noted by *The Atlantic Religion*:[383]

> Such empirical and spiritual symbolism pervades the legends and folklore of Celtish and related European cultures, and survives the Christian era—testament to the power of the [Pagan] old and mysterious worldview.

Some authorities consider the nickname to be purely arbitrary. Neither theory has much support, but the former is certainly more interesting.

# Meadowlark

Although they are not larks, the seven species of icterids with this name do, indeed, dwell in meadows. (See Lark for the explanation of that name.) In his *The Natural History of Birds* (1793), Buffon refers to an English *Alauda* lark as a "meadow-lark."[8] By 1828, Alexander Wilson et al. wrote about the New World bird in *American Ornithology; or The Natural history of the birds of the United States*:[98]

ALAUDA MAGNA MEADOW LARK.

Though this well-known species cannot boast of the powers of song which distinguish that "harbinger of day," the Sky Lark of Europe, yet in richness of plumage, as well as in sweetness of voice (as far as his few notes extend), he stands eminently its superior.

In a later volume, he refers to the same bird as "Sturnella Ludoviciana" [*sic*]. And in *The Birds of America, from drawings made in the United States and their territories* by John James Audubon (1844), the genus was labeled the "Meadow Starling" with the alternative name "Meadow Lark."[54] But, at least by the late 1800s, *Sturnella magna* was referred to variously as "Fieldlark" or "Meadowlark" (*Key to North American Birds* by Elliott Coues, 1903).[375]

## Melampitta

The Melampittidae is a very small family with only two species. The genus and common names of these little black, ground-dwelling birds are synonymous. The genus name was assigned by the German ornithologist Hermann Schlegl in 1871 from the Ancient Greek *melas-*, for "black," with the genus name "*Pitta*." As John Gould wrote in *The Birds of New Guinea and the Adjacent Papuan Islands* (1875–1888) under the heading of "MELAMPITTA LUGUBRIS, Black Ground-Thrush":[505]

> What are the natural affinities of this most curious bird? is a question which will exercise the ingenuity of ornithologists for some time to come. The generic appellation *Melampitta*, or "Black Ground-Thrush," bestowed upon it by Professor Schlegel, shows that by that eminent ornithologist the bird was evidently considered a near ally of the genus *Pitta*; and this is the position which I myself would assign to it.

(See also Pitta.)

## Melidectes

The chunky melidectes honeyeaters are all members of the synonymous genus in the Meliphagidae. The genus name derives from the Ancient Greek μέλι *meli*, meaning "honey," and δηκτής *dektes*, meaning "beggar or receiver," as explained by Sclater in the paper from 1873 that introduced the genus, with a description of the Ornate Melidectes.[278]

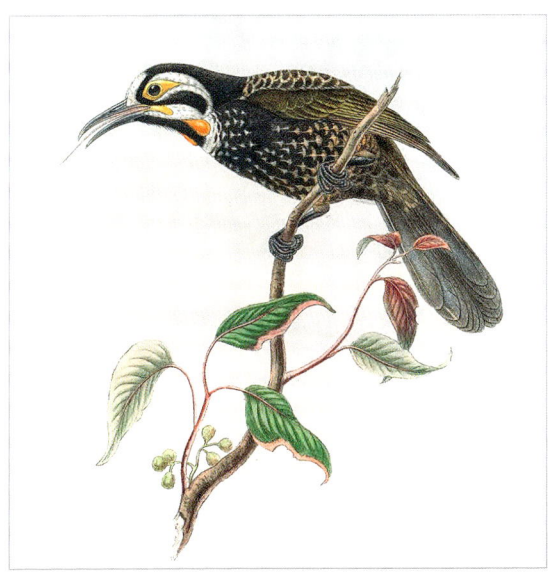

*Vogelkop Melidectes from Gould's* The Birds of New Guinea and the Adjacent Papuan Islands *(1875)*

## Meliphaga

A relatively large genus of fifteen species, the *Meliphaga* are sometimes referred to with the synonymous common name. The name is from the Ancient Greek *meli*, for "honey," and *phagos*, meaning "to eat," which is to say, they are honeyeaters. The name was assigned by John Lewin, the English-born artist and collector, in 1808.

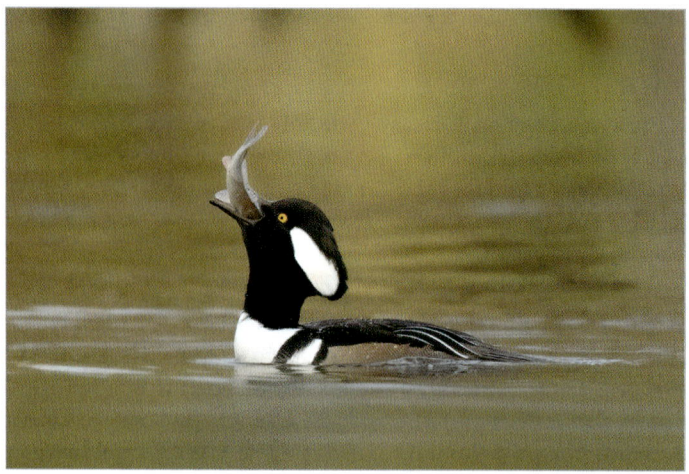

*Hooded Merganser (Washington State, USA)*

## Merganser

There are six species of mergansers, five of them in the genus *Mergus* and one in the *Lophodytes*. The name is an amalgam of the two Latin words *mergus* (an unidentified type of waterbird mentioned by Pliny, the name probably derived from the Latin *mergo*: to immerse, to plunge into water) and *anser*, meaning, respectively, "diver" and "goose."[26] The word *mergus* was in use from the time of Pliny in *Naturalis Historia* from AD77. Turner claimed it was a reference to a cormorant but this is unclear.[26] The common name merganser was used to refer to ducks in the genus *Mergus* in books dating back to the mid-1600s. In Ray and Willughby's *Ornithology* from 1678, the authors give alternative names to the Merganser as the "Gossander or Bergander," with the female being referred to as the "Dun Diver."[16] As well, a common name for *Mergus cinereus fuscus* (possibly a reference to *M. serrator*) is given as the "lesser tooth-billed Diver," while the female Smew was said to be called "*Mergus Glacialis*, which Mr. *Johnson* Englishes as the *Lough Diver*." The Common Merganser is still usually referred to as the Goosander in the United Kingdom.

## Merlin

The origin of the unique name of this small falcon is a journey from the Proto-German word *smiril*, for a falcon, through Old French *esmerillon*, to the Norman language *merilun*, and then into Middle English as variously *merlion*, *merlone*, or *merlinge*. And lastly, into English as *merlin*. When the Merlin was described by Catesby in *Natural History of Carolina, Florida and the Bahama Islands* (1729–1732), he called it the "Pigeon Hawk," reflecting the scientific name *Falco columbarius* meaning pigeon-like falcon.[70] It was first attested to in the Wycliffe's Bible, from the 1300s:

An Egle, & a griffyn, & a merlyon.

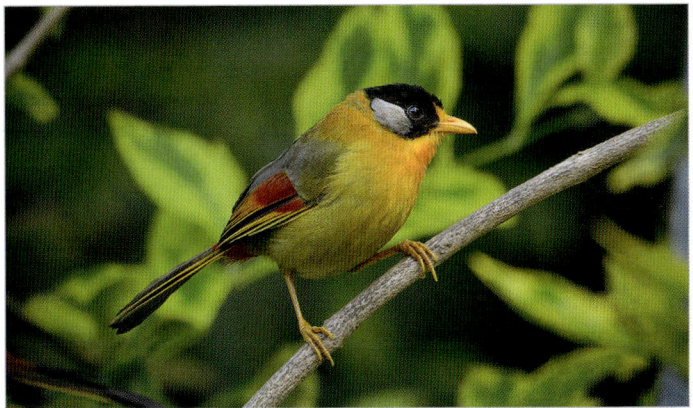

*Silver-eared Mesia (Peninsular Malaysia)*

# Mesia

The Silver-eared Mesia is a member of the Leiothichidae, the laughingthrushes and allies. The Nepali name is चाँदीकाने मिसीया *Cāṁdīkānē misīyā*. Some authors claim the word *mesia* is of Nepali origin, but a survey conducted in thirty-five districts of Nepal collecting genuine Nepali names for different birds found that not a single person was familiar with the name mesia in Nepali (pers. comm.), and it would appear the Nepali name is simply a transliteration of the English name.[2] That said, Hodgson (1841) did include *Mesia* in an article proposing the replacement of genera based on Nepali names:[322]

> Although I think the prevalent humour of the day, which cannot tolerate any other than Greek and Roman names of genera in Zoology, is, in good part, absurd and pedantic, yet as I am told that continued non-compliance therewith on my part will be considered by most persons as a sort of excuse for past and future appropriations of my discoveries in this branch of science, as described in your Journal, I have now the pleasure to transmit to you a series of classical substitutes for my previous local designations.

This suggests that Hodgson, who assigned the name *Mesia* without explanation in 1837, did base it on a Nepali name. He wanted to replace the genus name *Mesia* with *Philacalyx* (φιεος *philos*, "fond of," καλυξ *kalux*, "seedpods"). However, most authorities suggest *Mesia* is based on the Ancient Greek μέσος *mesos*, meaning "in the middle." How this concept of the middle might relate to this far-from-middling bird is unclear.

# Mesite

The Mesitornithidae are three species of mesites, endemic to Madagascar. The common name is based on the Ancient Greek word μέση *mesi*, meaning "in a middle position." In *Compléments de Buffon* in 1838, Lesson wrote about "the mesites":[432]

> They seem to form a curious type by their various points of analogy with several genera of birds; for the author [Geoffroy Saint-Hilaire] who was the first to make them known has said it to be analogous by the legs to pigeons, by the wings to most of the common gallinaceae, and by the characteristic shape of the beak and the cut of the nostrils to the finfoots. From these analogies derives the name of *mesite*, which recalls the mixed relations and the transitorial rank that the type species will have to occupy between several genera of very disparate birds.

# Metaltail

These nine species of hummingbirds are in the genus *Metallura* from the Ancient Greek *metallo* for "metal" and *ura*, "the tail." This is clearly a reference to one of the features of all members of the genus, their glittering tail plumage. The genus was established by John Gould in 1847; he never used a common name for the birds, although later in his *A Monograph of the Trochilidæ* he called them various names, including Coppery-tail, Brassy-tail, and Tyrian-tail.[6]

# Miner

There are two very different types of birds with the name miner—the eleven species of furnariids in the genus *Geositta* from the Neotropics, and the four species of *Manorina* honeyeaters from Australia. The origins of the common names differ:

*Manorina* honeyeaters—it's thought that the name "miner" is the legacy of a nineteenth-century spelling of the Hindi मैना mainā, "myna" for the *Acridotheres* starlings and the perceived similarity of the honeyeaters to them. In the 1800s there were a variety of phonetic spellings of the word, including "minah," "minor," "minar," and "miner." The Noisy Miner was the first to receive the appellation, and it may be that the black mask and yellow eyepatch and bill, along with its pugnacious behavior, reminded early observers of the Common Myna of Asia. By the time the latter spelling became the accepted form, the origin of the name was often misinterpreted as pertaining to "mining," with some authorities claiming that the clinking calls of the Bell Miners reminds the early colonists of Australia of the hammering of gold prospectors. There seems to be no evidence of this theory, however.

*Geositta* miners—the scientific name concurs with the common name. The genus name comes from the Ancient Greek *geo*, for "ground," and the genus name *Sitta*, for "nuthatch." In other words, a bird that looks like a "ground nuthatch." And the specific name *cunicularia* (of the Common Miner) is from the Latin *cuniculus*, meaning "mine or burrow." Their underground habits must have reminded the early European naturalists of the habits of human prospectors. These small South American birds of the Furnariidae ovenbird family excavate or co-opt long burrows for breeding.

# Minivet

The origin of this name for the fifteen species of *Pericrocotus* in the Campephagidae is a bit of a mystery. These slender, long-tailed forest birds are, with exceptions, red and black in color, yellow and black in the case of the females. The Latin *minium* means "red lead"—lead oxide, or minium, is the inorganic compound with the formula $Pb_3O_4$. The Dutch name Menievogel translates as "red lead bird," likewise the German name Mennigvogel. It seems likely that the English name is an anglicization of these words that reference the bird's unique plumage.

One of the first uses of the name is in the 1850 publication *Heineanum Directory of Ornithological Collection of Chief bailiff Ferdinand Heine, on Gut St. Burchard in front Halberstadt*:[63]

> Subfam. CAMPEPHAGINÆ. Raupenfresser.
> Gen. PERICROCOTUS Boie 1826. Mennigvogel.
> *Phoenicornis H. Boie* 1827. *Acis Less* 1831.

In *Birds of Asia* (1850–1883), John Gould has a number of entries for the *Pericrocotus* and gives various names, including "Hee-ah (little Gem), aborigines of Formosa," "Shah-Soki-kapir, Hindoos," "Sahelee, in the Himalayas," and "Nget-meng-tha-mee ('Princess Bird') of the Arakanese [*sic*]," but mentions the word minivet only once for *Pericrocotus cantonensis*:[33]

Of the well-defined groups of *Pericroci* or Minivets there are two very distinct sections, some species being, as in the present instance, of sober hue, whilst the others are unequalled in the brilliancy of their flame-coloured plumage. The birds must play an important part in nature, in keeping down insects and their larvae.

# Minla

The three species of minlas in the *Minla* and the *Actinodura* (one bird in the genus *Minla* has the common name sibia) are spritely members of the Leiothrichidae (laughingthrushes and allies), all found in Asia, where they all tend to favor higher altitude forests. Hodgson did include the name *Minla* in his list of "classical substitutes for [his] previous local designations," proposing the genus name *Proparus* in its stead.[322] The Nepali name for Red-tailed Minla is, in fact, मिन्ला *minla*. It is possible the name derives from the Jingpho (a language spoken in northern Myanmar and southern China) word *minla*, a "ghost," which comes from the Proto-Sino-Tibetan *m-hla*, meaning "beautiful" or "a god." But, if this is indeed the root of the word, how this relates to the bird name appears to be unrecorded. Numerous Himalayan languages use the syllable *lha* to refer to a god, a deity or a spirit.[330]

# Mistletoebird

The Mistletoebird, found in Australia and the Aru Islands, is the only flowerpecker that isn't called a flowerpecker. The original "Mistletoe-bird" was probably the Mistle Thrush, as mentioned by Newton:[7]

> MISSEL-BIRD or MISSEL-THRUSH, vulgar corruptions of Mistletoe-bird or Mistletoe-THRUSH.

The name for the flowerpecker, which, like all members of the Dicaeidae, preferentially feeds on mistletoe berries, has gone through a number of incarnations in the past. Lewin[355] called it the Crimson throat Fly-catcher, while Gould[103] called it the Swallow Dicæum and Leach[354] called it the Desmaretian Manakin, apparently named after the French zoologist Anselme Gaëtan Desmarest. The first use of the present-day name in print appears to have been in 1886, when Ramsay wrote:[356]

> This species is universally dispersed over the whole of Australia; feeds on berries and fruits of various kinds, but seems to prefer those of the *Loranthus*, of which we have in Australia so many varieties if not species, and of a *Viscum* (*V. aureum*), which is only found as a parasite on the *Loranthus*; this plainly accounts for the distribution of the *Loranthus* and *Viscum* [two species of mistletoe] all over the districts frequented by the *Dicæum*, and in which it is locally known as the Mistletoe Bird.

# Mockingbird

The sixteen species with the common name are all in the Mimidae, the mockingbirds and thrashers. Fourteen of these are in the genus *Mimus*, Latin for "mimic." The birds are renowned for their songs, which incorporate a lot of mimicry. These days, the word "mock" tends to mean to ridicule, but an alternative meaning is "to mimic or imitate." In the past the bird was often called the "Mock-Bird" as here, in Mark Catesby's *The Natural History of Carolina, Florida and the Bahama Islands*:[70]

> Hernandez justly calls it the Queen of all singing Birds. The Indians, by way of eminence or admiration, call it *Cencontlatolly*, or four hundred tongues; and we call it (though not by so elevated a name, yet very properly) the Mock-Bird, from its wonderful mocking and imitating the notes of all Birds, from the Humming Bird

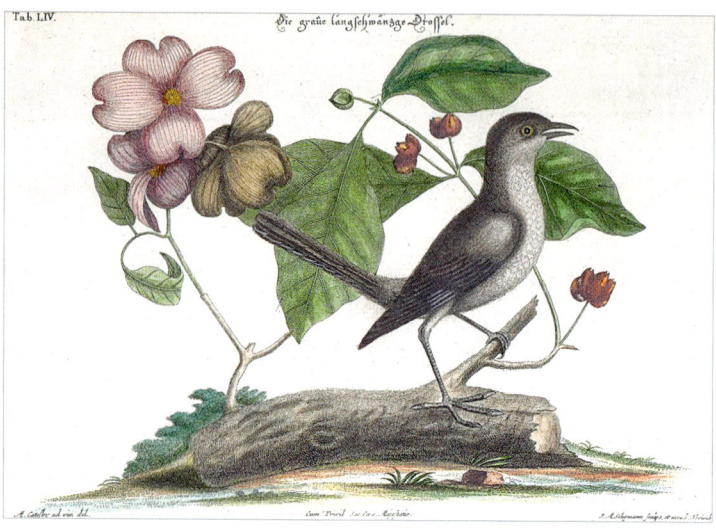

*Northern Mockingbird from Huth's* Recueil de divers oiseaux étrangers et peu communs *(1771)*

to the Eagle. From March till August it sings incessantly day and night with the greatest variety of notes; and, to compleat his compositions, borrows from the whole choir, and repeats to them their own tunes with such artful melody, that it is equally pleasing and surprizing. They may be said not only to sing but dance, by gradually raising themselves from the place where they stand, with their wings extended, and falling with their head down to the same place; then turning round, with their wings continuing spread, have many pretty antic gesticulations with their melody.

# Monal

The three monals are the members of the *Lophophorus* in the Phasianidae, the pheasants, grouse, and allies. These spectacular birds are found in the mountains of Asia. The name comes from the Hindi word मोनाल *monal*. To confuse matters, the Satyr Tragopan is actually the monal in the Nepali language, while the Himalayan Monal is called the डाँफे Ḍāṁphē. Blyth suggested that the word *monal* meant simply "fowl" and was possibly used for a number of game birds.[283] This is borne out by the fact that the names *monal* and *munal* are used for Snow Partridge, Himalayan Snowcock, Western Tragopan, and Satyr Tragopan, as well as Himalayan Monal in various regions of the subcontinent.[329] Blyth gave the Hindi names for Himalayan Monal as:

> *Monal*, or *Ghur Monal*; *Murgh i sari* ("Golden Fowl"); *Murgh Muhshor*: male, *Ratkap*; female, *Monali*.

Yule thought that the name might have its etymology in the Sanskrit word मुनि *muni*, meaning a "hermit" or a "sage," with the reasoning:[250]

> **MOONAUL**, s. Hind. *munāl* or *monāl* (it seems to be in no dictionary) . . . The *Lopophorus Impeyanus*, most splendid perhaps of all game-birds, rivalling the brilliancy of hue, and the metallic lustre of the humming-birds on the scale of the turkey . . . "In the autumnal and winter months numbers are generally collected in the same quarter of the forest, though often so widely scattered that each bird appears to be alone" (Jerdon). Can this last circumstance point to the etymology of the name as connected with Sanskrit *muni*, "an eremite"?

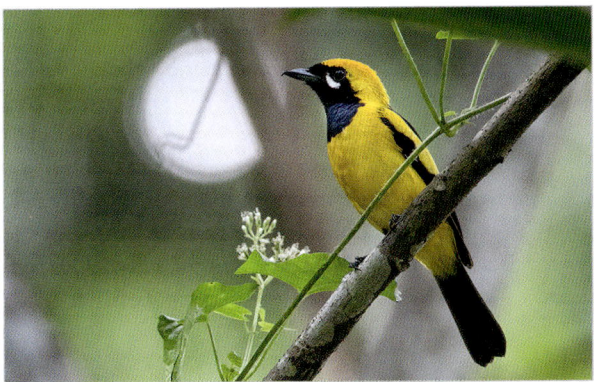

*Golden Monarch (West Papua, Indonesia)*

# Monarch

A family of lively, sallying, insectivorous birds, the Monarchidae, often erroneously called flycatchers, consists of 100 species, and 52 of these are called monarchs. The name comes from the Ancient Greek *monarkhes* (*monos*, "only" + *arkhos*, " leader") via the Latin *monarcha*, with the meaning "sole ruler or leader." When the genus name was first erected in 1827, the Black-faced Monarch was thought to be loosely allied with the Tyrranidae, hence the choice of the kingly title.[575] (See Tyrant.)

# Monjita

Seven species of *Xolmis* in the Tyrannidae (tyrant flycatchers) are named monjita, which means "little nun" in Spanish, a diminutive of *monja*. Monja or monjita seems to be a name that was used in Spanish to refer to a number of small birds with black and white plumage. In fact, one early scientific name for Blackpoll Warbler was *Monjita americana*. The name is no doubt the result of an analogy drawn between the black and white plumage of most of these birds and with the nuns' habits.

# Monklet

The Lanceolated Monklet is a small puffbird in the Bucconidae. The *-let* suffix is a diminutive, so a "small monk," and, in fact, the genus name *Micromonacha* means the same thing, from the Ancient Greek *mikros*, for "small," and *monakhos*, for a "monk." One can only surmise that the brown coloration and squat shape of the little bird reminded the early colonizers of the New World of the monks they were familiar with from Europe.

# Moorhen

There are seven species of moorhens, members of the Rallidae, the rails, gallinules, and coots. These wetland dwellers were known in Old English as *mōrhana*, from *mōr*, meaning "moor," but in Old English it also had the meaning of "marshland, bog, or swamp," and from *hen*, a "fowl." It is thought to perhaps be a corruption of *wōrhana*, meaning "pheasant." Swann wrote about the name Moorhen in 1913:[121]

> It is also commonly known as the Water Hen. Moor is from A. Sax. *mór*, and was anciently equivalent to morass or bog, the name having therefore much the same meaning as Water Hen. The name Moor Hen occurs in Merrett (1667). . . . Turner (1544) has "water hen, or Mot hen," and alludes to the bird as generally haunting "Moats which surround the houses of the great" and fish-ponds.

## Morepork

Described by Johann Friedrich Gmelin in 1788, and called by him "New Zeeland Owl," the Southern Boobook and Morepork owls were considered to be the same species until 1999.[223] But recent splits allow both common names to remain extant. Interestingly, Gould (in *Birds of Australia*, 1848) referred to the "*More-pork* of the Colonists" when discussing the Tawny Frogmouth.[40] This almost certainly arose from confusion, as the early colonists did indeed use the word, but for the Southern Boobook. The name is echoic of the bird's two-tone call, but, while Morepork fell out of use in Australia in the early 1900s in favor of the name boobook (see that entry), it had by then taken hold in New Zealand.

## Morningbird

The Morningbird is a whistler in the *Pachycephalidae*, found only on the tiny island of Palau. It owes its singular name to its loud morning songs. In a 1951 article entitled "The Avifauna of Micronesia, Its Origin, Evolution, and Distribution," Rollin H. Baker wrote:[433]

> It has a sweet song and may be considered as one of the finest singers in Micronesia. It heralds the break of day with its melodious carol, and its name is derived from its calling early in the morning. I heard the bird only infrequently in the hot part of the day, although it would sing when the skies were overcast. Its song could be heard also as evening approached.

## Morning-Thrush

The Spotted Morning-Thrush is an African member of the Muscicapidae in the genus *Cichladusa*. Though now known not to be a thrush, it was previously placed in the Turdidae. Ward explained the name in *The Avicultural Magazine* in 1973:[273]

> Spotted Morning Warbler, sometimes known as the Spotted Morning Thrush, is a small thrush-like bird, of six and a half inches . . . in a quiet sort of a way they are very attractive birds and have an extremely pleasant song. This is usually heard in the early morning and late evening.

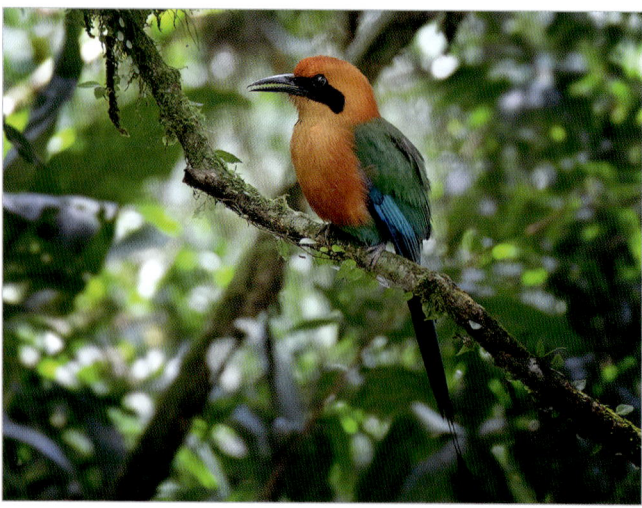

*Rufous Motmot (Costa Rica)*

# Motmot

The family of motmots, the Momotidae, is a small one with only fourteen species of colorful Neotropical birds. The word is an American Spanish one and is imitative of the bird's mellow *mot-mot* or repeated *mot . . . mot . . . mot* calls. In *Histoire Naturelle des Oiseaux* (1770), Buffon writes:[417]

> We confine to this bird the name of *Houtou* which the natives of your Guyana gave him, and which suits him perfectly, because he is the very expression of his voice: he never fails . . . we should have said, motmot from Mexico, because motmot is a Mexican name that Fernandez cited for this bird, while in Brazil it does not bear the name of motmot, but that of guiraguamumbi.

The Costa Ricans call the motmots *bobo*, which goes to show that onomatopoeia is in the ear of the beholder.

# Mountain-gem

The mountain-gems are seven species of hummingbirds in the *Lampornis* genus. Although Gould referred to the hummingbirds in general as "gems" many times in phrases such as "living gem," "gems of creation," and "flying gems," he didn't apply the common name mountain-gem anywhere in his *Monograph of the Trochilidæ*.[6] The name is clearly a reference to their brilliant plumage and preference for montane habitats. While it's not entirely clear where the name was first used, it seems to have first appeared in print in Ridgway's *Birds of North and Middle America* in 1911.[506]

# Mountaineer

The mountaineer is just one species of hummingbird, the Bearded Mountaineer, a Peruvian endemic and specialist of dry Andean valleys with scrubby slopes and open woodland. Gould was the first to use the common name Bearded Mountaineer in his *A Monograph of the Trochilidæ —Supplement* (1887) when describing a specimen collected by Whitely, the English naturalist and explorer.[507]

# Mourner

There are five species of mourners, three in the Tyrannidae and two in the Tityridae. All of them are named for their mournful vocalizations. The first use of the common name in the literature appears to have been in the 1912 book *The Birds of South America. Vol. 1* by Brabourne and Chubb.[45]

# Mousebird

The mousebirds are six species of African birds in the *Coliidae,* the mousebirds. They are so-called for their rodent-like habits, as they scurry about the ground searching through the leaf litter for fallen fruits. Additionally, as noted in an article in a 1896 volume of *The Ibis* entitled "Notes on the Ornithology of the Barberton District of the Transvaal," by Percy Rendall:[434]

> The colonial name for this species is the "Mouse-bird"—why I know not, but presume the decomposed feathers of the breast are responsible for the simile.

Note that the word "decomposed" in this case means feathers that have separate, loosely hanging barbs, not interconnected with barbules. The mousey brown and gray coloration of all species of mousebirds undoubtedly also contributes to the popular appellation.

# Mouse-warbler

Three species of mouse-warblers are endemic to New Guinea. They are members of the Australasian family, the Acanthizidae, a number of which are dubbed "warblers" or have been in the past. The mouse-warblers really are mouse-like in their appearance, including their colors and habits, and could easily be mistaken for the little rodents at first glance. This resemblance to mice is also referenced in the specific name of *Crateroscelis murina*, which is from the Latin for "gray-mouse-colored."

# Munia

In the Estrildidae (waxbills and allies) there are twenty-nine species of munias in the genus *Lonchura*. The name comes from the Hindi word मुनिया *muniya*, for the birds that are so familiar in the region. Hodgson was the first to use the name in the ornithological literature in 1836:[96]

> Munia, the name we have assigned to our new genus, is well known to the tarai [northern India and southern Nepal] and to the Hills as the generic appellation of several species of tiny gross-bills, distinguished for their familiarity with man, their gregarious habits, their depredations upon the rice crops, and their ingenious nests.

In a paper entitled "Hindustani-English Vocabulary of Indian Birds" published in 1908, the authors listed the following:[274]

> *Harī Lāl (male), Harī Muniyā (female): The Green Wax Bill (Estrelda Formosa)*
>
> *Lāl-muniyā, Lāl (the male), Munia (the female): The Red Wax-bill (Estrelda amandava).*
>
> *Teliyā muniyā. The Spotted Munia (Munia undulata).*

They referred to the names as being of Urdu or Hindi origin. The Hindu-English dictionaries define "munia" as "a beautiful rufous female bird" but also as a "common name used for a girl child," the former undoubtedly informing that latter.

# Murre/Murrelet

The two species of murres, and their diminutive cousins, the eight species of the murrelets, are members of the Alcidae. Early mentions of the name attach it not to the murres we know today but to Razorbills. In 1678, Ray and Willughby wrote an entry for Razorbill headed[16]

> The Bird called the Razor-bill in the West of England, the Auk in the North, the Murre in Cornwal: Alka Hoeiri [*Alca torda*].

Buffon made the same observation of the Razorbill in *Histoire Naturelle des Oiseaux* (1770):[417]

> In Norway, *alk*; in the Faroe Islands, *alck* or *alka*; in Gothland, *tord*; in Angermanie, *tordmulé*; in Ecoffe, *scout*; in northern England, *auk*; in western England, *razorbill*; in Cornwall, *murre*.

It would seem that murre was a local Cornish name for a different species that has now been misapplied to similar-looking species. The name is imitative of the calls of the juvenile birds.

# Myna

The Sturnidae is a big Old World family of starlings of 123 species, of which 26 are known by the common group name myna. The name came into English from the Hindi word for the *Acridotheres* mynas of the subcontinent, मैना *mainā*. When it was first adopted, it was transliterated phonetically with various spellings, such as "minah," "minor," "minar," "mynah," and "miner," finally settling as "myna." The root of the Hindi word is the Sanskrit मादन *madana*, meaning "maddening, gladdening, or delighting" which must surely refer to the birds' spirited vocalizations and feisty behavior.[336]

# Myza

The two species of myza honeyeaters are endemic to Sulawesi. The genus and common names are synonymous and derived from the Ancient Greek *muzao*, meaning "to suck," presumably a reference to their nectar-eating habits. In fact, the birds we know today as honeyeaters were called the honey-suckers at least into the late 1920s as in the *Systema Avium Australasianarum: A Systematic List of the Birds of the Australasian Region* by Gregory M. Mathews.[435] Although the common name didn't come into use until the 1980s, in *The Birds of Celebes and the Neighbouring Islands* by Meyer and Wiglesworth from 1898, we find one of the earliest uses of the name:[264]

The Honey-sucker, *Myza*, belongs to the purely Australasian family of the Meliphagidae.

# Myzomela

Another honeyeater with synonymous genus and common names, the myzomelas are thirty-five species of mostly brown or black sunbird-like birds with varying amounts of red in their plumage. The name comes from the Ancient Greek *muzao*, "to suck," and *meli*, meaning "honey," and was first given to the genus by Vigors and Horsfield in 1827.[169]

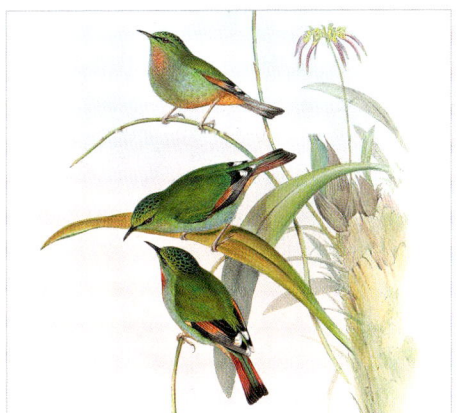

*Fire-tailed Myzornis from Gould's* Birds of Asia *(1850)*

# Myzornis

The single species of myzornis, the Fire-tailed Myzornis, is a member of the Sylviidae, the sylviid warblers, parrotbills, and allies, and can be found in the mountains of the subcontinent and southern China. This tiny, yet spectacular, bird was named in English by the great naturalist Edward Blyth in 1843, assigning the Ancient Greek μυζάω *myzáo*, "to suck," and ὄρνις *órnis*, "bird."[436] He commented:

I may notice here a beautiful little Nepal bird lately sent by Mr. Hodgson, which hardly seems to me to belong strictly to the Nectarinidae, though it is evidently a soft-billed honey-sucker, and I know not what else to approach it to.

The genus and common names are synonymous. Despite the name, they are not strictly nectarivorous; they also consume invertebrates, berries, and tree sap.

# N IS FOR NEEDLETAIL

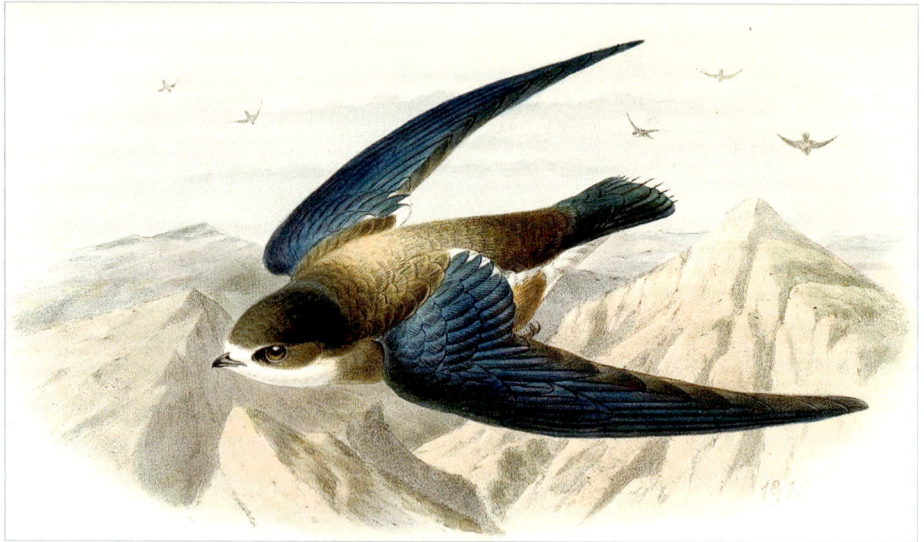

*White-throated Needletail in Lilford's* Coloured figures of the birds of the British Islands *(1885)*

## Needletail

Six species of needletails in three genera are exceptionally fast-flying swifts in the Apodidae, all found in Asia to Australasia. The White-throated Needletail is arguably the fastest bird recorded, having been clocked at speeds of up to 170 kph in horizontal flight. The name comes from the stiff feather shafts that extend beyond the end of the tail. Latham described the feature:[166]

> The tail feathers [are] furnished at the ends with projecting points, as sharp as a needle.

## Negrito

These two species of small tyrant flycatchers in the genus *Lessonia* were, until recently, considered subspecies of a single species. The name negrito is Spanish for "little black person" and was used in the literature for at least two other birds found in South America. In the *Histoire Physique, Politique et Naturelle De L'ile De Cuba* (1839) the Cuban Bullfinch *Melopyrrha nigra* was referred to as "Negrito."[437] And later, in *Contributions à la Faune Ornithologique de l'Europe Occidentale* by Léon Olphe-Galliard (1884), the Black Tern was also given the name.[438] And Wetmore gives the Spanish name for Blue-black Grassquit as "Arrocero Negrito" in 1958. The earliest use of the name for the *Lessonia* was in the 1975 *Wilson Bulletin*. Clearly, the name was used in American Spanish to refer to a variety of small black birds, which is a fitting description of these small, predominantly black-plumaged birds now known as Negritos in English. That said, the use of the epithet could be considered controversial.

## Newtonia

The four species of newtonias in the Vangidae have a synonymous genus and common name. The genus was erected by Schlegel in 1867, who named it after Sir Edward Newton, the British assistant colonial secretary on Mauritius from 1859 to 1877, who collected in Madagascar from 1861 to 1862.[508]

## Nicator

The three species in the small family the Nicatoridae are confined to Africa. The synonymous genus and common names derive from the Ancient Greek νικάτωρ *nikator*, meaning "conqueror," which is odd, given their shy and retiring habits. The name was coined in 1870 by Finsch and Hartlaub without explanation.[179]

## Nighthawk

The name nighthawk is now used only for the members of the Caprimulgidae of the Americas, but it was first used for the Eurasian Nightjar when the name was recorded in the King James Bible of 1611. In Leviticus 11:16:

And these are they which ye shall have in abomination among the fowls; they shall not be eaten, they are an abomination: the eagle, and the ossifrage, and the ospray, And the vulture, and the kite after his kind; Every raven after his kind; And the owl, and the night hawk, and the cuckow, and the hawk after his kind, And the little owl, and the cormorant, and the great owl, And the swan, and the pelican, and the gier eagle, And the stork, the heron after her kind, and the lapwing, and the bat.

Ingersoll wrote in 1937 that the nighthawks[9]

have received that name among us because they resemble small falcons when they swoop and dodge in the twilight as they pursue their insect prey through the air overhead. At such times they utter a whirring, booming sound, which has given them in the southern states the nicknames "bullbat" and "piramadig."

## Night-Heron

There are eleven herons in the Ardeidae that are known as night-herons, three of which are extinct. All of these herons, in three different genera, are crepuscular or nocturnal. An early use of the name was in 1781 in *Genera of Birds* by Thomas Pennant, who wrote under the heading of "Water-Fowl":[66]

All the Cloven-footed, or mere Waders, lay their eggs on the ground. Those with pinnated feet form large nests, either in the water, or near it. From the first, we must except the Heron and the Night-Heron, which build in trees.

## Nightingale

This old English name is given to two species of *Luscinia* in the Old World flycatchers, the Muscicapidae. The Old English *nehtægale* or *nihtegale* was borrowed from the Proto-Germanic *nahti-galōn*, from *nahti-* ("night") and *-galōn*, (*galan-* "to sing,"), the latter probably of onomatopoeic origin. The first known use of the name was in a poem from about 1250 entitled "The Owl and the Nightingale," in Old English "an hule and one niʒtingale."[416] The name later evolved to the Middle English *nyhtegale* or *nyghtgale*. It earns this compound name from its nocturnal vocal gymnastics. From *The Ornithology of Francis Willughby of Middleton in the County of Warwick, esq.* in 1678:[16]

> This bird is not remarkable for any variety or beauty of colours, but well known from its singing by night. And now that mention hath been made of singing, I cannot forbear to produce and insert the elegant words of that grave Naturalist Pliny, concerning the Nightingales admirable skill in singing, her study and contention, the sweetness of her accents, the great variety of her notes, the harmonious modulation and inflection of her voice.

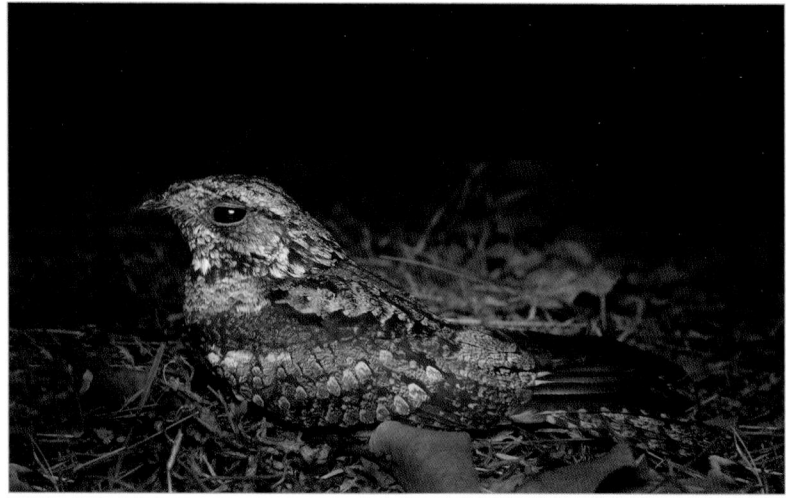

*Indian Nightjar (Goa, India)*

# Nightjar

The nightjars in the Caprimulgidae are named for their nocturnal habits in the first part of their name, but the second part is more interesting. The *-jar* suffix, meaning "discordant sound," as in "jarring," alludes to the harsh calls of the European nightjar, and, indeed, some other now outdated names include *nightchurr* and the Old English *nihthræfn*, "night raven." An early record of the word appeared in the French-language publication *L'Histoire Naturelle* by Salerne from 1767, in which he stated that the local *Caprimulgus* is called "in English the Fern-Owl, Goat Sucking-Owl, or Night jar."[382] Some of the other fabulous common names given for *Caprimulgus europaeus* in 1817's *Observations of the Natural History of Swallows* are Goatsucker, Dorhawk, Fernowl, Nighthawk, Churnowl, Goathawk, Wheelhawk, Wheelbird, Wheelowl, Spinningbird (the calls supposedly sounding like a spinning wheel), Goatmilker, and Evejar.[290]

# Nigrita

The common and genus names of these four small, mostly black species of Estrildidae finches from Africa are synonymous. The name is derived from the Latin *nigritia*, meaning "blackness" (the base word is *niger*, "black"). Formerly they were called Negro Finches, even up to the 1990s.[396] An illustration in the 1849 tome *Zoologia Typica, or figures of new and rare mammals and birds described in the proceedings, or exhibited in the collections of the Zoological Society of London* by Louis Fraser shows *Nigrita canicapilla*, but the common name given in the text is Grey-backed Finch.[397]

# Niltava

The niltavas, with their synonymous common and genus name, are six species of flycatchers with varying amounts of spectacular blue plumage. The genus name *Niltava* was erected by Hodgson in 1837 when he described four species in a paper in the *India Review* with the subheading "Genus Niltava nobis. Niltau of Nepal."[509] Jerdon also posted "Niltau, Nep." as the local name under the entry for *Niltava sundara* in *The Birds of India* (1862).[136] The modern-day Nepali word for blue, नीलो *nīlō*, is similar enough to *niltau* that one wonders if the words are related. In Nepali, the birds are called नीलतभा *neela tabha,* which can be directly translated as "splendid blue," a fitting description of all members of the genus.

# Noddy

The four species of noddies are delightful members of the Laridae, the gulls, terns, and skimmers, in the genus *Anous*, which is taken from the Ancient Greek word for stupid. I only mention this scientific name because it relates to the common generic name, which also means stupid. Noddy is probably a shortening of the obsolete words *noddypoll* or *hoddypoll*, which were Old English words for a "fumbling inept person."

The noddies gained these rather pejorative epithets from the ease with which they could be approached and killed; indeed, in the past sailors would approach nests and catch the birds by hand. In *The Natural History of Birds* (1793), Buffon wrote:[8]

> The Noddy, of which we now treat, has been termed the *foolish sparrow* (*passer stultus*); a very inaccurate denomination, since the Noddy is not a sparrow . . . We have adopted the name Noddy (*Noddi*) which occurs frequently in English voyages, because it expresses the stupidity, or silly confidence, with which the bird alights on the masts and yards of ships* and even on the sailors' hands.**
>
> *These are stupid birds, which, like the boobies, allow themselves to be caught by the hand, on the yards and the rigging of the vessel, on which they alight. *Catesby.*
> **The *Thouaroux* (the name of the Noddies in Cayenne) come to fish on very ample space, in company with the frigats; I never saw them alight on the water, like the gulls; but at night they come roving about the vessels to find repose, and the sailors catch them by lying on the top of the stern, and stretching out their hand, upon which the birds make no scruple to alight. *Memoirs communicated by M. de la Borde, king's physician at Cayenne.*

# Nothura

The nothuras are four species of Tinamidae tinamous in the synonymous genus *Nothura* from the Neotropics. The compound word is based on the Ancient Greek *nothos*, for "spurious or false," and *ouros*, meaning "tailed" (from *oura*, "tail"), which is fitting for a seemingly tailless, quail-like bird.

# Nukupuu

Contenders for one of the most unusual of bird names, the nukupuus are three species of congeners in the *Hemignathus* genus, one of which is extinct. The name is a compound of the Hawaiian words *nuku*, for "nose or bill," and *pu'u*, for "hill." The name literally means "hump bill," referring to the curved shape of the bill.

# Nunbird

There are five species of nunbirds in two genera in the Bucconidae, the puffbirds. One genus, the *Monasa*, the true nunbirds, derives its name from the Latin *monas*, meaning "solitary," which hints at their behavior, although this is not true for all species. The common name refers to their somber plumage and may also tie into their supposed retiring habits. As Frank H. Knowlton wrote in *Birds of The World: A Popular Account* in 1909:[298]

> Nunbirds (*Monasa*), so called from the somber black or slaty plumage, which, however, is somewhat relieved by a bright red or yellow bill, and in some there is white at the bend of the wing and about the bill.

# Nunlet

Six species in the Bucconidae (puffbirds) comprise the nunlets, a direct translation of their genus name *Nonnula*, a diminutive of the Latin *nonna* for "nun." These "little nuns" take their name from their plain attire, reminiscent of the plain habits worn by nuns. All of them are rufous, gray, and dull brown in various combinations, with no striking markings.

# Nutcracker

Three species of corvids in the genus *Nucifraga* are called nutcrackers. The name was first applied to the Eurasian Nutcracker because their main food throughout much of the year is pine nuts. They are specialized feeders on nuts, with an adapted physiology of the bill that serves as a nut-cracking tool. In 1544, Turner encountered the Eurasian Nutcracker in Europe, where it was called in German the "Nousbrecher," literally the "nut breaker."[119] He translated this in Latin as *Nucifraga*. At that time there was no English name, as the bird is a rare vagrant to the British Isles. One of the first uses of the name in print was in Ray and Willughby in 1678, although, strangely, they were using the name to refer to the "Virginia Nightingale" now known as the Northern Cardinal[16]

> which with its Bill cracks such kind of fruits, and other Grains or stones whence it is called *Nucifraga* or Nut-cracker. And that this Bird doth the like it is very probable, seeing it is likewise armed with a very thick and strong Bill.

# Nuthatch

The nuthatch family, the Sittidae, is made up of twenty-eight species, all with a similar morphology and coloring, and all sharing the same common generic name. Behaviorally, they are all very similar as well. They forage by hopping vertically up and down tree trunks, probing bark cracks for their invertebrate prey. In the winter months, though, they switch to a diet of seeds and nuts, and this is how they get their names. One of the earliest attestations of the name occurs in *The Promptorium Parvulorum* from around 1340 with an entry for the "Notehake, bryd."[261] Lockwood postulated the Old English word *hnuthæcca*; the second element of the name presumably from the verb *hæccan*, meaning "to hack."[84] He stated:

> In the subsequent evolution of the language, this verb came to sound like the unrelated (Middle English) *hachen* to hatch (of eggs) and passed out of use.

They literally hatch nuts, where hatch means "to crack." In *A History of British Birds* from 1871, William Yarrell threw light on the name:[343]

> The names of Nuthatch, Nuthack and Nut-jobber have been given to this bird from its habit of feeding at one season of the year on the kernels of hazel-nuts. These it

plucks from the bough or seeks when they have fallen to the ground, and flies with them between its mandibles to some tree or post, cleverly fixing them as though in a vice in some angular chink or crevice. It then hammers with the point of its bill at the shell until that is broken. Each stroke is delivered with the bird's full strength, and, working as it does from the hip-joint, the whole weight of its body is added to the force of the blow. While thus noisily employed—for the sound of the strokes can be heard at a considerable distance—it will often admit of very near approach, and the observer may then admire the skill with which the little workman will spring after any piece of the kernel that may be driven off by the violence of the blow, catch the morsel in the air, and quickly return to its occupation.

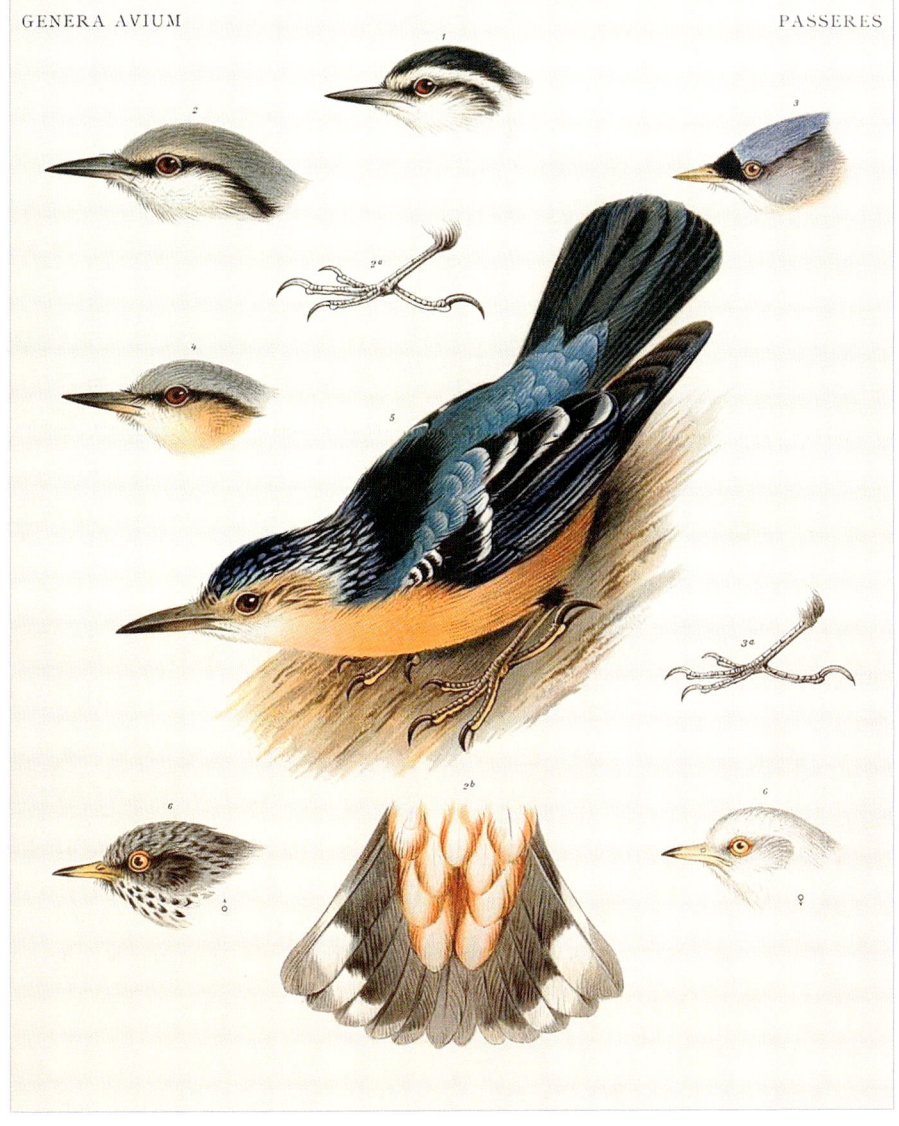

*Nuthatches in Wytsman's* Genera Avium *(1905)*

# O IS FOR OILBIRD

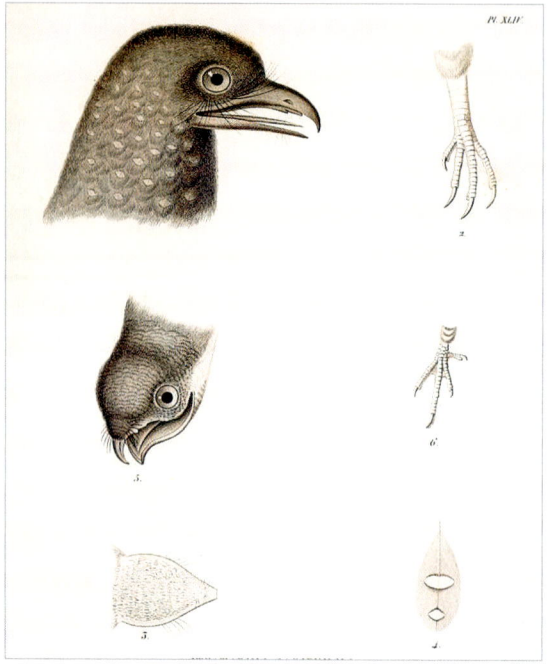

*Oilbird in Humboldt's* Recueil d'observations de zoologie et d'anatomie compar *(1833)*

## Oilbird

The remarkable Oilbird, the only nocturnal frugivore, is a member of the monotypic family, the Steatornithidae. Their diet of fruits with high lipid contents indirectly accounts for the name—the young have exceptionally high fat reserves as a result, and in the past, the birds were harvested for oil to light lamps. Bingley wrote in 1800:[73]

> The Indians enter the Cueva del Guachero once a year, near midsummer. They go armed with poles, with which they destroy the greater part of the nests. At that season several thousand birds are killed, and the old ones, as if to defend their brood, hover over the heads of the Indians, uttering terrible cries. The young, which fall to the ground, are opened on the spot. Their peritoneum is found extremely loaded with fat, and a layer of fat reaches from the abdomen to the vent, forming a kind of fatty cushion between the legs. At the period commonly called at Caripe the 'oil harvest,' the Indians build huts with palm leaves near the entrance and even in the porch of the cavern, where, with a fire of brushwood, they melt in pots of clay the fat of the young birds just killed. This fat is known by the name of butter or oil (*mantece* or *aceite*) of the Guachero. It is half liquid, transparent, without smell, and so pure that it may be kept above a year without becoming rancid. At the convent of Caripe no other oil is used in the kitchen of the monks but that of the cavern, and we never observed that it gave the aliments a disagreeable taste or smell.

# Oliveback

Three species of *Nesocharis* are very similar-looking estrildid finches with, not surprisingly, olive-colored backs. Early on they were called mannikins. In 1905's *The Birds of Africa, comprising all the species which occur in the Ethiopian region*, G. E. Shelley wrote:[139]

> These birds are well known as Mannikins, and I include under that English name the two species of *Ortygospiza* and the one of *Nesocharis*.

The present-day common name seems not to have come into use until the 1970s.

# Olomao

One of five Hawaiian thrushes in the genus *Myadestes*, the name is taken from the name *oloma'o*, given to it by the Indigenous inhabitants of the island of Molokai. There are theories that the Hawaiian name may be onomatopoeic or that the names of all the Hawaiian thrushes are metatheses of the name for the Hawaiian god Amaui, in honor of the beautiful songs. The first to describe the species was Scott B. Wilson in 1890. In *Aves Hawaiienses: The Birds of the Sandwich Islands*, he wrote: [47]

> It appears to be identical with the species inhabiting Molokai,—as was to be expected, seeing that the two islands are separated only by a narrow channel, some ten miles in width—and is known by the name of Olomao there: on Lanai, where I first obtained it, and after which island it is named, I met with no natives who knew the names of birds; indeed, in a few years there will not be many natives remaining.

# Omao

The Omao is another of the Hawaiian thrushes with a single name. According to Munro in *Birds of Hawaii* (2012):[265]

> The testimony of Bloxam (1823), Andrews (1865), Dole (1879) and the very old Hawaiian whom Perkins consulted on Kauai afford evidence that the original name of the thrushes on all the islands was Amaui (Manu a Maui), and that the different island names are corruptions of this.

And, indeed, in *Voyage of H.M.S. Blonde to the Sandwich Islands, in the years 1824–1825* by Byron and Bloxam in 1826, the *Myadestes* thrushes are treated as one species with the following entry:[510]

> 9.; native name, Amauii.
> Turdus Sandwichensis. Linn.
> Sandwich thrush.
> Found chiefly in Hawaii. There is a variety of the same at Oahu.

# Oo

Sadly, all four species of o'os in the Mohoidae, the Hawaiian honeyeaters, are now extinct. The birds were named in imitation of their calls. In one of the definitive books about Hawaiian birds, *Aves Hawaiienses: the birds of the Sandwich Islands* by Scott Wilson in 1890, this passage tells us something about the origin of the name:[47]

> Cassin [in 1858] goes out of his way to warn us that native names are not entitled to much consideration—a warning which, in the case of the Hawaiians (a people with a most accurate ear for sounds), is utterly uncalled for. He then proceeds to

observe that the name of this bird must sound quite different to different persons, and certainly—*Mo-hó, Hoohoo, Uho*—are strangely at variance; however we must ascribe this to the defects of ear of the individual explorers, since the Hawaiian gives to it but one name, *O-O*.

# Openbill

These two species of storks in the *Ciconiidae* possess remarkable bills with a noticeable space between the upper and lower mandibles, hence the name. Their diets consist mainly of large snails, which they are able to extract neatly from the shells using their specially adapted bills. Buffon wrote about the Asian bird as early as 1793 in *The Natural History of Birds*:[8]

> Such is the bird which we denominate Open-bill: in some respects, it resembles the Herons; and in others, it differs from them. It has, besides, one of those defects or natural imperfections, which we have remarked in a few species: its bill is wide open two-thirds of its length, both the upper and under mandible parting at that space and meeting again at the point. This bird is found in India, and we received it from Pondicherry.

# Orangequit

Another of the "quit" birds, this bird with a single name is a member of the Thraupidae, the tanagers and allies. It's a distinctive little dusky blue bird with an orange patch on the throat, from which it takes the first part of its name. As for the other birds with the suffix *-quit* in their names, it is derived from a Caribbean word applied to small birds. It is probably onomatopoeic, from the call note described as *guit-guit*. (See Bananaquit and Grassquit.)

# Origma

Australia's Rockwarbler is often also known as the Origma, also its genus name. This member of the Acanthizidae is found only in a very small area of eastern New South Wales, where it builds its hanging nest in rock crevices and under ledges. This brings us to the derivation of the word Origma—it comes from the Greek ορυγμα *orugma*, meaning "excavation, tunnel, or pit."

# Oriole

The common generic name oriole is used for two unrelated groups of birds, the Old World orioles, the Oriolidae, and the many species of New World orioles in the Icteridae. The word comes via the French *oriol* either from the Latin word *aureolus*, meaning "golden," an accurate characterization of the birds' bright yellow plumage, or as an imitative name for the call of Eurasian Golden Oriole. Albertus, the German Catholic friar, philosopher, scientist, and bishop, later canonized as a Catholic saint, wrote in 1250 that the name is onomatopoeic:[598]

> Oryoli aves sunt a sono vocis vulgariter sic vocati ut dicit Plinius. [Pliny says Oryol birds are so called for the sounds they commonly make.]

The name "oriole" was first used by Pennant in 1776 and is an adaptation of the genus name given to the *Oriolus* orioles by Linnaeus in 1766.[177] Pennant wrote:

> Its note is loud, and resembles its name.

Lockwood believed that Albertus's name was later misinterpreted as deriving from the Latin *aureolus* (the Latin name for oriole is *chlorion*).[84] Many authorities discount the onomatopoeia theory, and the true etymology of oriole may now be a matter of conjecture only.

Although they are unrelated, the thirty-two species of *Icterus* orioles do have similar plumage to the Oriolidae, as well as many similarities in behavior and diet, in an example of convergent evolution. The early European colonists noted the similarities and used the same vernacular name for these familiar birds.

# Oropendola

The remarkable oropendolas are nine species in the *Psarocolius* in the Icteridae, the troupials and allies. Despite the assertion by many that oropendola means "golden pendulum" (describing the courtship display when the male swings by his feet and flashes his bright yellow tail), the name is actually a Spanish one derived from the Latin *aurum*, for "gold," and *pinnula*, the diminutive of *pinna* for "wing." Originally, it was a word used by the Spanish for the Old World orioles, and, in fact, the modern-day Spanish name for Eurasian Golden Oriole is *Oropéndola Europea*. Brisson discussed the *Oiseau jaune de Bengale* (Bengal yellow bird) in *Ornithologie* in 1760, presumably an *Oriolus* oriole, using the Spanish names *Oroyéndola* and *Oropendola*.[76] These entries predate the earliest references in the European scientific literature to the *Psarocolius* oropendolas. It might be surmised that the Spanish colonizers of Latin America took the name with them and applied it to the similar-looking icterids in much the same way the English colonizers to the north appropriated the name oriole. Given that the oropendolas look very little like Old World orioles, it would seem likely that the name was first used for the *Icterus* orioles and later came into English in a case of mistaken identity.

# Osprey

The Pandionidae is a monotypic family with one species, the Osprey. The name in English is an old one, making one of its first appearances in the mid-1400s in John Russell's *Boke of Nurture*.[252] Under the heading "How to Carve a Large Roast," it appeared in the verse:

> Good soñ, of alle fowles rosted y telle yow as y Cañ,
>
> Every goos / teele / Mallard / Ospray / & also swanne,
>
> reyse vp þo leggis of alle þese furst, y sey the thañ,
>
> afftur þat, þe whynges large & rownd / þañ dare blame þe no man
>
> [Larger roast birds, such as the goose, teal, Mallard, *Osprey*, & also swan,
>
> raise up [? cut off] the legs, then the wings,
>
> lay the body in the middle,
>
> with the wings and legs round it.]

Shakespeare also mentioned the Osprey in *Coriolanus* from 1605 or thereabouts:

> I think he'll be to Rome / As is the Aspray to the fish, who takes it / By sovereignty of nature.

The name itself is possibly a misapplication. Livingstone noted:[511]

> The etymological dictionaries represent a tradition of uncertainty as regards its etymon and propose with caution and reserve L[atin] *ossifraga*, the "bone-breaker of the *Historia Naturalis* (x, 3) of the elder Pliny.

The Latin *ossifraga* is a compound of two words—*os*, for "bone," and *frangere*, "to break," and it was originally used by Pliny for the Lammergeier (see entry), which will fly up high to drop bones to the ground in order to break them open to eat the marrow. But wait, there's another theory—the Medieval Latin term for bird of prey was *avis praedae*, which may have morphed to the name osprey. In the same paper, Livingstone wrote:

> In spite of the efforts of etymologists to attach *osprey* at any cost to *ossifraga*, its etymon seems to me to lie in another direction: Latin *avis praedae* (V. L. *\*avisprede*) . . . The expression *oisel de proie* "hawk" is found in medieval texts . . . It might well represent a vernacular adaptation of an original *avis praedae*; cf. *avis maris = oiseau de mer*. The ordinary man doubtless had in the past, as he has today, a confused notion of hawks and their variety. It is entirely probable, however, that the *nisus*, the predatory fish-hawk, was early given the expressive and appropriate, although general, popular name *avis-prede* (= *avis praedae*) in a region frequented by it and where people were familiar with it and its habits and could distinguish it. . . . *Avis praedae* would have given quite regularly *ospreit* in Old French before the Conquest, . . . and later *osprei*, osprey.

In fact, the confusion between the similar-sounding words *ossifrage* and *avis prede* probably accounts for the present-day misapplied English name. (See Lammergeier.)

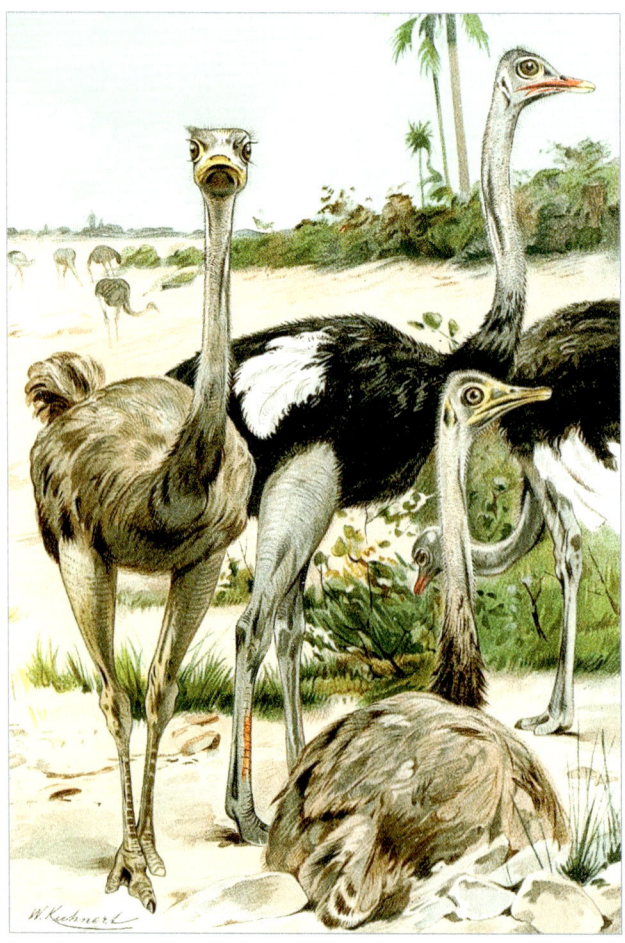

*Somali Ostrich from Lydekker's* The Royal Natural History *(1895)*

# Ostrich

Although long thought of as one species, recent molecular studies show there are in fact two species in the Struthionidae, the ostriches. The name has gone through various iterations since the 1200s, including *hostriche*, *ostrig*, and *estrich*, coming to English from the Latin *ostrica* via the French *ostruce*. The Latin *ostrica* came from the Vulgar Latin *avis struthio*, from *avis*, "bird" and *struthio*, "ostrich." The latter word ultimately derived from the Ancient Greek. The ostrich was well known to them, and they called it στρύθος *strýthos*, or, more poetically, *strouthokamelos*, "camel-sparrow," for its long neck. A very early reference to the ostrich appeared in the *Ancrene Riwle*, a book of rules for female monastic scholars or anchoresses written in the early 1200s:[301]

> Þe ostrice for his muchele flesch & oðer swuche fuheles maken semblaunt to fleon.. ah þe fet eauere drahen up o þe eorðe. Alswa fleschliche anker . . . hire fet eauer ase don þe ostrices þat arn hire lustes Drahen to þe eorðe. [The ostrich, because of his great flesh (big body), and other such fowls, make a pretense to fly, and beat the wings, but the feet always pull them to the earth.]

Apparently, the author drew an analogy with the bird's heavy body to represent the "lusts [that] . . . draw her to the earth," which would keep the sincere anchoress from flying like a true bird. Later, in 1521, Laurence Andrew wrote in *The noble lyfe*:[71]

> Stuciocamelo is a birde very gret & is moche in Ethiope & in Affrike & they be somwhat sibbe to the bestes & they be as hye as a hors / & they rōne moche faster throughe yᵉ helpe of their winges [The Ostrich is a great bird found in Ethiopia and in Africa, and they are kin to the beasts and as tall as a horse, and they run much faster with the help of their wings.]

# Ou

Originally called by Latham in 1822 the "Parrot-billed Grosbeak," this member of a monotypic genus, *Psittirostra*, in the Fringillidae, is yet another endangered Hawaiian honeycreeper.[17] In 1893 *The Avifauna of Laysan and the Neighbouring Islands: with a Complete History to Date of the Birds of the Hawaiian Possessions* by Walter Rothschild was published, with comments on the Hawaiian name:[37]

> Mr. F. Gay [a local collector] informs me that on Kauai the male is sometimes called "Ou poolapalapa" (Ou with the yellow head), while the female goes by the name of "Oulaevo" (the green Ou). Bloxam* also called the bird "Ohu."
>
> *In *Voyage of H. M. S. Blonde to the Sandwich Islands*, in the years 1824–1825. (1826)[510]

# Ouzel

These days there is only one ouzel, the Ring Ouzel, just one of 173 species in the Turdidae, the thrushes and allies. It is an old name with roots going back to the Old English for "blackbird," *osle*. It's thought that it came into our language from the Old German name for blackbird, *amsala*. However, Ray and Willughby claimed incorrectly in 1678:[16]

> This word Ouzel is undoubtedly of the same original with the Italian *Uccello*, and the French *Oiseau* signifying in general a Bird; however it be with us appropriated to this kind.

In the past, the word was used in many more common English names. For example, in Count de Buffon's *The Natural History of Birds* from 1793, there are multiple uses of the word, with entries such as the Little Crested Ouzel of China (in the chapter "Foreign Birds"), Rock Ouzel, the Great Mountain Ouzel, the Rose-Coloured Ouzel, Blue Ouzel, and Solitary Ouzel.[8] The word was often used interchangeably with "thrush," as seen in the chapter about Hermit Thrush which the author later referred to, in the same entry, as Hermit Ouzel.

# Ovenbird

The ovenbird of the Americas is a member of the New World warbler family, the Parulidae. Previously known as Golden-crowned Thrush or, more charmingly, the Golden-crowned Wood-Wagtail, as it appears in Audubon's *A Synopsis of the Birds of North America* from 1839, the alternative name "oven-bird" is the one that has stuck. It is so-called because of the domed shape of its nest of grasses, stems, and bark, made on the ground among thickets.[512]

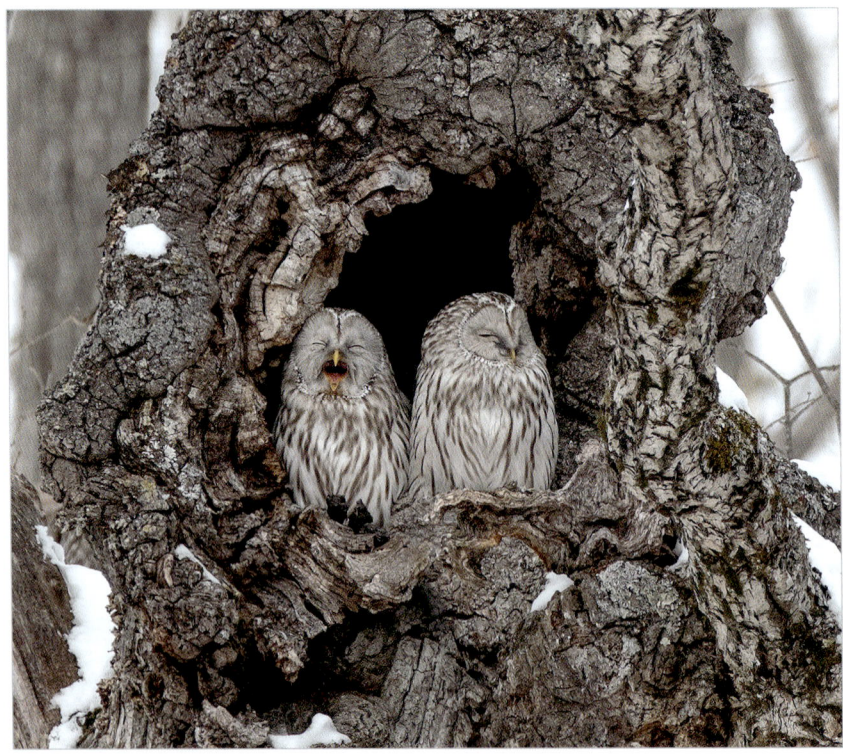

*Ural Owls (Japan)*

# Owl

Among the best-known birds in the world are the owls. Every English speaker knows the word. There are in fact two families of owls—the Tytonidae (with 18 species) and the Strigidae (with 225 species). Many names for owls are hyphenated, but I will treat them all here, as most are obviously indicative of morphology (Masked-Owl, Eagle-Owl, Hawk-Owl, Pygmy-Owl), behavior and diet (Fish-Owl, Fishing-Owl), or preferred habitat (Barn-Owl, Wood-Owl). The English name comes from the Proto-Germanic name for the birds, *uwwalon*. Even earlier, the root lies in the PIE word *uwal*, which was imitative of the owl's hooting call. It came into the English language as the Old English *ule*. There were many Middle English spellings of the word, including *uwile*, *oule*, *owell*, *hule*, *hoole*, and *howyell*.

"The Owl and the Nightingale" is a Middle English poem from around 1250, testifying to the importance of the owl in English culture.[416] The poem begins with the opening stanza:

> *Ich was in one sumere dale,*
>
> *in one supe dyele hale,*
>
> *iherde ich holde grete tale*
>
> *an hule and one nyhtegale.*
>
> *Þat plait was stif & starc & strong,*
>
> *[I was in a valley in springtime;*
>
> *in a very secluded corner,*
>
> *I heard an owl and a nightingale*
>
> *holding a great debate,*
>
> *their argument was fierce, passionate, and vehement,]*

(See Bay-Owl, Screech-Owl, and Scops-Owl for discussions of more obscure names.)

# Owlet

There are fifteen species of small owls in the Strigidae. The *-let* suffix is a diminutive derived from the Latin *-ettus* added to a noun to designate a smaller form.

# Owlet-nightjar

The Aegothelidae are the owlet-nightjars, a family of ten species all bearing the name. These small nightbirds, found only in Australia and New Guinea, are named for their coincidental resemblance to both small owls and nightjars. The name is first attested in 1840 in Gould's *The Birds of Australia*, in which he wrote:[40]

I also procured its eggs, and considerable information respecting its habits and actions, which differ most remarkably from those of the true *Caprimulgidae*, and on the other hand assimilate so closely to the smaller Owls, particularly those comprised in the genus *Athene*, as to form as perfect an analogical representative of that group of birds as can possibly be imagined, for which reason the English name of Owlet Nightjar has been assigned to it.

# Oxpecker

There are just two species in the oxpecker family, the Buphagidae, from Africa. They are a familiar sight over much of the continent, where they can be seen foraging on cattle and other African megafauna. As Buffon wrote in *The Natural History of Birds* in 1793:[8]

This bird is very fond of certain worms, or the larvæ of insects, which lodge under the epidermis in oxen. It alights on the backs of these animals, and pierces their skin with its bill, to extract these worms, and hence its name.

# Oxylabes

Two species in different genera but in the same family, Bernieridae, are called oxylabes. The common name (and the genus of White-throated Oxylabes) oxylabes comes from the Ancient Greek οξυλαβης *oxylabēs*, meaning "quick to catch," a reference to the active feeding habits of these small insectivores.

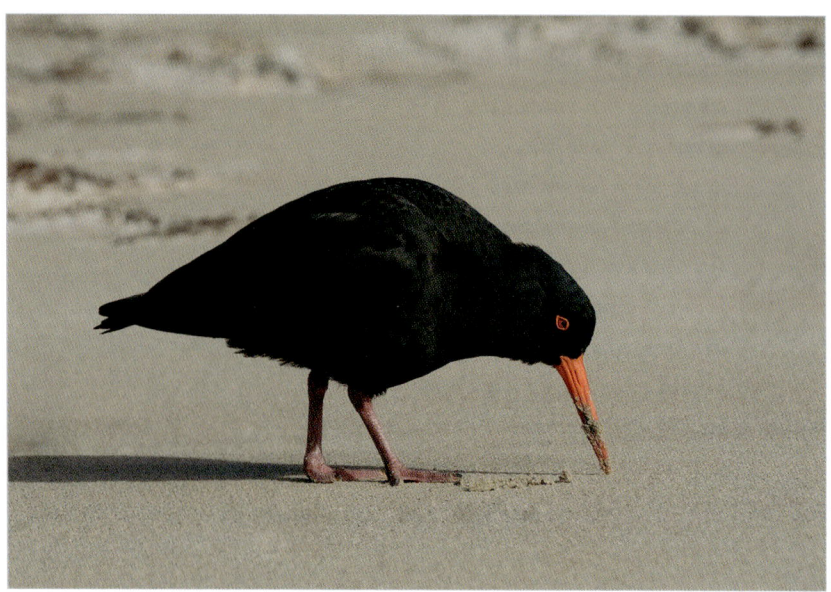

*Sooty Oystercatcher (Victoria, Australia)*

# Oystercatcher

In the Haematopodidae, there are twelve species of oystercatchers, all similar in appearance, with minor plumage differences being all-black or black with white underparts, red legs, long red bill, and a red eye-ring. These shorebirds have specially adapted bills that are laterally compressed with razor-sharp edges for cutting through the muscles of clams, oysters, and mussels with a lightning-quick jab in order to open them. An old and charming name for them was "Sea Pies," as mentioned in *British Zoology* by Thomas Pennant:[177]

> Their bills, which are compressed sideways, and end obtusely, are very fit instruments to insinuate between the limpet and the rock those shells adhere to, which they do with great dexterity to get at the fish. On the coast of France, where the tides recede so far as to leave the beds of oysters bare, these birds feed on them; forcing the shells open with their bills.

# P IS FOR PAINTED-SNIPE

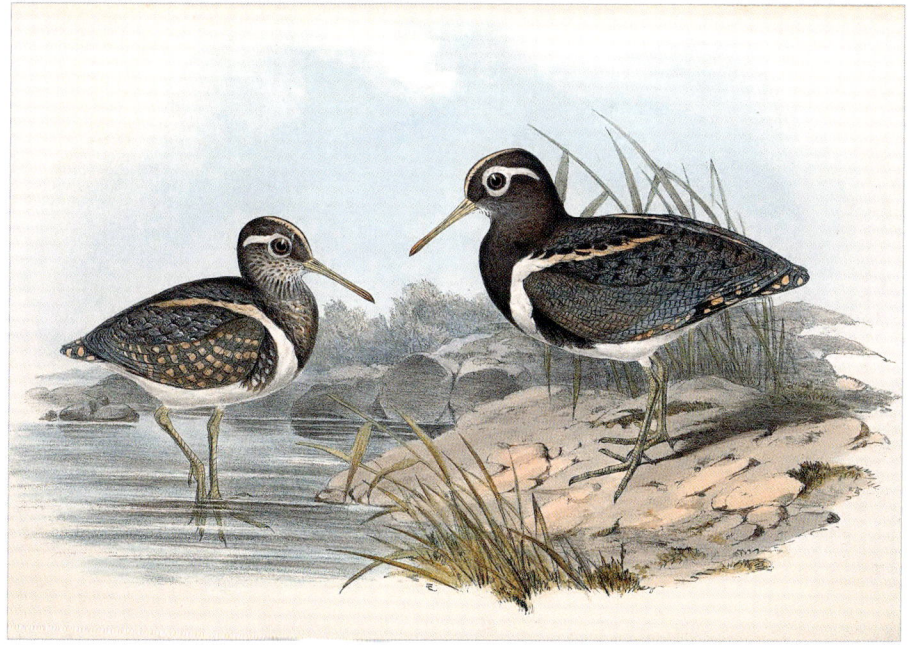

*Australian Painted-snipe in Gould's* The Birds of Australia *(1840)*

## Painted-snipe

Although they are not snipes, the three species of ornate, yet cryptic, painted-snipes in the Rostratulidae do have a similar appearance and habits. This, along with their brown, rufous, buff-and-gray-daubed plumage, is the explanation for the common group name. As Nicholson wrote in 1809, it[599]

> has a beautifully coloured wing; whence it is by many called the "painted snipe."

## Palila

The Palila is a species of Hawaiian honeycreeper in a monotypic genus, *Loxioides*, in the Fringillidae. It's probable that the name is simply the Hawaiian name for the bird. Pukui and Elbert list Palila in the *Hawaiian Dictionary* (1986):[234]

> An endangered gray, yellow, and white Hawaiian honeycreeper; endemic to the island of Hawaii.

They go on to note that there is a Hawaiian expression, "*pi'oloke ka leo o ka palila* frightened is the voice of the palila [bird]." (The word *pi'oloke* can be translated as "alarmed, startled, agitated.") In *Aves Hawaiienses: the Birds of the Sandwich Islands* from 1890, Wilson noted:[47]

> The Palila, as far as I know, has no song, but merely a very clear whistle-like note, which, when often repeated, is held by the natives to be a sign of approaching rain.

But in 1900, Rothschild had pointed out that there was some initial confusion between the Ou and the Palila.[37] Now, the Ou occurs on the island of Kauai, which is home to the mythological great warrior, Palila. This raises the possibility of some relationship between the name of the bird and the name of the warrior, who cried "I am Palila! I am from Kaua'i, raised at Alanapō, the temple of the gods!"[38]

# Palmchat

This member of the monotypic family the Dulidae is endemic to the island of Hispaniola. Palmchats make their enormous nests of sticks and twigs in the crowns of palm trees. (See also Chat.) In 1909, Knowlton and Ridgeway wrote in *Birds of the World: A Popular Account*:[298]

> This species, known locally as the Palm Sparrow, is a very common bird, breeding in colonies in nearly every grove of royal palms. . . . the nests are sometimes very large, quite an armful of twigs interwoven into a compact mass in the head of the palm or on the cluster of berries just below the head. In this ball of sticks there are generally three or four nests, merely burrows into the side of the mass, the end lined with finer twigs. Its habits are peculiar in many ways, and with its powerful legs, feet, and bill it breaks off and carries up into the palm tree sticks astonishingly large for its size.

# Palmcreeper

This member of the Furnariidae in the monotypic genus *Berlepschia* is strictly dependent on two species of *Mauritia* palm in the Neotropics. They build well-concealed nests in the crowns of the palms, where they also forage by gleaning invertebrates from the leaves and branches.

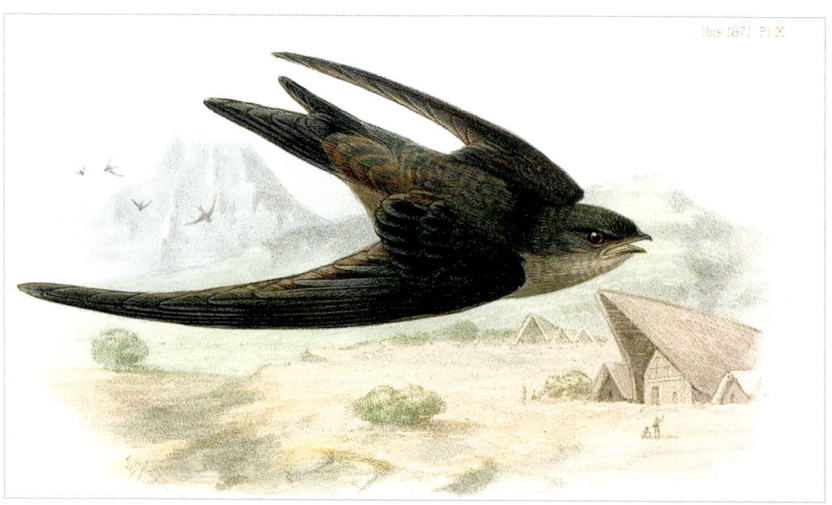

*Asian Palm-Swift in* The Ibis *(1871)*

# Palm-Swift

The five species of palm-swifts in the genera *Cypsiurus* and *Tachornis* are all closely associated with various species of palms, relying on them for roosting and nesting sites. One species, in particular, is also known to use thatched roofs, as related by Jerdon in tales of sightings of the Asian Palm-Swift, which he called the Palm-roof Swift, from Nagaland in a paper in *The Ibis* in 1871:[346]

The more highly civilized Khasi race have better houses than their neighbours on each side, who use huts thatched with palm-leaves. On these roofs this Palm-swift invariably builds its nest.

# Palm-Tanager

This small family, the Phaenicophilidae, known as Hispaniolan tanagers, consists of four species. Two of them, in the genus *Phaenicophilus*, are called palm-tanagers. They are quite similar in appearance and are sometimes considered conspecific. Despite the fact that both species can be found in a wide variety of habitats, early accounts of the bird—such as Brisson's[76] observation in 1760 that "It usually stays on Palm Trees, and makes its nest there" and Latham's[17] initial claim that it "is most frequent among the *palm-tree*" with consequent use of the name "Palm Thrush"—no doubt explains the present-day common name. Later, they were referred to as "palm tanagers"; the hyphen was subsequently added to differentiate them from the unrelated "Palm Tanager."

# Palm-Thrush

The palm-thrushes are two species of African muscicapids in the genus *Cichladusa*. Though now known not to be thrushes, they were previously considered to be in the Turdidae. Both species favor palm-dominated habitats, but not exclusively so.

# Pampa-Finch

These two species that are members of the Thraupidae, the tangers and allies, are not true finches but do bear resemblance to them. Both favor open grassland and savannas. The first of the two congeners in the genus *Embernagra* to be described was the Great Pampa-Finch. It is found in the grasslands of Argentina, the Pampas (the word is from the Indigenous Quechua *pampa*, meaning "plain").

# Paradigalla

The paradigallas are two species in the Paradisaeidae, the birds-of-paradise. The common and genus names are synonymous. The name was coined by Lesson in 1835, when he wrote in *Histoire Naturelle des Oiseaux de Paradis et des Épimaques*:[439]

> Oiseau de Paradis à pendeloques. Cette espèce, des plus rares, dont on ne connaît qu'un individu, n'est pas aussi grosse que le Sisilet. Son plumage est noir, le dessus de la tête excepté, qui est vert émeraude. Sa queue est composée de douze rectrices, régulièrement étagées. Les plumes des flancs sont molles, abondantes et laches comme celles des Sisilets. Une sorte de crête surmonte le devant du front, et deux pendeloques charnues s'attachent aux côtés de la mandibule inférieure. On pourrait la nommer *Paradigalla carunculata*.

> [Bird of Paradise with pendants. This species, of the most rare, of which only one individual is known, is not as large as the Sisilet. Its plumage is black except for the top of the head, which is emerald green. Its tail is composed of twelve rectrices, regularly stepped. The feathers of the flanks are soft, abundant and loose like those of the Sisilets. A sort of crest surmounts the front of the forehead, and two fleshy pendants attach to the sides of the lower mandible. It could be called *Paradigalla carunculata*.]

The name is a compound word combining the genus name *Paradisaea* (Linnaeus, 1758) and the Latin *gallus*, "cock or rooster."[14]

# Paradise-Flycatcher

There are fifteen species of *Terpsiphone* paradise-flycatchers in the Monarchidae, as well as one member of the Rhipiduridae, the fantails, which was, until recent molecular studies revealed otherwise, also classed as a monarch. In 1743, Edwards described the "Pyed Bird of Paradise" with accompanying illustration in *A Natural History of Uncommon Birds*:[78]

> Tho' the above Authors have placed this Bird with the *Manucodiata*,* I can by no Means agree with them; it being generally different, in having no Velvet-like Feathers about the Head, no gay or shining Colouring, and the Feet weaker and of a quite different Structure, as well as in having stiff long Bristles or Hairs above the Angles of the Mouth.

*An early name for the Paradisaeidae. (See Manucode.)

# Paradise-Kingfisher

There are nine species of *Tanysiptera* with the appellation paradise-kingfisher. In the 1860s, Gould used the common name kingfisher or tanysiptera for these birds that are found only in tropical Australasia.[513] Wallace wrote in *The Malay Archipelago* (1869) of birds he dubbed "racquet-tailed kingfishers":[205]

> The genus *Tanysiptera*, to which this bird belongs, is remarkable for the enormously lengthened tail, which in all other kingfishers is small and short. Linnæus named the species known to him "the goddess kingfisher" (*Alcedo dea*), from its extreme grace and beauty, the plumage being brilliant blue and white, with the bill red, like coral.

It may be for this reason that Rand and Gilliard coined the name paradise-kingfisher for the *Tanysiptera* in their *Handbook of New Guinea Birds* in 1967.[440]

# Parakeet

This widely used English bird name is applied to twenty-seven species in the Psittaculidae, the Old World parrots, and sixty-seven species in the Psittacidae, the New World and African parrots. Ray and Willughby explained in 1678:[16]

> Parrots *in respect of bigness may be divided into three kinds,* viz. *the* greatest, mean-sized, *and* least.

> The greatest *are equal in bigness to our common* Raven: *or (as* Aldrovandus *saith) to a well-fed* Capon; *and have long Tails: In English they are called* Macaos *and* Cockatoons. The middle *or meansized and most common* Parrots *are as big or bigger than a* Pigeon, *have short Tails, and are called in* English, Parrots *and* Poppinjayes. The least *are of the bulk of a* Blackbird *or a* Lark, *have very long Tails, and are called in English* Parakeetos. [*sic*]

The word may have come into English in the mid-1500s from the Middle French *paroquet* "parrot." (See Parrot.) Alternatively, it came from the Spanish *periquito*, a diminutive of *perico*, meaning "parrot," which came from the name Pedro. This name made an early appearance in Shakespeare's *Henry VI, Part 1*, thought to date to 1591, where it was used as an insult from Lady Percy to Hotspur:

> Come, come, you paraquito, answer me
>
> Directly unto this question that I ask.

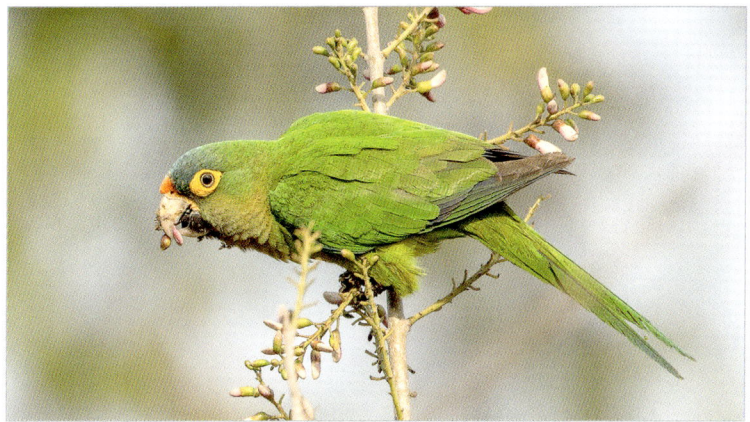

*Orange-fronted Parakeet (Mexico)*

# Pardalote

Pardalotes are endemic to Australia; there are four species in the Pardalotidae. The name was coined in the mid-nineteenth century in accordance with the genus name *Pardalotus*. One of the features of most members of the family is their spotted plumage, and this explains their name, which comes from the Ancient Greek πάρδαλις *pardalis*, meaning "leopard," from which we get the Latin *pardalotus*, "spotted like a leopard." Despite Vieillot's use of the common name, which appeared in the French publication *Dictionnaire d'Histoire Naturelle*[39] in 1816, Latham[17] used the name "manakin" for the birds, giving them names like Speckled Manakin, New Holland Manakin, and Supercilious Manakin. He did give a nod to the name Pardalote, attributing it to Temminck, who used the name in *Manuel d'Ornithologie* in 1820.[441] Gould was one of the first, if not the first, to use pardalote as a common name in English.[40] But he also lists some alternative names that are of interest, for example:

> *We-dup-we-dup,* Aborigines of the lowland districts of Western Australia.
> *Diamond Bird,* Colonists of New South Wales.

# Pardusco

This single-word bird is a member of the Thraupidae in the monotypic genus *Nephelornis*. It is remarkable for being described as late as 1973 but is otherwise a rather plain, small, brown tanager whose name means simply "brownish" in Spanish. In 1976, a paper entitled "A New Genus and Species of Nine-Primaried Oscine of Uncertain Affinities from Peru" appeared in *The Auk*. Its authors, Lowery and Tallman, wrote:[442]

> This paper describes still another recent Peruvian discovery, a bird that we call the "Pardusco," because that is the name applied to it by our Peruvian field assistants who live near the region where it is now known to occur.

# Parisoma

There are two parisomas in the Sylviidae, both in the genus *Sylvia*. This fairly diverse genus revels in possessing six different common group names. (See Babbler, Blackcap, Thrush-Babbler, Warbler, and Whitethroat.) The protonym *Parisoma Boehmi* [sic] was first assigned by Reichenow in 1882 when the Banded Parisoma *Sylvia boehmi* was considered to be aligned with the Paridae. The name is derived from the genus name *Parus* for the tits and the Ancient Greek *soma*, for "body."

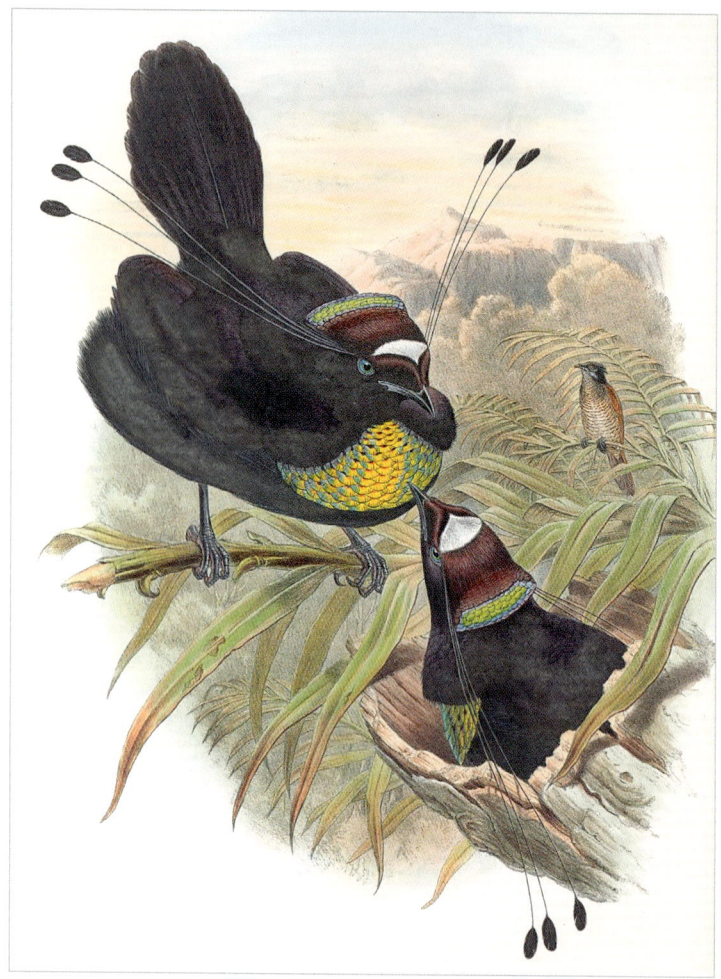

*Western Parotia from Gould's* The Birds of New Guinea and the Adjacent Papuan Islands *(1875)*

# Parotia

The five species of remarkable parotias in the Paradisaeidae share their common name and genus name. The genus name, which is an homage to the birds' elongated head plumes, was established by Vieillot in 1816.[514] The name is composed of two words of Ancient Greek origin—παρά *pará*, "beside, near, issuing from," and ὠτούς *otoús*, "ear." Vieillot described the head plumage in *La Galerie des Oiseaux* in 1826:[357]

> This beautiful bird wears a small crest, which extends over the crown of the head a little beyond the eyes; the feathers, which compose it, rise from the base of the beak, are fine, stiff, slightly barbed, and so mixed with black and white, that all of these colors present a very pearl tone. Three black quills, five to six inches in length, start from each side of the head, run backwards and are terminated by racquets that are longer than the others, and which, as they open out, have an oval shape. . . . Those of the throat are narrow at their origin, broad at their extremity, a beautiful velvet black in the middle, and of gold color changing to purple on the sides, with reflections of various green shades.

# Parrot

Recent molecular studies have now concluded that there are three families of parrots—the Psittaculidae, Old World parrots (seventy-four parrots*); the Strigopidae, New Zealand parrots; and the Psittacidae, New World and African parrots (eighty-eight parrots and parrotlets*). None of the New Zealand parrots has the common generic name, however. These birds are familiar the world over, not only as colorful and intelligent birds, but as popular pets, many of them with the ability to imitate human speech. They got their English name in the 1500s, probably from a corruption of the French name *Perrot* or *Pierrot*, which was a variant of the name Pierre. It is possible the name comes from that of a pet bird belonging to a character in Thomas de Saluces's *Le Chevalier Errant*, written in 1395.[443] In the story, the first known occurrence of the name "parrot," used as a proper name in the form Paroquet, appears in a tale within the allegory, which tells a parable of a lady who cheats on her older husband. The suspicious husband has her watched by his three pet parrots. But the wife remembers the parrots, who have seen and heard everything, after having received her lover during the night. She and her devoted servant question the youngest, who reports that he will repeat everything to his master upon his return, upon which the two women strangle the overly talkative parrot. The same goes for the second parrot who, besides being a spectator and witness of the night, also witnessed the death of his brother and will hasten to reveal everything to the master. He suffers the same fate. Finally, comes the turn of the oldest, who claims to have forgotten everything. He opens his speech with praise of the lady before concluding on the benefits of silence.[291]

In *Henry IV Part I*, written by Shakespeare in 1597, Prince Henry says of Francis:

> That ever this fellow should have fewer words than a parrot, and yet the son of a woman!

*These numbers include several hyphenated names—pygmy-parrot, king-parrot, tiger-parrot, shining-parrot, fig-parrot, hanging-parrot.

# Parrotbill

Previously placed in their own family, the twenty-one species of parrotbills are now placed in the Sylviidae, the sylviid warblers, parrotbills, and allies. The family they were formerly placed in was called the Paradoxornithidae—the paradoxical birds. For a long time after Gould first described the genus in 1836, they were known by the common name paradoxornis.[515] By the 1860s, Jerdon had dubbed them "finch-thrushes" in *The Birds of India*.[136] Later they became known in the literature as "crow-tits," but Baker started to use the name parrot-bill in his *Fauna of British India* in 1922 in reference to the stout, short bill.[444]

One other unrelated bird bears the name, also. The Maui Parrotbill is a Hawaiian honeycreeper in the Fringillidae named for its large, parrotlike bill.

# Parrotfinch

The twelve species of Estrildidae finches in the *Erythrura* all sport colorful plumage of reds, blues, and greens that are reminiscent of the gaudy parrots. An early use of the name was in 1783 when Latham described and illustrated the Red-throated Parrotfinch from New Caledonia, noting that "the body [is] parrot green."[218]

# Parrotlet

Nineteen species of tiny parrots are in the Psittacidae, the New World and African parrots. The -*let* suffix is a diminutive indicating their relatively small size compared to the generally larger parrots.

# Partridge

In the Phasianidae there are forty-five species of partridges, while in the Odontophoridae, the New World quails, there is but one partridge from Africa and three species of wood-partridges. If true, the etymology of the name is one of the most amusing in the English bird lexicon—ultimately the name comes from the Ancient Greek word πέρδομαι *pérdomai*, which means "to break wind," supposedly because the partridge makes the sound of a fart when it flies! Sadly, both Lockwood[84] and Liberman[384] found this explanation unconvincing, however, and the former thought the name is derived from the Middle English personal name *Partrich*. The Ancient Greek for partridge is πέρδιξ *pérdix*, which influenced the Old French *pertis*, coming into Middle English as *partriche* or *perdriz*, among other spellings. Not surprisingly, the first bird to be named "partridge" was the Grey Partridge of Europe with the scientific name *Perdix perdix*. This bird, of course, appeared in *The Boke of Nurture* in 1450 on the menu for the second course of the "dynere of flesche":[252]

> Partriche, wodcok / plovere / egret / Rabettes sowkere
>
> [Partridge, woodcock, plover, egret, suckling rabbit.]

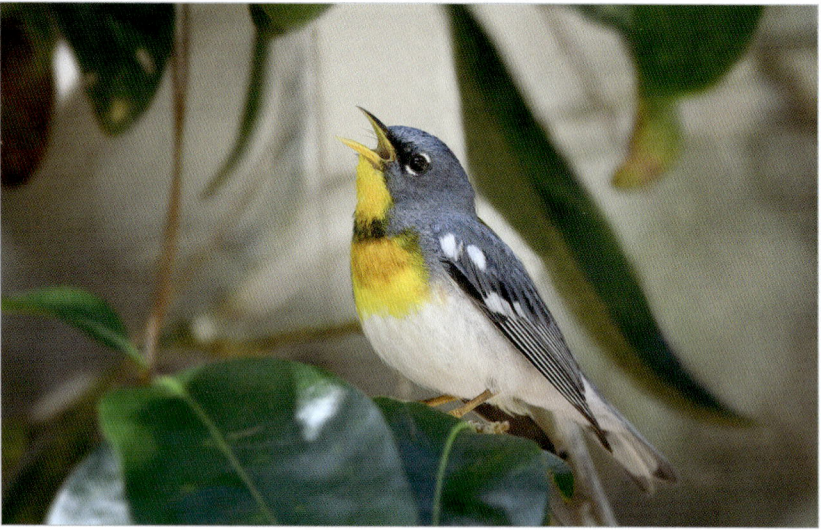

*Northern Parula (Florida, USA)*

# Parula

There are two species of parulas in the Parulidae in the genus *Setophaga*. They have gone through a variety of genus names, including *Parus* and *Parula* prior to the present-day one. Linnaeus assigned the protonym *Parus americanus* in the belief that the New World warbler was a tit and belonged in the Paridae. The Latin word *parulus* is simply the diminutive of the Latin *parus*, meaning "tit or titmouse." This in turn is believed to come from the unidentified bird of ill omen, the *parra*, mentioned in Latin in Pliny's *Naturalis Historia* (AD 77) 583 and Horace's *Odes* (23 BC) that has been variously and tentatively identified as an owl, a lapwing, a titmouse, and a jay. Buffon shed some light on the name in an entry headed "The Titmice":[8]

> In Greek the Titmouse is named Αιγιθαλος [Aigithalos], . . . in Latin, *Parra*, . . . in modern Latin, *Parus, Parix, Mefanga*. In Italy it is called *Parula*.

# Pauraque

The pauraques are three species of nightjars in the Caprimulgidae found in the Neotropics. The origin of the name is unsubstantiated, but the *American Heritage Dictionary* states that it is

American Spanish, perhaps from *pauraque*, *huaraque*, the name of a Native American people that once inhabited the Rio Grande region in Texas and northeast Mexico, or perhaps akin to dialectal Mexican Spanish (Rio Grande region) *parruaca*, *pauraque*, ultimately imitative of the bird's call.

The former theory is not particularly plausible given that, in 1926, Chapman suggested the name is an inaccurate rendering of the Spanish *para que*, which translates as "what for?"[300] He wrote:

I do not know the origin of this name. Possibly it is an inaccurate rendering of the Spanish words *para qué*, when it should be pronounced parakáy not "paurake."

This may work as a mnemonic, especially if one considers the upward inflection of the call and the vacillation between the spellings parauque and pauraque in the late 1800s. Ridgway first used the spelling pauraque in 1881, probably "a typographical error which has been perpetuated ever since" according to Stone, in *The Auk* in 1929:[516] In a 1930 volume of *The Auk*, a correspondent wrote:[614]

I am struck with the probability that the "u" may have been erroneously intruded in the first place and that Sennett made the mistake of writing the Spanish words "Para que?" (i.e. "what for?")—as they sounded when pronounced in the Mexican patois, Pau-ra-que. If the call of the bird is a three syllable call with a rising accent on the last syllable there would be little difference between "pow-rack-kee" and "pah-ra kay."

Skutch also asserted that it is a translation of one of its calls:[517]

The song of *Nyctidromus* is responsible for a number of its vernacular names, including *Pauraque* itself and *Cuyéo*, as this goatsucker is widely called in Costa Rica.

Taking into consideration that so many other local names throughout its range are imitations of its vocalizations, this seems likely. But if Chapman's claim is indeed the case, as he himself suggested, the present-day pronunciation [puh-RAH-kee] is in error, and should instead be something more like [pa-ra-KAY?].

# Peacock

While the peafowl are often colloquially called "peacocks," there is only one species with the official group name peacock,* the Congo Peacock in the monotypic genus *Afropavo*. The two-part name is composed of the elements, *pea* + *cock*. The *pea* in peacock, peahen, and peafowl came into Old English as *pāwa*, which changed over time to the Middle English *pe*, *po*, or *paue*, the derivation of which is the Latin word for "peacock," *pāvō*. The Latin peacock, the *pāvō*, is named for its "alarming" call, from the word *pavor*, meaning "alarm, fright, panic." Appropriate, given the bird's penetrating, trumpet-like vocalizations. The second element of the name, "cock," derives from the Proto-Germanic word *kukkaz*, probably an imitative word, which became *cocc* in Old English. The name has a long history in English literature, appearing in Chaucer's *Canterbury Tales* from the 1300s:

A sheef of pecok arwes, bright and kene,

Under his belt he bar ful thriftily,

(wel koude he dresse his takel yemanly:

His arwes drouped noght with fetheres lowe)

[A sheaf of peacock arrows bright and keen
Under his belt he bore very carefully
(Well could he keep his gear yeomanly:
His arrows had no drooped feathers low)]

The peacock also appeared on the menu in the *Boke of Nurture* as *Pecok* in the mid-1400s.[252] There were numerous spellings of the word at that time, including *pecok, pekok, pocok, pacok,* and *poucock.*

*According to the Clements taxonomy; IOC uses the name Congo Peafowl.

# Peacock-Pheasant

The eight species of peacock-pheasant in the *Polyplectron* genus are small pheasants of Asia with beautiful plumage covered with iridescent ocelli. Though these birds are not particularly peacock-like, Edwards first used the name in his *A Natural History of Uncommon Birds* in 1743.[78]

# Peafowl

The two species of peafowls are similar in appearance, but one of them, the Indian Peafowl, is familiar throughout the world and throughout history. The first element of the name is discussed under Peacock; the second element, *fowl*, evolved from the Old English *fugal*, which became the Middle English *fowel, foul,* or *foghel.* The origin of the word is the Proto-Germanic *fuglaz*, meaning "to fly."

The peacock of European antiquity was the Indian Peafowl; the presence of the species in Europe—and hence the antiquity of the English name—can be traced back to AD 900, as the bones of a bird have been found in a Viking burial site in Norway. In medieval England, it was costly to obtain a "peacock," so they were seen as a symbol of wealth in conspicuous displays of gluttony at feasts. At Richard III's coronation in 1483, forty-eight peacocks were bought for the royal banquet (as well as plovers, cranes, herons, woodcock, and pheasants).

# Pelican

The eight species of pelicans in the Pelecanidae are some of the most recognizable birds in the world. The word first appeared in Old English as *pellicane*, from the Latin *pelecānus*, which in turn was derived from the Ancient Greek πελεκάν *pelekan.* This was taken from the word πέλεκυς *pelekys*, meaning "axe or hatchet," it's thought due to the shape of the bird's bill. One of the first mentions of the name in the English literature was in an account of a voyage to "the Indies of Nova Hispania" in 1564:[175]

Of the sea-fowle above all other not common in England, I noted the pellicane . . . she is very deformed to beholde; for she is of colour russet: notwithstanding in Guinea I have seene of them as white as a swan, having legs like the same, and a body like a hearne, with a long necke, and a thick long beake, from the nether jaw whereof downe to the breast passeth a skinne of such a bignesse, as is able to receive a fish as big as ones thigh, and this her big throat and long bill doeth make her seem so ougly.

# Peltops

These two members of the Artamidae, the woodswallows, bellmagpies, and allies, from New Guinea have synonymous genus and common names. The name was coined by Johann Georg Wagler in 1829 when he disputed Lesson's placement of Lowland Peltops in the genus

*Eurylaimus* (broadbills). The name comes from the Ancient Greek words πέλτη *pelte*, a "small light shield of leather," and ὄψις *opsis*, "face," presumably a reference to the bird's conspicuous white cheek patches or possibly to the robust, hook-tipped bill.

# Penduline-Tit

The Remizidae is where the eleven species of penduline-tits reside. All but one of them (the Verdin) build suspended nests, which not only explains the present-day common name but the name *Picus nidum suspendens* (hang-nest woodpecker) which, according to Aldrovandi, was given to the group by Pliny.[101] These small birds (see Tit) get their names from the nature of these nests—they are "penduline," hanging down swinging from a branch, usually over water. Buffon gave a lovely account of the nest belonging to the bird he calls the Penduline Titmouse:[8]

> The most curious fact in the history of these birds is the exquisite art displayed in the construction of their nest. They employ the light down found on the buds of the willow, the poplar, the aspen, the juncago; in thistles, dandelions, flea-banes, cats' tails, &c. With their bill they entwine this filamentous substance and form a thick close web, almost like cloth: this they fortify externally with fibres and small roots, which penetrate into the texture, and in some measure form the basis of the nest. They line the inside with the same down, but not woven, that their young may lie soft: they shut it above to confine the warmth, and they suspend it with hemp, wild nettles, &c. from the cleft of a small plant branch, over running water, that it may rock more gently, assisted by the spring of the branch. In this situation the brood are well supplied with insects, which constitute their chief food; and they are protected from the rats, the lizards, the adders, and other reptiles, which are always the most dangerous and I am convinced that their conduct really proceeds from foresight; for they are naturally crafty . . . [The nest] of the penduline titmouse resembles sometimes a bag, sometimes a shut purse, sometimes a flattened bagpipe, &c. The aperture is made in the side, and almost always turned towards the water, and placed sometimes higher, sometimes lower; it is nearly round, and only an inch and an half in diameter, or even less, and commonly surrounded by a brim more or less protuberant.

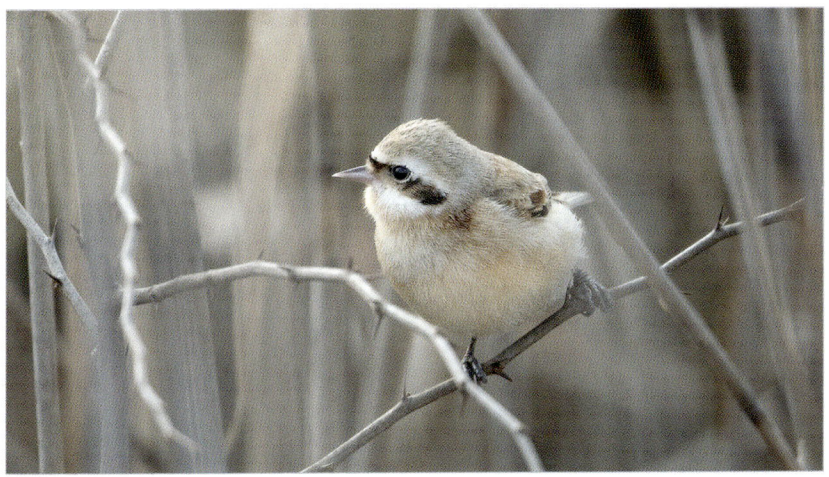

*Chinese Penduline-Tit (Japan)*

He goes on to note:

> The peasants regard them with superstitious veneration: one of these nests is suspended near the door of each cottage, and the possessors hold it as a protector from thunder, and its little architect as a sacred bird.

# Penguin

The Spheniscidae is composed of eighteen species, all with the common group name penguin. These iconic and well-loved birds got their names by default. The name was originally coined for the now-extinct Great Auk, which is the sole member of the genus *Pinguinus.* Richard Whitbourne wrote about the auks, calling them penguins, in his 1620 tome *A Discourse and Discovery of New-found-land*:[445]

> These Penguins are as bigge as Geese, and flye not, for they have but a little short wing, and they multiply so infinitly, vpon a certaine flat Iland, that men drive them from thence vpon a boord, into their boates by hundreds at a time; as if God had made the innocency of so poore a creature, to become such an admirable instrument for the sustentation of man.

The word predates Whitebourne's publication, however. In 1577, Drake's ship log reports seeing an "infinite number of fowle, which Welsh men named *pengwin*."[518] Despite this assertion, there is much debate about the etymology of the word. The word was first used in English and Dutch, possibly accounting for one of the most widely held theories that it comes from the Welsh *pen* for "head" and *gwyn* for "white." But as many linguists, including Yule, have pointed out:[250]

> Unfortunately for this etymology the head is precisely that part which seems in all species of the bird to be black!

Leading Skeat to speculate:[220]

> In that case, it must first have been given to another bird, such as the auk (the puffin is common in Anglesey), since the penguin's head is black.

Another theory is that the name derives from the flipper-like wings as in "pen-winged" or "pinioned." Yet another idea is that the name comes from the Latin *pinguis*, which means "fat" or "greasy," which the Great Auks certainly were, and sailors at the time would have been far more interested in the edibility of the birds than in their appearance. Others, including Liberman, consider an unknown folk etymology from a local North Atlantic Aboriginal language.[386] Finally, and possibly most plausibly, Sayers proposed that the name referred not to the bird's appearance but to its habitat:[385]

> Welsh or Breton *pen gwyn* "white headland," referencing the guano-covered cliffs at the northeastern corner of Funk Island, Newfoundland, a seamark in the Age of Discovery; later used of the black-and-white Great Auk, which bred on the island.

Citing references from the 1500s, Thier stated in 2000:[519]

> The frequent references to Penguin Island, some of which antedate use of the word as a noun denoting the bird suggest that the bird may have been named after the location, rather than vice versa.

When Drake and his sailors first encountered penguins in the Southern Hemisphere, they no doubt simply conflated the two groups of birds, the products of convergent evolution, and applied the name to what looked similar to them, as was common practice at that time.

# Peppershrike

The peppershrikes are two species in the genus *Cyclarhis* in the Vireonidae, two of the largest vireos found in the Neotropics. When the Black-billed Peppershrike was first described in the European literature by Le Vaillant (strangely in *Histoire Naturelle des Oiseaux d'Afrique*), he named it *Le Sourcirou* and referred to it as a "shrike."[93] The birds are still called *Sourciroux* in French, which translates as "red eyebrow." As Le Vaillant noted:

> Its forehead is of a ferruginous red which also occupies all the space between the eye and the nostrils, and then passes over the eyes, extending from behind in a kind of real eyebrow.

Buffon [417] had earlier called the same species the *verderoux*, the "green and red," a good name given that

> the plumage [is] of a more or less dark green, except for the forehead which is reddish on both sides of the head, in which extend two bands of this colour, from the forehead to the height of the neck behind the head.

With regard to the English name, the second element is explained by the early confusion caused by the short, stout, and hook-billed bill so like that of a shrike. The first element could be a corruption of the word *piper* for its melodious song, but more likely it is a reference to the thick, reddish eyebrow that could be said to be reminiscent of the color of a cayenne pepper.

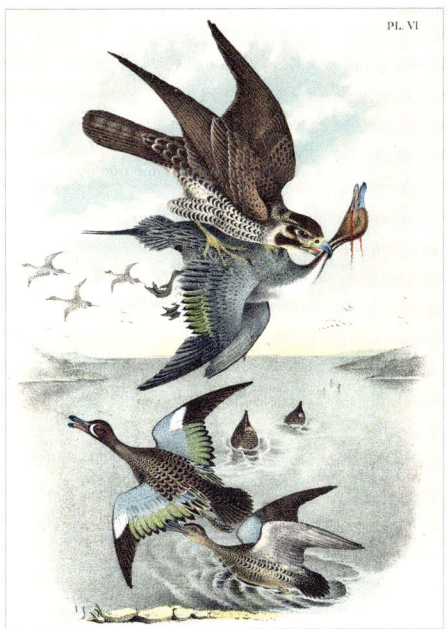

*Peregrine Falcon feeding on Northern Pintail from Studer's* The Birds of North America *(1903)*

# Peregrine

The Peregrine Falcon is often referred to simply as a Peregrine. The name was probably first used in print by Albertus in 1250 when he wrote a chapter entitled "*De falconibus qui peregrini dicuntur*" [*Of the falcons which are called peregrines*].[598] The name comes from the Latin word *peregrinus*, meaning "that which comes from foreign parts, strange, foreign, exotic," a reference to its migratory habits, at least in parts of its range, and widespread distribution.

# Petrel

The name petrel is widely used across three families; there are fifty-seven species, including the giant-petrels and diving-petrels, in the Procellariidae, the shearwaters and petrels; nine species of storm-petrels in the Oceanitidae, the southern storm-petrels, and eighteen species in the Hydrobatidae, northern storm-petrels. (See Storm-Petrel.) The most widely accepted theory for the etymology of the name comes from an entry in William Dampier's *A Voyage to New Holland, &c. in the Year 1699*:[41]

> In a storm they will hover close under the ship's stern in the wake of the ship (as it is called) or the smoothness which the ship's passing has made on the sea; and there as they fly (gently then) they pat the water alternately with their feet as if they walked upon it; though still upon the wing. And from hence the seamen give them the name of petrels in allusion to St. Peter's walking upon the Lake of Gennesareth.

Dampier was, in fact, writing about storm-petrels, as is clear from his description, but there is some debate as to the veracity of his claim regarding the source of the name, which, prior to his first use of the modern spelling, was spelled *pitteral* or *pittrel*. The allusion to St. Peter was, of course, to the apostle Peter's walk on the Sea of Galilee in Matthew 14:28, "petrel" being a diminutive form of the name Peter. Some doubt this conclusion, though, believing that the name *pitteral* predates the name petrel, at least in the spoken form via illiterate sailors of the 1600s. Lockwood maintained that the name is a diminutive formed from "pitter patter," the manner of movement on the water that Dampier refers to in the account of his voyage.[84] Examples of the use of the same diminutive include "dotterel," "whimbrel," and "cockerel."

# Pewee

All but one of the fourteen species of *Contopus* are known as pewees or wood-peewees. The name for these drab-plumaged, but much loved, members of the Tyrannidae is imitative of their vocalizations. Wilson described the behavior of the Eastern Wood-Pewee in *American Ornithology* in 1810:[28]

> It loves to sit on the high dead branches, amid the gloom of the woods, calling out in a feeble plaintive tone, *peto wăy; peto wăy; pee way*; occasionally darting after insects; sometimes making a circular sweep of thirty or forty yards, snapping up numbers in its way with great adroitness; and returning to its position and chant as before.

# Phainopepla

The charismatic Phainopepla in the small family the Ptiliogonatidae, or silky-flycatchers, has a synonymous genus name derived from the Ancient Greek φαίνω *phaínō*, "to shine," and πέπλος *péplos*, meaning "cloak or robe," appropriately for the male's glossy black plumage.

# Phalarope

Three very distinctive members of the Scolopacidae in the genus *Phalaropus* get their names from the Ancient Greek words φαλαρίς *phalarís*, meaning "coot," and πούς *poús*, meaning "foot." Brisson mistakenly concluded, on the basis of the structure of the feet, that they were related to coots and moorhens.[76] Despite Newton's protestations that[17]

> Brisson's maladroit rendering of the "Coot-footed Tringa" of Edwards who, in 1741, showed himself a better judge of its affinities than many others both before and after him, since for a long while some of the best authorities thought the Phalaropes allied

to the Coot, whereas they are unquestionably *Limicolae*, only somewhat modified in accordance with their habit of swimming

the name has stuck, as the scalloped toes of the feet are indeed coot-like.

# Pheasant

There are twenty-nine birds with the name pheasant in their group names, including peacock-pheasants and eared-pheasants. These iconic birds are all members of the Phasianidae, but the name originally applied to only one species, the Common (or Ring-necked) Pheasant. Although not originally found in Britain, they are believed to have been introduced there over two thousand years ago by the Romans, who bred them for the table, the bird having been brought to southern Europe from Asia. In Old English, they were called *wōrhana* from the Proto-Germanic words *wurzô*, for "grouse," and *hanô*, for "cock or rooster."[42] By the 1200s the bird had come to be known in Middle English by similar words of various spellings, such as *fesaunt, fesant, fesande,* and *feysaund,* due to the influence of the Old French *fesan.* Ultimately, the name came from the Ancient Greek φασιανός *phāsianós,* meaning "bird of the river Φᾶσις *Phâsis,*"* the area from which the pheasant was brought to Greece and then spread over Europe.

*The Phasis River, as it was known to the Ancient Greeks, is the present-day Rioni River that flows from the Caucasus Mountains, westward through Georgia to the Black Sea.

# Philentoma

Previously, these two members of the Vangidae, which share their common and genus names, were classified in the Muscicapidae and usually called by the name "flycatcher." These charming Asian forest dwellers derive their name from the Ancient Greek φιλία *philia,* for "love or affection," and ἔντομα *entoma,* "insects," which does indeed describe their feeding behavior.

# Phoebe

Three phoebes are found in the Tyrannidae. These familiar birds of the Americas are named for their vocalizations, which can be transliterated as *fee-bee.* Earlier it was often called the Pewit Flycatcher, as in 1810 in *American Ornithology,* in which Wilson wrote:[28]

> The favorite resort of this bird is by streams of water, under, or near bridges, in caves, &c. Near such places he sits on a projecting twig, calling out *pe-wee, pe-wit-titee pe-wee,* for a whole morning; darting after insects, and returning to the same twig; frequently flirting his tail, like the Wagtail, tho not so rapidly.

# Piapiac

The name of this unusual member of the Corvidae belonging to the monotypic genus *Ptilostomus* is onomatopoeic. The name was given to it by Le Vaillant, who wrote in *Histoire Naturelle des Oiseaux d'Afrique* in 1799:[93]

> I gave the name of Piapiac to the only species of magpie I found in my travels in Africa. . . . This bird seems to me to be of the same species as that which is described in Buffon under the name of the Senegal Magpie but I believe that [he described a juvenile bird]. So this bird is not only found in Senegal, so I thought I was allowed to give it another name, which suits it better and should be adopted, for the reason that it is analogous to the only cry I heard it make, and that he repeats as distinctly as we pronounce it.

## Piculet

The piculets are tiny woodpeckers, the Picidae. *Picus* is the Latin for woodpecker. The *-let* suffix is a diminutive, so they are "little woodpeckers." The name appears to have first been used in print in the 1830s.

## Piedtail

The Ecuadorian Piedtail was first described in 1860 by Gould, who gave it the name Pied-tailed Humming-Bird, stating that the[6]

> singularly coloured tail [renders] it quite distinct from every other member of the entire family.

He described the tail, from a specimen:

> Tail rounded; the four lateral feathers on each side white, with an oblique band of black or blackish purple occupying the centre of each, this band of black extending along the margin of the two outer feathers to the tip, so that the inner web only is white; not so on the next, which is terminated with a large spot or tip of white.

Another species of piedtail in the same genus, the *Phlogophilus*, was described in 1901; it is morphologically very similar to its congener.

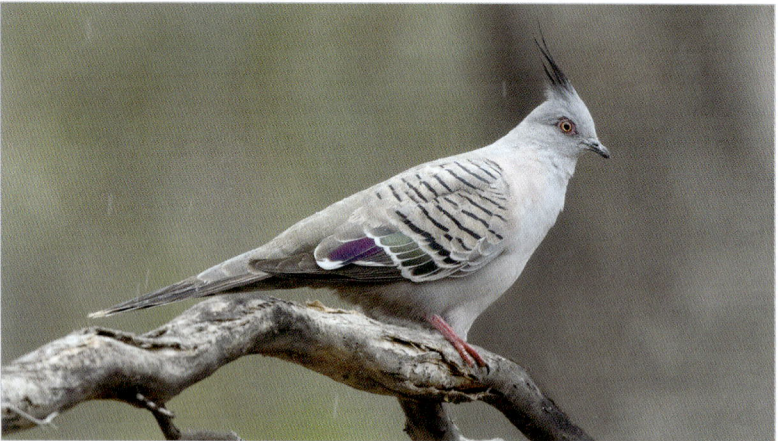

*Crested Pigeon (Victoria, Australia)*

## Pigeon

Everyone knows pigeons! The Rock Dove, usually called "the pigeon" colloquially, is the oldest domesticated bird and, and, thanks to human activities, now has a worldwide distribution. In the Columbidae, 151 species go by the name pigeon. Many of the names are in hyphenated forms, including the wood-pigeon, green-pigeon, mountain-pigeon, rock-pigeon, crowned-pigeon, blue-pigeon, and imperial-pigeon (see that entry), indicating appearance, morphology, habitat, or behavior. The journey of the name begins in the third century with the Latin word *pipionem*, the name for a pigeon derived from the verb *pipio*, "to peep, chirp"—literally a "piper." From the Old French *pijon*, the word arrived in Middle English in the 1300s as *pijoun* or variations on that spelling. Originally it was used as a term for a young bird, as seen in the *The Promptorium Parvulorum* with the entry for "*Pyion*, yong dove."[261] Pigeons appeared on the menu in the *Boke of Nurture* in 1315:[252]

Of quayle / sparow / larke / & litell*e* / m*e*rtinet,

pygeou*n* / swalow / thrusch*e* / osull*e* /

[Of quail, sparrow, lark, and little martin,

pigeon, swallow, thrush, ouzel.]

Prior to the arrival of the word in its Middle English form via the Normans, the Old English name for the Stock Dove, which was the familiar pigeon in England at the time, was *culfre*, which later became *culver*. This word had affiliations with the Latin *columba*, the genus name that Linnaeus introduced in 1758.

# Piha

The nine pihas in the Cotingidae, the cotingas of the Neotropics, are found in two genera. The name is echoic, with Indigenous language roots, of their extraordinarily loud vocalizations, although not all of them conform to this characterization. William Jardine was one of the first to use the common name in *The Naturalist's Library* in 1833 when he listed the Purple-throated Fruitcrow as Red-throated Piha with the observation:[58]

It seems to have no other note than that which resembles the words *pi-hau-hau*, uttered rather in an agreeable than a harsh tone.

He goes on to discuss "Grey Piha," the Screaming Piha, but makes no mention of its name or vocalizations. Later, in 1916, Charles Chubb wrote in *The Birds of British Guiana*:[176]

The "Warraus" called it *Paia-paia*, and the "Macusis" and "Arekunas" *Pai-paischo*.*

*The Warao (meaning "the boat people," after the Warao's intimate connection to the water), the Macushi, and the Pemón (also known as Arecuna, Aricuna Jaricuna, Kamarakoto, and Taurepang) are Indigenous groups in northern South America.

# Pilotbird

This unique member of the Acanthizidae belongs to the monotypic genus *Pycnoptilus*, the name by which it was known until the late 1800s. The name Pilotbird was almost certainly in use by local naturalists prior to this, but the first record of it in print was in an 1893 article in the newspaper *The Argus*, entitled "A Field Naturalists Collection," the author, identified only as D.M., wrote:[520]

Here, close together, are eggs of the lyre bird and the pilot bird—the last very rare, and only found quite lately in the Dandenong Ranges, where the lyre bird, too, has its home.

The bird receives its name from a back-to-front interpretation of its association with the Superb Lyrebird. The Pilotbird is often observed following the lyrebird in order to take advantage of the prey disturbed by its scratching of the soil. Early observers perceived that the Pilotbird "piloted"—in other words, guided—the Lyrebird. Littlejohns & Lawrence noted in *Birds of our Bush* from the early 1900s:[446]

The country beloved of the Lyre Bird is also the home of a species which, though not to the same extent remarkable, is almost as rare. It is said that the affinity of the species so far as locality is concerned is responsible for the name applied to the bird here described. Early observers who sought the Lyre Bird regarded the Pilot Bird as being an unfailing guide to the haunts of the former. Whether there is any truth in this explanation of the bird's name we know not, but we do know that in our experience the two species have invariably been found close together.

# Pinktail

Przevalski's Pinktail belongs to a monotypic family, the Urocynchramidae. Long regarded as either a bunting in the Emberizidae or as a rosefinch in the Fringillidae, recent molecular studies have shown that it is neither of these. Its name was changed in 2000, from the previously used Przewalski's Rosefinch and Pink-tailed Bunting, to Pinktail, in reference to the male's bright pink outer tail feathers.

# Pintail

There are four species of *Anas* ducks known as pintails, all with elongated central tail feathers giving the tail a pointed appearance, hence the common name, which was originally given to Northern Pintail. An earlier common name was "Sea-Pheasant," which John Hill explained in 1752 in *An History of Animals*:[447]

> The feathers which compose the tail are fifteen in number, and their proportions, as well as number and arrangement, are different from those of any other species: the two middle ones are considerably longer than any of the others, and run out into slender and sharp points; 'tis from this Angularity that the bird has obtained the name of the Sea-pheasant. . . . We call it the Craker and the Sea-pheasant.

The Northern Pintail, which would have been well known in Europe, had several now-obsolete names, including Springtail, Cracker, Winter duck, and Lady bird.

# Piopio

Two extinct New Zealand members of the Oriolidae go by the name piopio. These members of the *Turnagra* genus got their common name from the Māori name for the birds. In the 1906 article "Māori Bird Names," the Rev. H. W. Williams listed the Māori names for "*Turnagra tanagra* and *crassirostris*, (N.I. and S.I) thrush" as *korohea*, *koropio*, *piopio*, and *tiutiukata*.[44] Despite Williams's use of the name "thrush," the birds were being referred to as piopios as early as 1868 by authors such as Buller.[43] The Māori names are imitative of the bird's vocalizations.

# Pipipi

This member of the small Mohouidae of New Zealand is the congener of the Yellowhead and the Whitehead in the *Mohoua* genus. Williams listed the Māori names for "*Finschia novae zealandiae,* brown creeper" (the protonym and old common name) as *pipipi*, *pipirihika*, *titirihika*, and *toitoi*.[44] Other names include Brown Creeper, New Zealand Creeper, and New Zealand Titmouse. The Māori names are imitative of the bird's vocalizations.

# Pipit

In the Motacillidae, the wagtails and pipits, there are forty-four species with the common appellation pipit. Perhaps surprisingly, the word has a common origin with that of "pigeon." The Latin root is the word *pipio* or *pipire*, "to peep, pip or chirp." Going back even further, we can trace the word to the Ancient Greek πίπρα *pípra*, an imitative word used by Aristotle to name several small birds. Bewick wrote about the "Grasshopper Lark" in 1797, referring to the Tree Pipit, explaining:[149]

> It has been called the Pipit Lark from its small shrill cry, and in German *Piep-lerche* for the same reason. Mr White observes, that its note seems close to a person,

though at an hundred yards distance; and when close to the ear, seems scarce louder than when a great way off.

No doubt many modern-day observers can relate to Mr. White's frustrations!

# Piprites

These three small, chunky members of the Tyrannidae, with synonymous genus and common names, could be mistaken for manakins at first glance. This is reflected in the name, which is a compound of two Ancient Greek words, *Pipra*, one of the genera in the Manakin family, and the suffix -ίτης -*ites*, "those belonging to, resembling." The name πίπρα *pípra* can be found in Aristotle's *Historia Animalium* in about AD 610, referring to a small chirping bird of indeterminate species.[109]

# Pitohui

The seven species of pitohuis were previously all classified in the Pachycephalidae, the whistlers and allies, but recent studies have shown that four of those species belong in the Oriolidae. The pitohuis are unique among birds as they possess powerful neurotoxins in their skin and feathers, which are probably obtained from the birds' diet of *Choresine* beetles. It's believed these batrachotoxins provide a chemical defense against ectoparasites or against predators such as snakes, raptors, and humans. The name was coined by René Lesson, as the genus name, which still applies to the four members of the Oriolidae. In an entry about the bird now known as Northern Variable Pitohui in 1828's *Manuel d'Ornithologie*, Lesson mentioned the local vernacular name, which is thought to be onomatopoeic:[152]

> The Cape Gray Vanga inhabits the forests of New Guinea, to the Dorery Islands,* where the Papuans call it *pitohui*.

*Doréry or Dorey, now Manokwari harbor, West Papua.

# Pitta

For the keen birder, this family of forty-six species of jewel-thrushes, as they were once known, is considered to be one of the most charismatic and enigmatic families. All the members of the Pittidae are known by the common group name pitta. The name first appeared in

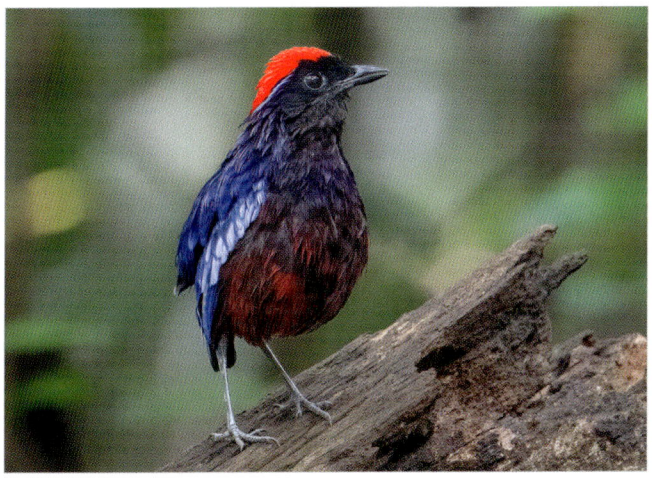

*Garnett Pitta (Peninsular Malaysia)*

the European literature in John Ray's *Joannis Raii Synopsis Methodica Avium & Piscium* in Latin in 1713.[521] In an entry entitled "Pica Indica *vulgaris,*" Indian Pitta, accompanied by Buckley's very unrealistic illustration of the bird, the "Madrass Jay," as it was then known, is remarkably likened to the Eurasian Jay. But Ray usefully informs us of the Telegu name (called Gentoo by Ray), listing it as "Ponnunky Pitta."[16] The present-day Telegu (the language of the Indian states of Andhra Pradesh and Telangana) name పొన్నంకి పిట్ట is usually transliterated as "Ponnangi Pitta." The word pitta is due to the romanization of the Telegu word and means simply "small bird." Compare this with some other names for birds in Telegu, such as "woodpecker," వడ్రంగిపిట్ట *Vaḍraṅgipiṭṭa*, and "kingfisher," లకుముకిపిట్ట *Lakumukipiṭṭa*. In *The Birds of India*, published in 1862, Jerdon's entry for "Pitta Bengalensis" [*sic*] gives the Telegu name as *Pona-inki* (as well as the Hindi name *Nourang*, meaning "nine-colored bird") with the common name "Yellow-breasted Ground-thrush."[136]

# Plains-wanderer

Gould appears to have been the first to have used the common name for this unusual sole member of the Pedionomidae. He described it in 1840 in his "The Birds of Australia," calling it the Collared Plain Wanderer. The word Plain was then pluralised and the descriptive "Collared" dropped possibly as late as the 1970s. Gould[40] wrote that:

> The structure of this singular little bird is admirably adapted for inhabiting those extensive and arid plains which characterize many of the central portions of Australia.

# Plantain-eater

The two species of plantain-eaters are in the same family as the turacos and the go-away-birds, the Musophagidae—the banana eaters. The family was established by Lesson in 1828, and the name derives from the Latin for "banana," *musa* (from the Arabic موزة *mawza*), and the Ancient Greek φάγος *phágos*, "eater or glutton."[152] Latham transliterated Isert's 1788 genus name *Musophaga* for Violet Turaco as plantain-eater.[17] It seems the common name was extrapolated to all members of the family before the other names came into use, leaving only the members of the genus *Crinifer* as plantain-eaters. In fact, these frugivorous birds are by no means exclusive banana eaters.

# Plantcutter

These three members of the remarkably diverse Cotingidae belong to the *Phytotoma* genus, which itself means "plant cutter" from the Ancient Greek name for "plant," φυτό *phyto*, and "cutting," Τομή *tomi*. These folivorous birds have robust, finch-like bills with finely serrated edges—an adaptation that allows them to cut leaves and stems, and it is from this behavior that they get their names. Latham was the first to use the common name in 1802 for the Rufous-tailed Plantcutter, making the observation:[17]

> Its food is vegetables, perhaps preferring the parts next the root, for with much pains, it digs about and cuts off the plants with its bill, as it were with a saw, close to the ground from this circumstance, it does much injury to the gardens, and is detested by the inhabitants.

# Ploughbill

This unique bird, the Wattled Ploughbill, from New Guinea belongs to the monotypic Eulacestomatidae. The name highlights the bird's very unusually chunky bill. Although

Sclater didn't use the common name in his article "On a Rare Passerine Bird from New Guinea" in the 1904 issue of *The Ibis*, he did give one of the first descriptions of the bird:[522]

> Easily recognised by its excessively compressed bill (which is somewhat like that of the Neotropical form *Cyclorhis*\*) and the curious rictal wattles of the male.

\**Cyclarhis* peppershrikes

# Plover

The name plover spans three families, the monotypic Pluvianellidae, the Magellanic Plover; the monotypic Pluvianidae, the Egyptian Plover; and the Charadriidae, the plovers and lapwings, with thirty-five species known as plovers. The name is often hyphenated, as in golden-plover, sand-plover, sandpiper-plover, and so on. This English common bird name has a long history, first being recorded for an unknown species\* in 1312 as a "pluver" when fifty of them were listed for purchase for the table. In 1315, the bird appeared in the *Boke of Nurture*:[252]

> Sawce gamelyň to heyroň-sewe / egret / crane / & plovere;
> [Cameline sauce for common heron, egret, crane and plover;]

There are three theories as to the origin of this old name, which was spelled in various ways in Old and Middle English, including *plouier, ploware, plowere, pluwer, plovere, plower*, and *pluuer*. The first is that the Anglo-Normans brought the name from the Old French *plovier*, which was derived from the Latin *plovarius*, which is of disputed origin; perhaps from the Latin for rain, *pluvia*. This theory is reinforced by the German name for the Charadriidae, *Regenpfeifer*, which translates to "rain piper." Skeat wrote:[220]

> Formed as if from a Low Lat. *pluviarius*, equivalent to Lat. *pluvialis*, belonging to rain, because these birds are said to be most seen and caught in a rainy season.

The possible association between plovers and rain may be due to the timing of the arrival of migratory birds in Europe coinciding with the rainy season, or possibly from the supposed restlessness they display when rain is approaching. Another theory, the least plausible, is that the name refers to the spotted plumage, which looks like raindrops, of the Golden-Plover. And lastly, Lockwood (and others) maintain that the name is imitative of the bird's drawn-out, clear *plō* whistles and that it was[84]

> needless now to add that the many attempts to find a rational connection between the Plover and rain, by writers ancient and modern, have necessarily been in vain.

\*The lapwing, which was often called a plover colloquially, is also mentioned in the *Boke of Nurture* as the "lapewynk."

# Plovercrest

These two hummingbirds in the genus *Stephanoxis* are named for their long, upright crests. The crest is very similar to that of the Northern Lapwing, which was often called simply "the Plover" in England in the 1800s. Gould described the crest in *A Monograph of the Trochilidæ*:[6]

> The lengthened crest, which adds so much to the beauty and elegance of the *Cephalepis DeMandi*, terminates in a single elongated plume,—a very remarkable circumstance, since the feathers of birds are usually arranged in pairs.

## Plumed-Warbler

The two species of *Micromacronus* plumed-warblers in the Cisticolidae are found only in the Philippines. Long treated as members of the Timaliidae, they were previously known as "miniature-babblers." Oliveros et al. introduced the name in 2012 with the comment:[601]

> The term "Plumed Warbler" is a more appropriate English name for this genus because of its distinctive extended plumes on the flanks and back.

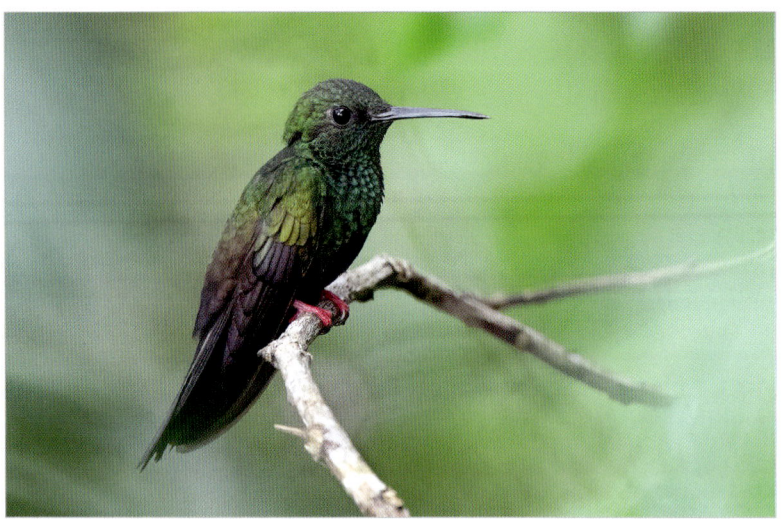

Bronze-tailed Plumeleteer (Costa Rica)

## Plumeleteer

These two large hummingbirds in the *Chalybura* genus receive their names from their distinctive undertail coverts, which, according to Gould, are[6]

> snow-white, and so largely developed as to form conspicuous plumes.

The name is a three-part composition from *plume* and the diminutive *-let* plus a suffix of *-eer*, denoting something associated with, in this case, plumes.

## Plushcap

This atypical tanager in the Thraupidae belongs to the monotypic genus *Catamblyrhynchus*. Its namesake feature is the bright yellow patch of short, dense feathers on the forecrown. Initially, it was placed in the Fringillidae, the finches, and for a long time was known as the "Plush-capped Finch." It is thought the unusual plumage may be an adaptation to its specialized feeding mode, in which it probes dense whorls of bamboo for its insect prey, which would otherwise damage less robust feathers.[615]

## Plushcrown

The Orange-fronted Plushcrown is a unique member of the Furnariidae in the monotypic genus *Metopothrix* with distinctive yellow and green plumage, the only bird in the family so adorned. The soft, bright yellow plumage on the forecrown accounts for the name, although it was previously called the "Manikin Woodhewer" in *The Birds of South America* for unexplained reasons.[45]

# Pochard

The original pochard, in the sense of the use of the name, was the Common Pochard, which is widespread in Europe as well as parts of Africa and Asia. The name has been in use since the 1500s, when variations on the name were used, including *pockard* and *poker*. There are now six species of pochards in the Anatidae, the ducks, geese, and waterfowl, following the European expansion across the globe. Newton was uncertain of the origin of the name:[7]

> The derivation of these words . . . is very uncertain. Cotgrave has *Pocheculier* which he renders "Shoveler," nowadays the name of a kind of Duck, but in his time meaning the bird we commonly call SPOONBILL. Littré gives *Pochard* as a popular French word signifying drunkard. That this word would in the ordinary way become the English Pochard or Poker may be regarded as certain; but then it is not known to be used in French as a bird's name.

The true etymology lies in the bird's feeding behavior as it upends in the water, seemingly to "poke" the lake bed. "Poker" was an English nickname for various species of ducks when Ray and Willughby wrote about the "Poker, or Pochard" in 1678.[16] The latter is from *poach*, "to poke, thrust," from the Middle French *pocher*, in turn from the Proto-Germanic *puk*. The *-ard* suffix is a pejorative signifying someone who is in a certain condition (like a drunkard).

# Pond-Heron

The pond-herons are four species of *Ardeola* in the Ardeidae. They are often found on small ponds, giving rise to the English name, probably coined by the English colonists in the subcontinent in the 1800s, and shared by four of the six species in the genus.

# Poorwill

There are five species of poorwills, small caprimulgid nightjars from the Americas. Like the closely related whip-poor-wills, the name is imitative of the vocalizations, in this case of the Common Poorwill, having been originally given to this well-known North American bird. Audubon, who called the bird the "Nuttall's Whip-poor-will," wrote of his experience of the bird in the field, describing the calls:[54]

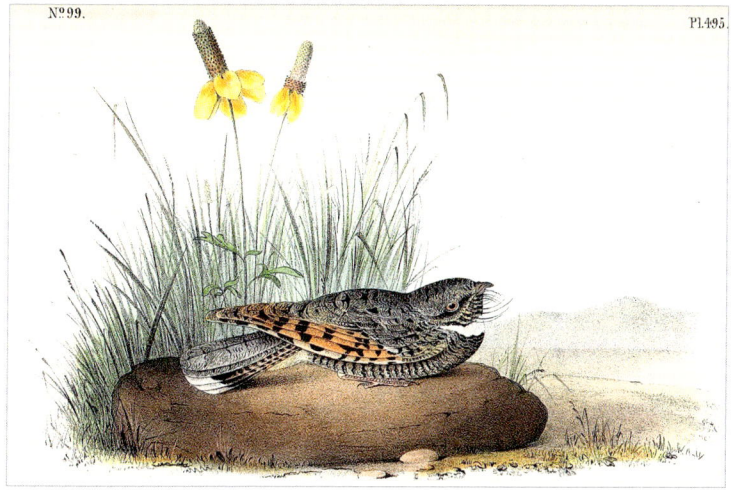

*Common Poorwill in Audubon's* The Birds of America *(1839)*

The sounds we heard, were indeed those of a Whip-poor-will, cut short of much of their compounds, for it was reduced to the syllables *Oh-will, Oh-will, Oh-will*, repeated often and as quickly as is the fashion of our own common species.

Later, while still calling it the Nuttall's Whippoorwill [*sic*], Baird wrote:[21]

This species is said to have a note somewhat similar to that of the whippoorwill, except that the first syllable is omitted leaving the sound something like that of "poor-will."

# Poo-uli

This member of the Fringillidae is a Hawaiian honeycreeper in the *Melamprosops* genus that, amazingly, was only discovered in 1973 by college students on a University of Hawaii expedition to the Haleakala Volcano on Maui Island. The name, given to it by the renowned authority on Hawaiian culture, Mary Kawena Puku'i, is Hawaiian for "black-faced" or "black head," referencing its striking bandit mask.

# Potoo

The potoos are a family of seven species, the Nyctibiidae, all going by the name. The name was first used in the mid-1800s and derives from Jamaican Creole, which was heavily influenced by the Akan language of Ghana. Although it's often said that the name is of imitative origin, the Akan word for owl is *Patuo*, which undoubtedly is the true etymology.

# Prairie-Chicken

The impressive prairie-chickens are two species of grouse in the Phasianidae found in the vast grasslands of North America. Members of the Phasianidae are often referred to as "chickens," but this is the only bird in which the name appears in the official common English name (it is probably a Proto-West Germanic *kiukīn*, also meaning "chicken"). The word "prairie" is the name for the American flat grassland that is their home. The word is a loanword from the French *pré* or *prairie*, meaning "meadow, grassland, or pasture," itself derived from the Latin word *pratum*, for "meadow."

# Pratincole

The small Glareolidae, the pratincoles and coursers, contains eight species called pratincole. The name was first coined by the German naturalist Wilhem Heinrich Kramer in 1756, gaining currency when Pennant adopted it in his influential *Genera of Birds*.[281] He wrote about the Collared Pratincole, which he called simply "Pratincole":

Pratincola, or inhabitant of meadows, a name given it by Dr. Kramer, and adopted by me; placed by Linnaeus with the Hirundo, by Brisson among his Glareolae.

The Latin root of the word, which is echoed in the Collared Pratincole's scientific name, *Glareola pratincola*, is *pratum*, for "meadow," and *incola*, "inhabitant."

# Prickletail

The Spectacled Prickletail is a single species of *Siptornis* in the Furnariidae of the eastern Andes. The bird's namesake feature is the graduated tail with stiff central rectrices, the shafts of which are virtually devoid of barbs.

# Prinia

The prinias are found in Africa and Asia, with thirty species in the Cisticolidae, the cisticolas and allies. The name was first used by Thomas Horsfield in 1821 when he erected the genus name with the comment that the name is "*Prinya* Javanis" [Prinya in Javanese].[46] This is a transliteration of the Javanese word *prenjak,* said to be the name of the Bar-winged Prinia but actually a name used in Indonesia for a number of species of small bird. In colloquial Indonesian, *prenjak* is used for other birds, including the tailorbirds, and Latham actually lists "Batavian Barbet" with the comment that it "inhabits the Island of Java, called there Prinya."[17]

# Prion

These dwellers of the southern oceans are six species of *Pachyptila* in Procellariidae, the shearwaters and petrels. The name refers to the comb-like lamellae on the upper mandible (for filtering plankton) that looks like a saw, which in Ancient Greek is πρίων *priōn.*

Note: the etymology of the microbiological prion is unrelated; it is a portmanteau coined in 1982 from the words "protein" and "infection" and should properly be pronounced [PREE-on], while the bird name is pronounced [PRY-on].

# Ptarmigan

The three species of *Lagopus* ptarmigans are members of the Phasianidae. These lovely birds, with their "rabbit feet" (*lago,* "hare," *pus,* "feet"), were known in Scottish Gaelic as *tarmachan,* simply their name for the Willow Ptarmigan, with the meaning of "croaker," for its low, growling calls. The *pt-* was added later in the mistaken belief in a Greek origin, as if related to *ptero-* ("wing"). There is still a mountain in Scotland called Meall nan Tarmachan, meaning "Hill

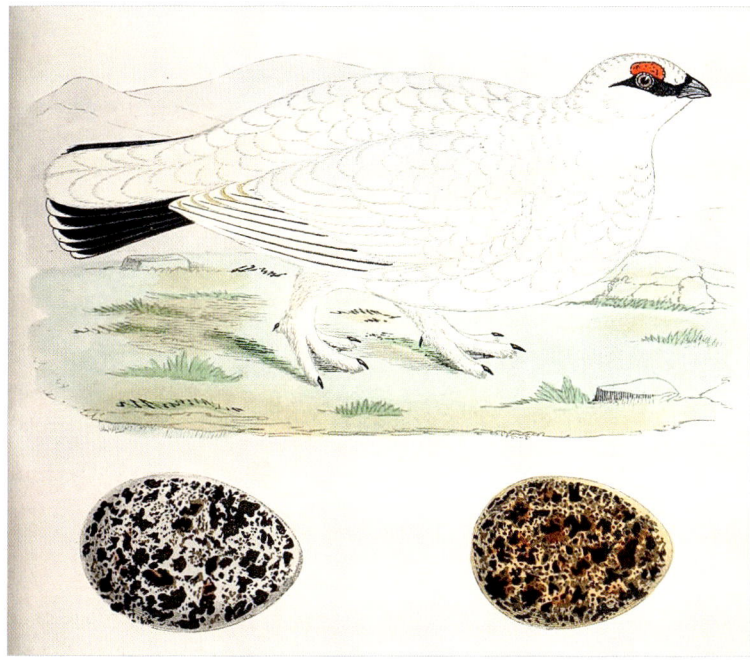

*Rock Ptarmigan from Des Murs's* Oiseaux d'Europe *(1886)*

of the Ptarmigan." Oddly enough, despite the origin of the name, the subspecies of Willow Ptarmigan, which is native (endemic) to the British Isles and now lumped in with the more widespread/Holarctic Willow Ptarmigan, is known as the Red Grouse in the United Kingdom.

# Puaiohi

The Puaiohi is the odd one out of the Hawaiian thrushes, although it, too, belongs to the genus *Myadestes*. It not only looks noticeably different from its four congeners, the Kamao, Omao, Olomao, and Amaui, but it is the only one whose name is not a corruption of manu-a-Maui, Maui's bird. (See Hawaiian Thrushes.) In *Aves Hawaiienses* the Kamao is listed with a note:[47]

> Our natives always said there was a different variety [on Kauai] called the *Puaiohi*, which they said had a different note from the common Kamao. I never believed much in what the natives said about it, as the Kamao varies so much in colour and spots. This bird may be more common on the windward side of the island, as the name of Puaiohi is more commonly used there than here.

As it turned out, the "natives" were correct, and the Hawaiian name Puaiohi is onomatopoeic for the bird's song.

# Puffback

The six species of *Dryoscopus* puffbacks in the Malaconotidae, the bushshrikes and allies, bear quite a similarity to the shrikes of the Lanidae and in the earlier literature were called "puff-back shrikes." As Hopkinson noted in *The Birds of Gambia* in 1909:[49]

> The feathers of the rump are long, white and downy, and form the "puffback."

# Puffbird

The Bucconidae, the puffbirds, are made up of thirty-six species and ten genera; of these, twenty-four are known as puffbirds (including two striated-puffbirds). According to Swainson in *Zoological Illustrations* from 1820:[50]

> At such times the disproportionate size of the head is rendered more conspicuous by the bird raising its feathers so as to appear not unlike a puff ball; hence the general name they have received from the English residents in Brazil; of which vast country all the species, I believe, are natives. When frightened, their form is suddenly changed by the feathers lying quite flat.

# Puffin

The three species of much-loved puffins are members of the Alcidae, the auks, murres, and puffins. But the name didn't always belong to them. It was originally used for the Manx Shearwater in the Isle of Man and North Wales (Ray and Willughby[16]) and this name survives in its scientific name, *Puffinus puffinus*. The word came into English from the Middle Cornish *pópa* and Breton *poc'han*; in Middle English, the shearwaters were called *poffin, poffoun*, or *puffon*. The accepted etymology is that the word comes from *puff*, meaning "swollen" or "fat," with the suffix *-ing*, in reference to either the shearwater's fat, fluffy chicks or to their cured carcasses. The name was subsequently applied to any sort of seabird with similar appearance and habits, giving rise to the present-day use of the name.

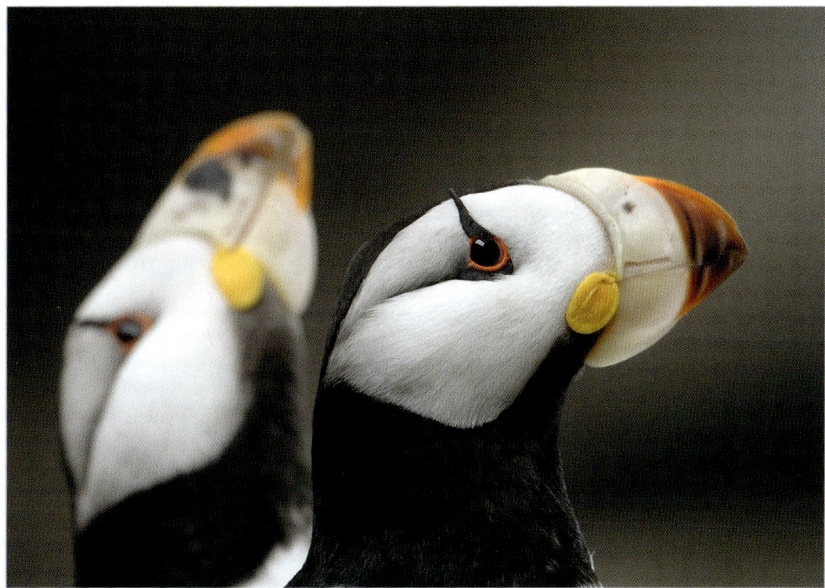

*Horned Puffin (Alaska, USA)*

# Puffleg

The fourteen pufflegs are hummingbirds in two different genera, all sharing the distinctive fluffy plumage around the legs. As Wood wrote in *The Illustrated Natural History* in 1862:[85]

> Several of the Humming-birds are remarkable for a tuft of pure white downy feathers which envelop each leg, and which has obtained for them the popular title of puff-legs, because the white tufts bear some resemblance to a powder-puff.

# Purpletuft

The *Iodopleura* are three species called purpletufts in the Tityridae, the tityras and allies. Prior to recent molecular research, they were classified as cotingas. Previously, as in *The Birds of South America* in 1912, they were called "chatterers."[45] The present name is relatively new and refers to the purple tufts of feathers on the male's flanks that are prominent only during displays of courtship or aggression.

# Pygmy-Goose

The three species of pygmy-geese in the genus *Nettapus* in the Anatidae are in fact ducks and not geese. Two are found in the Asia-Australasia region and one in Africa. The genus was erected by Brandt in 1836 from the Greek ναττα *nētta*, for "duck," and πους *pous* for "foot." He described the birds in Latin:[608]

> Quae quidem rostro et collo Anserem, pedum structura autem et corpore Anatem se praebet [It is with the beak and neck of a goose, however the structure of the body and feet is that of a duck.]

The common group name appears to have been first used by Gould in *The Birds of Australia* in 1848,[40] but prior to that he presented a specimen of a Cotton Pygmy-Goose at the Zoological Society in 1842 as "a new species of goose."[347] It seems his equivocation has given us a wholly unsuitable common name.

# Pyrrhuloxia

This is a unique name for a cardinal in the Cardinalidae that was coined by Bonaparte in 1850 when he introduced it as the genus name, now obsolete. In his *Monographie des Loxiens* [*Monograph of the Crossbills*], Bonaparte wrote:[51]

> By its remarkable beak, the *Cardinalis sinuatus* . . . which in truth could constitute a separate genus (*Pyrrhuloxia*?), which the hues might include among the crossbills, here attaches the genus *Cardinalis*, whose loose plumage also resembles that of the genus *Paradoxornis*, and which includes two other certain species, besides a few doubtful ones.

Some theorize that the name is from the Greek, πυρρός *purrhos*, meaning "fire red" and λοξος *loxos*, meaning "oblique or crooked," referring to the peculiar short, crooked bill of this bird of the American Southwest desert. But the evidence from Bonaparte's writings suggests this is not the case; rather it is portmanteau of the genera *Pyrrhula*, bullfinch, and *Loxia*, crossbill. He considered the bird to be intermediate between two taxa based on the bills and plumage colors.[52]

> Medius quasi inter Paradoxornitheos et Loxiinas. [It is midway between the parrotbills or finches and the crossbills.]

Ridgway used the common names Arizona Pyrrhuloxia and St. Lucia Pyrrhuloxia in *The Auk* in 1887.[523]

# Pytilia

Five species in the Estrildidae have synonymous common and genus names. The name is believed to have been coined by Swainson in 1837 by constructing a diminutive from the genus *Pitylus*, the protonym of the *Pheucticus* grosbeaks, a reference to their short, stout bills no doubt.[524] (This genus name *Pitylus* in turn is based on the name of a type of small bird mentioned by Hesychius, the Greek grammarian who compiled the richest lexicon of Ancient Greek words in the fifth century.) But right from the start, the name caused ornithologists to scratch their heads. A report from the German-language *Journal für Ornithologie* of a meeting held in Berlin 1877 stated:[525]

> Mr. Cabanis takes this opportunity to discuss the etymology of the genus name *Pytelia* by Swainson. Because he could not find a classical root for the word, the lecturer suggested the generic name *Zonogastris* in 1851 and this has also been accepted on various occasions. Swainson first wrote *Pytilia*, but then because the English have to write an "e" if they want to speak "i," they wrote Pytelia. If one wants to correct the name absolutely, one could assume that it was supposed to denote a diminutive of *Pitylus*, (in the figurative sense about the same as our German "Ruderfink"). The latter name was erroneously spelled *Pythilus* instead of *Pitylus* by Boie (1826). So it is possible that Swainson actually meant to say *Pitylia*. At least the name would no longer be completely nonsensical.

# $Q$ IS FOR QUAIL

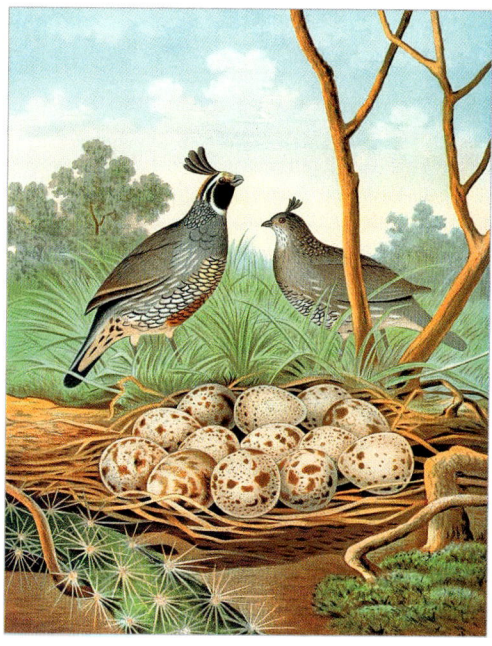

*California Quail in Gentry's* Nests and eggs of birds of the United States *(1882)*

## Quail

The name quail is used for birds in two families. In the Phasianidae (pheasants, grouse, and allies) there are eleven species (also four species of bush-quail), while in the New World quail family, the Odontophoridae, there are ten species of quails (also fifteen wood-quails). The name came into English in the 1300s, replacing older ones such as *ersc hen* (stubble-field hen) and *edisc hen* (field hen), via the Old French *quaille*, possibly from the Medieval Latin *quaccula* or from the Frankish (Old German) *kwahtila* or *quahtala*.[84] These names were imitative of the bird's cry. In Middle English, there were various spellings, including *quayle*, *quaile*, and *quaille*. The quail is mentioned as part of a royal feast in the *Boke of Nurture* from 1460 by John Russell:[252]

> Of quayle / sparow / larke / & litelle / mertinet,
>
> pygeoun / swalow / thrusche / osulle / ye not forgete,
>
> þe legges to ley to your souereyne ye ne lett,
>
> and afturward þe whyngus if his lust be to ete.

> Of quail, sparrow, lark, and little martinet
>
> pigeon, swallow, thrush, ouzel, and don't forget
>
> give your lord the legs first.
>
> and afterward the wings if his lust is to eat.

## Quail-Dove

There are nineteen quail-doves in four different genera in the Columbidae. All are found in the Americas. The combination of their small size, shy terrestrial habits, and plump body accounts for the common name, but it may also have been influenced by its noisy flight when flushed. Bendire wrote in the *Life Histories of American Birds* in 1892:[348]

> Its flight is noisy when starting, similar to that of the European Partridge, from which it receives its misleading name "Perdiz." [closely related to quails]

## Quailfinch

This is a single species of estrildid finch in the genus *Ortygospiza* found in Africa. The name is a compound of the two names "quail" and "finch." The quail part of the name probably comes from its habits, which could be said to be quail-like, in that they are said to spend most of their lives on the ground and will flush when approached. In earlier literature, it was referred to as the "Quail-Finch."

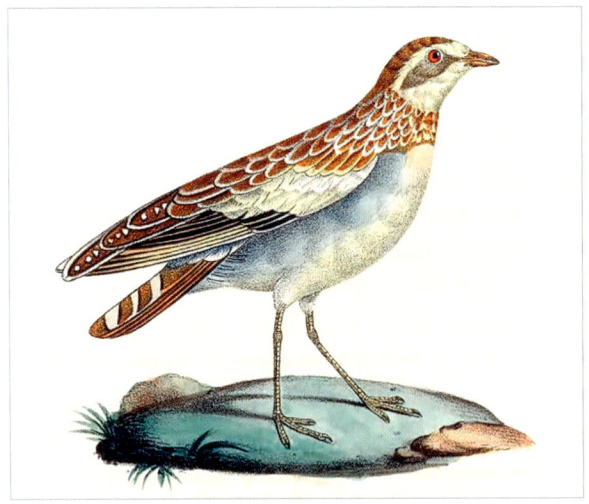

*Quail-plover in Vieillot's* La Galerie des Oiseaux *(1825)*

## Quail-plover

This unusual bird is the only member of the Turnicidae, the buttonquails, outside the genus Turnix. The sole member of the *Ortyxelos*, the Quail-plover is found in tropical Africa and was first described by Vieillot from a specimen in 1819. Vieillot himself was confused about the bird's affinities, a state of affairs that is still reflected in its common name as it is neither a quail nor a plover but has a superficial resemblance to both. Vieillot wrote in 1825:[357]

> We have no positive information on the instinct and habits of the unique species contained in this division; we only know that it is in Senegal. We notice in it some connection with the *Turnix* . . . but having since examined it with the greatest attention, we have determined that it was misclassified . . . , since it differs from it mainly by its otherwise shaped beak, by the length of its feet, and especially by its semi-bare legs. These last two attributes bring it closer to the *Waders* than to the *Gallinaceae*, all of which always have their legs completely covered with feathers; and we suspect from these characters that the *Ortyxeles* inhabits the marshes not

frequented by the Turnixes, as far as I know. It is based on this suspicion that we imposed the generic name by which we describe it. If we are in error, it is up to the naturalist who will observe it in living nature to correct us.

# Quail-thrush

The quail-thrushes belong to the small Australasian Cinclosomatidae of quail-thrushes and jewel-babblers. There are eight species, all in the genus *Cinclosoma*. Their unimaginative name given to them by the European colonists of Australia derives from their superficial similarity to both quails and thrushes, both in appearance and habits. Gould named the birds ground-thrushes but did make allusions to quail, writing "When suddenly flushed it rises with a loud burring noise, like the Quail or Partridge."[40] He put it in the Merulidea, the former name for a family of thrushes. The name quail-thrush arguably makes its first appearance in a 1911 issue of *The Victorian Naturalist*, in which H. B. Williamson wrote in an article called "A Cycle Trip Through East Gippsland":[143]

> A new acquaintance I also made in the Spotted Ground-bird—suitably named, for it rarely leaves the ground and logs. When it does fly, it rises with a noise like that of a quail's flight, hence one of its local names—"Quail Thrush."

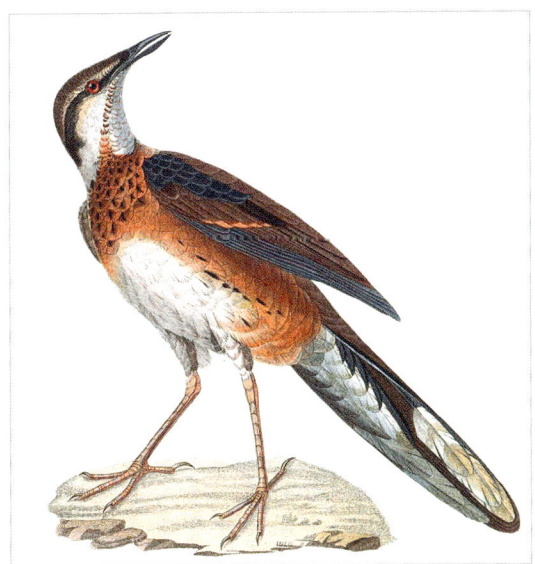

*Painted Quail-thrush from Temminck's* Tableau Méthodique *(1838)*

# Quelea

The quelea are three species of weavers in the Ploceidae, one of which is said to be the most numerous bird species in the world. The synonymous common and genus names probably derive from the Latin word for quail, *qualea*, which is odd given their distinctly un-quail-like appearance. Linnaeus named the Red-billed Quelea in 1758 in his landmark *Systema Naturæ*, recognizing it as a bunting and giving it the binomial name *Emberiza quelea*, but he gave no explanation.[474] In a 1973 issue of *Bokmakierie*, the journal of the South African Ornithological Society, M. W. Jeffreys made an etymological connection between the plague-like swarms of the queleas that devastated crops in modern Africa and the huge numbers of quail (qualea) that invaded Israelite camps.[399] He quoted the fourth book of Moses 11:31:

A wind arose, sent by the Lord, and made quail come from the sea and let them fall on the camp, a day's journey around the camp, two cubits high on the earth.

Deepening the mystery, Gotch, in his *Birds: Their Latin Names Explained* believed that the name comes from an unspecified African language name for the bird.[398] But there appears to be little support for this, and it seems likely he confused this with another colloquial English name, "Dioch," a Wolof and Nigerian name for the Red-billed Quelea.[400]

# Quetzal

The quetzals are six species of spectacular green and red members of the Trogonidae (trogons) in two different genera, *Euptilotis* (one species) and *Pharomachrus* (five species). In 1896 Newton wrote:[7]

QUESAL or QUEZAL the Spanish-American name for one of the most beautiful of birds, abbreviated from the Aztec or Maya *Quetzal-tototl*, the last part of the compound word meaning fowl, and the first, also written *Cuetzal*, the long feathers of rich green with which it is adorned. . . . the Mexican deity *Quetzal-coatl* had his name, generally translated "Feathered Snake," from the *quetzal*, feather or bird, and *coatl*, snake . . . *Quetzal-itzli* is said to be the emerald.

The name came into English from the Nahuatl (Aztec) language *quetzalli* via the Spanish. The Nahuatl word *quetza* means "to raise or lift"; could the Nahuatl name literally mean "flying emerald"?

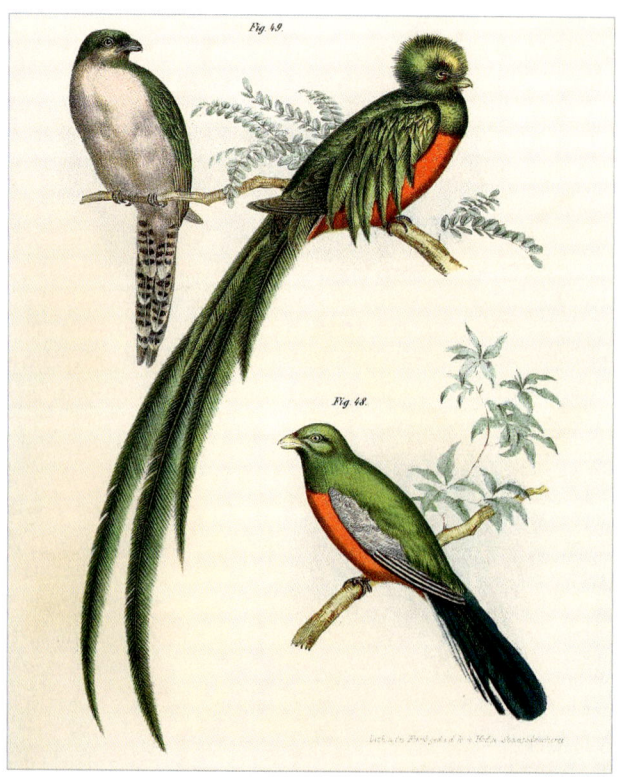

*Resplendant Quetzal (and Narina Trogon) in* Fitzinger's Bilder-atlas zur Wissenschaftlich-populären Naturgeschichte der Vögel *(1864)*

# R IS FOR RACKET-TAIL

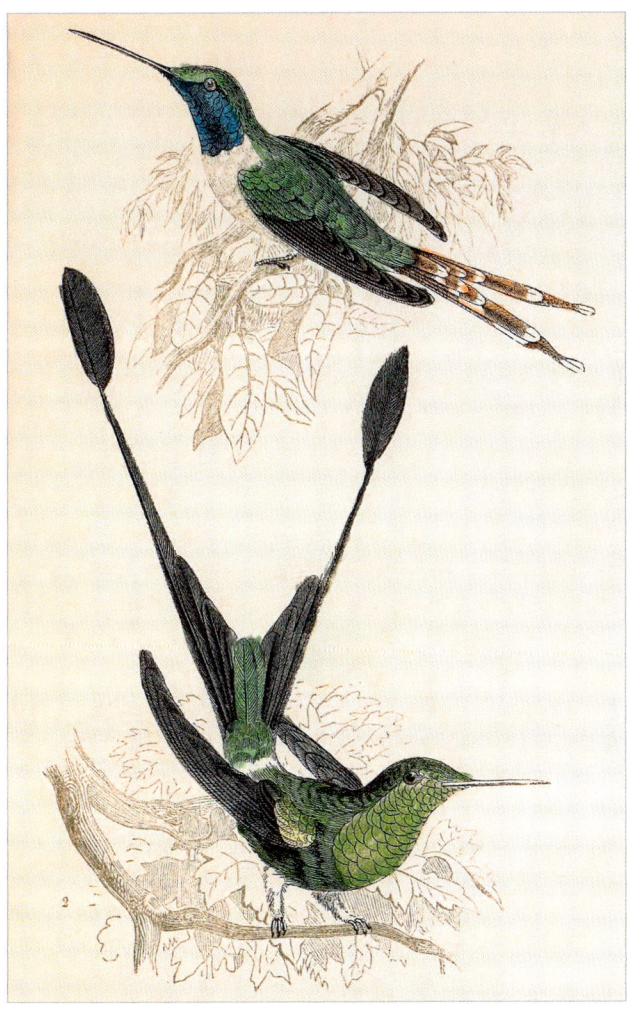

*Sparkling-tailed Hummingbird and Booted Racket-tail in*
Adams's Humming Birds, described and illustrated *(1862)*

## Racket-tail

The Booted Racket-tail is the only bird, a Trochilidae hummingbird, to own this common name. That said, there are other birds with the term used as a descriptor in the common name, for example, Greater Racket-tailed Drongo, Racket-tailed Treepie, and Racket-tailed Roller. In all cases, the name refers to their unique elongated outer tail feathers with spatulate tips. In 1862's *The Illustrated Natural History*, Wood stated of the Booted Racket-tail:[85]

It is chiefly remarkable for the curious formation from which it derives its popular and appropriate name.

# Racquet-tail

These ten species of *Prioniturus* parrots all possess tails of the type mentioned above. As described by van der Hoeven in *A Handbook of Zoology* (1856):[526]

> Tail [is] moderate, even, broad, with the two middle feathers elongate, the shaft naked, vaned at the extremity, discoidal.

In the legendary *The Malay Archipelago* published in 1869, Alfred Russel Wallace wrote about the birds:[205]

> Of the ten parrots found in Celebes, eight are peculiar. Among them are two species of the singular raquet-tailed [*sic*] parrots forming the genus *Prioniturus*, and which are characterized by possessing two long spoon-shaped feathers in the tail. Two allied species are found in the adjacent island of Mindanao, one of the Philippines, and this form of tail is found in no other parrots in the whole world.

The spelling of the word "racquet" has been inconsistent over the years, variously being transcribed as "racket" and "raquet," in addition to "racquet," and, indeed, the IOC and some other taxonomies call these parrots racket-tails. And in Wallace's same volume, he refers to the "Racquet-tailed Kingfisher" (*Tanyseptera* paradise-kingfishers).

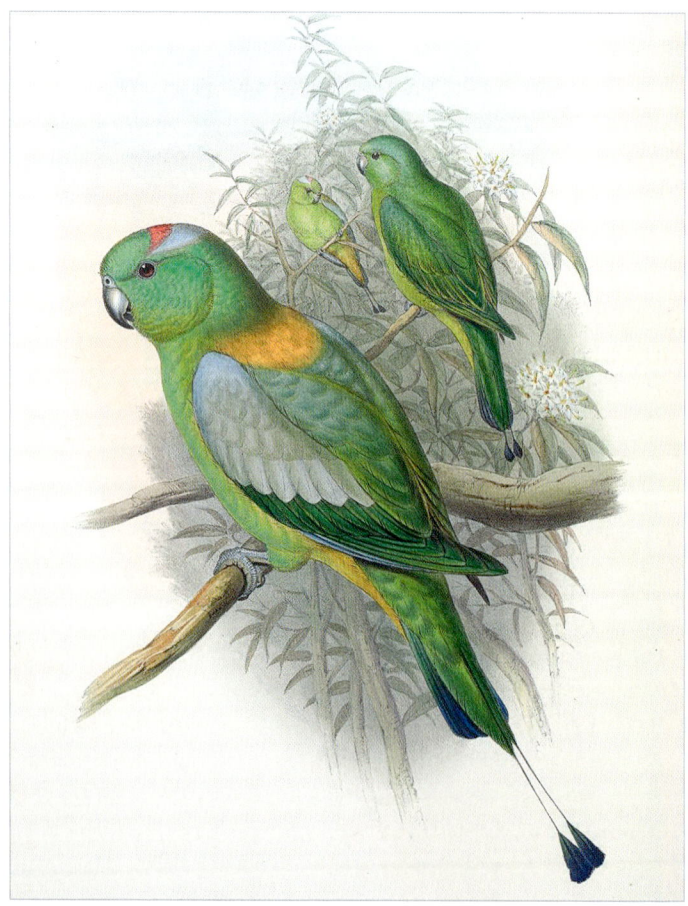

*Golden-mantled Racquet-tail from Gould's* Birds of Asia *(1850)*

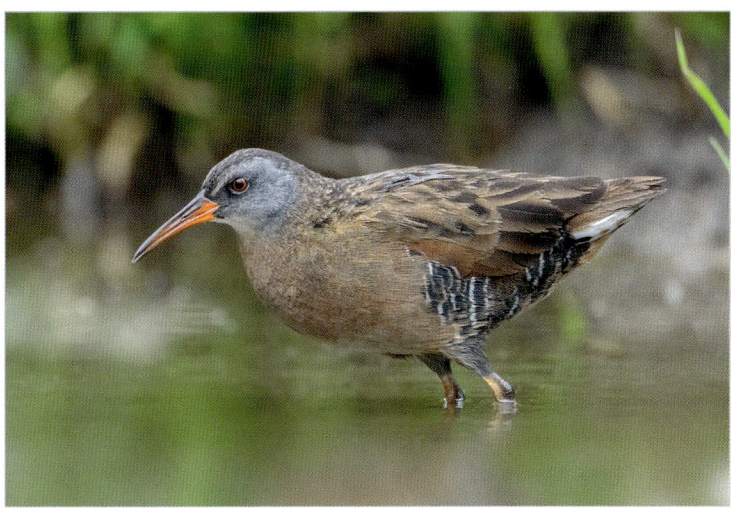

*Virginia Rail (Washington State, USA)*

# Rail

This well-known, albeit very secretive group of birds, the rails, are members of the Rallidae, the rails, gallinules, and coots. There are seventy-three species with the group appellation, including a few with compound names such as wood-rail and forest-rail. Despite Ray and Willughby's conjecture that:[16]

> It is called Rallus or Grallus perchance from its stalking (à gradu grallatorio) [Latin: one who steps on stilts] or perchance from Royale, because it is a Royal or Princely dish.

the name is actually an anglicized respelling of the Old French *rasle*, "to rattle," which in turn comes from the Latin *rallus*. It is probably a metonym from the Latin *radere*, "to scrape," applied to the original rail, the Water Rail, in imitation of its harsh vocalizations.

Despite the yarns, the expression "thin as a rail" has nothing to do with the bird. The phrase was first used by Mark Twain in *Roughing It* in 1872. According to authorities such as the *Oxford English Dictionary*, it is not a comparison to the bird, however, but rather a comparative reference to a straight stick—a fence rail—from the Latin *regula*, replacing earlier forms of the idiom such as "thin a lathe."

# Rail-babbler

This bird in the monotypic Eupetidae is neither a rail nor a babbler and really doesn't resemble either very closely. Its furtive habits in dense vegetation may be said to be somewhat rail-like, but the habitat preference is quite different—they can be found on the rainforest floor in Borneo, Sumatra, and Peninsular Malaysia. In 1935, Frederick Chasen seems to have been at least one of the first to use the name in his *A Handlist of Malaysian Birds*.[296]

# Raven

The original raven was the Common Raven in the Corvidae, the crows, jays, and magpies. These days there are nine species with the common name. The raven has held a place in European mythology since antiquity. The bird made an appearance in *Beowolf*, the Old English epic poem from the tenth century:

Ve´steinn Vali, enn ViWll Stu´fi,

Meinþjo´fr Mo´i enn Morginn Vakri,

Áli Hrafni [raven], til´ıss riðu

en annarr austr und Aðilsi,

gra´r hvarfaði, ge[i]ri undaðr.

[Ve´steinn on Falcon, and ViWll on Stub,

Meinþjo´fr on Heath and Morginn on Waker,

A´li on Raven, they rode to the ice,

and another (horse) east under Aðils,

grey it wandered, wounded with a spear.]

The Old English name had various spellings: *hræfn*, *hrefn*, *refen*, as well as the aforementioned *hrafni*. It came to England from the Proto-Germanic *hrabnaz*, which was then shortened to *hrabn*. All that said, it is believed the name is imitative of the bird's harsh calls, *hrabn* deriving from the PIE *krep-* as in the Latin *crepere*, "to creak." Contrary to some accounts, the noun *raven* and the verb *raven* ("to plunder, to eat greedily, to be ravenous") are completely different words, with different roots, different pronunciations, and different meanings.[18]

# Rayadito

The two species of rayaditos are closely related in the Furnariidae with a Spanish name meaning "striped or lined" (*rayado*) in the diminutive, probably referring to the black-and-orange-striated plumage. In *The Birds of Chile* in 1932, Hellmayr referred to the bird with the name "Thorn-tailed Warbler" given by Latham but later in the text wrote:[448]

The "Rayadito" is a characteristic bird of the southern forests.

# Razorbill

This auk is the sole member of the genus *Alca*, in the Alcidae. It has been known by many names in the past. In his *Natural History of Birds* in 1793, Count de Buffon under the heading "The Penguin" noted that they were called:[8]

In the north of England the Auk: in the west of England the Razorbill: in Cornwall the Murre: in Scotland the Scout.

The obvious etymology of the name is the bird's remarkable bill, which

is two inches long, arched, very strong and sharp at the edges; the colour black: the upper mandible is marked with four transverse grooves; the lower with three; the widest of which is white, and crosses each mandible [from *British Zoology,* Thomas Pennant[177]].

# Recurvebill

These two species of foliage-gleaners from the Neotropical forests in the Furnariidae are so named for their unusual, upturned bills. The Peruvian Recurvebill was first described by Chapman in 1928 but appears not to have received a common name until much later.[299] According to Parker in a 1982 issue of the *Wilson Bulletin*, the bird[402] seems to favor bamboo thickets, but it is not restricted to them. They hop along branches and trunks of fallen trees and understory palms, and [have been seen] hammering dead bamboo stalks, presumably to dislodge prey items.

The bill is surely an adaptation for this hammering manner of feeding.

# Redhead

Like many of the pochards, the Redhead has a rufous head and neck. For quite some time, the Redhead of the Americas was not split from the Old World Common Pochard. In 1784, Pennant wrote about the Common Pochard:[19]

Inhabits North America, in winter, as low as Carolina; and, I believe, is the Red-headed Duck of Lawson.

While Alexander Wilson wrote in *American Ornithology* (1808):[28]

Anxious as I am to determine precisely whether this species be the Red-headed Wigeon, Pochard, or Dun bird of England, I have not been able to ascertain the point to my own satisfaction; though I think it very probably the same, the size, extent, and general description of the Pochard agreeing pretty nearly with this.

The first use of the name in its present form seems to have been in Coues's *Key to North American Birds* in 1903:[375]

FULIGULA PERINA (L.) Sw.,
var. AMERICANA (Eyton) Coues.
Redhead; Pochard.

# Redpoll

The three species of redpolls in Fringillidae are also redheads. All sport bright red forecrowns, so the first part of the compound name is obvious. The second part, *-poll*, may be less clear to us in the modern era. The word is a Middle English one, *pol*, *polle*, meaning "scalp, pate," from the Proto-Germanic *pullaz*, for "round object, head, top." It was even an old name for red-headed people. Newton wrote:[7]

The lively colours which glow upon the cock-bird at liberty are in confinement lost at the first moult and never resumed, so that the very name Redpoll becomes a misnomer—the top of the head changing to dark orange, hardly visible in some lights.

# Redshank

The two shorebirds that are the redshanks are named for their long, red legs. The word *shank*, meaning the part of the leg between the knee and the ankle, has largely fallen out of favor in Modern English. It came from the Middle English *schanke*, and ultimately from Ancient Greek σχάζω *shkázō*, "to limp." The name has been in use at least since the 1500s, and maybe even from the late 1400s, when it appeared in one of the Chester Mystery Plays:

Heare are doves, digges, drackes, [Here are doves, ducks, drakes,]

Red-shonckes roninge through lackes. [Redshanks running through lakes.]

Another old name for the bird mentioned by Ray and Willughby was "Pool-Snipe."[16] As an aside, in the sixteenth century, Redshank was also a nickname for Scottish mercenaries. They were called redshanks because they hitch up their kilts and wade bare-legged through rivers in the coldest weather.

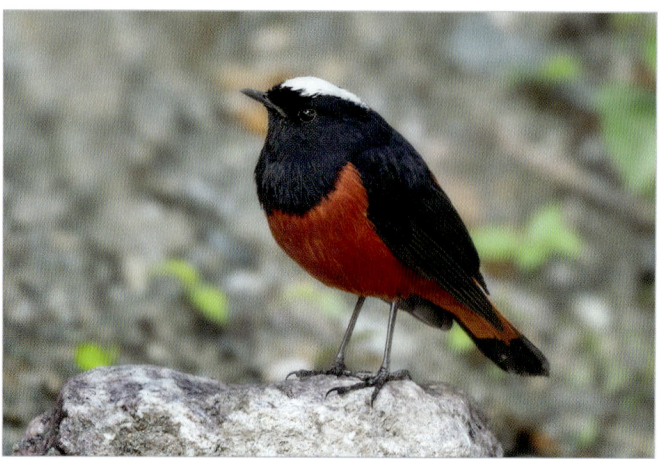

*White-capped Redstart (India)*

# Redstart

Although the name is now applied to birds in two different families—fifteen species in the Muscicapidae, Old World flycatchers, and thirteen species in the Parulidae, New World warblers—redstart was first used for the familiar little Common Redstart in Europe. With its habit of shivering its red tail, it earns the second part of its name, *-start*, an Old English word for "tail." This probably came into English from the Proto-Indo-European *ster-*, meaning "stiff." In a fabulous little book from 1791 entitled *The Natural History of Birds: containing a variety of facts selected from several writers, and intended for the amusement and instruction of children*, there is an entry about the "Red Start":[388]

> He flies lightly, and when perched, has a little cry, which he repeats, moving his tail with a kind of tremulous motion, not up and down, but from one side to the other, like a dog when it fawns.

# Redwing

The Redwing is a species of *Turdus* thrush found across Europe and the Palearctic that would have been very well known to the inhabitants of England from ancient times. It gets its name from the rusty-colored underwings that are conspicuous in flight. Ray and Willughby wrote about the Redwing in 1678, mentioning its namesake feature:[16]

> The covert-feathers of the underside of the Wings, and of the sides of the body under the Wings, which in the *Mavis* [Song Thrush] are yellow, in this kind are of a red Orange-colour, by which mark it is chiefly distinguished from it.

# Reedhaunter

The remarkably diverse Furnariidae, ovenbirds and woodcreepers, has three species called reedhaunters. These skulking and shy birds inhabit thick vegetation in marsh and other wetlands—in other words, they haunt the reeds.

# Reedling

The Bearded Reedling belongs to the monotypic family the Panuridae. Due to confusion about its taxonomic affinities in the past, it has gone by a number of different names, including Bearded Tit and Bearded Parrotbill. The present name could be described as meaning simply "little reed dweller." The birds are found strictly in reed-like vegetation, and *-ling* is an Old English diminutive.

## Reeve

(See Ruff.) The female to the Ruff is the Reeve, possibly a result of a transformation of the older name "Ree" after Ruff was adopted for the male birds. Lockwood believed that the name predated ruff and referred to the male's aggressive behavior on the breeding grounds, citing an Old English word, *hrēoh* or *ree*, meaning "frenzied."[84] He goes on to note that the later modified name may have arisen by analogy with the ornate plumage of the bird and the livery of the government officials known as reeves.

## Rhabdornis

The rhabdornises are four species of creeper-like birds endemic to the Philippines, now placed in the Sturnidae, the starlings. The synonymous common and genus name references their striped plumage. The Ancient Greek ῥάβδος *rhabdos* means "stripe" and the Latin *ornis* is for "bird." Reichenbach coined the name for the genus in 1853 but gave no explanation other than to say that Stripe-sided Rhabdornis, which he described, has the[527]

upper head, neck and upper back black, with white shaft stripes on all feathers.

## Rhea

There are two rheas in the rhea family, the Rheidae. The bird was named the rhea by Möhring, a European in the mid-1700s, after a character in Greek mythology, Rhea, daughter of the earth goddess Gaia and the sky god Uranus. Her name may or may not be a metathesis of the

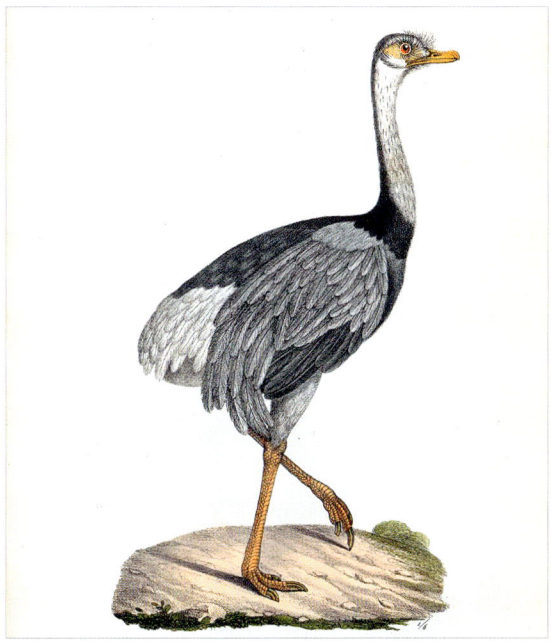

*Greater Rhea from Schinz and Brodtmann's* Naturgeschichte und Abbildungen der Vögel: nach den neuesten Systemen bearbeitet *(1836)*

Ancient Greek word for "ground," so it possibly relates to the bird's earthbound flightlessness. In *The Animal Kingdom, or zoological system, of the celebrated Sir Charles Linnæus* from 1792, a common name is listed as "Tougai" without explanation.[223] And then in 1793's *The Natural History of Birds*, Buffon makes an observation that could well be made of many bird names to this day, albeit with less pejorative language:[8]

> Moehring and Brisson prefer the Latin name *rhea*, to which the latter annexes the American epithet *Touyou*, formed from *Touyouyou*, by which it is generally known in Guiana. The savages settled in other parts of the continent have given it different names: *Yardu, Yandu, Andu* and *Nandu-guacu* in Brazil; *Sallian* in the island of Maragnan; *Suri* in Chili, &c. So many names have been bestowed on an animal with which we were so lately made acquainted! For my part, I shall readily adopt that of *Touyou*, which Brisson has applied or rather retained, and I shall not hesitate to prefer this barbarous word, which has probably some analogy with the voice or cry of that bird, to the scientific terms, which only serve to convey false notions, and to new names, which mark no character, no essential property of the animal on which they are bestowed.

# Riflebird

The four species of *Ptiloris* are all called riflebirds, members of the Paradisaeidae, the birds-of-paradise. There are two theories behind the name. One is that it describes the bird's call, the other that it is named for the male's black plumage. Under the entry of "Rifle-bird," Edward E. Morris wrote:[20]

> Sometimes called also Rifleman . . . ; a bird of paradise. The male is of a general velvety black, something like the uniform of the Rifle Brigade. This peculiarity, no doubt, gave the bird its name, but, on the other hand, settlers and local naturalists sometimes ascribe the name to the resemblance they hear in the bird's cry to the noise of a rifle being fired and its bullet striking the target.

Unless the early settlers were suffering from a case of mistaken identity, the former explanation seems most plausible, as the vocalizations certainly do not resemble the sound of a rifle.

# Rifleman

This little green, almost tailless passerine is a member of the Acanthisittidae, the New Zealand wrens. This bird might win the prize for the most common names over the course of history! Since it was first described in 1787, it has been known in English as Creeper, Wren, Citrine Warbler, Green Wren, Stripe-faced Wren, and, finally, Rifleman, as well as a plethora of Māori names, including *Pi Wau Wau, Titipounaniu, Mirumiru, Tititipounamu, Kikimutu, Kikirimutu, Pihipihi, Piripiri, Tokepiripiri*, and *Moutuutu*—all listed in various publications up to the late 1800s. The present-day name supposedly stems from an imagined resemblance of its plumage to the uniform of an early New Zealand colonial regiment.

# Ringneck

The Ringneck is a single species of Australian parrot in the genus *Barnardius*. There are four subspecies that are sometimes recognized as distinct, but all possess a conspicuous yellow collar that accounts for the name. Called the "Ring-neck Parrakeet" by Morris in the *Dictionary of Australasian Words, Phrases and Usages* from 1898, this was a name already in use for the bird now called Rose-ringed Parakeet.[20] In the ornithological literature, the bird was called the Barnard's Parrakeet or the Mallee Parrot, but in an issue of *The Emu* in 1911, the author of an article wrote that it is[528]

> *Barnardius* [*zonarius*] *semitorquatus*—the common "Ring-neck" of settlers—was distinctly rare.

# Roadrunner

There are two species of ground cuckoos in the *Geococcyx* genus, an iconic bird thanks to the Road Runner's long-running feud with Wile E. Coyote.[449] The name comes from its habits, as described by Baird in 1858:[21]

> This remarkable genus is represented in the United States by a single species known as the Paisano, Chapparal Cock, or sometimes Road Runner, on account of its frequenting public highways. Its very long legs enable it to run with very great rapidity, faster even than a very fleet horse.

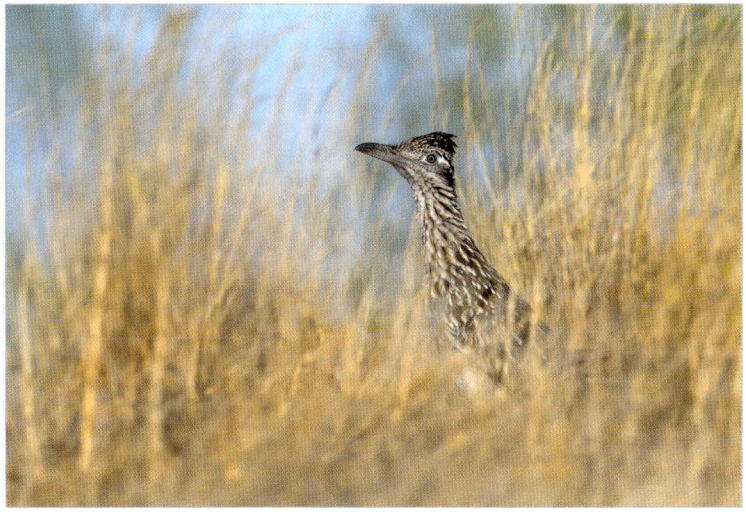

*Greater Roadrunner (Arizona, USA)*

# Robin

This very widely used bird name has somewhat mundane beginnings. It started with the European Robin in the fifteenth century, when a fashion started of giving people's names to familiar birds and other animals of the British countryside. The little bird gained the nickname Robin Redbreast, a cute alliteration, Robin being a diminutive of the name Robert. (Other examples include Jenny Wren, Billy Goat, and even Magpie from Margaret). Eventually, the "redbreast" part of the name was dropped, leaving just "robin," a word that eventually traveled around the world with the expansion of the British empire. There are now so many bird names that include the word "robin" that it's hard to keep track. In the Muscicapidae (Old World flycatchers), to which the original Robin Redbreast belongs, there are over fifty species, including bush-robin, scrub-robin, magpie-robin (see that entry), three species of Turdidae (thrushes) robin, and at least thirty-six species of Petroicidae (Australasian robins) including, of course, robins, as well as ground-robin and scrub-robin. These names were given to similar-looking birds that the English diaspora encountered, at times out of a sense of nostalgia or homesickness, at others arguably just from a lack of imagination.

## Robin-Chat

All but one of these fifteen members of the Muscicapidae, the Old World flycatchers, are in the same genus. A previous name was Chat-Thrush, suggesting uncertainty about which English appellation to apply to unfamiliar birds with a mix of characteristics from the, at that time, more familiar English birds. But they are noisy birds, like chats, and most sport reddish plumage, like the original robins. (See Robin and Chat.)

## Rockfowl

These two species of charismatic birds from Africa are the only members of the Picathartidae, the rockfowl. Previously known as Bald Crow or Bald-headed Crow, due to its unusual bare head and nape, it is still sometimes referred to colloquially by the genus name *Picathartes*, which would translate from the Latin as "magpie vulture." These forest dwellers are usually found near cliffs, outcrops, or caves due to their penchant for building their mud nests attached high on vertical rocky surfaces. This accounts for the present-day English name of the bird. Interestingly, though, Lieutenant Boyd Alexander gives us some insight into local names for chicken-like birds in his book *From the Niger to the Nile* when he wrote about the Stone Partridge:[450]

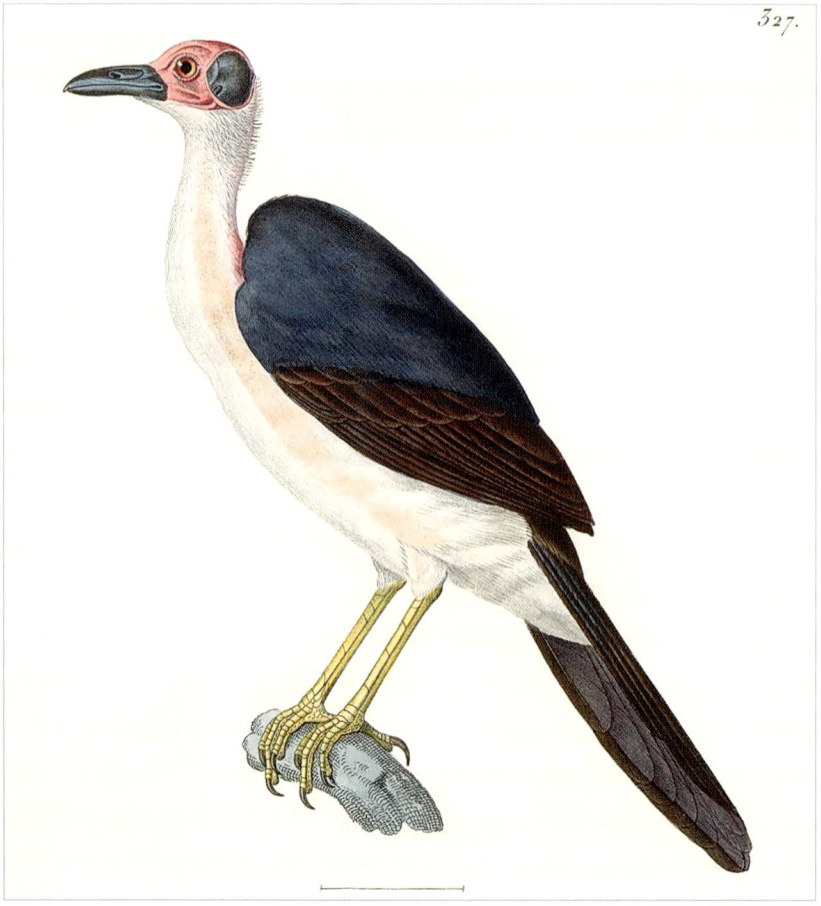

*White-necked Rockfowl in Temminck's* Nouveau recueil de planches coloriées d'oiseaux *(1838)*

Where there are rugged hills and kopjes, the graceful little rock-pheasant (*Ptilopachys fuscus*) is found, or *Casa duci** (rock-fowl) as it is called by the Hausas; I have seen as many as twenty together taking refuge, when disturbed, in the crannies of the rocks.

The Hausa language is spoken through large areas of West Africa. It seems likely that the Hausa name was one used for all large, ground-dwelling birds.

*The present-day spelling of *Casa duci* would be *Kaza dutzen*.

# Rockjumper

These two species in their own family, the Chaetopidae, live up to their names, finding their homes in rocky outcrops, where they jump and bound from rock to rock in search of their invertebrate prey. In 1906, William Beebe described a rockjumper in *The Bird; its Form and Function*:[99]

> Of the Rock-jumper, it is said: "These curious birds are only to be found on the rock-strewn slopes and summits of mountain-ranges where they are able to hop from rock to rock for a distance without having to cross level or open ground; . . . at the slightest alarm they either drop into a crevice or bound from rock to rock with extraordinary speed, looking more like india-rubber balls than birds, for there is no perceptible interval between the end of one leap and the beginning of the next, and the distance they can clear at a single hop must be seen to be believed. Should they have to cross a piece of level ground between two rocks which they cannot clear with a single bound, they run across it with great speed and usually with outspread wings. So feeble are their powers of flight that they seldom attempt to fly, and never when in a hurry or alarmed; at the most they flutter feebly for a few hundred yards down hill."

# Rockrunner

The Rockrunner is the sole representative of the genus *Achaetops* in the Macrosphenidae, the African warblers. When it was first described in the 1920s, it was known as the Damara Rock-jumper, and it shares many habits of those birds. They inhabit rocky outcrops, remaining hidden in crevices and grassy tufts, until the breeding season, when they will run among the rocks and sing from prominent positions.

# Rock-Thrush

Although the twelve members of the *Monticola* genus have the word "thrush" in their common group names, they are not thrushes but members of the Muscicapidae, the Old World flycatchers. That said, they were formerly included in the Turdidae and have a similar gestalt to the thrushes. And despite the first part of the name, not all of them are associated with rocky habitats by any means. The name was first used for the Rufous-tailed Rock-Thrush *Monticola saxatilis*—that is, "the mountain-dwelling rock lover" (*montis*, "mountain," *colere*, "to dwell," and *saxatilis*, "rock-frequenting"). In the past, it was also colloquially known as Rock Crow and Rock Shrike.

# Rockwarbler

This unique member of the Acanthizidae in the genus *Origma* is strongly associated with the sandstone rock formations that are such a feature of the land around Sydney, Australia. The bird was first described in 1808 by Lewin, the English artist and collector, who described it as a warbler in the genus *Sylvia*. It seems the early European naturalists could not help but assume everything was related to something back home. (See Origma.)

# Roller

The twelve species of rollers are in their own family, the Coraciidae, and most authorities believe they got their name from their manner of flight. They are known for their conspicuous and acrobatic display flights, in which they dive rapidly in a roller-coaster manner with wings and body rocking back and forth. Gessner was the first to use the name in 1555 when he wrote in *Historia Animalium*:[178]

> Avis hæc cuius figuram ponimus, circa Argentoratum Roller uocatur per onomatopœiam, ut audio, in aere perquam alte uolat. [This bird is called the Roller in Strasbourg, the name is onomatopoeic and it is reported to be flying very high up in the air.]

Although very few others concur, Lockwood, citing Gessner, believed the name arose from the German word *rollen*:[84]

> A verb imitative of sounds made in rapid succession and particularly applicable to the cries of this noisy bird.

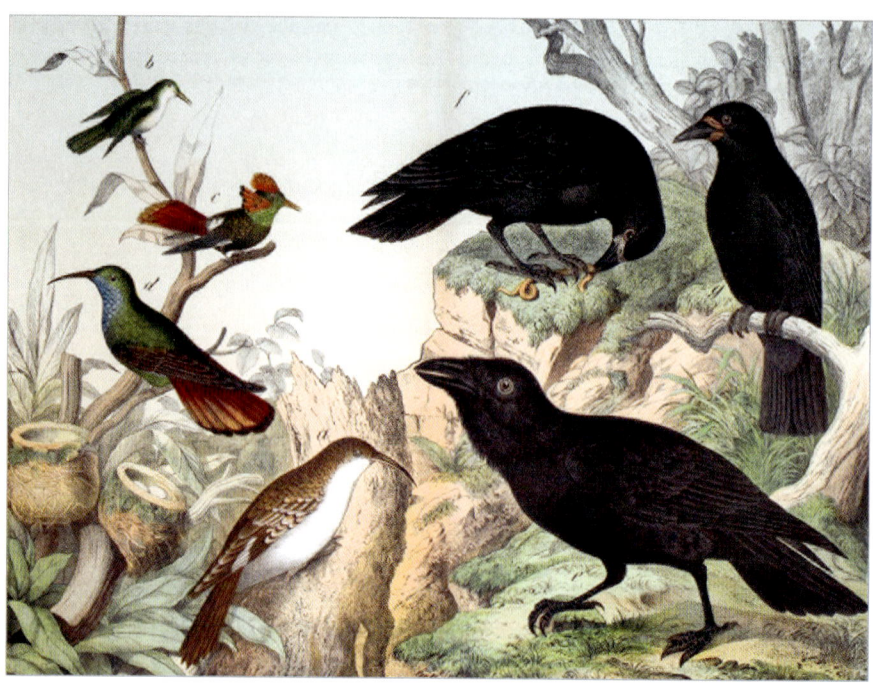

*Rook and others from Kirby's* Natural History of the Animal Kingdom for the Use of Young People *(1889)*

# Rook

Just one member of the large *Corvus* genus is known by the name Rook. It has a long history in the English language, first appearing as the word *roke* closest to its present form in the 1400s. In 1581, in his Latin book *Historiæ Animalium*, Conrad Gessner noted:[178]

> Among the English, too, I hear, the bird is called the rook, the plain beak is black, but the posterior is whitish.

The English-language name is believed ultimately to be derived from the bird's harsh call, with its roots in the Old English *hroc*, which came from the Proto-Germanic *hrōkaz*. But some evidence suggests *hroc* was originally used for any of the corvids found in the British Isles. In a fascinating study of the etymology entitled "When is a *hroc* not a *hroc*? When it is a *crawe* or a *hrefn*," Eric Lacey stated:[389]

> Close examination of the linguistic evidence, both in Old English and in later developments, however, . . . reveals that these three birds were not distinguished on visual grounds and were prone to be confused with one another. Indeed, the most significant criteria for differentiating the three birds are those sounds embedded in their onomatopoeic names: *hroc* ("rook"), *hrefn* ("raven") and *crawe* ("crow"). . . . The aural data suggests that it [*hroc*] was originally a raven term that was transferred to rook . . . [and] it is difficult to avoid concluding that Proto-Germanic *xrōkaz* was originally a term referring to the raven rather than the rook. It then probably became a vertically polysemous term referring to the three largest *Corvidae* collectively, before narrowing semantically to cover the rook and other crows, and then eventually just the rook.

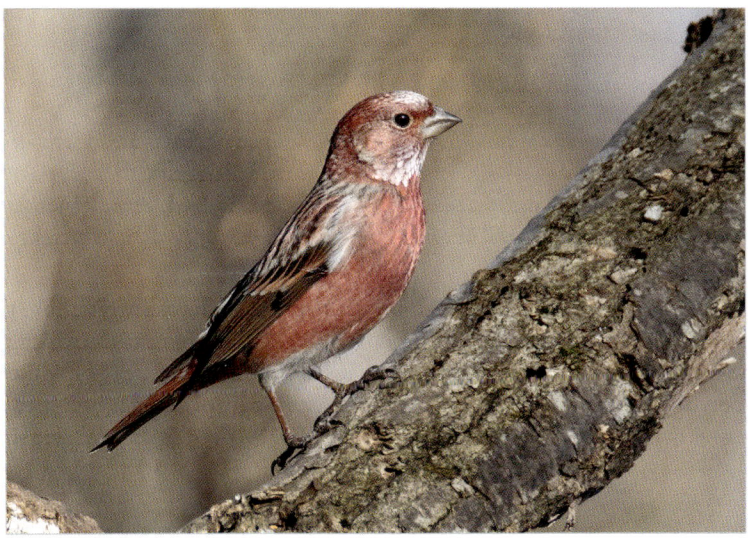

*Pallas's Rosefinch (Japan)*

# Rosefinch

The rosefinches are twenty-six species of finches in the Fringillidae, all but two of them in the *Carpodacus* genus. All of them have varying degrees of a rosy-pink hue, their namesake feature. Although one might think the name is quite old, it doesn't appear in the English literature until the early 1800s; even Newton makes no mention of it.

# Rosella

These familiar and much-loved parrots of the Australian bush belong to the genus *Platycercus*. The original rosella was the Eastern Rosella, a common sight around Sydney, where the first English colonists settled. As Gould noted in his *Birds of Australia* (1848), it is the "Rose-hill of the Colonists"; he, however, dubbed it the Rose-hill Parakeet.[40] The British Museum has an 1822 sketch of the bird with an attached explanation:

> This beautiful, and in England, scarce bird, took its name from a hill, called Rose Hill, in New South Wales, on account of the numbers of them resorting to it.

Rose Hill, the fortified camp built around Government House near Sydney, would have been a conspicuous landmark in the settlers' eyes. The colloquial name for the bird, "Rose-hiller," eventually morphed into the present-day "rosella."

# Ruby

The only ruby, the Brazilian Ruby is a hummingbird endemic to southeastern Brazil. Buffon described the bird as having a[8]

> throat of a bright ruby or rosette color, following the aspects; head, neck, front & outline of body, emerald green with golden highlights; the tail is red

and called it Le Rubis Émeraude, the Ruby Emerald, in 1780. In 1849, Gould noted under the heading "Brazilian Ruby":[6]

> Besides being one of the commonest birds in our collections, the *Clytolcema rubinea* is also one of the most beautiful members of its lovely family, and is rendered eminently conspicuous in the group by its rich ruby-coloured throat, surrounded with equally glittering green.

# Rubythroat

The three species of very similar-looking *Calliope* rubythroats are members of the Muscicapidae, the Old World flycatchers. And as the name suggests, they all possess bright ruby-red throats. One of the earliest uses of the name and, indeed, earliest mentions of the bird in the English literature, is in Latham's *A General Synopsis of Birds*, in which he described the Siberian Rubythroat:[218]

> Plumage above, the colour of a Nightingale: beneath, yellowish white: throat the colour of vermilion, bounded on each side by a black and white streak, which takes rise at the bill: between the bill and eye black: over the eye a streak of white: tail rounded, the colour of the back. Inhabits the east part of Sibiria; first met with about the river Jenisei. Seen mostly on the tops of the trees. Is an excellent singer, chiefly in the middle of the night.

# Ruff

This *Calidris* shorebird is remarkable in many ways—not only for its unusual name, but also for its extreme sexual dimorphism, as well as for the annual transformation of the male, the Ruff, into striking breeding plumages that may include bare bright orange facial skin, a red bill, fluffy black breast feathers, rufous head plumes, and a large collar of ornamental feathers. The Ruff exhibits one of the most extreme cases of polymorphism known among birds. It is from the latter feature that the male bird derives its name—a ruff was a projecting starched, circular, fluted frill worn around the neck in Europe in the sixteenth and seventeenth centuries.

The Ruff and the Reeve are among only a very few birds with widely used different names for the males and the females, and this probably derives from the aforementioned extreme sexual dimorphism, which is unusual in shorebirds. That said, the name Reeve is not now used except colloquially. (See Reeve.)

In fact, the earlier name for the Ruff was "Rey" or "Ree." A very old reference to the "Ree" was in Leland's "Collectanea" with a list of food to be served at Neville's Banquet in 1465, which included not only an extraordinary "4000 Mallards and Teal, 12 Porpoises and Seals and 6 Wild Bulls" but also "2400 of The Foules called Rees."[529] Later, the bird is mentioned in *The Regulations and Establishment of the Household of Henry Algernon Percy* from 1512:[325]

AT PRINCIPALL FEESTS.

Item REYS to be hadde for my Lordes owne Mees at Princypall Feestes and at ij *d.*\* a pece.

Thomas Percy, who compiled the *Northumberland Houshold Book* [*sic*] in 1770 from the original manuscript of the early 1500s noted that the "Reys" were the Ruffs and Reeves.[530] This original name, Ree, is possibly derived from the Middle English *rei*, *reh*, *reoh*, from Old English *hrēoh*, meaning "fierce, wild, angry, tempestuous," which in turn is from the reconstructed Proto-Germanic *hreuhaz* for "bad or wild." Presumably, this was an allusion to the males' combative behavior on the breeding grounds. Under an entry for "Ruff and Reeve" Swann wrote:[121]

[The] Derivation of Reeve is thought to be from Anglo Saxon *gerefa*, literally one in authority, perhaps so called from the pugnacious habits of the males. A wood-reeve was anciently the overseer of a wood. The name is found in Willughby as "The Ruff, whose female is called a Reeve"; in Merrett as "Rough and Reev," perhaps a mere phonetic spelling . . . The name Ruff is invariably applied to the male bird, the female being called Reeve. According to Willughby, "They breed in Summer time in the Fens of Lincolnshire about Crowland," but it is, alas, now nearly a thing of the past for them to breed anywhere in England.

The Middle English word "reeve" for a Crown official, such as Oswald the Reeve in Chaucer's *Canterbury Tales*, may have arisen as a secondary form of the word for the bird, drawing a parallel between the male bird's ornate breeding plumage and the official's fancy garb. As Swann noted, Reeve comes from the Middle English *reve*, from Old English *rēfa*, an aphetism of *gerēfa* from the Proto-West Germanic *garāfijō*, meaning an "officer or official."

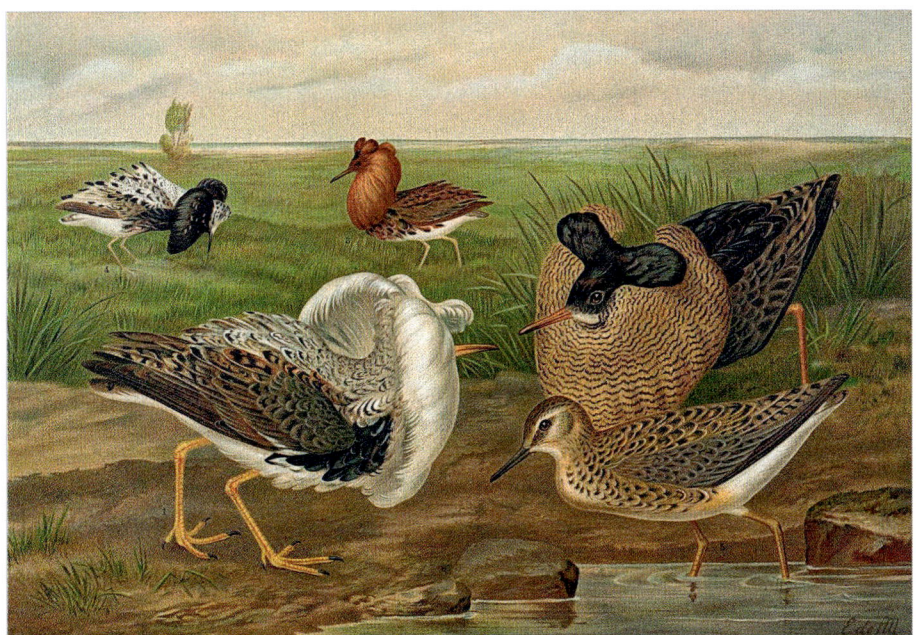

*Ruffs and a Reeve from Naumann's* Naturgeschichte der Vögel Mitteleuropas *(1905)*

At first blush it would seem that "Ree" and "Reeve" are merely two forms of the same word, but that is not necessarily the case.

In the case of Ruff, Newton was unsure which came first—the chicken or the egg, or in this case the ruff or the rough:[7]

> It seems to be at present unknown whether the bird was named from the frill, or the frill from the bird. In the latter case the name should possibly be spelt Rough (cf. "rough-footed" as applied to Fowls with feathered legs), as in 1666 Merrett (Pinax, p. 182) had it.

Lockwood says that the present-day names were first noted in print in 1634 in what was essentially a shopping list: "12 Ruff and reeve 3 dozen."[84] And Willughby and Ray wrote about them in 1678:[16]

> The Ruffe, "*Avis pugnax*" The female of this is called the Reeve. These Birds differ wonderfully in colours, so that scarce can there be found any two alike.

How or why the names morphed from Ree to Reeve and Ruff is something of a mystery, but the old name survives in the female form. To finish off, Gurney gives us an amusing account of another early reference to the bird.[324] Although he doesn't elaborate on which name was used at the time, it does provide some insight into the analogy between the bird and the clothing item:

> Aldrovandus gives as many as seven illustrations of the Ruff and Reeve, and of these, one has been discovered by Mr. W. H. Mullens to have been copied from an anonymous pamphlet, printed in England soon after 1586. The cut, which is quite meritorious, represents in very characteristic fashion, one of a flock of Ruffs, which were "intangled and caught" at Crowley in Lincolnshire. Mr. Mullens is of opinion that in this pamphlet, which was evidently written as a protest against the extravagant fashions of the day, we have the earliest mention of the Ruff as a British bird, together with the first published figure and description of it.

(*In the Middle Ages, a "j" was sometimes substituted for the final "i" of a "lower-case" Roman numeral, such as "ij" for 2.)

# Rushbird

The wren-like Rushbird is a member of the Furnariidae in the genus *Phleocryptes*. As the name suggests, this species is strongly associated with rushes in freshwater wetlands. In one of the earliest accounts of the species, Sclater (1888) used the common name "Rush-loving Spine-Tail" and wrote that it is[22]

> very abundant in the rush-beds growing in the water, where alone it is found. [It] spends the warm season secluded in its rush-bed: and when disturbed flies with great reluctance, fluttering feebly away to a distance of a few yards, and then dropping into the rushes again, apparently quite incapable of a sustained flight . . . This Spine-tail seldom ventures out of its rush-bed, but is occasionally seen feeding in the grass and herbage a few yards removed from the water.

Curiously, in the same publication, Sclater used the common name "Curved-Bill Rush-Bird" for *Limnornis curvirostris*, the Curve-billed Reedhaunter.

# S IS FOR SABREWING

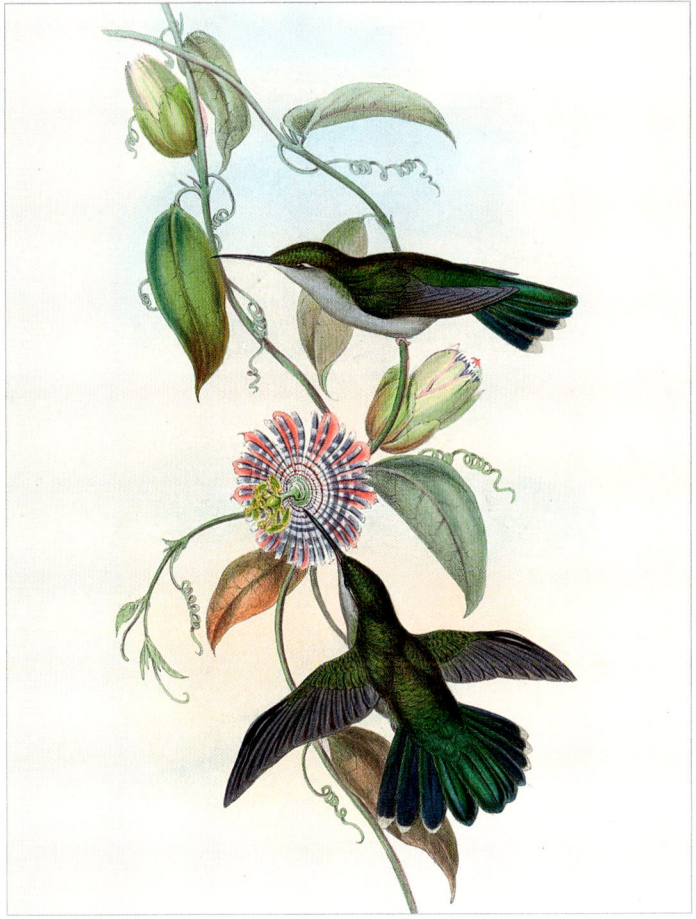

*Grey-breasted Sabrewing in Gould's* Monograph of the Trochilidæ *(1849)*

## Sabrewing

The sabrewings are a genus, *Campylopterus*, of thirteen species of hummingbirds. The common name is mirrored in the genus name, which is derived from the Ancient Greek καμπυλος *kampylos*, "bent," and πτερον *pteron*, "wing," for the distinctive thickened and curved shafts of the outermost primary flight feathers of the male's wings. An older name was "Sickle-winged Humming-bird." In his introduction to the hummingbirds in *The Naturalist's Library* (1833), Jardine described the namesake feature of the sabrewings:[58]

> In all, the shafts of the quills are remarkably strong and elastic but in a few species, known under the denomination of sickle or sabre-winged humming-birds, and forming the genus *Campylopterus* of Swainson, they are developed to an extraordinary degree at the base, and nearly equal the breadth of the plume.

## Saddleback

The two very similar saddlebacks in the New Zealand family, the Callaeidae wattlebirds, sport a singularly characteristic rufous "saddle" from the mantle to the rump that contrasts against the predominantly black plumage. Reischek wrote about the saddleback, giving the Māori name as "Tieke," in 1886:[287]

> This bird derives its popular name from a peculiarity in the distribution of its two strangely contrasted colours, uniform black, back and shoulders ferruginous, the shoulders of the wings forming a saddle.

## Saltator

There are seventeen species of saltators in the eponymous *Saltator* genus and one in the *Saltatricula*, all are tanagers in the Thraupidae. In 1816, the French ornithologist Louis Pierre Vieillot introduced the genus for the Buff-throated Saltator without explanation, but the name is from the Latin *saltō* and *tor*, meaning "one that jumps or dances" or "dancer."[248] This name alludes to the way the Buff-throated Saltator hops on the ground or on tree limbs with both feet together. Buffon and other French naturalists earlier referred to the bird as the "Grand Tangara," *Tangará* deriving from the Tupi Guarani *atá cará*, "to walk in leaps and bounds, the jumper."[133]

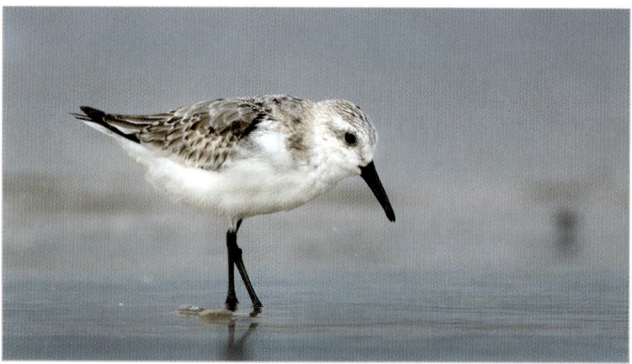

*Sanderling (Georgia, USA)*

## Sanderling

The sanderling, a small member of the Scolopacidae, the sandpipers and allies, has had other names in the past, as witnessed by Ray and Willughby, who mentioned both "sea-dotterel" and "curwillet."[16] Two possible etymologies are advanced, one that it is from the word *sand* with the suffix *-ling*, signifying the diminutive or "manner, direction, or position," in other words, "associated with sand." The other is that it derives from Old English *sand-yrðling*, "sand-ploughman." The word *yrðling* translated as ploughman literally means "earthling," so the *-ling* suffix serves the same function here.

## Sandpiper

The sandpipers are often referred to colloquially, in North America at least, as "peeps," which actually is a clue to the etymology of the name. Clearly, it is a compound of two words— "sand" from the Proto-Germanic *samdaz*, and "piper" from the Latin *pīpiō*, "to chirp, peep," as so many sandpipers do when they take flight. There are twenty-six sandpipers in the Scolopacidae. The word first entered the written literature in Ray and Willughby in 1678.[16]

## Sapayoa

An enigmatic bird of the Neotropics, the Sapayoa has long been a taxonomic mystery (as reflected in the scientific name *Sapayoa aenigma*). It's now thought to be most closely related to the Old World broadbills but is placed in a monotypic family, for now. It was first described by Hartert, who named it for the place it was first collected, the Rio Sapayo (Rio Zapallo Grande) in Ecuador. In 1903, he wrote in *Novitates Zoologicae*:[426]

> A single example . . . evidently fully adult, was obtained on the Rio Sapayo in N.W. Ecuador, November 2nd, 1901, by Mr. Miketta, one of Mr. F. W. H. Rosenberg's correspondents in South America, who has discovered several other fine novelties in Ecuador.

## Sapphirewing

One of the largest of the hummingbirds, the Great Sapphirewing was named by Gould, who opined:[6]

> This is perhaps without exception one of the finest species of the family yet discovered.

He named it "Temminck's Sapphire-wing" in honor of Coenraad Jacob Temminck, the Dutch zoologist. The males are iridescent, green with bright, sapphire-colored wings, and, as Gould wrote:

> The brilliant colouring of its wings renders it conspicuously different from all other known species.

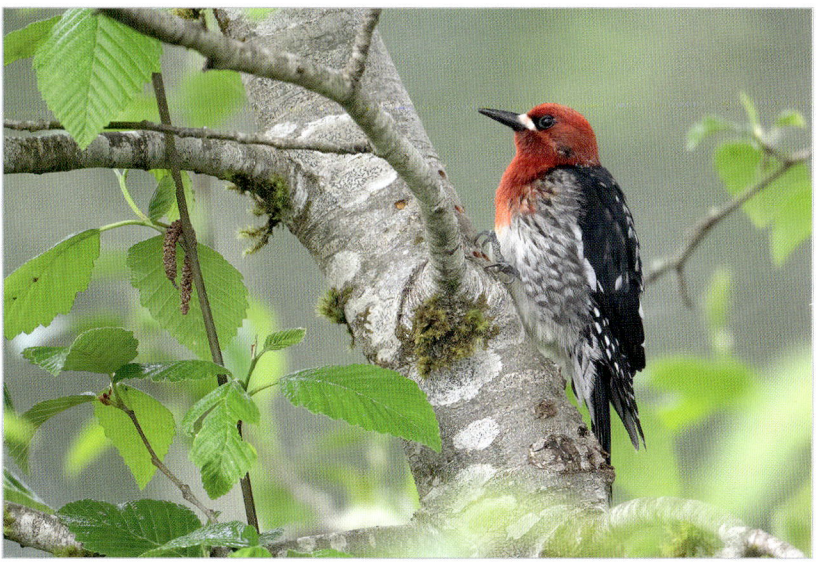

*Red-breasted Sapsucker (Washington State, USA)*

## Sapsucker

Four species of sapsuckers in the Picidae, the woodpeckers, are found in North and Central America. All of them belong to the genus *Sphyrapicus* and derive their common name from their mode of feeding, which is different from that of other woodpeckers. They drill neat rows of shallow holes in tree bark, from which they lap up the exuding sap with their specialized brush-tipped tongues. In the past, the name was used for other woodpeckers, including "Big Sapsucker" for Hairy Woodpecker and "Little Sapsucker" for Downy Woodpecker,[452] and even for White-breasted Nuthatch in Barton's *Fragments of the Natural History of Pennsylvania* in 1799.[451] Newton wrote that it was[7]

> a common name in North America for many of the smaller Woodpeckers . . . but strictly only applicable to *Sphyropicus varius*, which . . . has a lingual structure, first described by Macgillivray for Audubon . . . very different from that of most Picidae and a mode of feeding to correspond.

# Satinbird

Initially thought to be bowerbirds, the three members of the Cnemophilidae were placed in the Paradisaeidae until Cracraft and Feinstein's 2000 paper showed a marked divergence from that family.[53] As they wrote:

> The removal of the cnemophilines and *Macgregoria* from the paradisaeids necessitates a change in their vernacular names since all are referred to as birds of paradise. The cnemophilines have no readily available vernacular name, hence it is proposed here that the three biological species be called Loria's cnemophilus (*Cnemophilus loriae*), crested cnemophilus (*C. macgregorii*), and yellow-breasted cnemophilus (*Loboparadisea sericea*). The word 'cnemophilus' refers to being a lover of the mountain slope.

Cracraft and Feinstein's suggestion was not taken up, and, instead, the more easily pronounced "satinbird" was chosen to reference the velvety or silky plumage that all three species possess.

# Saw-whet

The saw-whets are three species of small owls from North and Central America. The most widely advanced theory of the etymology of this unusual name is that the bird's calls resemble the sound of a saw being sharpening with a whetting stone. Audubon wrote:[54]

> The Little Owl is known in Massachusetts by the name of the "Saw-whet" the sound of its love-notes bearing a great resemblance to the noise produced by filing the teeth of a large saw. These notes, when coming, as they frequently do, from the interior of a deep forest, produce a very peculiar effect on the traveller, who, not being aware of their real nature, expects, as he advances on his route, to meet with shelter under a saw-mill at no great distance.

However, despite the fact that the Northern Saw-whet Owl has up to eleven different vocalizations, nobody is quite sure which one sounds like a saw being sharpened. Another theory that more closely fits the law of parsimony is that it is an anglicization of the French word *chouette* for "owl," a word that would have been widely used in French-speaking parts of North America around the time the species was first described in the English language literature.

# Sawwing

The five species of sawwings from Africa, previously known as "rough-winged swallows," as in Sharpe's *A Monograph of the Hirundinidae* in 1885, are named for the rough outer primary feathers on the wings.[453] The function of these feathers, with tiny recurved hooks on the leading edges, is not yet known, but it's believed by some authorities that they may produce a sound during territorial or courtship displays.

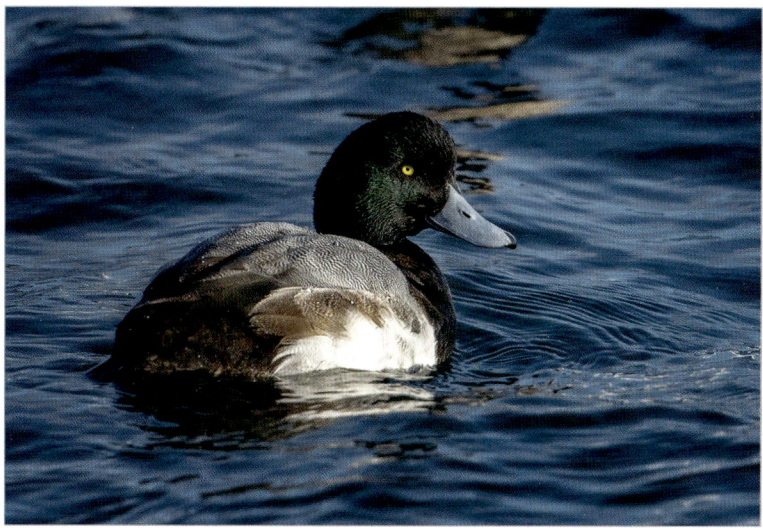

*Greater Scaup (Japan)*

# Scaup

In 1678, Ray and Willughby briefly explained the name for these ducks, of which there are three species:[16]

> It is called Scaup-duck from its feeding upon Scaup, i.e. broken Shelfish.

The word is a Scots language one meaning a bed or stratum of shellfish in the sea, especially a bed of mussels or oysters. It was variously spelled *scaup*, *scawp*, *skaap*, *scape*, or *scalp*. There are other claims that the name comes from the duck's vocalizations, but this can probably be discounted. The name started out as "scaup-duck" to signify that it is "a duck that feeds on scaup"; as the meaning of the word "scaup" was lost on most English speakers, the second element of the name was dropped.

# Schiffornis

These are seven species in the synonymous genus *Schiffornis*, in the Tityridae. Bonaparte coined the name for the genus in 1854, naming it in honor of Moritz Schiff, the German surgeon, pioneer anatomist, and physiologist. In 1846, Schiff became director of the ornithology department of the Frankfurt Zoological Museum, where he had worked in his youth. Schiff classified the birds of South America and collaborated during his time at the museum with Charles Bonaparte. The latter component of the name is, of course, the Greek *ornis* for "bird."

# Scimitar-Babbler

Following the shake-up of the Timaliidae, many of the babblers have been shifted to new or different families. The scimitar-babblers now span two families with two species of *Napothera* in the Pellorneidae, ground babblers and allies, and fifteen species of *Pomatorhinus* and *Megapomatorhinus* in the Timaliidae, the tree-babblers, scimitar-babblers, and allies. All of them have long, decurved bills ranging from the remarkably long bill of the Slender-billed Scimitar-Babbler to the shorter and straighter bill of the Streak-breasted Scimitar-Babbler. These specialized bills are used for various foraging methods, including for working through the leaf litter, probing flowers for nectar, and probing in the substrate for invertebrate prey. Jerdon seems to have been the first to use the vernacular name in *Illustrations of Indian Ornithology* in 1847 when he described the Slender-billed Scimitar-Babbler, calling it the "Scimitar Billed Babbler" with an account of the bill:[284]

> The bill much longer and more slender, and very thinly compressed throughout its length, widening only at the extreme base, and describing a considerable incurvation.

The long bills are clearly likened to a scimitar, a sword of Persian origin that features a curved blade. The word came into English from the Middle French *cimeterre*, ultimately derived from the Persian شمشیر *shamshir*, "sword."

# Scimitarbill

There are three species of scimitarbills in the Phoeniculidae, the woodhoopoes and scimitarbills. These members of the genus *Rhinopomastus* were initially placed with the hoopoes in the Upupidae and were called "Scimitar-billed Hoopoe" in 1903 in *The Birds of South Africa* by Sclater.[454] They are named for their long, decurved bills. (See Scimitar-Babbler.)

# Scops-Owl

The name of these fifty-five members of the Strigidae, the owl family in the genus *Otus*, has one of the oldest traceable etymologies. The first element of the name, scops, is from the Latin *scops*, which in turn came from the Ancient Greek σκώψ *skōps*, "a type of small owl." Pliny wrote in *Natural History* that "Homer maketh mention of a kind of Bird called Scopes."[583] And, indeed, he did. In *The Odyssey*, written in the eighth century BCE, we can find around Nymph Calypso's cave abode:[55]

> A thick wood of alder, poplar, and sweet smelling cypress trees, wherein all kinds of great birds had built their nests—owls [σκῶπες *skópes*], hawks, and chattering sea-crows that occupy their business in the waters.

The root of the Ancient Greek word is from σκοπός *skopós*, "watcher, guardian, spy," which seems appropriate for these ever-alert but secretive birds.

# Scoter

The scoters, with their predominantly black plumage, are six members of the *Melanitta* genus, from the Ancient Greek *melas*, "black," and *netta*, "duck," in the Anatidae. The etymology of their common name is not as straightforward, however. Newton stated that it is[7]

> a word of doubtful origin, perhaps a variant of SCOUT—one of the many local names shared in common by the distinct genus.

(Other black waterbirds, such as guillemots and coots, were also known as "scouts" along with a plethora of other unusual local epithets.*) Ray and Willughby seem to have been among the first to use the name in print, heading an entry "The black Diver or Scoter."[16] The best clue to the source of the name may be provided in a letter from John Ray received by Tancred Robinson, which he related in *Some Observations of the French Macreuse*:[56]

> The first knowledge of this Bird, we had from Mr. Jessop who sent us the skins of this among others stuft [stuffed], from Sheffield in Yorkshire, by the name of Scoter as it seems they call it thereabouts whether from the dark or black colour of it, or (which is more likely,) from Scotland, whence they might suppose it to come.

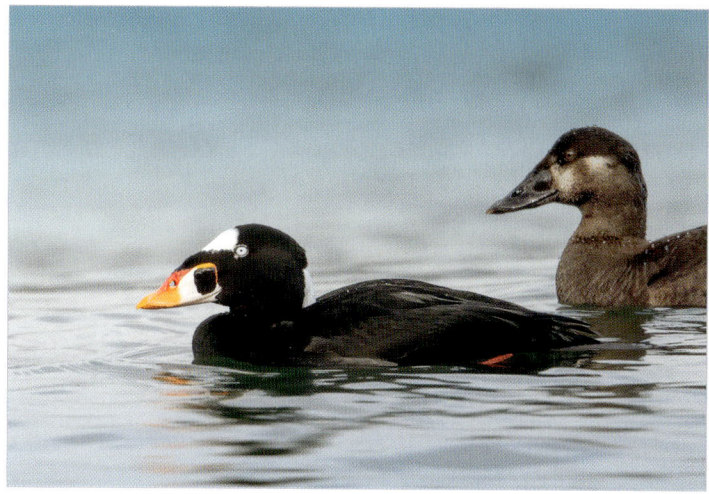

*Surf Scoter (Washington State, USA)*

Buffon reinforced Ray's views in *The Natural History of Birds* in 1793:[8]

> It has been pretended that the Scoters are engendered, like the barnacles, in shells or in rotten wood. Hence the name Scoter; Scotland being the principal scene of this fabulous transmutation of the barnacles.

In summary, the common name is either a corruption of the name "scout" (Newton); a geographical epithet, that is, Scoters from Scotland (Ray and Buffon); or, in some way, a reference to the black plumage, as per Ray and Willughby. Lockwood believed the name to be a corruption of an unrecorded word, *sooter*, in reference to the bird's notably black plumage.[84] The true etymology of this unusual bird name will probably never be revealed, but parallel names in other languages, for instance the German *Russente* (soot duck) and Dutch *Zwarte Zee-eend* (black sea duck), support the latter hypothesis.

*Including Frowl, Kiddaw or Skiddaw, Langy, Lavy, Marrock, Scout, Scuttock, Strany, Tinker or Tinkershire, and Willock.

# Screech-Owl

There are twenty-three small screech-owls in the *Megascops* genus of the Americas. As Ingersoll noted:[9]

> "Screech-owl" is an ancient but very indefinite designation. Certainly the term ought not to be used in this country as it constantly is for our little red and gray crooner, because the gentle, wavering serenade with which he lulls us into sleep is the very antithesis of a "screech."

In fact, "screech owl" was a name widely used for the Barn Owl in Europe. For example, in Bewick's *History of British Birds* in 1797, the "White Owl" is listed with alternative names "Barn Owl, Church Owl, Gillihowlet or Screech-Owl" with the observation:[149]

> Its flight is accompanied with loud and frightful cries, from whence it is denominated the Screech Owl.

Then Ray and Willughby wrote this for the Tawny Owl:[16]

> The Brown or Ivy-Owl, and from its schreeking [*sic*] noise, the Screech-Owl.

Even Audubon wrote about "*Strix Americana*, Aud. American Screech-Owl.—Barn Owl" in *A Synopsis of the Birds of North America* in 1839.[512] Although in 1828, Wilson wrote in *American Ornithology* that *Strix asio* (Eastern Screech-Owl *Megascops asio*)[531]

> is another of our nocturnal wanderers, well known by its common name, the Little Screech Owl; and noted for its melancholy quivering kind of wailing in the evenings.

# Scrub-bird

This Australian family, the Atrichornithidae, contains only two species of scrub-birds. These exceptionally secretive birds live in dense, scrubby vegetation and can be almost impossible to see even when seemingly calling at one's feet. Gould related a letter he received from the Australian ornithologist Edward Ramsay:[532]

> I have frequently stood on a log waiting for it to show itself from among the tangled mass of vines and weeds at my feet, when all of a sudden it would begin to squeak and imitate first one bird and then another, now throwing its voice over my head, then on one side, and then again apparently from the log on which I was standing. This it will continue to do for hours together; and you may remain all day without catching sight of it.

In 1848 Gould used the name "brush-bird" but by 1865 he was calling the birds "scrub-birds." In an 1849 volume of *The Tasmanian Journal of Natural Science, Agriculture, Statistics, &c.*, the common name for *Psophodes crepitans* [Eastern Whipbird *P. olivaceus*] is listed as "Crested scrub bird," so the term was probably used by local people for many types of skulking bird.[455] As Newton observed:[7]

> The name [is scrub-bird] (for want of a better, since it is not very distinctive).

The word "scrub," although used widely, is particularly applied in Australia to describe dense plant communities dominated by low shrubs and grasses. The word is a variant of "shrub," a variant of *shrobbe*, perhaps influenced by a Scandinavian word, such as the Danish *skrub* or Norwegian *skrubbe*, meaning "a small, stunted tree."

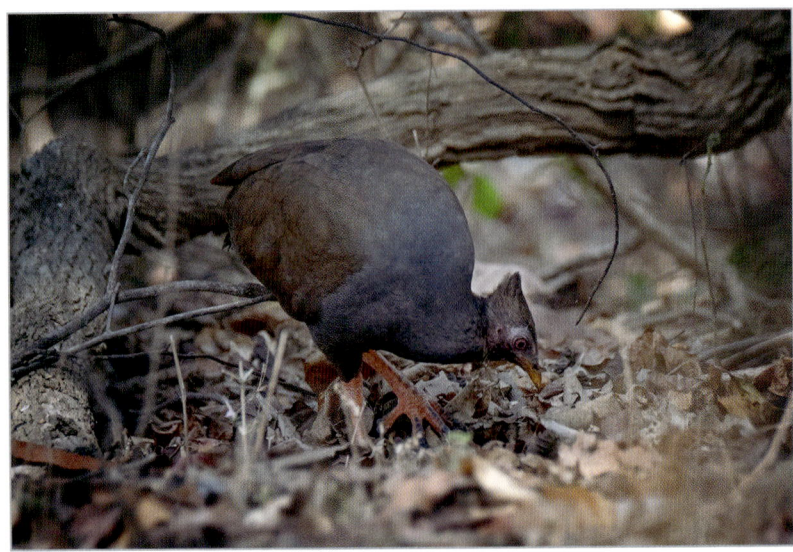

*Orange-footed Scrubfowl (Komodo Island, Indonesia)*

## Scrubfowl

Fourteen species of scrubfowls in the Megapodidae, the "big footed" mound-builders of Australasia, are close relatives of the brushturkeys, and their names mean much the same thing. The name was coined in Australia in the late 1800s as an allusion to their somewhat chicken-like appearance (as Fraser and Gray noted, "familiarity was more important than systematics"[130]) and the fact that they dwell in lowland tropical "scrubby" forests.

## Scrub-Jay

The four species of scrub-jays in the Corvidae are confined to North America. The scrub-jays were called simply "jays" in the literature up until the 1940s, when "scrub-jay" was first used in print by Grimes in *Bird Lore* in a 1940 article entitled "Scrub Jay Reminiscences."[533] In *The Wilson Bulletin* in 1946, Eisenmann and Poor wrote:[57]

> In some instances it will . . . be necessary and desirable for the Committee to adopt a name . . . For example, . . . "Florida Jay" (n)or "California Jay" would be suitable for the *Aphelocoma coerulescens* group. . . . "Scrub Jay" (is a) possible suggestion here.

## Scrub-Robin

These secretive members of the Petroicidae, the Australasian robins, are three terrestrial species that dwell in dense thickets and understory in forests and semiarid scrublands. There are also ten species of unrelated African Muscicapid scrub-robins in the genus *Cercotrichas*. The English colonists used the name "robin" for any Australian bird that bore even a passing resemblance to the unrelated birds of their homeland. Gould introduced the name in 1848 and wrote:[40]

> I discovered this singular bird in the great Murray Scrub in South Australia, where it was tolerably abundant; I have never seen it from any other part of the country, and it is doubtless confined to such portions of Australia as are clothed with a similar character of vegetation.

## Scrubwren

Up until the late 1800s, the scrubwrens, of which there are thirteen species in the *Sericornis*, were referred to simply by the genus name. In 1897, North used the name "Scrub Wren" for the Chestnut-rumped Heathwren.[534] Then in 1898, Morris[20] listed "Scrub-Wren" as "any little bird of the Australian genus *Sericornis*." This compound word derives from the preferred scrubby habitat of all members of the genus in the first element and its superficial similarity to the true wrens in the second element.

## Scythebill

With their remarkably long, curved bills, the scythebills are aptly named. Five species are so named in the Furnariidae, the ovenbirds and woodcreepers. The name comes from the similarity of the shape of the bill to that of the grass-cutting instrument with a long blade in a sweeping curve. The Old English word *sīþe* is from the Proto-Germanic *segitho*, "sickle" (curiously, the *sc-* spelling was adopted in the early 1400s, due to the mistaken belief that there was an association with the Latin *scindere*, "to cut."). As for the second part of the name, it came into Old English as *bile*, and, surprisingly, is of unknown origin. It's thought it might come from an adapted use of Old English *bil, bill*, meaning "a hooked point; curved weapon," from the Proto-Germanic *biljǫ*, "axe; sword; blade."

# Secretarybird

The strange Secretarybird is the sole member of the Sagittariidae. The first European account of the bird came in 1769 from Vosmaer, who received a specimen in the Netherlands from the Cape of Good Hope and was told that the bird was known as the "Sagittarius," or Archer, from its striding gait thought to resemble that of a bowman advancing to shoot. He noted:[535]

> Les Païsans du Cap, au lieu de Sagittaire, l'ont appelé par corruption, Secrétaire. [The Cape townspeople give the bird—surely by corruption of the word Sagittarius—the name of Secretarius.]

Later, Le Vaillant would note that the traditional Indigenous name for the bird was "snake-eater" and would call the bird in French "Le Mangeur de Serpents."[93] Both of these early observations are enshrined in the scientific name *Sagittarius serpentarius*.[86] In 1785 Sparrman used the name "the secretaries bird" and in French "l'Oiseau des secretaries," both meaning "belonging to secretaries," which is odd.[87] The theory proposed by Fry[88] that the name is derived from the Arabic phrase "*saqr et-tair*" has been dismissed by Glenn[86] as historically implausible and linguistically illogical in 2018. As Glenn suggests in his appropriately entitled paper "Shoot the Messenger? How the Secretarybird *Sagittarius serpentarius* got its names (mostly wrong)," it seems likely that

> the bird named Secretarius in Africa had its name corrupted or changed to Sagittarius en route to Holland—presumably someone misheard or mistranscribed the name.

Ultimately, as Buffon surmised in 1783 when he named the bird "Le Secretaire ou Le Messager" [The Secretary or the Messenger], the bird was named Secretary from the fancied resemblance of the plumes on the nape to the pens clerks would stick behind their ears.[417]

> Comme il doit sans doute celui de *secrétaire* à ce paquet de plumes qu'il porte au haut du cou, quoique M. Vosmaër veuille dériver ce dernier nom de celui de *sagittaire*. [It undoubtedly owes the name *secretary* to this bundle of feathers which he wears at the top of his neck, although M. Vosmaër wishes to derive this last name from that of *Sagittarius*.]

# Seedcracker

These three members of the Estrildidae, the waxbills and allies, have remarkably stout, strong bills even for a group of birds known for this feature. These birds can really crack seeds! In *The Naturalist's Library* from 1862, Jardine called the Crimson Seedcracker the "Crimson Nut-Cracker" and wrote:[58]

> It may safely be affirmed that this extraordinary bird has the thickest and most massive bill in the feathered creation. . . . What are the nuts or seeds, the breaking of which requires such an amazing strength of bill, is perfectly unknown; but they must be of a stone-like hardness.

# Seedeater

This catch-all name applies to thirteen species in the Fringillidae, the finches, euphonias, and allies; three species in the Cardinalidae, the cardinals and allies; and the thirty-eight species in the Thraupidae, the tanagers and allies. All of them feature distinctively conical bills and, not surprisingly, feed on seeds.

# Seed-Finch

Six species in the Thraupidae belong to the same genus, the *Sporophila*, as the seedeaters of that family. These finch-like birds are also seed eaters.

# Seedsnipe

The South American family the Thinocoridae are four species of shorebirds all called seedsnipes. Although they could hardly be said to resemble snipes, in fact looking more like small sandgrouse, they do live up to the first part of their name—they feed on seeds, albeit not exclusively; they also feed on buds and leaves. However, when we look at the etymology of the word "snipe" (see entry), the name is perfectly logical as a reference to the bird as a shorebird and, as we shall see, Sclater did see similarities with the true snipes.[22] Initially called quail-plovers, the name seed-snipe seems to have been first used by Sclater in 1888:

> This curious bird has the grey upper plumage and narrow, long, sharply-pointed wings of a Snipe, with the plump body and short strong curved beak of a Partridge . . . When alighting the Seed-Snipe drops its body directly upon the ground and sits close like a Goatsucker; when rising it rushes suddenly away with the wild hurried flight and sharp scraping alarm-cry of a Snipe. It is exclusively a vegetable-feeder. I have opened the gizzards of many scores to satisfy myself that they never eat insects, and have found nothing in them but seed (usually clover-seed) and tender buds and leaves mixed with minute particles of gravel.

# Seriema

The Cariamidae is a family consisting of only two species of seriemas, found in the savanna, woodlands, and grasslands of South America. The name is taken from the South American Tupi-language word for the bird, which is variously transliterated as *siriema*, *sariama*, or *çariama*. The name is a compound of the Tupi words *çaria*, "crest," and *am*, "raised," which very aptly describes the frontal crest of both species, especially of the more widespread Red-legged Seriema. In 2011, Costa wrote of the name:[456]

> Toponym of Tupi origin defined . . . as a form corresponding to *ceri-eim*, "that which flies or rises a little"; corresponding to *çariama*: *çaria*, meaning "crested" and,—*am* meaning "uplifted," that is, "uplifted crest or armed with a crest."

# Serin

Eleven birds in the Fringillidae, the finches, euphonias, and allies, derive their names from the French name, which is used in that language for all the members of the two genera *Crithagra* and *Serinus*. In English, the members of the genera are either "canary" or "serin." In 1555, Pierre Belon wrote in *L'Histoire de la Nature des Oyseaux*:[378]

> The Serin takes its French appellation for the excellence of its song: because just as it is said that the Syrenes lull the sailors to sleep with the sweetness of their songs, apparently because this little bird, of stoutness almost comparable to a small kinglet, sings so softly, he took the name of the Serin.

If Belon is correct, the etymology of the bird name "serin" is therefore taken from the Sirens (Σειρήν, *Seirín*) of Greek mythology, the dangerous creatures who, with their enchanting music and singing voices, would lure sailors to their deaths on the rocky coast of their island.

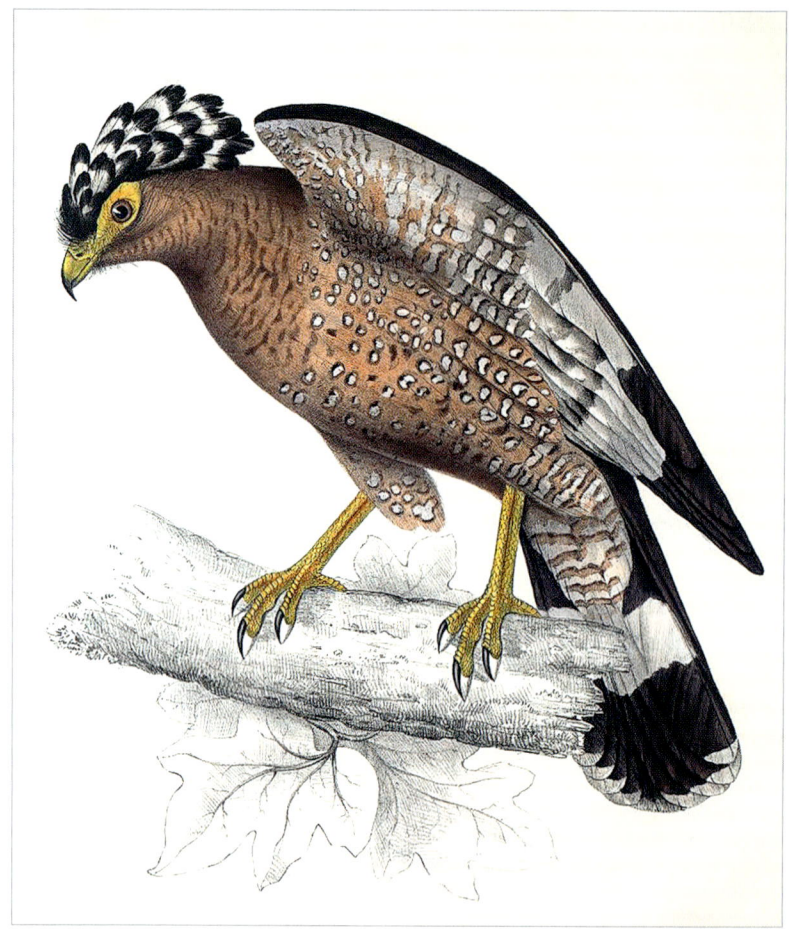

*Crested Serpent-Eagle from Gould and Gould's* A Century of Birds from the Himalaya Mountains *(1831)*

# Serpent-Eagle

The serpent-eagles are eight species of raptors in three genera that all prey on snakes to lesser or greater degrees. Jerdon used the name in 1839, saying that its "most favourite food of the *samp-mar* is, as its Indian name implies, snakes," but he was referring to the Short-toed Snake-Eagle. (See Snake-Eagle.)[275] Later, in 1862, the same author used the name for the birds of the *Spilornis* genus:[136]

> The birds of this genus differ from those of *Circaetus* in being more arboreal, and much less on the wing, darting on their reptile prey from the bough of a tree. They are confined to the tropical parts of Asia.

# Shag

At present there are fifteen species officially known as shags. They are all in the genus *Phalacrocorax* and all possess shaggy crests, which is how they got their names. Lockwood calls these types of names "part for whole," giving examples of other bird names, such as Ruff (collar), auk (neck), and snipe (bill).[84] The Old English name for the "sea ravens" was *scega*, derived from the Scandinavian word for a beard, *skjegg*. Other old names for European Shag, such as *Skart* (Orkney Isles) and *Scarf* (Shetland Isles), give some hints as to the evolution of the word, especially when you consider that the sounds *sk* and *sc* were pronounced as *sh* in Old English.

## Cormorant or Shag?

In England, where these English names came from, there are found the Cormorant and the Shag—known everywhere else as the Great Cormorant and the European Shag, respectively. They are easily told apart on the British Isles, but worldwide there are forty species of cormorants and shags in the Phalacrocoracidae. Is there a rule for what makes a cormorant and what makes a shag? The quick answer is no, the more expansive answer is "sort of"—much like the bird names "dove" and "pigeon," there is no hard and fast rule for what is a "shag" and what is a "cormorant." The names have been used interchangeably over many years and in many places, but the name shag has tended to be assigned to smaller members of the family that possess shaggy crests. To understand the often-arbitrary nature of the assignation of common generic names, just look at the South Georgia Shag and the Imperial Cormorant, which are often treated as conspecific.

# Shama

There are seven species of shamas in the *Copsychus* genus in the Muscicapidae, the Old World flycatchers. The name is taken from the Hindi name for the White-rumped Shama, शामा *shaama*, and it derives from the Sanskrit *syāma*, meaning "black, dark-colored." In what appears to be a case of some sort of miscommunication, when writing of the Blue Rock-Thrush, Jerdon said:[284]

> Its name in Hindustani is *Shama*, and it is said to be sometimes caged by Faqueers and others for its song, which is highly prized in the north of India, but its musical qualities appear unknown in the south, and the *Copsychus macrourus* [White-rumped Shama] appears known as the Shama or Shahmour in Bengal.

# Sharpbill

This species is a member of the small Oxyruncidae, the Sharpbill, royal flycatchers, and allies. The Sharpbill is in the monotypic genus *Oxyruncus* and is named for its conical, very pointed bill which Shaw described in 1826 as[81]

> short, very straight, trigonal at the base, attenuated beyond the base, the tip very acute.

# Sheartail

The sheartails are three species of hummingbirds with deeply forked tails, described by Montes de Oca in *Ensayo Ornitológico de los Troquilideos ó Colibríes de México* in 1875:[457]

> The tail is purple-black, open in the shape of scissors, decreasing the size of its feathers two by two towards its center, having the second and third on each side, about a third of its width, light brown, without reaching up to the tip, giving them at first sight the figure of a spatula.

The birds are clearly named for the shape of their tails, which resemble shears, the tool like scissors used for cutting cloth or for removing the fleece from sheep.

# Shearwater

Shearwaters, whose skimming, twisting flight accounts for their name, are thirty-one species in the Procellariidae, the shearwaters and petrels. The name was first introduced by Ray and Willughby in *Ornithology* in 1678:[16]

> The Shear-water is a Sea-fowl, which fishermen observe to resort to their Vessels in some numbers, swimming swiftly to and fro, backward, forward, and about them, and doth as it were, *radere aquam*, shear the water, from whence perhaps it had its name.

# Sheathbill

The Chionidae contains only two species, the Snowy Sheathbill and the Black-faced Sheathbill. These birds of the Antarctic are named for the unusual horny sheath that covers the upper mandible. Latham was one of the first to describe the bird 1785:[17]

> Over the nostrils a horny appendage, which covers them, except just on the forepart; and descends so low on each side, as to hang over part of the under mandible; this is moveable, and may be raised upwards, or depressed so as to lay flat on the bill.

The marine biologist Carl Chun explained the function of the sheath in 1902:[282]

> A mother penguin cannot leave her egg for a moment but that a sheath-bill, *Chionis*, dashes its beak into it. The sheath that lies over this bird's upper mandible, like a saddle with the pommel tilted up into 'the air,' has been explained by Studer as a benevolent provision to prevent its nostrils from being stuffed up by the contents of the egg—the very thing which the enraged parent penguin would like to see happen to the piratical *Chionis*.

# Shelduck

Nine species of shelducks span three genera, but the name was originally coined for the familiar Common Shelduck in England. The bird was known by many local names, such as Shieldduck, Sheldrake, Shelduck, Burrowduck, Skygoose, Skeelduck, and Skeelinggoose, before settling on the more familiar name of today. The name comes from the word *sheld* in a British dialect meaning "spotted, piebald, or speckled." Ray and Willughby mentioned the bird in 1678:[16]

> The Sheldrake, or Borough-Duck: *Tadorna Bellonii*. It is called *Sheldrake* from its being particoloured, Sheld signifying dappled or spotted with white; and *Burrow-duck* from building in Coney-burrows [rabbit burrows].

# Shikra

Unlike most of its congeners, which are known as either sparrowhawks or goshawks, the Shikra has earned a unique name. The name is from the Hindi word शिकरा *shikara* and the Urdu word شکاری *shikari*, which both mean "hunter." This beautiful and sleek bird of prey obtained its name from having long been a favorite with falconers in India and Pakistan. Shikras are known for being easy to train and for having high success rates, so that they were often kept for the sole purpose of acquiring food to feed larger falcons. The males and females were given different names by the Indian falconers, and Blyth explained the names in 1849:[283]

> *Shikra*, (from *shikárkardan*, to pursue game), female; *Chippuck* (or *Chipka*, Jerdon, from the voice), male.

Jerdon wrote in *The Birds of India* in 1862 that the Shikra[136]

> is more commonly trained than any other hawk in India. It is very quickly and easily reclaimed, and, though not remarkable for speed, can yet seize quails and partridges if put up sufficiently close. It is, however, a bird of great courage, and can be taught to strike a large quarry, such as the common crow, the small grey hornbill, the crow pheasant, (*Centropus*), young pea fowl, and small herons.

# Shining-Parrot

The shining-parrots are three species endemic to Fiji. They are brightly colored, large members of the Psittaculidae and may have first received their name from British cage bird enthusiasts. One of the first mentions of the birds appears in 1887's *Parrots in Captivity* by W. T. Greene, who asserted:[458]

> There is a blue band at the neck, and the back is dark shining green, from which it may get its name of Shining Parrakeet.

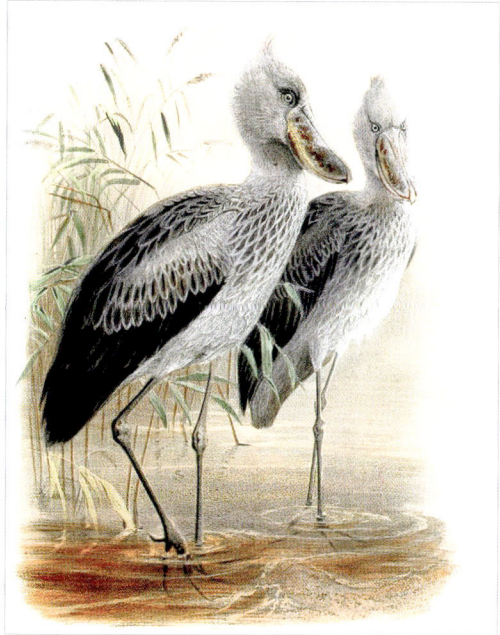

*Shoebill by Joseph Wolf in* Transactions of the Zoological Society of London *Vol. 4 (1862)*

# Shoebill

There is no mystery about the origin of the name of this amazing bird. The sole member of the Balaenicipitidae, the Shoebill possesses one of the most unusual bills of any bird. The enormous boot-shaped appendage gives rise to another common name, the whalebird, which is reflected in the family name. Newton wrote about the bird and its name in 1896:[7]

> SHOE-BILL or SHOE-BIRD, renderings of the Arabic name *Abu-markub* (Father of a Shoe) that have been given by travellers to one of the most remarkable-looking of Central-African birds, *Balæniceps rex*, also called by some writers the Whale-headed Stork—the bird's huge bill, in shape not unlike a whale's head, and tipped with a formidable hook, suggesting all these names.

## Sholakili

This is a new name for the Southern Indian birds that were until recently called either shortwings or blue-robins. The genus *Sholicola* was established in 2017, and at the same time, the common name Sholakili was coined.[59] It is a combination of two Tamil or Malayalam words, the languages of the two Indian states in which the birds are found. In Tamil it is சோலை *cōlai* (pronounced *SO-lay*), for "grove or small forest" and கிளி *kili*, for "bird." In Tamil it would be written சோலைக்கிளி *cōlaikkili*. In Malayalam it is ഷോല *sēāla* and കിളി *kili* (ഷോലകിളി *sēālakili*).

## Short-tail

The Sao Tome Short-tail in the genus *Motacilla* is hardly the only bird to possess a short tail, but when compared to other members of the genus, all known as wagtails, its shortened tail is indeed worthy of mention. This wagtail is more wagger than tail. It was previously known as "Bocage's Longbill" after José Vicente Barbosa du Bocage (1823–1907), the Portuguese zoologist and politician. It was never a well-known species, and it disappeared from view—presumed extinct after 1928—before its rediscovery in 1990. The present-day name, Sao Tome Short-tail, appeared in the 1970s.

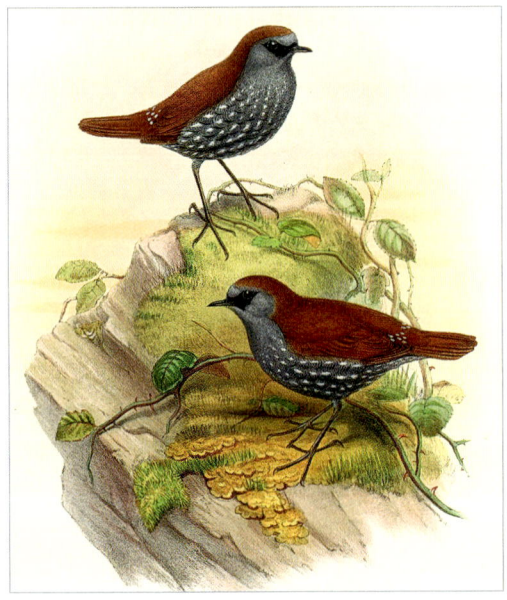

*Gould's Shortwing from Gould's* Birds of Asia *(1850)*

## Shortwing

Eight species, seven in the *Brachypteryx* and one now in *Heinrichia*, belong to the Muscicapidae, the Old World flycatchers. These charming but shy ground dwellers are dumpy little birds with long legs, short tails, and short wings perfectly adapted for their lives in the dense undergrowth of the Asian forests. Thomas Horsfield introduced the genus name *Brachypteryx* in 1822, taking the Ancient Greek words *brakhus* for "short" and *pterux* meaning "wing."[46] He made the observation in *Transactions of the Linnean Society* of London that "it possesses . . . peculiarities in the shortness of the wings." Gould was the first to use the common name, listing Gould's Shortwing as "Stellated Shortwing."[33]

# Shoveler

There are four species of shovelers, but the name was first used for the Northern Shoveler, which is widespread in the Northern Hemisphere, including in England. These unusual ducks are most notable for their large, spatulate bills that resemble a shovel, hence the name. Their bills are adapted for straining food items from water. Newton noted:[7]

> Shoveler, formerly spelt Shovelar, and more anciently Shovelard, a word by which used to be meant the bird now almost invariably called Spoonbill,* but in the latter half of the 17th century transferred to one hitherto generally, and in these days locally, known as the Spoon-billed Duck.

*Which went extinct in Britain in the late 1600s.

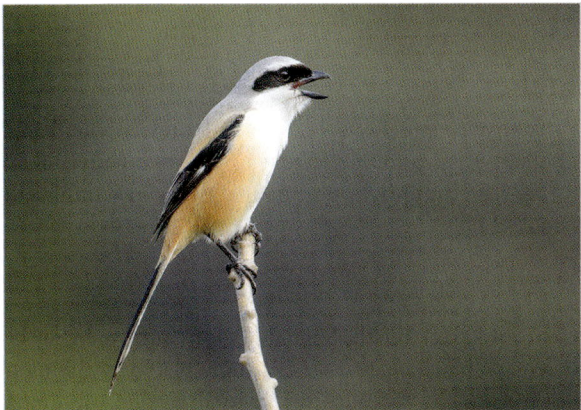

*Long-tailed Shrike (India)*

# Shrike

The Shrike family, the Laniidae, contains thirty-three species, twenty-seven of which are known as shrikes. Also sometimes known in the past as "butcher-birds" (a name now used for the unrelated Australasian *Cracticus* genus), the name shrike is thought to be derived from the Icelandic word *skríkja*, literally "a shrieker." The name is first found in Turner in 1544, when he wrote about "a shrike, a nyn murder" (Nine Killer being a German name for the Great Grey Shrike from the superstition that the bird slays nine birds every day).[119] Ray and Willughby wrote in 1678:[16]

> Of Butcher-Birds or Shrikes called in Latine *Lanii* or *Colluriones*.

> The new name of Lanius or Butcher was by Gesner imposed on this bird . . . In English it is called a Shrike.

But the vocalizations of the shrikes could hardly be described as a shriek, leading Newton to believe that the name was misapplied to the Great Grey Shrike and that Turner was misinformed in 1544:[7]

> There can be little doubt that Turner's informant was mistaken, and that the name, signifying a bird that screeches or shrieks (A.S. *Scríc*, old Norsk *Skrikja*, mod. Scand. *Skrika*—a Jay) probably applied originally to the Mistletoe-THRUSH, known to Charleton in 1668 as SHREITCH, and to Willughby as SHRITE—a name it still bears in some parts of England, to say nothing of cognate forms such as *Screech-bird* and *Shirl*.

# Shrike-Babbler

It came as a surprise to many to find, following molecular studies, that the nine birds known as shrike-babblers were not actually babblers in the Timaliidae, but rather members of the Vireonidae. With this discovery, they became the first recognized vireonids in the Old World. They acquired the second element of their name through their previous classification in the Timaliidae, the Old World babblers (now the tree-babblers, scimitar-babblers, and allies). The first element of the name refers to the thickish, hook-tipped bill—like a shrike's.

# Shrikebill

The shrikebills are five species of monarch flycatchers in the genus *Clytorhynchus* endemic to Melanesia. The rather unimaginative name is taken from their shrike-like, long, laterally compressed bills.

# Shrike-Flycatcher

Two species of shrike-flycatchers from Africa are found in the Vangidae, the vangas, helmetshrikes, and allies. They combine features of the two better-known birds they take their name from—they have an upright posture and behavior reminiscent of flycatchers, as well as shrike-like hooked bills.

# Shrikejay

The enigmatic Crested Shrikejay from the Sundaic region (Borneo, Malay Peninsula, Sumatra, and Java), with its remarkable elongated crest, is now placed in the monotypic family the Platylophidae. Until recently, it was called the Crested Jay and subsumed in the Corvidae. But recent morphological and molecular studies show that it is quite distinct from Corvidae and may actually be closer to the Laniidae, the shrikes. This is a newly coined name; the jay-like appearance along with the shrike-like hook-tipped bill explains the compound name.

# Shrikethrush

Eleven species of shrikethrushes in the *Colluricincla* belong to the Pachycephalidae, the whistlers and allies. The name is another example of British bird names being transferred to unfamiliar Australian birds by the English colonizers. In 1890 Broinowski wrote:[60]

> Each part of Australia possesses at least one species of this form, which both in structure and habits is a medium between the Shrikes and Thrushes.

# Shrike-tit

The Crested Shrike-tit of Australia is the sole member of the monotypic family the Falcunculidae. As is so often the case for the Australian birds, the early English colonizers named the bird for ones they were familiar with back home. It has the large head and heavy hooked bill of a shrike and bears a more than passing resemblance in appearance and behavior to the Great Tit. Gould was the first to use the name, at least in print, and he gave a wonderful description of the bird:[40]

> It is very animated and sprightly in its actions, and in many of its habits bears a striking resemblance to the Tits, particularly in the manner in which it clings to and climbs among the branches in search of food. While thus employed it frequently erects its crest and assumes many pert and lively positions: no bird of its size with

which I am acquainted possesses greater strength in its mandibles, or is capable of inflicting severer wounds, as I experienced on handling one I had previously winged, and which fastened on my hand in the most ferocious manner.

# Shrike-Vireo

There are four species of robust vireos in the genus *Vireolanius.* The genus name, too, attests to the shrike-like appearance of the bird, being *Vireo* and *lanius*, the genus name for the shrikes. (See Vireo and Shrike.)

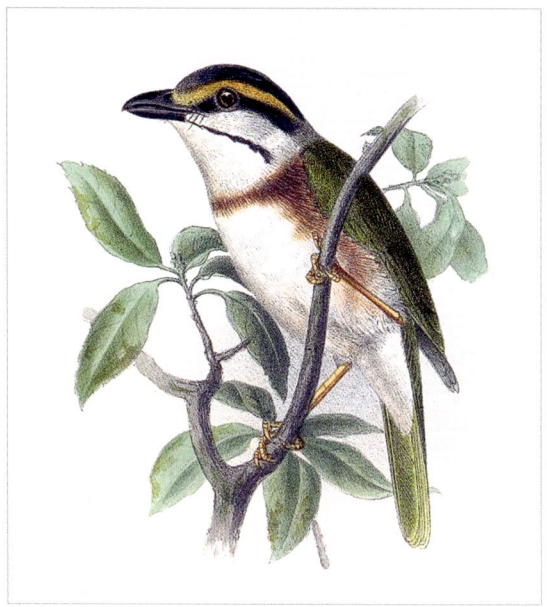

*Chestnut-sided Shrike-Vireo in Sclater's* Exotic ornithology *(1869)*

# Sibia

The delightful sibias found in Asia are seven species in the *Heterophasia* genus, and one that is now in the *Minla* genus, of Leiothrichid. The name comes from the Nepali name for the bird, सबिया *sibiyā*, originally for the Rufous Sibia *Heterophasia capistrata*.[329]

# Sicklebill

There are two groups of sicklebills—four birds-of-paradise in the genera *Drepanornis* and *Epimachus* and two hummingbirds in the genus *Eutoxeres*. All the sicklebills have long, decurved—in other words, sickle-shaped—bills. In the case of the hummingbirds, the declination of the bill is remarkably sharp.

# Sierra-finch

There are eleven sierra-finches in the Thraupidae, the tanagers and allies. Though not true finches, these tanagers superficially resemble the members of the Fringillidae. That, along with their mountain-dwelling habits, accounts for their name, a relatively recent one first appearing in the literature in the 1970s. The word *sierra* is Spanish for "mountain range" or "mountain chain."

# Silktail

The silktails are two members of the Rhipiduridae, the fantails, endemic to two tiny islands in Fiji and named for the conspicuous white patch on the rump and upper tail. They were first chronicled in 1873 by Otto Finsch, who described their most striking feature:[61]

> Quite singular are the tail-feathers as regards the loose and separated disposition of their radii, and the splendid shining silky white coloration which they show on their basal portions, like the rump and upper tail-coverts.

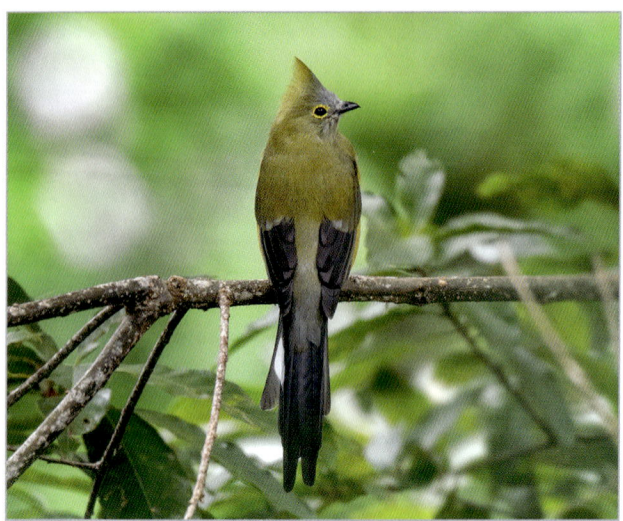

*Long-tailed Silky-flycatcher (Costa Rica)*

# Silky-flycatcher

The Ptiliogonatidae is a small family of only four species. The name silky-flycatcher references their smooth, silky plumage. The initial conflation of these mostly frugivorous birds with the tyrant-flycatchers explains the second element of their name. Swainson, who called Grey Silky-flycatcher by the name "Yellow-vented Short-foot" in *Zoological Illustrations* in 1833 claimed, incorrectly as it turns out, that "By viewing this as the type of the Scansorial group of the Tyrant-flycatchers, (Tyranninae) every circumstance, even the most minute, regarding its structure and its colours will be explained."[50] Howell and Webb drop the second element of the name, calling it the Silky, believing that the flycatcher part is both superfluous and confusing.[323]

# Silverbill

Three species of silverbills in two different genera, *Euodice* and *Odontospiza* (one found in India, two in Africa), in the Estrildidae, the waxbills and allies, have silvery-gray bills. The name appears to have originated in the early 1800s, chiefly among the cage bird enthusiasts.

# Silverbird

The Silverbird is a member of the Muscicapidae in the genus *Melaenornis*, although it is often put in the monospecific *Empidornis*. Found in East Africa, the bird is named for its predominantly slaty-gray, some might say silver, plumage.

# Silvereye

Although most other members of the *Zosterops* genus are known by the name "white-eye," the much-loved Silvereye of Australia and New Zealand has retained its local appellation given to it by the English colonizers. Gould reported in 1848 that the "Colonists of New South Wales" called the Silvereye "White-eye," which he called "Grey-backed Zosterops."[40] The first use of the present-day name may have been in New Zealand. Walter Buller wrote in *Transactions and Proceedings of the New Zealand Institute* in 1870:[62]

> By the settlers it has been variously designated as Ring-eye, Wax-eye, White-eye, or Silver-eye, in allusion to the beautiful circlet of satiny-white feathers which surrounds the eyes; and quite as commonly the "Blight-bird," or "Winter-migrant."

# Sirystes

These four members of the Tyrannidae share their common and genus names. The name was given to Sibilant Sirystes by Cabanis and Heine in 1859 with the explanation:[63]

> Σιρύστες = sibilator, von σιρύςω (zischen) [*Sirýstes* = sibilant, from σιρύςω *sirýso* (hiss)].

The four species are differentiated mostly on the basis of vocalisations but all of them could be said to call with sibilant whistles. The Ancient Greek word συρίζω *syrízo* can be defined as "play the pipe, make a whistle or hissing sound, or catcall."

# Siskin

The nineteen species of siskins in the Fringillidae, the finches, euphonias, and allies, are small green, yellow, and black birds (with the exception of the Red Siskin). The first bird to receive the name was the Eurasian Siskin, when Turner wrote in 1544 of the bird:[26]

> In English a siskin, in German eyn zeysich: The Luteola is much smaller than the Lutea above described, and with a colour tending more to green. It has a yellow breast, a longish, slender, pointed bill, like that in Aurivittis, and two spots of black, one on the forehead, one beneath the chin; it warbles with some sweetness. In England it is rare, and scarcely to be seen elsewhere than in cages.

It is thought the name came from a German dialect *Sisschen* or *Zeischen*, a diminutive form of the Middle High German *zisec*. Others claim it is derived from a Scandinavian word. Swann wrote in *A Dictionary of English and Folk-names of British Birds*:[121]

> The derivation is probably from the Dan[ish] *Sidsken*, or Swedish *Siska*, a chirper.

But Siskin is usually etymologized as a Slavic loanword in German, from the Czech *čížek* via the Middle Low German *sisek* and the Middle Dutch *siseken*. As a German loanword, the "-kin" part of the name comes from the German "-chen," a diminutive. Webster's dictionary dates the first occurrence of the English name at 1544 (Turner) and calls it a German dialect word. The *Oxford English Dictionary* also states that the name is based on an Middle High German word *zisec*, being of Slavic origin. Vasmer says that the name "čížek," also known as the undiminutized "číž" in some Slavic languages, is onomatopoeic.[326]

> Звукоподражательное; ср. передачу крика чижа в укр.: чий ви! чий ви! "чей вы?" [Onomatopoeic; cf. transmission of the cry of a siskin in Ukrainian: chiy vi! chiy vi! "whose are you?"]

## Sittella

The three species in the small Australasian family the Neosittidae are called sittellas. The name is a diminutive form of the genus *Sitta*, the nuthatches of the Northern Hemisphere. The Ancient Greek name σίττη *sittē* was mentioned by Aristotle and Hesychius in reference to a climbing woodpecker-like bird and adopted by Linnaeus for the Eurasian Nuthatch in 1758. The genus *Sittella* was established by Swainson in 1837 on the basis of the shared characteristics with the nuthatches, another example of convergent evolution.[64] This wasn't recognized, however, until the 1960s, prior to which the sittellas were placed in the Sittidae, alongside their doppelgängers.

## Siva

The Blue-winged Minla in the Leiothrichidae is sometimes called the Blue-winged Siva, which was also the now disused genus name coined by Hodgson in 1837.[536] The name is possibly from the Hindu god शिव *Shiva*, which is derived from the Sanskrit word *śiva* and means "auspicious, kind, benevolent, friendly," which could describe these charming birds.

## Skimmer

Pennant was the first to coin the name skimmer for the three species of *Rynchops* in the Laridae. He wrote in 1773:[66]

> I call it Skimmer, from the manner of its collecting its food with the lower mandible, as it flies along the surface of the water.

Prior to this, in 1730, Catesby had called it the "Cut Water," commenting that "The bill, which is the characteristic note of this Bird, is a wonderful work of nature."[70] Much later, in 1877, Holder gave a wonderful description of the bird:[65]

> The Skimmer darts swiftly along the surface of the ocean, dipping the extremity of its curious bill into the water as it moves along, for the purpose of capturing the small fishes and crustacea upon which it chiefly feeds. This, however, is not, according to Lesson, the only use of the bill; that writer states that on the coast of Chili the Skimmers insert knife-like extremity of the lower mandible into the gaping shells of the bivalve mollusca left nearly dry by the retreating tide; the mollusc, objecting to this treatment immediately closes his shell, and in so doing of course seizes the bill of his enemy, who then drags him from his retreat amongst the sand, carries him up to the beach, breaks his shell open by a few blows, and speedily devours its contents.

## Skua

Four members of the Stercorariidae, the skuas and jaegers, get their names from Faroese, a Scandinavian language. The word "Faroese," *skúgvur* for the "Great Skua," from Old Norse *skúfr*, is of unknown origin but is possibly echoic of the birds' vocalizations. The island of Skúvoy in the Faroe Islands is home to one of the largest colonies of the Great Skua; it is debatable if the island was named after the bird or the bird after the island. Newton wrote of the word:[7]

> Thus written by Hoier (circa 1604) as the name of a Færoese bird (*hodie Skúir*), an example of which he sent to Clusius. The word being thence copied by Willughby has been generally adopted in English, and applied to all the congeners of the species to which it was originally peculiar.

# Skylark

Three species in the Alaudidae are called by the name skylark. (See Lark.) These larks are known to sing as they fly high into the air. The English name is a direct translation of the German *himmellerch* (*himmel* meaning "sky," *lerche* meaning "lark") mentioned by Gessner in 1555.[178] In 1678 Ray and Willughby referred to

> the common Skie-Lark.

The present-day name and spelling was subsequently popularized by Pennant in 1766, with a description of its namesake habits, when he wrote:[342]

> This is the only bird that sings as he flies; raising its note as it soars, and lowering it till it quite dies away as it descends. It will often soar to that height that we are charmed with the music when we lose sight of the songster.

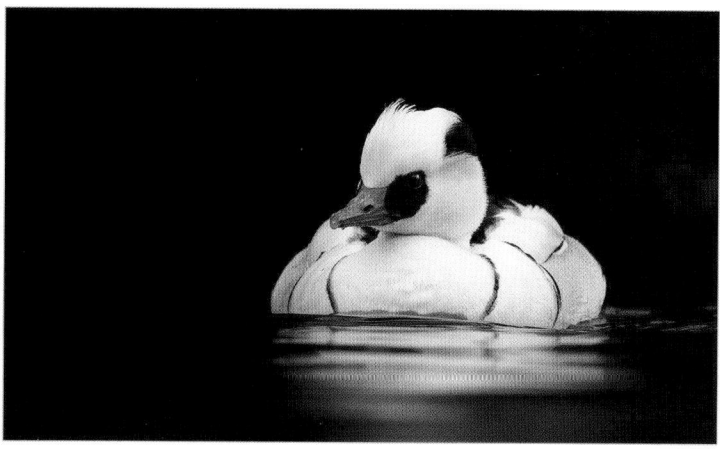

*Smew (Japan)*

# Smew

The Smew is a unique merganser in the monotypic genus *Mergellus*. This unusual name is an early provincial name from Norfolk, also recorded as *Smee*, and is derived from the Dutch name *smient* for Eurasian Wigeon and a German dialectical word for any small wild duck, *Schmeiente* (*-ente* or *-ent*, meaning "duck").[75] Ray and Willughby wrote about the Smew in 1678 under an older name, the "White Nun":[16]

> The Male and Female in this and the precedent differ so much in colour that they have been even by the best Naturalists described and figured for diverse Species. I had the Female of this latter lately sent me from *Cambridge*, by the title of a *Smew*. I suppose the name is originally *High Dutch*; for I find in *Baltner* our common *Wigeon* intitled *Ein Schmey*.

Newton also wrote about the origin of the name:[7]

> The commonly-accepted name for the smallest of the Mergansers . . . though not unfrequently applied in this country to some other *Anatidae* as the WIGEON and Pochard; but then generally in the form of SMEE-DUCK (cf. Dutch Smiente = Wigeon) or SMETHE, while in America one or other of these variants is locally used for the PINTAIL . . . Originally it would seem to have been used for the female of the species to which it is now ordinarily applied, while the male was the Nun.

According to Trumbull, in the 1800s the Northern Pintail was known by the names *Smee*, *Smees*, and *Smeethe* in the New Jersey area, a region that historically was heavily influenced by Dutch immigration.[276] There is a widely disseminated opinion that the name is "perhaps for ice-mew," but there appears to be no evidence for this.

# Snake-Eagle

These six species of *Circaetus* raptors, found predominantly in Africa, are all specialist herpetivores—that is, as their name suggests, they feed on snakes, but they also feed on other reptiles, such as large lizards. The feeding habits of the Banded Snake-Eagle, which are typical of the genus, were described in *Raptors of the World*:[277]

> Still-hunts from tree perch, taking prey on ground and also snatching snakes from tree. Holds writhing snake in bill and carries it to perch, there crushing neck until dead and often holding it for some time before swallowing whole.

# Snipe

The snipes are twenty-four species in the Scolopacidae, the sandpipers and allies. The name has been used since the 1400s in Britain as *snype* or *snyte* as, for example, in the *Boke of Nurture*, in which the ingredients of the third course of "A dynere of flesche" [A dinner of flesh] include:[252]

> Curlew / brew / snyt*es* / quayles / sp*a*rows / m*er*tenett*es* rost.

The word is believed to be derived from the Old Norse *snípa*, as in the Faroese name for Common Snipe, *mýrisnípa*. *Snípa* in Norwegian and *snäppa* in Swedish both mean simply "sandpiper" or "shorebird," and the Scolopacidae are the *Snipefamilien*. Meanwhile, the name *mýrisnípa* ("marsh sandpiper") is used only for the Common Snipe in Faroese; the Norwegian name for Dunlin is *myrsnipe*; while the name for Marsh Sandpiper is *Damsnipe*. The root word is likely the Proto-Germanic *snipōn*, meaning "a long, thin object," descriptive of the bills of many of the Scolopacidae.

# Snowcap

The Snowcap is a hummingbird in the monotypic genus *Microchera*. This lovely bird is characterized by a brilliant white forecrown that contrasts with the darker purple-bronze plumage of the male. The common name was coined by Gould in 1849 when he called it "Snow-cap."[6]

# Snowcock

The five species of snowcocks in the genus *Tetraogallus* are all high-altitude dwellers of the mountain ranges of Asia up to the snow line, where they inhabit scree and rocky outcrops, generally above the tree line. Although these members of the pheasant family, the Phasianidae, tend to avoid snowy areas, they are all found in habitats where snow is a permanent feature. In an 1847 article entitled "Extract of a letter from W. Jameson, Esq., Superintendent Botanic Gardens, N.W. Provinces," the author related a journey over the Himalayas via the Niti La, which connects India and Tibet with China:[350]

> The Rasores [an obsolete order of birds comprising the Gallinae and the Columbae] are so regularly distributed in the Himalayas as to present an excellent barometer to mark heights: thus from 1,800 feet to 5,000, we have the Gallus bankiva [Red

Junglefowl], and Phasianus pucrasia [Koklass Pheasant], &c.: from 5,000 to 8,000 we have the Phasianus Stacii [Ring-necked Pheasant], Tetrao chukor [Chukar], &c.: from 8,000 to 12,000 Lophophorus refulgens [Himalayan Monal], &c.: and from thence upwards to the snow the Tetraogallus, or snow pheasant of travellers.

# Snowfinch

There are eight snowfinches in the genus *Montifringilla* that belong to the Passeridae, the Old World sparrows. These sparrows superficially resemble finches, and their high-altitude habitats account for the "snow" element of the name.

# Softtail

The softtails are five members of the *Thripophaga* genus in the Furnariidae, the ovenbirds and woodcreepers. All of them have long tails with soft, loose feathers. Darwin wrote of a member of the *Synallaxis* spinetails, with which the softtails were previously lumped:[36]

> The tail-feathers were . . . so loosely attached, that I seldom procured a specimen with all of them perfect; and I saw . . . flying about with no tail.

# Solitaire

The eleven species of solitaires are medium-sized insectivores in the genera *Myadestes*, *Cichlopsis*, and *Entomodestes* of the thrush family, the Turdidae. As the name suggests, they are usually seen alone. It derives from the French *solitaire*, a name perhaps borrowed from Vieillot, who gave the name "La Grive Solitaire" to Hermit Thrush in 1807.[68] The root word is the Latin *solitarius* for "alone, lonely, or isolated." As Dawson noted of Townsend's Solitaire in 1909:[67]

> The bird thus . . . has long been a puzzle to naturalists. It has been called Flycatcher, Thrush, and a combination of the two; but the name Solitaire seems to express both our noncommittal attitude toward the subject, and the demure independence with which the bird itself proceeds to mind its own affairs.

Another bird known as Rodrigues Solitaire, very famous but extinct since the 1770s, predated the thrushes. It was named by French explorer François Leguat, referring to its solitary habits. Even earlier than that, the name may have been given to the Reunion Ibis, extinct since the early 1770s.

# Songlark

The two species of songlarks, endemic to Australia, were formally described in the 1820s. Gould wrote about the Rufous Songlark in 1848, under the heading "Rufous-tinted Cincloramphus" but noted that it is the "Singing Lark of the Colonists."[40] In 1877 Richardson rather unfairly described the same species:[65]

> The Australian Singing Lark is one of the few Australian birds which deserve notice on account of the sweetness of their song. It is found in all parts of Australia, and dwells principally on the ground, from which it ascends perpendicularly to a great height in the air, singing both in its ascent and descent in the manner of our skylark.

The name evolved so that by the late 1800s the bird was called the "Song-Lark." This is another example of inappropriate names being applied by English settlers to unrelated birds that vaguely resemble those of the homeland. Gould mentioned another name, "*E-role-dél*, Aborigines of the Mountain districts of Western Australia," a far more evocative name.

# Sora

This little *Porzana* crake is the only member of the genus to boast a unique name. The origins of the name are uncertain, but it's probable that it is taken from a Native American word that is echoic of the bird's call, which has been described as "similar [to its song] in its bright quality but sharper and ascending, with a whiplike accent on the second syllable: *so-REE, so-REE*."[537] The first written description of the bird appears in 1731, when Catesby wrote of the "Soree":[70]

> The Bird, in size and form, resembles our Water-Rail. The whole body is cover'd with brown feathers; the under part of the body being lighter than the upper. The bill and legs are brown. These Birds become so very fat in Autumn, by feeding on Wild Oats, that they can't escape the Indians, who catch abundance by running them down. In Virginia (where only I have seen them) they are as much in request for the delicacy of their flesh, as the Rice-Bird is in Carolina, or the Ortolan in Europe.

Catesby's description sounds much more like a Virginia Rail, however. And in *Burnaby's Travels through North America* in 1775, the author wrote:[69]

> The night I spent here [a "Pamunky Indian town"], they went out into an adjoining marsh to catch soruses; and one of them, as I was informed in the morning, caught near a hundred dozen. The manner of taking these birds is remarkable. The sorus is not known to be in Virginia, except for about six weeks from the latter end of September: at that time they are found in the marshes in prodigious numbers, feeding upon the wild oats.

In 1785 Latham referred to the "SOREE G[allinule]":[17]

> These inhabit Virginia, at certain seasons, in vast plenty. Burnaby mentions catching one hundred dozen of Sorusses in one.

So, the initial forms of the word were *soree* and *sorus* in the singular, with the plurals *soruses* and *sorusses* apparently modified to sora in the early 1800s..

# Spadebill

In the Tyrannidae, the seven members of the neotropical *Platyrinchus genus*, which translates from the Latin as "broad bill," all have wide, flat, triangular bills. The genus was erected in 1805 by Desmarest, who called them the Platyrinques.[164]

# Sparrow

The original sparrows belong to the Passeridae, the Old World sparrows, of which there are thirty-three species with the name. But there are also sixty-five species called "sparrow" (including ground-sparrows) in the New World sparrows, the Passerellidae. Additionally, there are the two *Lonchura* sparrows, Timor and Java Sparrows, which aren't really sparrows but finches in the Estrildidae. The name is probably one of the oldest for a bird in the English language. The Old English word was *spearwa*, from Proto-Germanic *sparwan*. It even appeared in *The Apocryphal Gospels of Mary* from the fifth century:[609]

> Hēo ġeseah spearwan nest on ānum lāwertrēowe. [She saw a sparrow's nest in a laurel tree.]

Later, in 1527, in *The noble lyfe*, Laurence Andrew, wrote about the "natures of man of bestes, serpentys, fowles":[71]

Passer / The Sparowe is a lytell byrde . . . The sparowes be wylye & they make theyr nestes in the holes of the walles or onder the rydges of ehe howses / the he is somwhat blacke about the bylle [The sparrow is a little bird . . . The sparrows are clever and they make their nests in the holes of the walls and under the ridges of the houses, the head is somewhat black about the bill.]

As Newton wrote, the word from the Proto-Indo-European *spér* or *spor-wo-* for "sparrow" is "perhaps (like the equivalent Latin *Passer*) originally meaning almost any small bird, but gradually restricted in signification and nowadays in common English applied to only [the Old World sparrows]."[7]

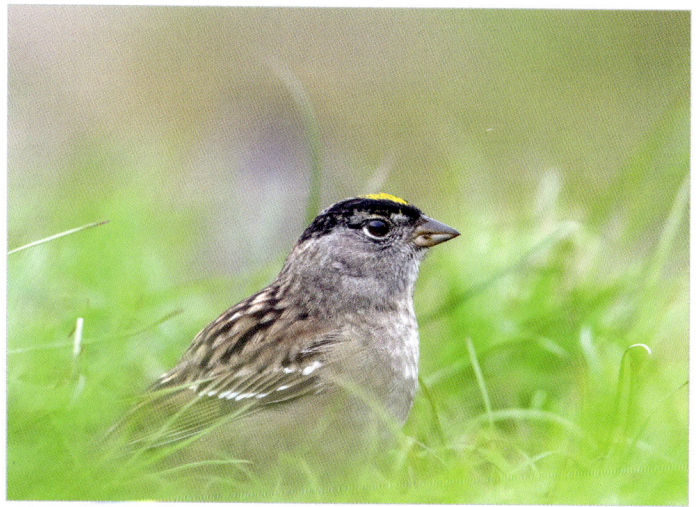

*Golden-crowned Sparrow (Washington State, USA)*

# Sparrowhawk

There are eighteen small accipiters with the sparrowhawk moniker, but the bird to first be called by the name was the Eurasian Sparrowhawk. The Middle English name *sperhauk* dates from the 1400s, replacing the Old English *spearhafoc*, a hawk that hunts sparrow-sized birds, from *spearwa*, meaning sparrow. By the 1500s, the bird was called "sparow hawke" as in *The noble lyfe*:[71]

Nisus is a sparow hawke / & it is a gentyll byrde & is federed like a goshawke / & whan his felowe sitteth vpō their egges than hath he a place where he plucketh hys byrdes that he taketh / & they be clene whā he bereth it to the neste & geueth it his felowe sitting on the egges. [Nisus is a sparrowhawk and it is a gentle bird and is feathered like a goshawk, and when this fellow sits upon their eggs then he has a place where he plucks his bird that he has taken, and they be clean when he takes it to the nest and gives it to his mate sitting on the eggs.]

# Sparrow-Weaver

The four species belonging to the genus *Plocepasser* in the Ploceidae, the weavers and allies, are all known colloquially as sparrow-weavers. Although they are weavers, they are sparrow-like in appearance, and in the past they have been placed in the Passeridae, the Old World sparrows. Both families are part of the Passeroidea superfamily.

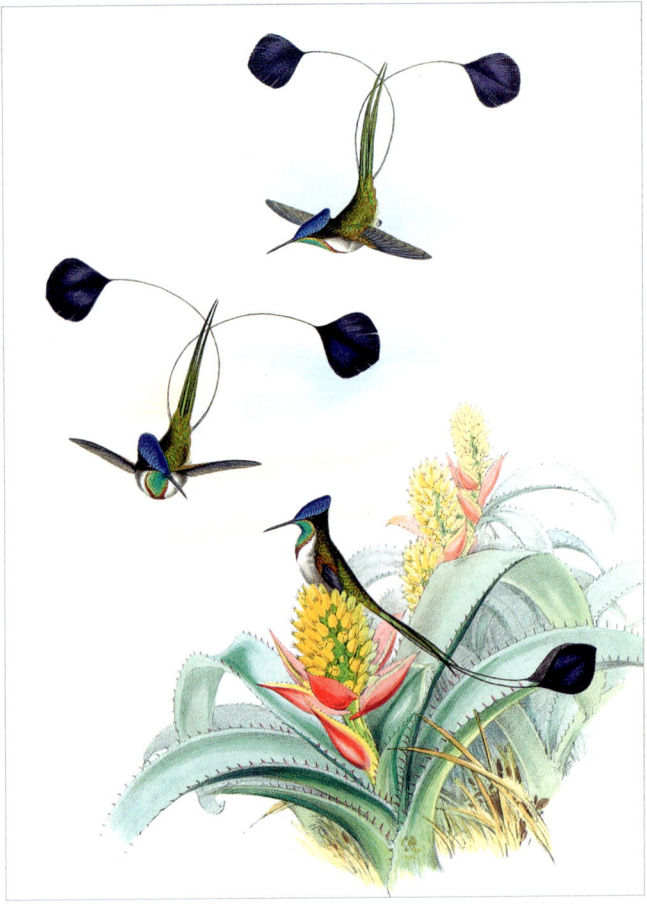

*Marvelous Spatuletail from Gould's* Monograph of the Trochilidæ *(1849)*

# Spatuletail

The astounding Marvelous Spatuletail is a hummingbird with a remarkable tail. Gould, who did not use the name, did, however, describe the unique tail:[6]

> These [outer tail] feathers cross each other twice, first near the base, and secondly towards the middle; consequently each spatule, as represented in the drawing, belongs to the feather of that side. How very remarkable is this arrangement, and how different from what is found to occur in any other known species!

# Speirops

The four species of white-eyes in the *Zosterops* genus were previously in a synonymous genus, *Speirops*, which is formed from the Ancient Greek words σπεῖρα *speira*, "a ring or coil" and οψ *ōps*, eye. The name was first proposed for Black-capped Speirops in an 1836 work by Reichenbach:[72]

> Wreath eye: ἡ σπεῖρα the wreath, and ἡ οψ the eye, so *Speirops*.

The species in question has a particularly broad, pronounced eye-ring that could almost be said to be bulbous.

# Spiderhunter

The thirteen species of *Arachnothera* spiderhunters, and now one species of *Kurochkinegramma*, do hunt spiders, but they also frequently consume other types of invertebrates as well as nectar, in much the same way of their close relatives the sunbirds. Remarkably, Swainson called the *Arachnothera* "spider-suckers" in his *On the Natural History and Classification of Birds* in 1836.[64]

# Spindalis

The Spindalidae contains four species of similar-looking birds in the synonymous *Spindalis* genus. Despite the fact that they are confined to the Caribbean Islands, an examination of the name suggests it comes from the Ancient Greek σπίνδαλος *spindalos*, an Indian bird akin to ἀτταγᾶς *attagas*, Black-bellied Sandgrouse. The genus name was coined by Jardine and Selby in 1837 without explanation.[538] Newton stated in 1896, also without explanation, that the name is an error for *Spindasis*, which would be from the genus *Spinus*, for "siskin," and the Ancient Greek *dasos*, for "thicket."[7] As a name for these birds that are nothing like a sandgrouse, this would make more sense.

# Spinebill

These beautiful honeyeaters are two species in the genus *Acanthorhynchus*, which derives from the Ancient Greek ακανθα *akantha*, "spine," and ρύγχος *rýnchos*, "bill." Interestingly, the Eastern Spinebill's scientific name is a tautology: *Acanthorhynchus tenuirostris*—"spine bill slender bill" and, in fact, Gould gave it the name "Slender-billed Spine-bill."[40] He gave the local names as:

> Cobbler's Awl, Colonists of Van Diemen's Land.
>
> Spine-bill, Colonists of New South Wales.

Not surprisingly, the latter is the one that has stuck, and they do, indeed, possess slender, spine-like bills.

# Spinetail

The name spinetail is used for birds in two families—there are seven species in the Apodidae, the swifts; and sixty-one spinetails and ten tit-spinetails in the Furnariidae, the ovenbirds and woodcreepers of the Neotropics. In both groups, the tail feathers are modified so that the ends of the shafts are bare, producing a "spiny" appearance. These tails feathers are modified for bracing support. In the case of the swifts, they are unable to land on horizontal surfaces—in most swifts, the hind toe is rotated forward as an aid in gripping vertical surfaces, but the spinetails gain support from the short needle-tipped tail feathers, as their feet are not modified in this way. The short tails of the spinetail swifts allow them to roost on vertical cliffs and rock walls. The tail performs a similar support function in the furnariids, but their tails are longer and graduated, with sharply pointed individual feathers, the central pair thinning toward the tip, again giving a spiny appearance to the tips. They use their tails in a similar manner to woodpeckers, as support when feeding on invertebrates they take from bark crevices on tree trunks and branches.

# Spinifexbird

The uniquely named Spinifexbird is a member of the Locustellidae in the genus *Poodytes*. This Australian outback dweller favors habitats with stands of *Triodia* grasses, known colloquially as spinifex (the *Spinifex* genus are coastal grasses), where they feed on insects and seeds. When alarmed, the birds will seek cover in the notoriously spiny stands of spinifex grass clumps. The word "spinifex" is formed from the Latin *spīna*, "a thorn," and *-fex*, "maker," from *facere*, "to make." The bird was first called the "Desert-Bird" in the English literature, but not everyone agreed with the name, as reported by Whitlock in 1913:

> When observing the Desert-Bird, I always think that 'Spinifex-Bird' would have been a more appropriate trivial name. Our north-west coast at its worst can hardly be regarded as a 'desert,' considering its numerous rivers and creeks, the majority of which are well-timbered with eucalypts and other trees of respectable dimensions.

## Spoonbill

These iconic members of the Threskiornithidae, the ibises and spoonbills, are six birds in the *Platalea* genus. The first to receive the English name was the Eurasian Spoonbill despite the fact that it went extinct in Britain in the 1600s. At that time the bird was variously known as the "Shovelard" and "Shovelar." As Newton observed:[7]

> The bird now so called was formerly known in England as the Popeler, Shovelard or Shovelar, while that which used to bear the name of Spoonbill is the Shoveler [duck] of modern days—the exchange of names having been effected about 200 years ago, when the subject of the present notice, the *Platalea leucorodia* of ornithology, was doubtless better known than now, since it evidently was, from ancient documents, the constant concomitant of Herons, and with them the law tried to protect it.

The namesake feature of these birds could not be described any better than Ray and Willughby did in 1678:[16]

> The Bill is of a singular and unusual figure, plain, depressed, and broad, near the end dilated into an almost circular figure, of the likeness of a Spoon whence also the Bird it self is called by the Low Dutch, *Lepelaer*, that is, *Spoon-bill*.

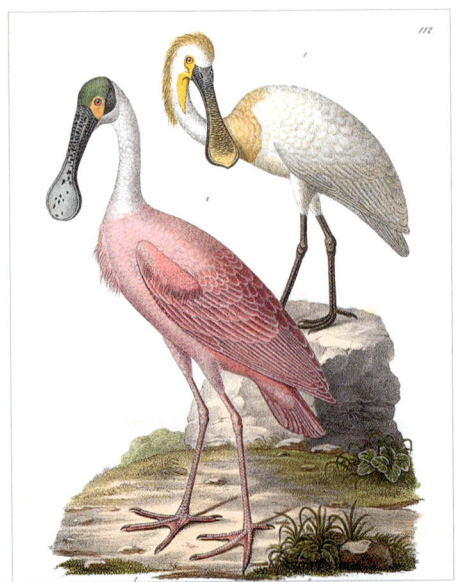

*Eurasian Spoonbill and Roseate Spoonbill from Schinz and Brodtmann's*
Naturgeschichte und Abbildungen der Vögel: nach den neuesten Systemen bearbeitet *(1836)*

# Spot-throat

The Modulatricidae consists of only three species, including the Spot-throat, which is found in a monotypic genus, as are its two congeners. This small babbler-like bird found in Africa is uniformly brown except for the spotting on the buffy throat.

# Spurfowl

The genus *Galloperdix* in the Phasianidae is composed of three species of spurfowls found on the Indian subcontinent. Males uniquely possess two to four prominent tarsal spurs; females can also have one or two spurs. The spurs are used in mating competition and territory defense.

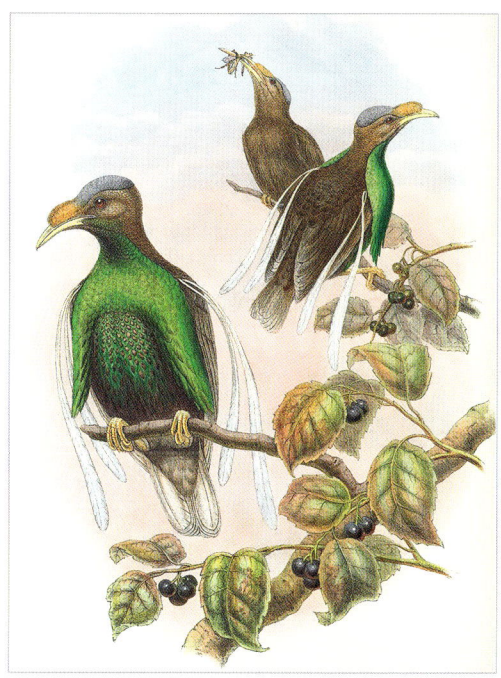

*Wallace's Standardwing from Sharpe's* Monograph of the Paradiseidae

# Standardwing

Wallace's Standardwing, now often referred to as Standardwing Bird-of-Paradise, is the westernmost representative of the Paradisaeidae, being endemic to the Indonesian island of Halmahera. It is named for its unique plumes, which, to Wallace, brought to mind a "standard" or long-poled "battle flag." According to Wallace, the bird is[205]

> especially distinguished by a pair of long narrow feathers of a white color, which spring from among the short plumes which clothe the bend of the wing, and are capable of being erected at pleasure. . . . The most curious feature of the bird, however, and one altogether unique in the whole class, is found in the pair of long narrow delicate feathers which spring from each wing close to the bend. On lifting the wing coverts they are seen to arise from two tubular horny sheaths, which diverge from near the point of junction of the carpal bones. They are erectile and when the bird is excited are spread out at right angles to the wing and slightly divergent. They are from six to six and a half inches long, the upper one slightly exceeding the lower.

## Starfrontlet

The whimsically named starfrontlets are seven species of hummingbirds in the *Coeligena* genus. The common name was given to them by Gould, who referred to them as "starfrontlets" in honor of the glittering metallic plumage on their forecrown:[6]

> the forehead only is decorated, with a star brighter than Venus, the queen of planets.

## Starling

The Sturnidae is a large family of 123 species, with 92 members known by the common name starling. The bird was originally called the *staer* in Old English, to which a diminutive suffix was added, giving us *starlinc* or *stærlinc* in the eleventh century. Originally both *staer* and *staerling* were used independently, as the adult and juvenile European Starlings differ so markedly, but as Lockwood pointed out, *"the original diminutive sense was later lost."*[84] In Middle English the name took various forms, such as *starling*, *sterling*, and *sterlinge*. There were numerous variations throughout the islands, such as *Stare* in Ireland and Cornwall, *Starnel* in the North, *Starn* in the Shetland Isles, and *Staynil* in Yorkshire. The name spread throughout the world as early European naturalists and colonists encountered similar species in their travels. The name probably came into English from the Latin name for the bird, *Sturnus*. The word is often associated with the Ancient Greek ψάρ *psár*, "starling"; it is probably an etymon, and, ultimately, the name is derived from the Proto-Indo-European *storo-* or *(s)tern-* for "starling."

Romantic interpretations of the name abound, such as that " 'Sturnus' and 'starling' mean the same thing: 'little star' "[539] for European Starling and that "In flight their wings are short and pointed, making them look rather like small, four-pointed stars (and giving them their name)."[540] There is no support for these theories, and as Anderson wrote in an analysis of the poetry of Colergidge:[327]

> How many English-speakers, when they see or say the word "starling," have in their mind's eye the starry spots of the starling's plumage? It could—though it doesn't—mean "little star."

## Starthroat

These four species of hummingbirds are all named for the bright, metallic plumage on their throats. As is so often the case with the hummingbirds, it appears that Gould coined the name in the 1850s in his *Monograph of the Trochilidae*, calling them the "Star-throats."[6]

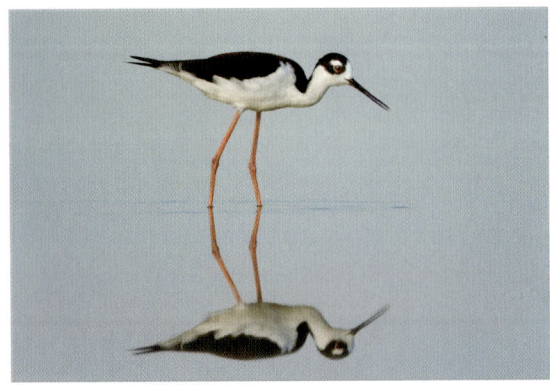

*Black-necked Stilt (Arizona, USA)*

# Stilt

Five members of the Recurvirostridae, the stilts, are named for their remarkably long legs, which bring to mind the long poles that people use to stand high above the ground. The name is taken from a direct translation of the French *Échasse* first used by Brisson in 1760.[77] He wrote:

> Echasse, nom que j'ai donné aux especes de ce genre à cause de la lougueur de leurs jambes. [Stilt, the name I gave to the species of this genus because of the length of their legs.]

The English word for the birds was believed to have first been used in the early 1800s. Bingley wrote in 1800 that the "Stilt, or Long-legged Plover" has[73]

> habits [that] are altogether maritime, and it is said to feed on the spawn of fish, tadpoles, gnats, flies and other aquatic insects. The legs of this bird are remarkably slender, and longer, perhaps in proportion, than in any other known bird, it consequently staggers and reels in its gait, while balancing itself on its stilt-like legs.

# Stint

There are four species of small sandpipers in the genus *Calidris* called stints. As Newton mentioned, the name is "akin to Stunt," appropriate given the bird's small stature.[7] The two words *stint* and *stunt* appear to have a common etymological derivation; both derive from the Old English words *styntan* and are cognate with the Old Norse *stynta* or *stytta*, "to shorten, to make short." In Middle English, the words *stont* and *stunt* meant "short or dwarfed." Ultimately, the words are from the Proto-Germanic *stuntaz* for "short, compact, stupid, dull." One of the earliest records of the word occurred in 1465, when the birds appeared on the menu at the "Feast Neville" a days-long feast marking the installation of Archbishop Neville of York, held at Cawood Castle. For the the "thirde course," the guests were served "Plovers rost; Quayles and Styntes rost, [etc.]."[390]

# Stipplethroat

The eight species of *Epinecrophylla* in the Thamnophilidae, the typical antbirds, have throats with a black and white "checkered" pattern; in other words, the plumage looks stippled. They were previously included in the genus *Myrmotherula* (when they were known as antwrens), but recent molecular studies show them to be from a separate clade, in the genus *Epinecrophylla* with the note:[610]

> The stipple-throated assemblage is distinguished morphologically by at least one sex having a black throat stippled white or buffy white combined with a comparatively long, unmarked tail.

The new name Checker-throated Stipplethroat, for Checker-throated Antwren, certainly seems clumsy.

# Stitchbird

The Stitchbird of New Zealand is the sole member of the Notiomystidae. According to Buller:[74]

> The male bird utters at short intervals and with startling energy a melodious whistling call of three notes. At other times he produces a sharp clicking sound like the striking of two quartz stones together: the sound has a fanciful resemblance to the word "stitch," whence the popular name of the bird is derived.

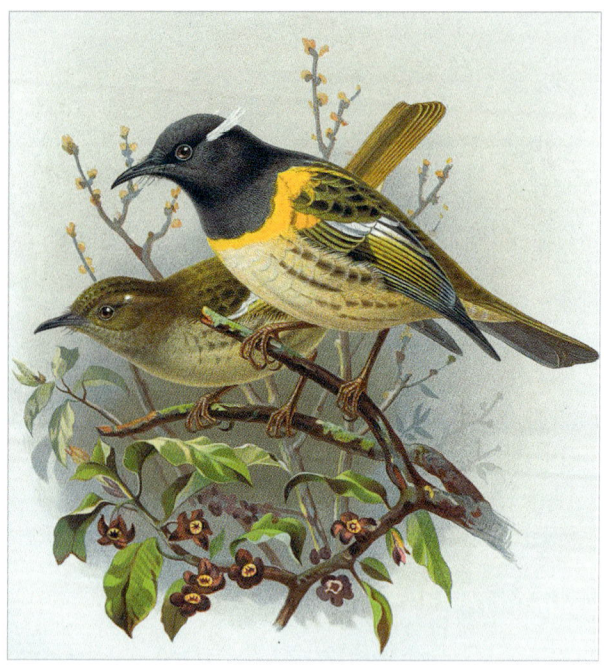

*Stitchbird from Buller's* A History of the Birds of New Zealand *(1873)*

One can't help but wonder if one of the Māori names for the bird might be more suitable. Buller also mentioned the names *Hihi, Tihe, Kotihe, Kotihewera, Tiora,* and *Tiheora,* and the name *Hihi* is in fact widely used in New Zealand.

# Stonechat

There are seven species of stonechats in the Muscicapidae. The genus name *Saxicola* is a compound of the Latin words *saxum,* for "rock" and *colō,* "to inhabit," and, indeed, most of them live in barren, rocky, or, should we say, stony, landscapes with shrubby vegetation. Both Turner[26] and Merrett[122] use the name "Stone-chatter," while Ray and Willughby[16] list the "Stone-smich or Stone-chatter." (See Chat.)

# Stone-curlew

While most of their congeners are called thick-knees, the two species of stone-curlews retain the name, although the names are used interchangeably for other species in the Burhinidae. The first element of the name relates to the habitat of the Eurasian Thick-knee, still locally called the "Stone Curlew" in Britain. The second element is based on their perceived similarities with unrelated true curlews. (See Curlew.) Swainson wrote in 1886 that the bird in England is[75]

> so called from its frequenting stony localities and uttering a cry resembling the sound of the word "curlui."

The name has been around for a long time as can be seen in Ray and Willughby from 1678:[16]

> The Chin, Breast, and Thighs are white: The Throat, Neck, Back, and Head covered with feathers, having their middle parts black, their lateral or borders of a reddish ash-colour, like that of a *Curlew:* Whence they of Norfolk call it, The *Stone-Curlew.*

# Stork

The stork family, the Ciconiidae, contains nineteen species, fourteen of which are known by this common name. The name is thought to have arrived in English from the Proto-Germanic word for the birds, *sturkaz* probably from the PIE word *ster-* for stiff, in reference to the stiff or rigid posture of the bird. In Old English the name was *storc*, evolving to *storke* in Middle English. According to the *The Booke of Demeanor* from 1619 one must[541]

> Nor practize snufflingly to speake,
>
> for that doth imitate
>
> The brutish Storke and Elephant,
>
> yea and the wralling cat.

The White Stork, which was the originally named bird, is the subject of many myths and fables. One of the first accounts in English appeared in *The noble lyfe* in 1510:[71]

> A Storke is a byrde w$^t$ whyte and black feders / & it clappeth wyth his byll & maketh gret noyse / and he is mortall ennemye of the serpentes for he eteth thē & other venymous bestes also but he eteth no todes excepte great hūger dryue him to it / and in the londe of thessaly who so sleeth a storke must nedys dye hȳ selfe as yff he had slayne a man. & these byrdes renewe eueri yere theyr neste / & whan theyr iōges be full growen & federed thā they caste one of their ionges out of the nest for a tribute vnto y$^e$ lorde of the groūde and some say that they geue it god for the tythe / & therfore in the lond of Turingia where as ther is no tythe geuē ther they cometh nat.

> [A stork is a bird with white and black feathers, and it claps with his bill and makes a great noise, and he a mortal enemy of the serpents for he eats them and other venomous beasts but eats no toads except if driven by great hunger, and in the land of Thessaly (Greece) whoever slays a stork will himself die as if he had slain a man and these birds renew their nests yearly, and their eggs be grown and feathered then they cast one of young out of the nest for a tribute to the lord of the ground and some say that they give it to god for a tithe, and therefore they turn to the land of Turingia (Germany) where they need not give a tithe.]

# Storm-Petrel

These are nine species in the Oceanitidae, the southern storm-petrels, as well as eighteen species in the Hydrobatidae, the northern storm-petrels. Various names preceded the present-day one—Little Petrel, Stormy Petrel, and Storm-Finch. The earliest mention of the birds appears to have been in Høyer's letter to Clusius in 1605, when he referred to the "Stormfinck," the Faroese name for the birds.[77] Catesby[70] mentioned them in 1731 under the heading the "Storm-Finch, or Petrel":

> Their appearance is generally believed by Mariners to prognosticate a storm, or bad weather; and I must confess I never saw them but in a troubled sea. They use their wings and feet with surprising celerity. Their wings are long, and resemble those of Swallows, with which they are equally swift, but without making such angles or short turns in their flight, as Swallows do, but fly in a direct line. Though their feet are formed for swimming, they are likewise so for running, which use they seem most to put them to, being oftenest seen in the action of running swiftly on the surface of the waves in their greatest agitation, but with the assistance of their wings.

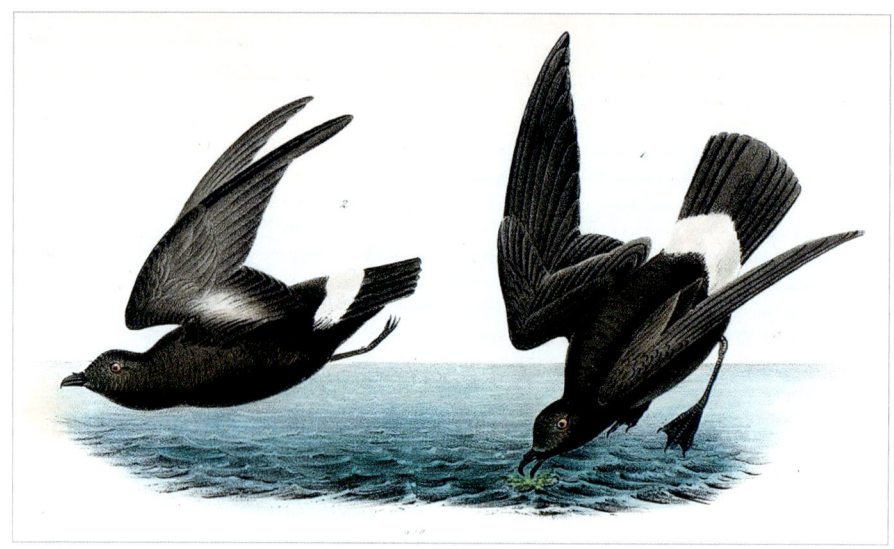

*Least Storm-Petrel from Audubon's* Birds of America *(1839)*

Even earlier, in 1729, Dampier wrote of the storm-petrels but referred simply to "petrels":[41]

> They are not so often seen in fair weather; being foul-weather birds, as our seamen call them, and presaging a storm when they come about a ship; who for that reason don't love to see them.

Pennant referred to the "Stormy Petrel" in 1781 for *Procellaria*, the genus in which most of the storm-petrels were placed, with the petrels, at that time.[66] The birds got their name because the sailors of yore believed their appearance foreshadowed a storm. (See also Petrel.)

# Straightbill

The straightbills are two species of honeyeaters in the Meliphagidae endemic to the island of New Guinea. For a long time they were referred to simply by the common name "honeyeater," as the present-day common name wasn't in use until the mid-1900s. It refers to the straight, sharp bills.

# Streamcreeper

The elusive Sharp-tailed Streamcreeper is in the monotypic genus *Lochmias* in the Furnariidae. It takes its name from its habitat and habits, being closely associated with rocky streams and rivers, where it forages along the edges and in shallow water. Although the species was described in 1823, a common name wasn't used until 1835 when Swainson, a follower of the Quinarian classification system that divided all taxa into five subgroups, wrote under the heading "Climbing Birds—Scansores":[603]

> Lochmias *Sw.* Creeper

However, he used this name for many of his so-called Climbing Birds. In 1841 Swainson called it the "Sharp-tailed Wren," and then in 1888, Sclater called it the "Brazilian Lochmias."[22] And then in 1921, Chubb chose to use the remarkable name Chestnut-backed Sharp-tailed Creeper.[176] The present-day name first gained favor in the 1950s. These examples illustrate well the need for standardized English names for birds!

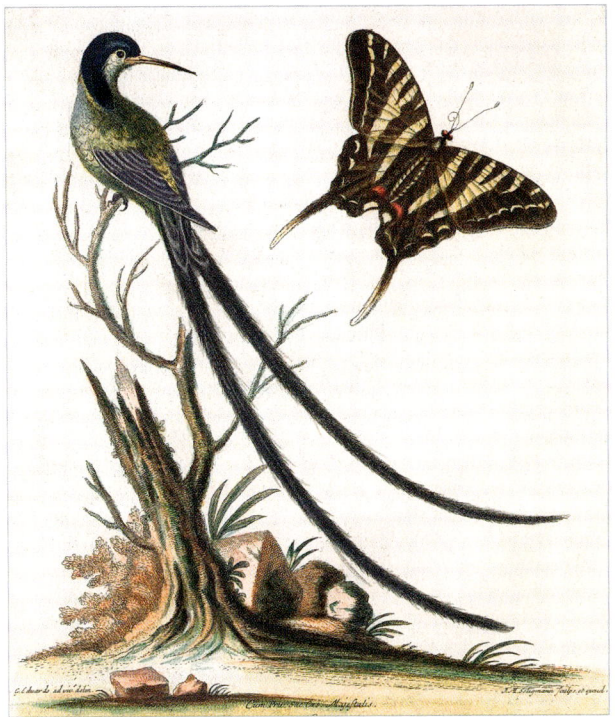

*Streamertail from Edwards's* A Natural History of Uncommon Birds *(1743)*

# Streamertail

The Streamertail is a spectacular hummingbird endemic to Jamaica with an exceptionally long, forked tail in the male. One of the first mentions of the species is in Edwards in 1751 with an account of the "Long tailed black cap humming bird."[78] The bird's name has evolved over the years from "Long-tailed Hummingbird" in the 1800s to "Streamer-tailed Hummingbird" in the early 1900s, then to "Streamertail Hummingbird" in the 1970s. The stand-alone name Streamertail appeared in the 1990s.

# Striped-Babbler

Four species in the *Zosterornis* were previously considered to be members of the Timaliidae, the Old World babblers, but are now placed in the Zosteropidae by many authorities. They are small babbler-like insectivores endemic to the Philippines with striped plumage on the underparts.

# Stubtail

There are three stubtails, tiny members of the Scotocercidae, the bush warblers and allies, in the *Urosphena* genus. These little, largely terrestrial warblers are almost tailless, hence the name.

# Sugarbird

The sugarbirds are a family, the Promeropidae, of only two species. They are long-tailed and long-billed, feisty birds that feed on nectar (but also insects), accounting for the vernacular name. They are particularly fond of the nectar of *Protea* flowers, known locally as sugarbushes.

For a long time, they were thought to be representatives of the Nectarinidae and were called "long-tailed sun-birds." Burchell wrote in 1822:[79]

> The *Nectariniae*, here called by the Dutch colonists *Suiker-vogels* (sugar-birds), from having been observed, at least in the neighbourhood of Cape Town, to feed principally on the honey of the flowers of the *Suiker-bosch* (sugar-bush).

# Sunangel

Seven species of sunangels belong to the genus *Heliangelus*, which is derived from the Ancient Greek "Ηλιος *helios*, for "sun," and ἄγγελος *ángelos*, for "angel." The genus was established by Gould in 1848, who also gave it the common name "Sun Angel"; the common name is a direct translation of the genus name.[542]

# Sunbeam

The four species of *Aglaeactis* hummingbirds were named by Gould who wrote:[80]

> These extraordinary birds, to which I have given the trivial name of Sunbeams, are among the most wonderful of the Trochilidae.

He probably did so in homage to what he called their "rich and glittering hues."

# Sunbird

In the Nectariniidae there are 129 species of sunbirds found in Africa, Asia, and Australasia. Stephens in Shaw seems to have been the first to use the name in 1826 in *General Zoology, or Systematic Natural History*.[81] Swainson explained the name in 1836:[64]

> So called by the natives of Asia in allusion to their splendid and shining plumage . . . those gay and beautiful tints which are so strikingly developed in the sunbirds: a rich golden green, varied on the under parts with steel-blue, purple, bright orange, or vivid crimson, decorates nearly all the species, and produces a brilliancy of colours only rivalled by those of the humming-birds.

# Sunbird-Asity

Two asities in the small Philepittidae are named sunbird-asity due to their superficial resemblance to the true sunbirds. They were classified in the Nectariniidae, sunbirds and spiderhunters, until relatively recently. (See Asity and Sunbird.)

# Sunbittern

Often likened to either a heron or a bittern, the remarkable Sunbittern is neither. It belongs to the monotypic family the Eurypygidae. Brehm, who called it the "Peacock Heron" in *Cassell's Book of Birds* (1875), tells us that "it is a favourite pet-bird of the Brazilians, who call it Pavao (pronounced 'Pavaong'), or Peacock."[106] Figuier gave an account of the bird in 1873:[82]

> A Bird about the size of the Partridge, with a large and fan-like tail. Its brilliant hues have obtained for it in Guinea the name of the Little Peacock or Sun Bittern. It is very wild in its nature.

One of the first people to describe the bird in the European literature, in 1781, was Pallas, who called the bird "Sonnenreyger," which translates from the German to "Sun heron":[543]

> In the first . . . collection of l'Abbé Rozier you can find an illustration of it under the name *Petit Paon des roseaux de Cayenne*, which does not make our [name

Sonnenreyger] superfluous . . . Fermin mentions it very vaguely and imperfectly, according to his judgment. The name *Sun Bird*, which is customary in Surinam and which he has kept (*Oiseau du Soleil*), I would never have guessed from his description that he was talking about the bird described here. . . . it is only found in a few collections, under this and the name given above; and so the little that Fermin has recorded of the customs of his sunbird can certainly be drawn here. . . .

I was also told that he sometimes spreads the tail feathers like a peacock; and the above-mentioned French designation, which is common in Cayenne, seems to confirm this: as a result of which the sun heron became even more different from the other herons.

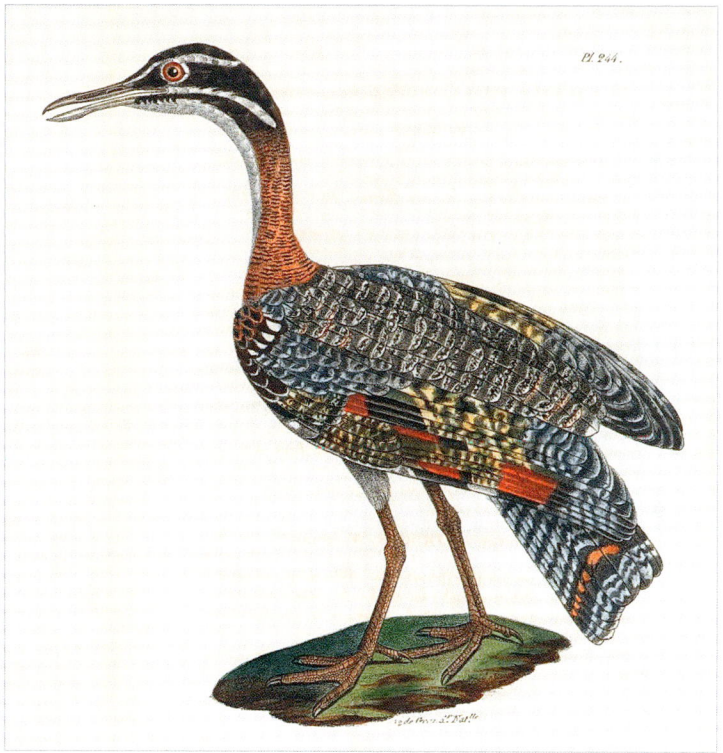

*Sunbittern in Vieillot's* La Galerie des Oiseaux *(1825)*

# Sungem

The Horned Sungem is a single species of hummingbird in the genus *Heliactin*. The common name of this remarkable little bird with its red, blue, and gold horn-like crest was coined by Gould, who commented:[6]

> well does this elegant little bird represent in the air the brilliants which lie hidden in the deep primitive rocks over which it flies: fairy-like in form and colour, we might easily imagine that one of the jewels had become vivified and had taken wing.

In his *Illustrated Natural History* in 1862, Wood poetically paraphrased Gould, describing the bird as[85]

> A veritable gem sprung out of the mountain and suddenly gifted with life.

# Sungrebe

This Neotropical member of the small finfoot family, the Heliornithidae, belongs to the monotypic genus *Heliornis*. In 1776, the bird was mentioned in *New Illustrations of Zoology* under the heading "The Surinam Tern" with the comment:[83]

> Called by the Dutch *sonne-vogel*, which signifies bird of the sun, or sun BIRD; feeds on flies, is often domesticated: its head and body is in a continual motion.

Of course, it is neither a tern nor a grebe, but the latter name has stuck. The name "Sun-Grebe" was also given as an alternative name to African Finfoot in the 1800s, along with "Water Treader." There is a fair description of a Sungrebe in the 1817 volume of *Nouveau Dictionnaire d'Histoire Naturelle* under the heading "L'Héliorne d'Amérique," with an alternative name given as "le Grèbe-foulque," the "Grebe-coot."[328] But while the physical description matches the bird in question, the description of its behavior and the allusion to them being kept as pets seems far more likely to be a reference to Sunbittern, with which the author, Virey (whose "lack of scientific spirit" was criticized at the time), may have been confused:

> It is also seen in dwellings in Surinam, where it is known as the *sun-berd* (bird of the sun). This name comes from the fact that by continually moving the head and the neck, and at the same time extending the wings and the tail, it appears, it is said, to resemble this star.

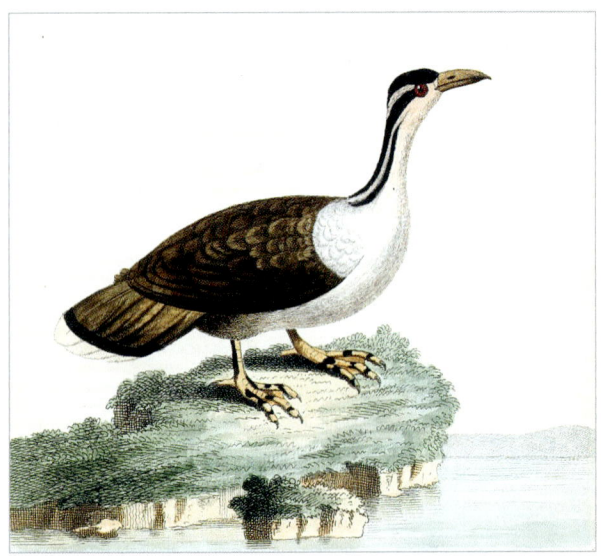

*Sungrebe, "The Surinam Tern," in Brown's* New Illustrations of Zoology *(1776)*

# Surfbird

Audubon assigned the Surfbird to the genus *Aphriza* from the Ancient Greek ἀφρος, "foam," and ζαω, "to live," and he also gave it its unique name.[54] It is now in the genus *Calidris* but retains its unique common generic name, of course. He mentioned that Townsend, the American naturalist who collected the bird, told him:

> It was sitting on the edge of the steep rocks, and the heavy surf frequently dashed its spray over it as it foraged among the retreating waves. When it started, it flew with a quick, jerking motion of its wings, and alighted again at a short distance.

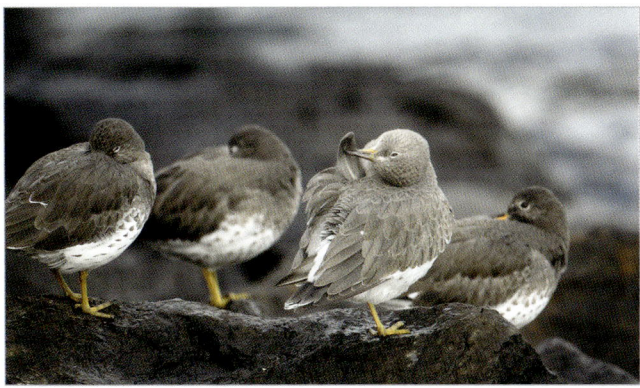

*Surfbirds (Washington State, USA)*

# Swallow

The Swallow family, the Hirundinidae, is a large one of eighty-six species, of which fifty-five are called by the common name swallow. Ultimately, the name comes from the Proto-Germanic word *swalwō*, believed to be the name for the Barn Swallow, which would have been a familiar migratory bird to early Europeans. It came into English as the Old English *swealwe*, evolving in Middle English to *swalwe, swalewe, or swalowe*. Instructions on how to serve are given in *The Boke of Keruynge* (*The Book of Carving*) in 1481:[483]

> Quayle, sparow, larke, martynet, pegyon, swalowe, & thrusshe, yᵉ legges fyrst, than yᵉ wynges. [Quail, sparrow, lark, martin, pigeon, swallow and thrush, the legs first, then the wings.]

The Proto-Germanic *swalwō* meant "cleft stick," an allusion to the shape of the tail. (The verb "to swallow" is unrelated with a different root word—the Proto-Germanic *swelgana*, "to swallow or devour.")

# Swamphen

The generally similar-looking members of the genus *Porphyrio* are all known as either gallinules or swamphens. The name gallinule (see entry) derives from the Latin for "little hen," an analogy that can, of course, be seen in the name swamphen as well but with the added reference to its favored habitat. Gould claimed in 1848 that the name was[40]

> Swamp-Hen, [given by the] Colonists of Western Australia.

But he called it the "Azure-breasted Porphyrio," writing that it inhabits "the thick reed-beds and swampy districts of the lakes and rivers round Perth and Fremantle."

# Swan

The six members of the genus *Cygnus* are the much-loved swans, birds of legend wherever they occur. It is the European legend that gives the swan its common English name. The word is derived from an Indo-European root, *swen*, which means "to sound or to sing," despite the fact that the namesake Mute Swan, while not entirely mute, is not terribly vocal. A legend, dating at least from Ancient Greece, is that swans would sing beautifully just before their death, since they have been silent for most of their lifetime, the so-called swan song. From the Proto-Germanic *swanaz*, meaning "the singing bird," the bird's name came into Old English as *swañ*, and *swannes* in the plural. As Ray and Willughby wrote:[16]

> For my part, those stories of the Ancients concerning the singing of Swans, viz, that those Birds at other times, but especially when their death approaches; do with a most sweet and melodious modulation of their voice, sing their own Naenia or funeral song, seemed to me always very unlikely and fabulous.

It was proposed by early naturalists, such as Pallas, that the legend might actually pertain to Whooper Swans since they are winter visitors to the eastern Mediterranean, but Lockwood, ignoring the legend, pointed out that a bird making a sound is hardly a unique characteristic.[84] On the other hand, when a swan takes flight, it

> is remarkable for its singing, metallic throb, which, in favourable conditions, is audible a good mile off and more. Such a unique phenomenon is more than likely to inspire a name.

# Swift

There are fifty-five species of swifts in the Apodidae. Early on the name was used interchangeably with "martin" due to confusion with the Hirundidae. Ray and Willughby wrote of "The *black Martin* or *Swift*" in 1678:[16]

> We have observed four sorts of Swallows in England, and not more elsewhere. Those are, 1. The common or House-Swallow: 2. The Martin, or Martinet, or Martlet. 3. The Sand-Martin or Shore-bird. 4. The black Martin or Swift. Of this last we have seen a sort painted with the whole Belly white. And Julius Scaliger affirms, that he hath seen one of this kind as big as a Buzzard [!]

Newton observed later:[7]

> Swift, a bird so called from the extreme speed of its flight, which apparently exceeds that of any other British species, the *Hirundo apus* of Linnaeus and *Cypselus apus* or *murarius* of most modern ornithologists, who have at last learned that it has only an outward resemblance but no near affinity to the Swallow or its allies.

The word swift is derived from the Proto-Indo-European word *sweyp*, meaning "to twist or wind around," which became *swiftaz* in Proto-Germanic, meaning "fast, quick."

# Swiftlet

The thirty-nine species of swiftlets, also in the Apodidae, are so-called for their relatively diminutive size, hence the *-let* suffix.

# Sylph

The sylphs are three species of beautiful, long-tailed hummingbirds in the genus *Aglaiocercus*. Wood wrote in 1862 that these hummingbirds are[85]

> one of the most beautiful of the birds which are called by the name of Sylphs, in allusion to their beautiful form and graceful movements.

The Sylph is a mythological spirit of the air. The name was coined by Paracelsus, the Swiss physician, alchemist, theologian, and philosopher, in the 1500s, possibly from the Latin *sylvestris* and *nympha*—"nymphs of the woods." Paracelsus described four elemental or mythic beings, corresponding to the classical elements—fire (salamanders), earth (gnomes), water (undines), and air (sylphs).

# T IS FOR TACHURI

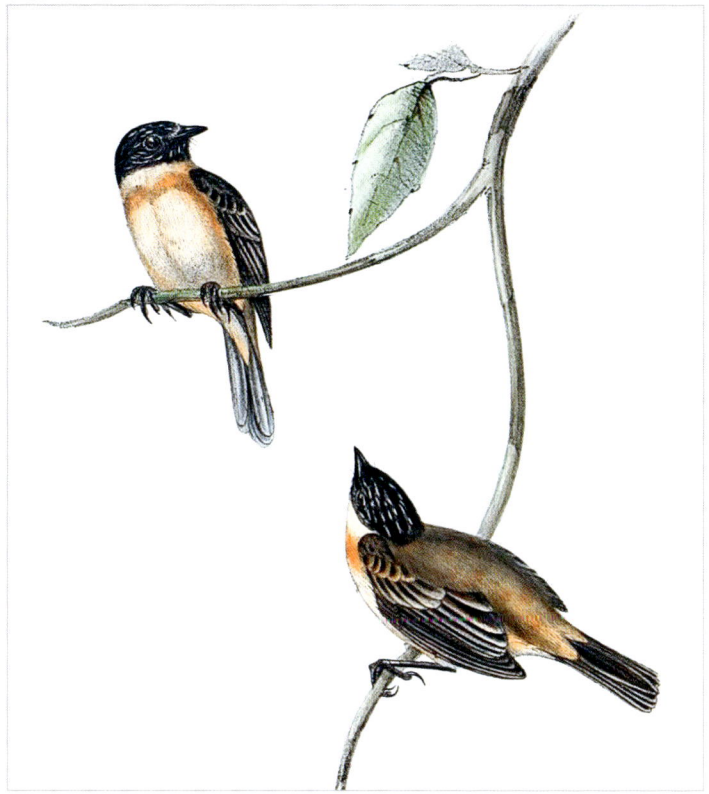

*Bearded Tachuri in Darwin's* The Zoology of the Voyage of H.M.S. Beagle ...
during the years 1832–1836 *(1838)*

## Tachuri

These two South American insectivorous birds of the grasslands are members of the Tyrannidae. The name is coined in English via Spanish from the transliterated Guarani name *Tachurí* or *Tarichú*, meaning "grub-eater" for several small birds. Félix Manuel de Azara, the Spanish naturalist, introduced the name to European ornithologists in 1805 when he wrote:[188]

> Incluye esta familia muchos paxarillos, que los Güaranís llaman en general *Tachurís*, y *Tarichús*, sin que yo sepa por que. [This family includes many birds that in general the Guaranís call *Tachurís* and *Tarichús*, I don't know why.]

Later, Lafresnaye, who was annoyed with Vieillot for, in his view, appropriating Azara's work on the species he named *Sylvia rubrigastra*, assigned the genus name *Tachuris* to the Many-colored Rush Tyrant *Tachuris rubrigastra*. Jean Cabanis used the common name extensively in his German-language *Museum Heineanum* (1859) for the present-day tachuris as well as for the *Euscarthmus* pygmy-tyrants, *Serpophaga* tyrannulets, and the *Anairetes* tit-tyrants.[63] And, in fact, the name is still used colloquially for the pygmy-tyrants in Paraguay.

# Tailorbird

The fifteen species of tailorbirds in the Cisticolidae share a number of characteristics—long, pointed bills; cocked tails; and the ability to "sew" remarkable nests. And this is how they get their very appropriate name. A tailorbird's nest cannot be described any better than Thomas Pennant does in his 1769 volume *Indian Zoology:*[35]

> It picks up a dead leaf, and surprising to relate, sews it to the side of a living one, its slender bill being its needle, and its thread some fine fibres, the lining, feathers gossamer and down: its eggs are white, the color of the bird light-yellow, its length three inches, its weight only three-fixteenths [*sic*] of an ounce, so that the materials of the nest, and its own size, are not likely to draw down a habitation that depends on so slight a tenure.

# Takahe

There are two species of takahes—one is very rare, the other is, sadly, extinct. The name is a Māori one for the bird. In 1850 a notice appeared in the *Proceedings of the Zoological Society of London* with the long-winded title "Notice of the discovery by Mr. Walter Mantell in the Middle Island of New Zealand, of a living specimen of the Notornis, a bird of the Rail family, allied to Brachypteryx, and hitherto unknown to naturalists except in a fossil state."[459] In it, Mantell wrote:

> This bird was called *Moho* [by the North-Islanders], or *Takahe* [by the South-Islanders], and was of a black colour, destitute of wings, and had a long bill, which, as well as the legs, was of a bright red colour. No traces of the "Moho" had been discovered since the arrival of any of the English colonists.

The Māori name itself is said to be *takahea* and means "to stand up tall and stamp one's foot on the ground" and may have come from the word *takahi*, meaning "to trample or to press down."[23]

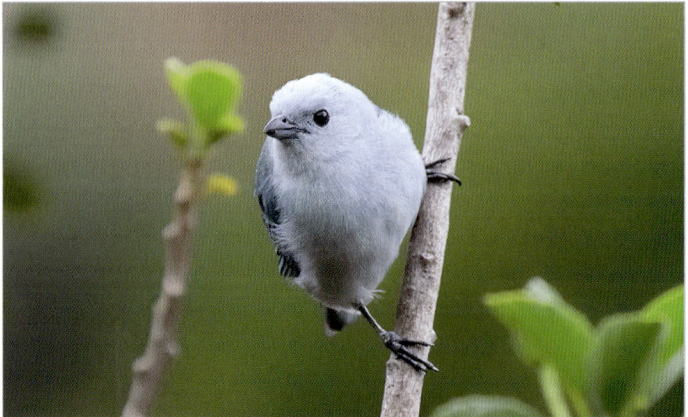

*Blue-gray Tanager (Costa Rica)*

# Tanager

This widely used common name can be found in the names of birds from eight different families, albeit often in hyphenated forms, including thrush-tanager, chat-tanager, palm-tanager, ant-tanager, shrike-tanager, and mountain-tanager. The prefixes in each case describe the birds' appearance, habitat or behavior. (See entries.) The name first appears in the European literature in Marcgrave's *Historia Naturalis Brasiliae* from 1648 under the heading (in Latin) "TANGARA Brazilian birds; several colour variants were found."[90] Brisson proposed the genus *Tangara* in 1760 writing it is "the name given in Brazil to some species of this genus." However, *tangará* is the Brazilian name for manakins not tanagers, known there as *saíras*. The *Dicionário de Palavras Brasileiras de Origem Indígena* 25 states "Tangará: Bird of the family of piprids…" Garcia confirms *tangará* comes from the Tupi words *atá*, to walk or jump, and *carã*, go around—especially applied to the Swallow-tailed Manakin, known colloquially as the Dançador meaning dancer[133].

# Tapaculo

The tapaculos are a family, the Rhinocryptidae, of sixty-five species, fifty-seven of which go by the name. These incredibly furtive forest ground dwellers scurry about under the cover of dense vegetation, and when one is lucky enough to see them, they usually have their tails cocked. The name has the literal meaning in the Spanish *tapar culo* of "cover your ass"; according to folklore, this is a suggestion that the bird should be a little more discreet. However, it's probable that rather than an admonition, the name is, in reality, descriptive. Tapaculos often cock their tail forward of vertical, thus covering the "ass," which would be the rump in human terms (like knee in thick-knee, not anatomically correct), thus it is descriptive, as opposed to a reprimand to "cover your ass," given most people don't think of the cloaca as a bird's "arse" (cf. wheatear, etc.). Darwin wrote:[295]

> It is called Tapacolo [*sic*], or "cover your posterior"; and well does the shameless little bird deserve its name; for it carries its tail more than erect, that is, inclined backwards towards its head.

On the other hand, Johnson claimed in *The Birds of Chile* that the name is onomatopoeic of the call of the White-throated Tapaculo:[316]

> As a rule, however, it is to be located not by sight but by the call notes which obligingly are different from those of all other members of the family. The call most frequently used is a musical tri-syllabic "Tá-pa-koo-Tá-pa-koo," pronounced with a short "a" and the accent on the first syllable. In our opinion the adoption of the English name "Tapaculo" to represent the entire *Rhioncryptidae* family is unfortunate and misleading. It is derived directly from the "Tá-pa-koo" call of this one species whereas the calls of all other members of the family are completely different.

The former interpretation seems more likely, while that latter may be coincidental, but the true derivation of the name may well be lost in the mists of time.

# Tattler

The two species of tattlers, shorebirds in the *Tringa* genus, earn their names from their noisy dispositions. In 1842, William MacGillivray wrote in *A Manual of British Ornithology*:[460]

> The Tattlers are birds of small or moderate size, which frequent the shores of the sea, lakes, marshes, and rivers; feed upon worms, insects, Crustacea, and mollusca; have a rapid, light, rather wavering flight, and in their habits [are] remarkable for their timidity, vociferousness, and the balancing motion of their bodies.

Interestingly, the name was also used for the American birds now known as Solitary and Spotted Sandpipers, being referred to as the "Solitary Tattler" and the "Spotted Tattler" in "Reports on the Fishes, Reptiles and Birds of Massachusetts" in 1839.[544]

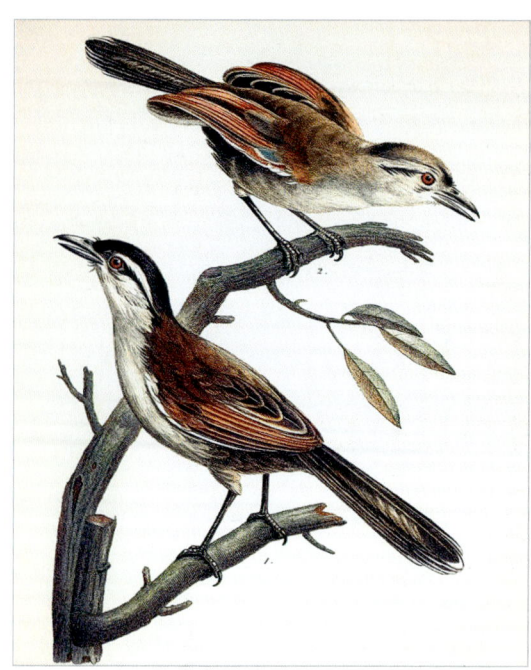

*Southern Tchagra from* Le Vaillant's Histoire Naturelle des Oiseaux d'Afrique *(1799)*

# Tchagra

The tchagras are five species of striking Malaconotidae bush shrikes in the synonymous *Tchagra* genus from Africa. Francois Le Vaillant was one of the first to describe the birds in his *Histoire Naturelle des Oiseaux d'Afrique* in 1799, and he took it upon himself to name the Southern Tchagra after its vocalizations:[93]

> The male has a cry that can be heard from afar, and which is expressed very well by *tcha-tcha-tcha-gra*, from which I took the name of this species.

# Teal

Although there are now twenty-two species of birds known as teals, the bird that originally owned the name was the Eurasian Teal, which was not only very familiar to early Britons but was regularly hunted and consumed. It is mentioned in the *Boke of Keruynge,* (the *Book of Carving and Arranging; and the Dishes for all the Feasts in the year)* from the mid-1400s. The second course of the "feest of saynt Myghell unto the feest of Chrystynmasse" must have been truly amazing:

> potage, mortrus, or conyes, or sewe / than roste flesshe, motton, porke, vele, pullettes, chekyns, pygyons, teeles, Widgeon, wegyons, mallardes, partryche, woodcoke, plouer, bytture, curlewe, heronsewe / venyson roost, grete byrdes, snytes, Fieldfares, Chewets, Beef, with sauces Gelopere and Pegyll, feldefayres, thrusshes, fruyters, chewettes, befe with sauce gelopere, roost with sauce pegyll, & other bake metes.

Notwithstanding the antiquity of the name, the etymology of it is unclear. It is thought that it derived from an unrecorded Old English word, *tæle*, from the Proto-Germanic *tailijaz*, the meaning of which is unknown. Lockwood proposed that a syllable, *tēl*, imative of the chuckling sounds the feeding birds make, is the true etymology.

# Tern

There are forty species of Laridae known as terns. The word has a long history in the English language, but there are two theories regarding the origin of the word. According to the *Oxford English Dictionary*, the name comes from an East Anglian dialect of the 1670s, from the Old Norse *þerna*, meaning "tern or maidservant," ultimately from the Proto-Germanic þewernǭ ("maidservant; servant's daughter"). It is difficult to determine if there is any association between the seabirds and maidservants that could explain the use of the name. There is a possibility it relates to a tern that dives into the ocean to tell Venus about the fate of Cupid, calling her "mistress," in Apuleius's "Tale of Cupid and Psyche" from the second century AD. But the Old English cognate *stearn* predates the 1600s as it had already appeared in an Anglo-Saxon poem, "The Seafarer," written by an unknown author sometime between *circa* 450 and *circa* 1100:

Stormas þær stanclifu beotan, [Storms there beat the stony cliffs,]

þær him stearn oncwæð, [where the tern spoke,]

Following his usual form, Lockwood attributed the name to onomatopoeia.[84] In 1544, William Turner published *De Historia Avium* in which he Latinized a local English name for the Black Tern:[26]

huius generis est & alia parua auis, no strati lingua **sterna** appellata, quae marinis laris ita similis est, ut sola magnitudine & colore ab illis differre uideatur: est enim iste larus, marinis minor & nigrior.

[There is another small bird of this kind, called Stern (the Black Tern *Sterna nigra*) in local dialect, which is so like the sea Lari that it seems to differ from them only in its size and color.]

Linnaeus adopted *Sterna* as the genus name for the terns in 1758, either from the Scandinavian or from the Anglo-Saxon *stearn*.

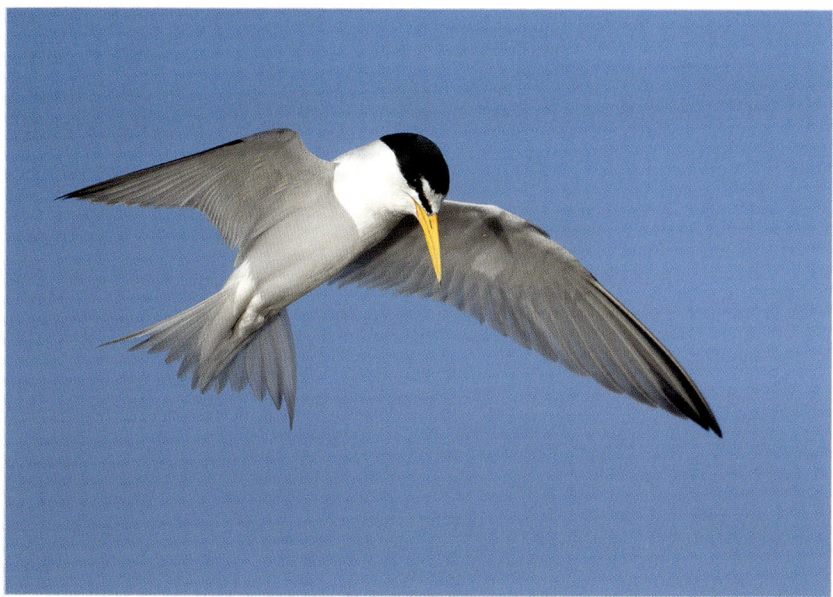

*Least Tern (Florida, USA)*

## Ternlet

There is just one ternlet, which somehow managed to avoid the epithet noddy, like its four congeners. It is simply a diminutive form of the name *tern*.

## Tesia

These five species of little, almost tailless ground dwellers are denizens of the Asian rainforests in the Scotocercidae, the bush warblers and allies. Hodgson was the first to describe them to Western science, citing the Nepali name for the birds as his justification. Under the heading "On three new Genera or sub-Genera of long-legged Thrushes, with descriptions of their species" in *The Journal of the Asiatic Society of Bengal* (1837), Hodgson wrote "Tee-see of the Nipalese" as a heading for his description of Grey-bellied Tesia.[545] The Bombay Natural History Society lists the transliterated Nepali name as *Tisi*. The Nepali name for the birds is टिसिया, transliterated as "Tisia."

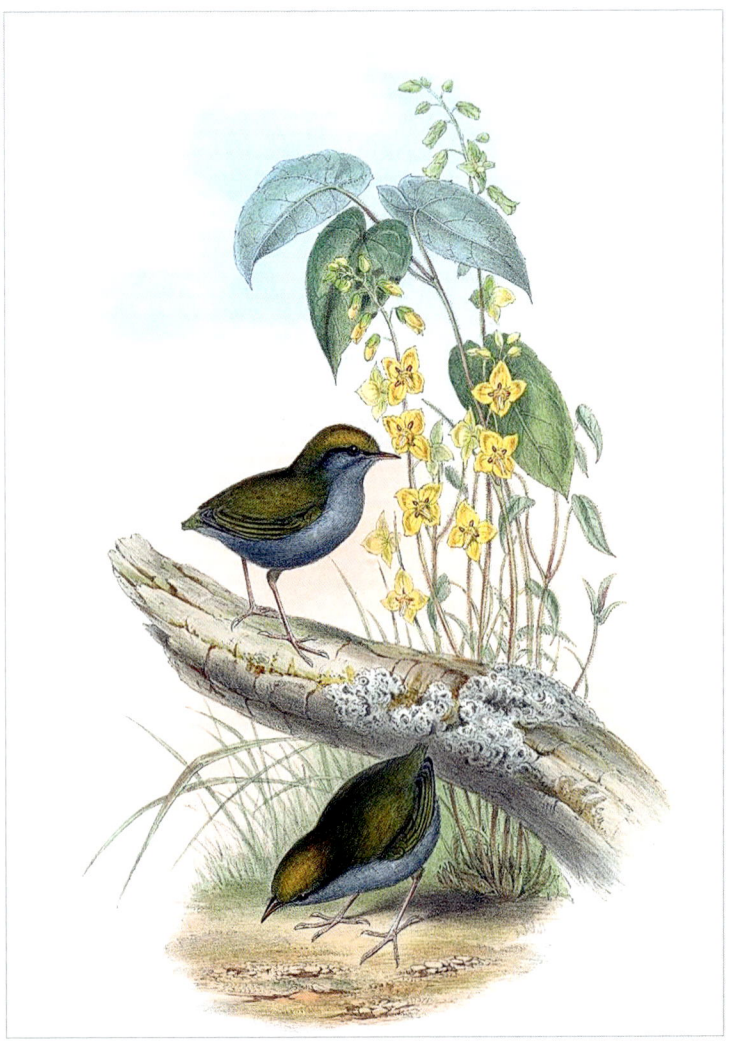

*Grey-bellied Tesia from Gould's* Birds of Asia *(1850)*

# Tetraka

Four species of Bernierid Malagasy warblers, the tetrakas, were previously classified as *Phyllastrephus* bulbuls in the Pycnonotidae. When molecular studies in 2005 revealed them to be endemic Malagasy "warblers" in the genus *Xanthomixis*, the common generic name tetraka was widely adopted. The name had already been in use as an alternative name in the English literature since the 1990s, no doubt based on Alfred Grandidier's entry for *Bernieria zosterops* (now Spectacled Tetraka *Xanthomixis zosterops*) in *Histoire physique, naturelle, et politique de Madagascar* from 1885:[461]

> Les Malgaches nomment cet oiseau Tetekâ, comme son congénère, par imitation de son cri. [The Malagasy people name this bird Tetekâ, like its congener, by imitation of its cry.]

Richardson, however, listed a slightly different rendering of the name as *Te'kite'ky*, "a Sakalava name applied to at least four species of birds, all different kinds of warblers" with alternative names, including *Boretiky*, *Parety*, and *Peretika*, illustrating the fact that there are actually many Malagasy dialects.[231]

# Thamnornis

An endemic Malagasy warbler in the Bernieridae, Thamnornis is named for its genus name, which is simply the Ancient Greek θάμνος *thamnos*, meaning "bush," and *ornis*, for "bird." The name was applied in the 1880s by Grandidier, who wrote:[461]

> The only known Thamnornis to date has been killed by one of us in the dry, arid plains that stretch north of Cape St. Mary, the southernmost tip of the island of Madagascar, and which are covered with small shrubs, thorny bushes and nopals [prickly pears]. This bird lives in the bushes and undergrowth, looking for the insects that make up its food . . . The Antandroys refer to it by roughly the same name as the Sakalavas give them: Kiritikâ (lit.: small).

# Thicketbird

There are five thicketbirds in the Australasian *Cincloramphus* genus in the Locustellidae. The common name acknowledges the birds' preference for dense undergrowth in their forest habitats, which differs from other members of the genus variously known as grassbirds, bushbirds and songlarks.

# Thicket-Fantail

The three species of thicket-fantails (a name that appears to have been coined in the 1960s) in the *Rhipidura* genus favor areas of particularly dense vegetation—in other words, thickets. (See Fantail.)

# Thick-knee

The thick-knees are a small but very distinctive family of ten species. Eight of them in the genus *Burhinus* are known as thick-knees, ostensibly in homage to their prominent tibiotarsal joints, which in birds is not in fact the knee, but the ankle. The name first appeared in print in Thomas Pennant's 1776 work *British Zoology*, in which he discussed the Eurasian Thick-knee under the heading "Thick-kneed Bustard."[177] He described the feature in a vivid manner:

The knees thick, as if swelled, like those of a gouty man; from whence Belon gives it the name of *Oedicnemus*.*

* From οἰδέω, and κνήμη (from *oidéo*, for "swollen," and *kními*, for "leg, shank").

In many places the birds are still called "stone-curlews" and Newton clearly didn't approve of the alternative name:[7]

Stone-Curlew—called also, by some writers . . . most wrongly and absurdly the Thick-Knee or Thick-Kneed Bustard.

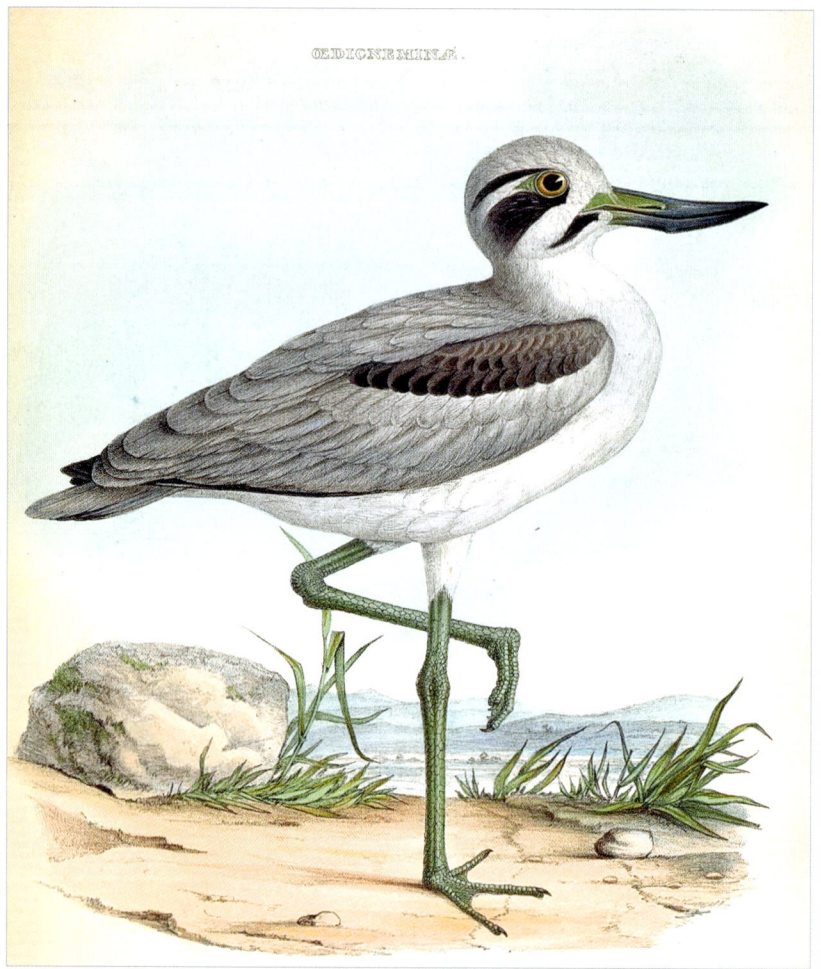

*Beach Stone-Curlew from Studer's* The Birds of North America *(1873)*

# Thistletail

The thistletails, of which there are nine species in the South American Furnariidae, were for a long time colloquially referred to as "spine-tails," a name still in use for many of the furnariids. In a *Catalogue of Birds of the Americas and the Adjacent Islands* published in 1925, the justification for the name is laid forth:[456]

The tail feathers are decidedly attenuated or acuminate terminally.

# Thornbill

Two families feature birds known as thornbills—the Trochilidae hummingbirds and the Acanthizidae, thornbills and allies. Both, very different, groups of birds share one thing in common—they have unusually short bills. It would seem that Gould coined the common name for the hummingbirds in 1846 in his *Monograph of the Trochilidæ*.[6] Commenting on Purple-backed Thornbill, which he called "Thorn-bill," he wrote:

> As its name implies, the bill is remarkably small, and in fact, there is no species yet discovered that has this organ so diminutive: how minute then must be the insects upon which it feeds, how small must be the flowers from which they are obtained, and how active must it be to procure a sufficient supply of these microscopic creatures for the sustenance of life!

Remarkably, one of the first names given to an *Acanthiza* thornbill (Yellow Thornbill) was Dwarf Warbler, a name that appeared in *The Animal Kingdom* by Georges Cuvier in 1829.[407] Later they went by names such as Yellow-rumped Tit, Yellow-tail, Tom-Tit, Little Tit, Yellow Tit, and the like in publications such as *The Insectivorous Birds of Victoria* by Robert Hall, in 1900.[359] It was Alfred J. North in *The Destruction of Native Birds in New South Wales* in 1901 who first used the colloquial name thornbill, presumably in accordance with the name favored by laypeople.[547] Despite this, the birds continued to be referred to in the literature with the Anglo-centric name "tit" until 1926.

# Thornbird

The nine species of furnariid thornbirds are famous for their massive stick nests up to two meters long that are suspended from the ends of tree branches and often contain many compartments. Skutch made the observation that

> the majority of the sticks brought to the nest were not thorny. Probably in more arid districts, where the vegetation bristles with spines, our bird uses enough thorny twigs to justify its name.[27]

He also noted that up to "nine species of birds are known to make more or less use of thornbirds' nests for breeding."

# Thrasher

The thrashers belong to a number of different genera in the Mimidae; there are fourteen species with the common name. The name arose locally in England for the Song Thrush and was carried to North America by the colonists. Until the mid-1800s, the American thrashers were treated as *Turdus* thrushes. Newton discussed the origins of the name in 1896:[7]

> These words are doubtless derived from Thrush, if they be not corruptions of it. An esteemed American correspondent has suggested to me that Thrusher originated in the wish to indicate that the bird so called was bigger than an ordinary Thrush, of which word it might be said to be (if the expression be allowable) the "comparative degree." In that case the other two must be regarded as corruptions. They have nothing to do with threshing.

Prior to this, the common generic name cycled through the names mocking-bird, thrush, and thrasher. Wilson amusingly recounts an anecdote about the "French Mocking-bird," a name by which the Brown Thrasher was known in the 1800s:[28]

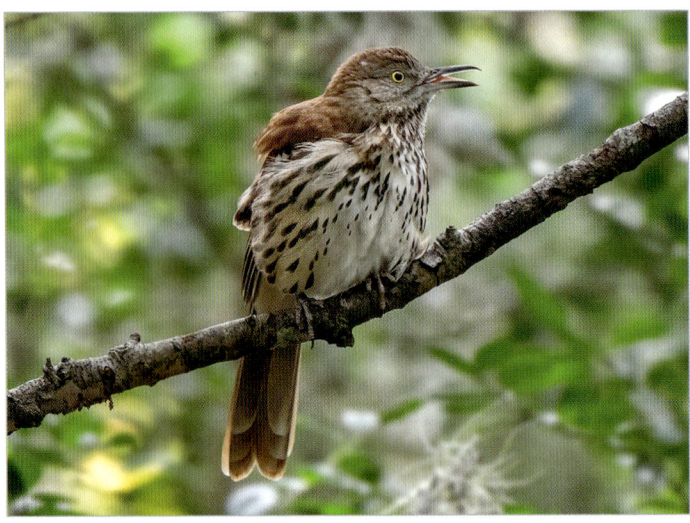

*Brown Thrasher (Georgia, USA)*

The native notes of the Mocking-bird have considerable resemblance to those of the Brown Thrush, but may easily be distinguished by their greater rapidity, sweetness, energy of expression and variety. Both, however, have in many parts of the United States, particularly in those to the south, obtained the name of Mocking-bird. The first, or Brown Thrush, from its inferiority of song being called the French, and the other the English Mocking-bird. A mode of expression probably originating in the prejudices of our forefathers; with whom every thing French was inferior to every thing English.

## Thorntail

The thorntails are five species of hummingbirds in the genus *Discosura*. All have notably long forked tails, which explains the common group name conferred upon them by Gould in 1849, with the unelaborated comment

> *Gouldia, Popelairia, Gouldomyia* and *Prymnacantha* are all generic terms proposed for the four species I have called by the trivial name of Thorn-tail.[6]

## Thrush

This very old English word arrived in the Old English as *þrysce* or *þræsce*, probably from the Proto-Indo-European *trosdos*, the name for the thrush. In Middle English, many different spellings were used, including *thrusche*, *thrusche*, and *thryshe*. Another old name for the Song Thrush was the *throstle* or *throstelle*, from the Proto-Germanic *þrustlō*. The bird names appeared on the menu in Middle English publications like *The Boke of Nurture* in 1450 ("Sparows / thrusches / alle þese .vij. wit*h* salt & synamome")[252] and later in *The Boke of Keruynge* in 1513 ("sparowes & throstelles with salte & synamo*n*").[483] In 1544 William Turner gave "Thrusche, Thrushe, Throssel, Mavis, or Wyngthrushe" as names for the Redwing (Thrush) *Turdus iliacus*.[119] Then in 1678, Ray and Willughby listed under the heading "The Thrush Kind":[16]

- The Missel-Bird, or *Shrite*, and in the North the Thrush simply without addition, *Turdus viscivorus major*.

- The Throstle, *Mavis*, or *Song-Thrush*, T*urdus simpliciter dictus sen viscivorus minor.*
- The Redwing, *Swine-pipe*, or *Wind-Thrush*, *Turdus iliacus, Illas sen Tylas.* It should rather be written and pronounced, The *Wine-thrush*).

This appears to be one of the first examples in print of the use of the present-day bird name. Although Shakespeare did write in *A Winter's Tale* in 1623:

> The lark, that tirra-lyra chants,
>
> With heigh! with heigh! the thrush and the jay.

Today there are 135 species in the Turdidae known as thrushes, some of which have hyphenated names—flycatcher-thrush, ant-thrush, nightingale-thrush, and ground-thrush. There are three other hyphenated bird names—the quail-thrushes, thrush-babbler, and thrush-tanager—belonging to thrush-like, but unrelated, birds (see those entries).

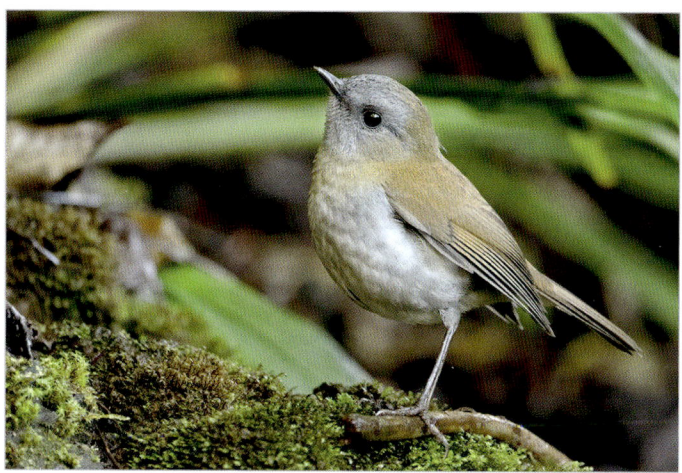

*Black-billed Nightingale Thrush (Costa Rica)*

# Thrush-Babbler

Dohrn's Thrush-Babbler is an unusual sylviid warbler in the genus *Illadopsis,* found only on the tiny island of Príncipe, in the Gulf of Guinea. Dohrn first collected the species in 1866 but gave it no common name. It is babbler-like in its behavior and has a rich, some might say thrush-like, song. Up until the 1930s, and maybe later, other members of the genus of eight species were also called thrush-babblers, for example as in Sclater's *List of the Birds of the Ethiopian Region*, with a reference to Usambara Thrush-Babbler, now called Pale-breasted Illadopsis.[604]

# Thrush-Tanager

The Rosy Thrush-Tanager is a member of a monotypic family, the Rhodinocichlidae. In 1874's *Catalogue of the Birds in the British Museum* the common name given was Rose-breasted Wren.[131] By the 1930s, the name thrush-tanager was being used in the literature. As Winkler et al. stated:[29]

> Few common names do as serviceable a job as this one. This seeming cross between the two birds is suggested not only by its morphology but also by its song, which incorporates the simple structure of many tanager songs and the ringing whistled quality of many thrush songs.

# Tiger-Heron

The three Neotropical *Tigrisoma* herons are well named, although they were known as "tiger-bitterns" in the 1800s. In fact, the name echoes the genus name, which combines the Ancient Greek *tigris*, meaning "tiger" and *somā*, meaning "body." Buffon wrote of the Rufescent Tiger-Heron in his 1793 book *The Natural History of Birds*:[8]

> All its plumage is agreeably marked, and widely intersected, by black cross bars in zigzags, on a rufous ground on the upper side of the body, and of a light gray on the under.

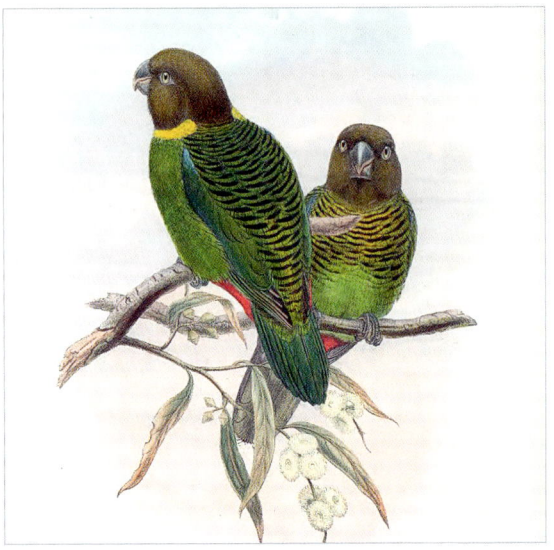

*Brehm's Tiger-Parrot in Gould's* The Birds of New Guinea and the Adjacent Papuan Islands *(1875)*

# Tiger-Parrot

All four species of New Guinean tiger-parrots have varying degrees of tiger-like striping in their plumage. Gould did not call them tiger-parrots in his *The Birds of New Guinea and the adjacent Papuan islands* in 1875, but rather by the plain names of "parrot" and "parakeet."[505] The name tiger-parrot appears to be a relatively new one from the 1970s.

# Tinamou

The tinamous are a family of forty-six species, all but four of them—including two "crested-tinamous"—with the name. The name comes from the Carib-language (a family of languages indigenous to northeastern South America) names for these pheasant-like birds that were widely hunted by the Indigenous people of the region. The name in Surinam and Guyana for Great Tinamou *Tinamus major* is *inamû* in Wayampi, *mamu* in Arawak, and *anamu* in Surinamese.[30] In 1779 Buffon introduced "Les Tinamous" in *The Natural History of Birds*:[8]

> The Creoles of Cayenne call it the *Pintado Tinamou*; but this appellation is improper, for it bears no resemblance to the Pintado, and its striped plumage is not dotted.

Later, in 1783, Latham wrote:[17]

> To this is given the name of Tinamou . . . This name has been given to them by the natives in the parts where they are found.

# Tinkerbird

These nine members of the genus *Pogoniulus* in the Lybiidae are small African barbets. As Gurney noted in "A Sixth additional List of Birds from Natal" in *The Ibis* from 1864 about the "Little Tinker Barbet" (Red-fronted Tinkerbird *Pogoniulus pusillus*):[548]

> The note of this curious little bird so much resembles the tapping of a hammer on an anvil (having that peculiar metallic ring), that it is called in Natal the Tinkerbird. It is silent during the winter months commencing its monotonous cry in the spring, and continuing it throughout the summer.

# Tit

The tits are found in two families—the Paridae (tits, chickadees, and titmice) and the Aegithalidae (long-tailed tits), although the word appeared in a number of hyphenated forms for small birds of other families. In the 1600s the name "titmouse" was more often used, such as in Ray and Willughby in 1678:[16]

> Of Titmice: Of these we have observed in England five kinds, viz. I. The great Titmouse or Oxe-eye. 1. The Colemouse. 3. The Marsh-Titmouse or Blackcap. 4. The blue Titmouse or Nun. 5. the long-tail'd-Titmouse. The crested Titmouse and Wood Titmouse of Gefner, we have not yet found in England.

The word came into English in the 1500s from the Icelandic *tittr*, the Norwegian *tita*, or the Swedish *tätt*, originally for something small, as a horse or a child.[31, 32] (See Titmouse.) Where the Scandinavian words came from is unclear; they probably appeared spontaneously as low-energy words used as diminutives.

Note that the word tit has no connection to the vulgar word for "breast," which is a corruption of the word *teat* from the Proto-Germanic *titta*.

# Tit-Flycatcher

The tit-flycatchers are two small African flycatchers in the Muscicapidae, their diminutive size recognized in the first part of their compound name. (See Tit.)

# Titmouse

There are now five species of titmouses from the Americas in the Paridae, although in the past the name was much more widely used and has a long history in the English language. The word *mase* was an Anglo-Saxon word probably from the Ancient Greek μειος *meios*, for "less, small," and came to mean "small bird." An Anglo-Saxon name for the wren was *hise mase*, and the original form of titmouse was *tit mase*. So, contrary to popular opinion, the "mouse" part of the compound word has nothing to do with mice*. Over time the word *mase* morphed into the word *mouse* (derived from *us*), probably as a result of people mistakenly conflating the two small animals.

*So, the correct plural is titmouses, rather than titmice, but good luck with that!

# Tit-Spinetail

These ten members of the Furnariidae are small spinetails, the first part of their name signifying their relatively diminutive size. (See Tit and Spinetail.)

# Tit-Tyrant

Another example of the use of the name "tit" in a hyphenated name to signify a diminutive version of a related bird, in this case, the tyrants. There are eight species in the Tyrannidae, the tyrant flycatchers.

# Tit-Warbler

The delightful Central Asian tit-warblers are members of the Aegithalidae, the long-tailed tits. In this case, the birds are tits, as other members of the family, but not warblers. The songs do not suggest warblers, so the explanation could lie in this Gould comment:[33]

> The bill is slenderer than in any known Tit; it is, however, entire at the tip, and very hard and very sharp-pointed. I think that we must accept this as a sort of link between the Warblers and the Long-tailed Tits.

# Tityra

Three species of tityras in the relatively small Tityridae are said to have been named in homage to *Tityrus*, the shepherd in Virgil's *Eclogues*, due to the bird's noisy and pugnacious behavior. However, Virgil's *Tityrus* was anything but aggressive, preferring to make cheese! The tityrus or musimon is also a fictitious and raucous creature, a cross between a ram and a goat, and it seems more likely this is the true allusion to the bird.[605, 14]

# Tody

The Todidae are five species of todies—tiny, brilliant green insectivores related to motmots and kingfishers. The name is a modification of the French name *todier*, given to it by Brisson in 1760.[7, 76] It comes from the Latin *todus*, the name given to an unknown species of small bird by the Roman grammarian Sextus Pompeius Festus. Newton noted that the name comes from "the French *Todier* of Brisson of the somewhat obscure Latin word *Todus*."[7] But although the genus was erected by Brisson in 1760, the name was first used in 1756 in *The Civil and Natural History of Jamaica* by Patrick Browne, who wrote under the heading "Todus The Tom-tit":[146]

> This little bird is hardly larger than the green Humming-bird; but its legs and thighs are longer, and the bill more compressed and quite straight. It is a very familiar and beautiful bird, and will often let a man come within a few feet, and before it moves. It keeps much about houses in country parts, flies very slow, and probably may be easily tamed.

# Tody-Flycatcher

Nineteen members of the Tyrannid tyrant flycatchers go by this name. They are all relatively small, hence the first part of the name from the Latin *todus*, meaning "a small bird."

# Tody-Tyrant

Eighteen members of the genus *Hemitriccus* in the Tyrannidae, as with the tody-flycatchers, are relatively small birds, which earns them the epithet tody, *todus* for "small bird." The word *todus* is listed in *A Dictionary of English and Latin Words* from 1570 as "A Titmose, bird, *todus*."[549]

## Tomtit

While this name is now in use only for the New Zealand endemic the Tomtit in the Petroicidae, it is a name that has been in use in Britain for many centuries, mainly, but not exclusively, to refer to the Blue Tit. Originally, the bird was nicknamed "tom titmouse," perhaps in accordance with a fashion of the time of giving people's names to familiar birds and other animals of the British countryside (see Robin) and as an alliteration. Over time it was shortened to the "Tom Tit." The British colonizers of New Zealand brought the name with them and used it for a bird that reminded them of the tits from home (as it was in Australia for many of the small thornbills).

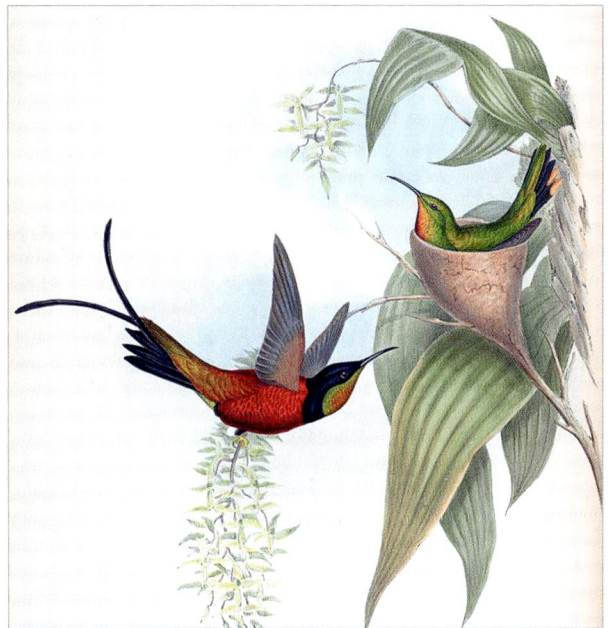

*Fiery Topaz in Gould's* A Monograph of the Trochilidæ *(1849)*

## Topaz

Two spectacular species of hummingbirds in the genus *Topaza* are known by the common group name topaz. They are named for the color of their iridescent plumage, the color of topaz. Gould wrote:[6]

> This gorgeous species, which may be regarded not only as one of the gems of Ornithology, but as one of the most beautifully adorned species of the Trochilidae.

## Torrent-lark

The Torrent-lark is a member of the Monarchidae in the genus *Grallina* along with its only congener, the Magpie-lark. Gould, who named the bird "Bruijn's Grallina" after the collecter, wrote:[550]

> The discovery of a species of *Grallina* in New Guinea is of great interest, as the genus was until recently considered to be entirely peculiar to Australia . . . the native name is *Tada . . .* These birds are found flying about creeks and hopping about stones; they seem to feed on insects obtained there.

Sims noted Shaw-Mayer's field notes in a 1956 issue of the *Bulletin of the British Museum*:[271]

> Native name, "Notechll" ("ll" as in Welsh). This is a bird of the mountain stream, and one was seen by a rushing torrent at about 6,500 ft. When it alighted on a rock it fanned out its tail in the same manner as the Plumbeous Red-start, *Pheonicurus fuliginosus*, of the fast-flowing streams of the Himalayas.

The name Torrent-lark appears to have been first used in the 1970s, bringing the name in line with its congener.

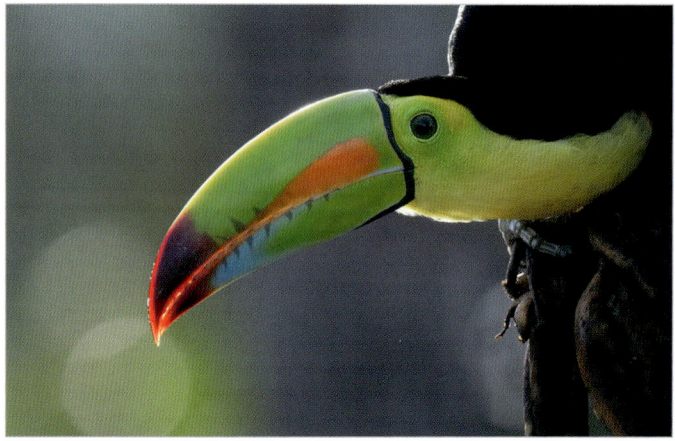

*Keel-billed Toucan (Costa Rica)*

# Toucan

There are seven toucans and four mountain-toucans in the Ramphastidae, the toucans. The name was used in print in 1560 in *Icones Avium Omnium* by Conrad Gessner with the entry "*Toucan ab Americae incolis uocatur* [Toucan from the American natives]."[266] These remarkable birds got their English names in the mid-sixteenth century via the French and Portuguese from the Tupi-language *tucano* or *tukano*, which was their name for the bird. It is said to probably be imitative of its call but could also mean "that which beats or claps strongly," which could reference the loud bill-clapping that some species are known to perform.[25]

# Toucanet

The name toucanet, for the thirteen toucanets and two emerald-toucanets, signifies a smaller toucan, the suffix *-et* being a diminutive, while the name emerald-toucanet is self-explanatory for these lovely green birds.

# Towhee

The nine towhees are birds of the Americas in two different genera in the Passerellidae, the New World sparrows. Some early names for the Eastern Towhee included Ground Robin, Marsh Robin, Towhee Bunting, and Chewink, as Coues listed in *Birds of the Northwest* in 1874.[551] He went on to describe an encounter with the bird:

> The jaunty Towhee, smartly dressed in black, white, and chestnut, comes into view, flying low, with a saucy flirt of the tail, and dashes again into the covert as quickly as it emerged, crying "*tow-hée*" with startling distinctness.

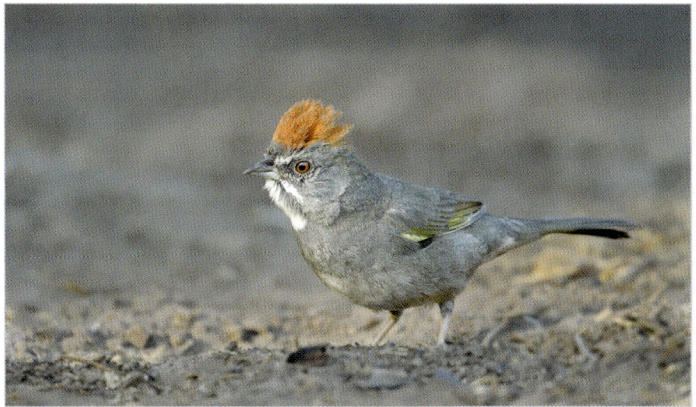

*Green-tailed Towhee (Mexico)*

But the name, which is clearly imitative of the bird's vocalizations, predates this entry, appearing in *The Natural History of Carolina, Florida and the Bahama Islands* by Mark Catesby in 1729.[70] He spelled or misspelled the name under the heading "The Towhe Bird" but went on to list it as "Towhee bird" in the index.

# Tragopan

These five species of pheasants, the tragopans, are some of the most spectacular birds in the world. The common and genus names are synonymous and appear in Aldrovandi's Latin tome *Orithologiae hoc est de avibus historiae* in 1599:[101]

> Sola quidem Auiu Tragopanas veteribus cornuta putata est, Plinio teste, sed qui fabulosam esse arbitretur. [The only old skin of the bird Tragopan is owned by Pliny, but this is treated as a myth.]

The name is a Latinized word from the Ancient Greek τράγος *trágos*, "a billy-goat," and Πάν *Pán*, "Pan" (the Greek god of the wild, shepherds, and flocks, who had the hindquarters, legs, and horns of a goat). This is in reference to the two brightly colored, fleshy horns on the male's head that can be erected during courtship displays

# Trainbearer

In 1862's *The Illustrated Natural History*, Wood wrote:[85]

> Remarkable for the length of its tail is the Trainbearer, a native of Quito. This bird, although a small creature, possesses a long and very straight tail, something like that of the *Polytmus* [goldenthroats] or the *Sappho* [Comet], but much larger in proportion, the length of the elongated feathers being nearly six inches.

The males of the two species of trainbearers do indeed possess exceptionally long tails, like the train of a wedding gown or ceremonial robe.

# Treecreeper

There are two families of treecreepers, the Certhiidae treecreepers and Climacteridae Australasian treecreepers. Even up to the 1960s, the Australasian treecreepers were placed in the Certhiidae, which explains the name they share. But they are not related; rather, they are a remarkable example of convergent evolution. They are clearly named for their habits, although they have gone by many names in the past. Forster listed the colloquial names for *Certhia familiaris* in *Observations of the Natural History of Swallows* (1817):[290]

> Creeper, Treecreeper, Insecthunter, Timber-climber, or Treeclimber.

Earlier in 1781, Latham described the behavior of Eurasian Treecreeper:[17]

> It may be thought more scarce than it really is, by the less attentive observer; for, supposing it on the body or branch of any tree, the moment it observes any one, it gets to the opposite side, and so on, let a person walk round the tree ever so often: the facility of its running on the bark of a tree, in all directions, is wonderful, doing this with as much ease as a fly on a glass window.

In Latham's same volume, he remarkably calls the bird now known as White-throated Treecreeper by the names *Certhia leucoptera* and "Dirigang Honey-eater," writing of it:

> Inhabits New South Wales; called there a Woodpecker, from its being frequently seen running up the trees in the manner of that bird; is most frequently found in the thick forests, chiefly on oak trees, and is named by the natives, Dirigang.

(See Creeper.)

# Treehunter

The ten treehunters are found in three different genera in the Furnariidae. All of them forage by moving along epiphyte-covered limbs and branches in the Neotropical forests, gleaning and pecking their invertebrate prey from clumps of bromeliads, mosses, bark, and dead leaves. In one of the first descriptions by Sclater in *Exotic Ornithology* (1869), he gave the common name "Striped Bush-Hopper" to Flammulated Treehunter *Thripadectes flammulatus*.[34]

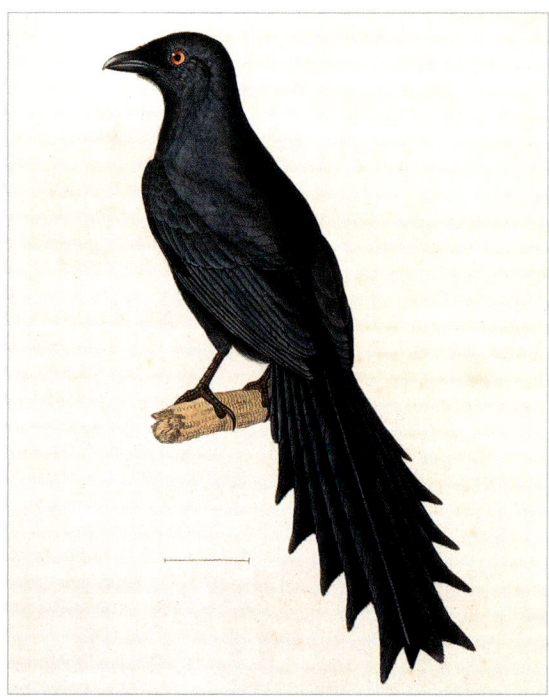

*Ratchet-tailed Treepie in Temminck's* Tableau Méthodique *(1838)*

# Treepie

This common name rhymes with magpie, and, in fact, the two words have a common origin. There are ten birds in two genera, the *Dendrocitta* and the *Crypsirina* in the Corvidae, which includes the crows, jays, and the magpies. The treepies have cycled through a number of common names and were for a while also known as magpies. In 1821, Latham referred to Rufous Treepie as the "Grey-tailed Roller" or the "Rufous Crow";[17] in 1850 Gould[33] called Hooded Treepie *Crypsirhina cucullata*, the "Hooded Crypsirhina"; and then in 1866 Bulger[552] called White-bellied Treepie *Dendrocitta leucogastra*, the "Long-tailed Magpie." One of the first uses of the name appears to have been in the 1867 issue of *The Ibis* with an entry for Andaman Treepie *Dendrocitta bayleii* as "Bayley's Tree-Pie."[553]

The first part of this compound name refers, of course, to the birds' arboreal habits. The source of the second part of the name is the same as that for magpie. It comes from the Middle English *pye*, from Latin *pīca*, meaning "woodpecker," and ultimately from the Proto-Indo-European *(s)peyk-*, which would have been a word used for "woodpecker or magpie." (See Magpie.)

# Treerunner

The five species of treerunners in the Furnariidae are named for their foraging habits. These scansorial birds hunt for invertebrates on epiphyte-covered branches, often hanging upside down or performing other acrobatic feats. The name treerunner or tree-runner has appeared in the ornithological literature since the 1700s as names for *Certhia* creepers and Australian sitellas, among others.

# Trembler

Two members of the Mimidae occur on tiny islands in the Caribbean. According to Sclater in 1869, the Grey Trembler was known as "La Merle Trembleuse" on Martinique.[34] He was unfamiliar with the bird, as he was working from skins, as he commented:

> As to the habits of this bird we have unfortunately no information, but from the singular name which it bears in Guadeloupe, and which is shared by its representative in the island of Sta. Lucia, we cannot but suppose that its mode of life exhibits some peculiarities.

He was correct:[3]

> Tremblers are well named, for they indeed tremble as they sing. The trembling is mainly a quavering of the wings, but also a vibration of the cocked tail, giving the effect that the entire bird is reverberating as it sings.

# Triller

The trillers are all members of the *Lalage* genus in the Campephagidae, and they are closely related to the cuckooshrikes and cicadabirds. In the 1800s, the genus name *Campephaga*, which means "caterpillar eater" from the Ancient Greek κάμπια *kámpia*, for "caterpillar," and φαγα *faga*, "to eat," was used interchangeably with *Lalage* from the Ancient Greek λαλαγή *lalagē*, "to prattle." In 1859's *Catalogue of the Birds of the Tropical Islands of the Pacific Ocean, in the collection of the British Museum*, the common group names flycatcher;* thrush;** and, for the triller*** found on Fiji, "*Karou or Caru* of the natives" are all listed as trivial names for the genus.[462] Especially in Australia, the common names "caterpillar-catcher" and "caterpillar-eater" were the names used until the 1920s for the birds we now know as trillers. One of the first uses of the name triller appeared in *The Emu* in 1924:[554]

> *Lalage tricolor.* White-winged Triller (Caterpillar-eater, Peewit Lark).—Only one seen here, and that a female or male in immature plumage. This is a sprightly little bird, the male having a warbling song, often uttered on the wing. A migrant, and fairly constant in times of arrival, which in western districts is about the middle of November.

\**Campephaga* (*Lalage*) *nævia.* Nævous Flycatcher
\*\**Campephaga* (*Lalage*) *pacifica.* Pacific Thrush
\*\*\**Campephaga* (*Lalage*) *karu.* "Karou" or "Caru" of the natives.

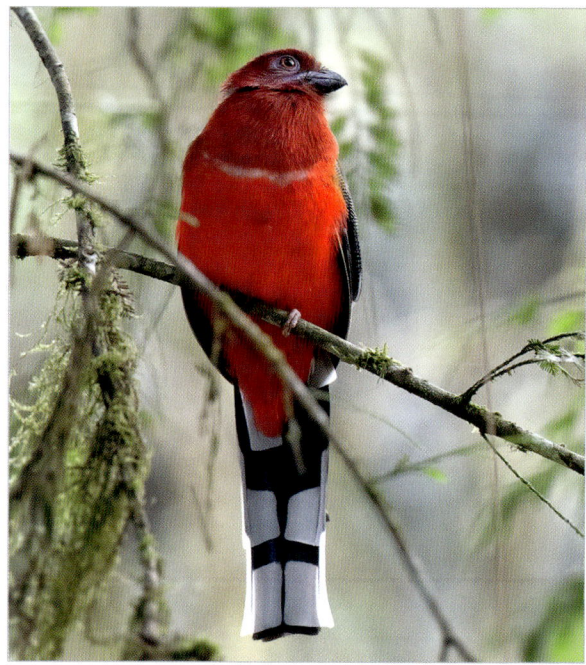

*Red-headed Trogon (Peninsular Malaysia)*

# Trogon

In the Trogonidae, there are thirty-seven birds with the common group name trogon. These pantropical birds were known in the English literature initially as *curucui(s)* or *couroucou(s)*, a name taken from the Brazilian Tupi word for the bird, *surucuí* or *surucuá*. Pennant even used the name for the Malabar Trogon of India in the late 1700s, despite the origins of the word on a different continent.[35] Buffon wrote in 1793:[8]

> Such is the name which these birds bear in their native climate of Brazil. This word imitates their cry so exactly, that the natives of Guiana have omitted only the first letter, and call them *Urucoos.*

The common name trogon is also used in the genus name for the trogons of the Americas, but it wasn't until the mid-1800s that the name started to be used in the colloquial form. It comes to English from the Ancient Greek word τρώγω *trógō*, to "to chew or gnaw," and supposedly refers to the fact that these birds gnaw holes in trees to make their nests. Pliny was the first to mention the name in his *Natural History, in thirty-seven books* (a translation from 1601):[583]

> Of foreigners, one who is called Hylas is thought to have written learnedly concerning Auguries. He reporteth that the Noctua Bubo, the Picus that pecketh Holes in Trees, the Trogon[2] and the Comix come out of their Shells with their Tails first; because through the Weight of their Heads the Eggs are turned (with the wrong End down-ward), and so the hinder part of their Bodies lieth next under the Hen to cherish with her Heat.

with the footnote (²) from the translator:

> Perhaps Trygon; Columba turtur, the Turtle-dove. Wern. Club.

Newton also noted:[7]

> *Trogoneem* . . . occurs in Pliny as the name of a bird of which he knew nothing, save that it was mentioned by Hylas, an augur, whose work is lost; but some would read *Trygonem* (Turtle-Dove). In 1752 Mohring applied the name to the "Curucui" (pronounced "Suruquá") of Marcgrave, who described and figured it in 1648 recognizably. In 1760 Brisson adopted Trogon as a generic term, and, Linnaeus having followed his example, it has since been universally accepted.

So, it would appear that Pliny's reference to a "trogon" was not to the birds we now know as trogons. Even so, in 1781 Pennant wrote in *Genera of Birds* that "The reason for the name Trogon seems to be, because Pliny has such a name after the Picus."[66]

# Tropicbird

The gorgeous tropicbirds are a family, Phaethontidae, of three species; these seabirds are found in tropical or subtropical oceans. Dampier encountered them in the 1600s, writing that "They are never seen far without either Tropick, for which reason they are called Tropick-Birds.[555] They are very good food." As Buffon so poetically noted in *The Natural History of Birds* (1793):[8]

> We have seen birds travel from north to south, and with boundless course traverse all the climates of the globe: others we shall view confined to the polar regions, the last children of expiring nature, invaded by the horrors of eternal ice. The present, on the contrary, seems to attend the car of the sun under the burning zone, defined by the tropics: flying perpetually amidst the tepid zephyrs, without straying beyond the verge of the ecliptic, it informs the navigator of his approach to the flaming barriers of the solar track. Hence it has been called the *Tropic Bird*, because it resides within the limits of the torrid zone.

# Troupial

Only three species retain the name troupial in the Icteridae; most of their congeners are now called orioles. Originally called troupiales, the name first appears in Brisson's *Ornithologie* in 1760 with the note:[76]

> Troupiale, nom qu'on donne en Amérique à quelques especes de ce genre. [Troupiale, name given in America to some species of this kind.]

The French name *troupiale* comes from the French *troupe* ("troop") and was given to them because they can occur in vast flocks.

# Trumpeter

The trumpeters are a family of only three species, the Psophiidae, medium-sized but rather long-legged chicken-like ground dwellers from the Neotropics. They are named for their very loud vocalizations. An eye-opening account of the Grey-winged Trumpeter appeared in *A Dictionary of Arts and Sciences* in 1806:[280]

> The most characteristic and remarkable property of these birds consists in the wonderful noise they make either of themselves, or when urged by the keepers of the menagerie. Some have supposed it to proceed from the anus, and some from the belly. It is now certain, however, that this noise proceeds from the lungs.

# Tuftedcheek

The two species of tuftedcheeks in the Neotropical Furnariidae both have prominent buffy tufts on the cheek and the sides of the neck.

# Tui

This striking honeyeater is endemic to New Zealand and belongs to a monotypic genus, the *Prosthemadera*. In the 1800s, the English colonists of New Zealand called the Tui the "Parson-bird." Newton wrote that it was[7]

> so-called by the English in New Zealand from the two tufts of curled and filamentary white feathers hanging beneath its chin, which were supposed to resemble the bands worn until lately by clerics.

The present-day name comes from the Māori *Tūī*, the earliest attestation of which was probably in Augustus Earle's *A Narrative of a Nine Months' Residence in New Zealand, in 1827* from 1832 when he wrote:[361]

> All was quiet, beautiful, and serene; the only sounds which broke the calm were the wild notes of the *tooe* (or New Zealand blackbird), the splashing of our own oars, or the occasional flight of a wild duck (or shag), disturbed by our approach.

The etymology of the Māori name is harder to ascertain, but, according to *The Te Aka Māori-English, English-Māori Dictionary, korokoro* [throat] *tūī* means "melodious singer."[165] Could *tūī* allude to the bird's extraordinary vocalizations, about which Buller wrote:[362]

> Many of the passages in the Tui's ordinary song are of surpassing sweetness, and so rapid is the change from one set of notes to another that the naturalist never tires of listening to the wild melody . . . the musical cadence of the note is exquisite, as all who are familiar with it will readily admit.

# Turaco

Originally called "Crown-birds" (and mistakenly thought to be from Mexico)[78], the turacos are a family, the Musophagidae, of twenty-three species, eighteen of them with the common group name turaco. Newton noted in 1896:[7]

> It is the *Cuculus persa* of Linnaeus, and *Turacus* or *Corythaix persa* of later authors, who perceived that it required generic separation. Cuvier, in 1799 or 1800, Latinized its native name.

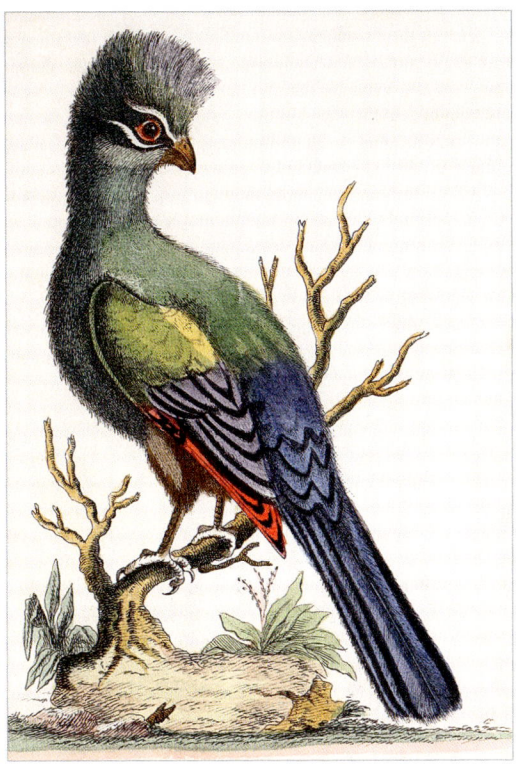

*Guinea Turaco in Edward's* A Natural History of Uncommon Birds:
and of some other rare and undescribed animals *(1743)*

The origins of the "native name" are unrecorded. Turacos also lend their name to the two unique pigments that they alone possess. *Turacoverdin*, responsible for the birds' bright green coloration, is the only true green pigment in birds known to date. Further to that, *turacin* is a red pigment seen in the wings of the turacos.

## Turca

The Moustached Turca is in the same genus with the huet-huets, possessing a similarly singular name. In *The Zoology of the Voyage of H.M.S. Beagle . . . during the years 1832–1836,* Darwin made the observation:[36]

> The *P. megapodius,* is called by the Chilenos, "El Turco"; it lives on the ground amongst the bushes which are sparingly scattered over the stony hills.

Why the Chileans called these birds "the Turks" does not appear to be documented, but almost certainly it is a parallel drawn between the bird's distinctive broad, white moustachial stripe and the luxuriant mustaches typically worn by Turkish nationalists as political statements since the final days of the Ottoman Empire.[403]

## Turkey

The two species of turkeys in the genus *Meleagris* are members of the Phasianidae. Both the genus and common names of these game birds came into English through a series of mistakes. The Helmeted Guineafowl was first brought to Europe in the mid-1500s from

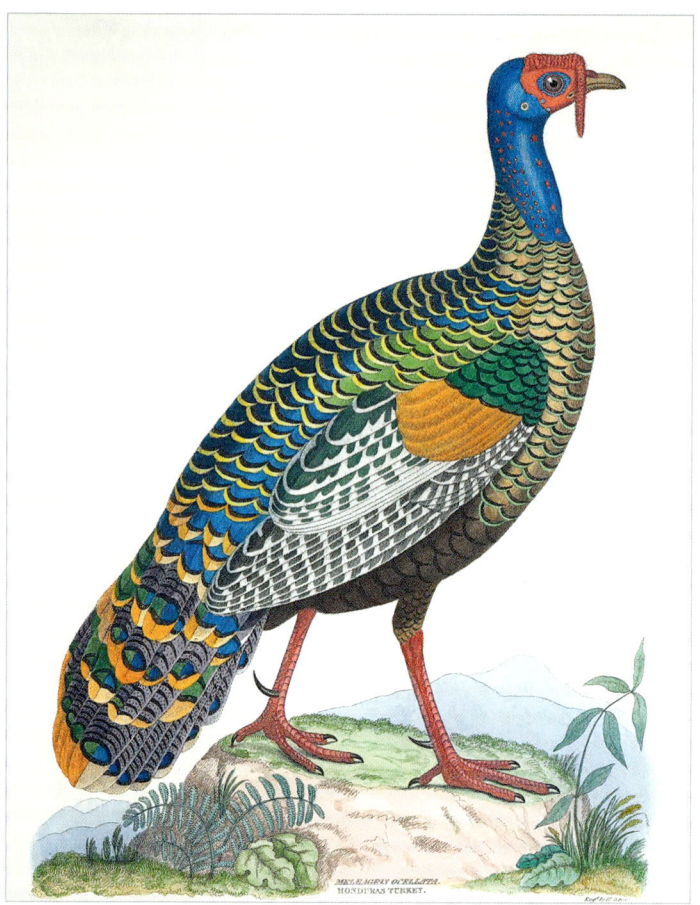

*Ocellated Turkey in Brown's* Illustrations of the American ornithology of Alexander Wilson and Charles Lucien Bonaparte *(1835)*

Africa via Turkey but was call "guinea-fowl" by the Portuguese in honor of their colony in West Africa. Around the same time, the Spanish imported another game bird, *Meleagris gallopavo*, that had been domesticated by the Aztecs. At first, Europeans didn't differentiate between the two species, calling them both turkeys, presumably by association with the erroneously perceived geographic origins of the birds. Later, when it came to be more widely understood that the two birds were, in fact, different, the name "turkey" stuck for the New World birds. Even Linnaeus may have been confused—he assigned the Ancient Greek name for guineafowl *Meleagris* to the New World turkey. Pennant described the confusion in his publication from 1785:

> Turkies are natives only of *America*, or the *New World*; and of course unknown to the antients. Since both these positions have been denied by some of the most eminent naturalists of the fifteenth century, I beg leave to lay open, in as few words as possible, the cause of their error. . . . Belon, the earliest of those writers who are of opinion that these birds were natives of the old world, founds his notion on the description of the *Guinea* Fowl, the *Meleagrides* of *Strabo, Athenaeus, Pliny*, and others of the antients . . . Gesner; [*sic*] who falls into a mistake of another kind, and wishes the Turkey to be thought a native of *India*.

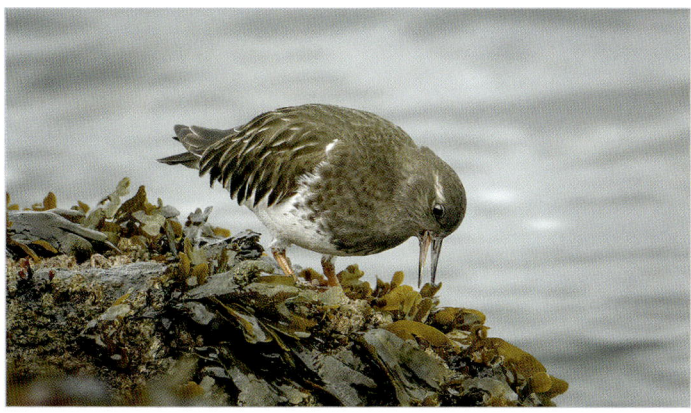

*Black Turnstone (Washington State, USA)*

# Turnstone

Two species of unusual sandpipers in the Scolopacidae are distinguished by their stout bills with a slightly upturned upper mandible, giving them a singular appearance. Catesby was the first to use the name in *The Natural History of Carolina, Florida and the Bahama Islands* in 1729, when he wrote about an unfortunate bird he had captured:[70]

> It was very active in turning up stones, which we put into its cage; but not finding under them the usual food, it died. In this action it moved only the upper mandible; yet would with great dexterity and quickness turn over stones of above three pounds weight. This property Nature seems to have given it for the finding of its food, which is probably Worms and Insects on the Sea-shore.

The name persisted with Buffon, in 1793's *The Natural History of Birds*:[8]

> Adopt[ed] the name Turn-Stone given by Catesby, because it indicates the singular habit which this bird has of turning over the stones at the water's-edge, to discover the worms and insects that lurk under these; whereas all the other shore-birds content themselves with searching in the sand or mud.

# Turtle-dove

Seven species of Columbidae are called by the name turtle-dove. The "original" turtle-dove, though, is the European Turtle-Dove. The Romans called this bird the *turtur*, the word appearing in its scientific name as *Streptopelia turtur*. The Latin word is onomatopoeic for the bird's cooing *turr turr turr* song. There is no connection at all with the reptilian turtle. The mispronounced name came into Old English as *turtle* or *turtla*, evolving over time to *turtledoufe* or *turtildove* in Middle English. Chaucer wrote in *The Parliament of Fowls* in 1381:

> The wedded turtel, with hir herte trewe. [The wedded turtledove with her heart true.]

# Twinspot

These spotted estrildids are six species in four different genera, all found in Africa. They are named for the numerous pale twin spots on the underparts. As Shelley pointed out in *The Birds of Africa* (1896),

> In adults, many of the feathers of the underparts have rounded pale subterminal twin spots, one on each web.[139]

# Twistwing

The two species of Neotropical tyrannids are named for their very unusual modified outer primary feathers, which possess thickened and twisted shafts.[3] Up until the early 2000s, *Cnipodectes subbrunneus* was known simply as Brownish Flycatcher. It wasn't until Rufous Twistwing was described in *The Wilson Journal of Ornithology* in the late 2000s, that the name twistwing came into use.[556]

# Twite

This little congener of the linnets in the Fringillidae has a unique name that is imitative of its distinctive *twit* call. The name first appeared in Albin's *A Natural History of English Song-birds* in 1737, but he gives no explanation of the name.[391] It is thought to be a colloquial name known to the "bird-catchers" that Albin mentions in the text, imitative of the bird's calls, the original form of which was probably *tweet*.

# Tyrant

The Tyrannidae, or tyrant flycatchers, is a huge family of well over 400 species; in fact, it is the largest of all bird families, and the name tyrant, in one form or another, appears in the epithets of 120 species. The attribution of the name lies in the supposed tyrannical behavior toward other birds. Catesby, who is credited with coining the name, albeit for the bird now known as Eastern Kingbird, wrote in 1731:[70]

> The courage of this little Bird is singular. He pursues and puts to flight all kinds of Birds that come near his station, from the smallest; to the largest, none escaping his fury: nor did I ever see any that dar'd to oppose him while flying; for he does not offer to attack them when sitting. I have seen one of them fix on the back of an Eagle, and persecute him so, that he has turned on his back into various postures in the air, in order to get rid of him, and at last was forced to alight on the top of the next tree, from whence he dared not move, till the little Tyrant was tired, or thought fit to leave him. This is the constant practice of the Cock while the Hen is brooding: he sits on the top of a bush, or small tree, not far from her nest; near which if any small Birds approach, he drives them away; but the great ones, as Crows, Hawks, and Eagles, he won't suffer to come within a quarter of a mile of him without attacking them.

The hyphenated forms of the names of species with the Tyrannidae include Black-Tyrant, Bristle-Tyrant, Bush-Tyrant, Chat-Tyrant, Ground-Tyrant, Manakin-Tyrant, Pygmy-Tyrant, Shrike-Tyrant, Tit-Tyrant, Tody-Tyrant, Wagtail-Tyrant, Water-Tyrant. The first element of each name indicates appearance, morphology, habitat choice, or similarity to other species.

# Tyrant-Manakin

Seven species of Pipridae manakins are likened to tyrants in their compound names. Like many of the tyrants, they have rather drab olive plumage and are often mistaken for those birds.

# Tyrranulet

These small tyrant flycatchers, of which there are sixty-seven representatives, are named for just that—*tyrant* with the diminutive suffix -*let*.

# U IS FOR ULA-AI-HAWANE

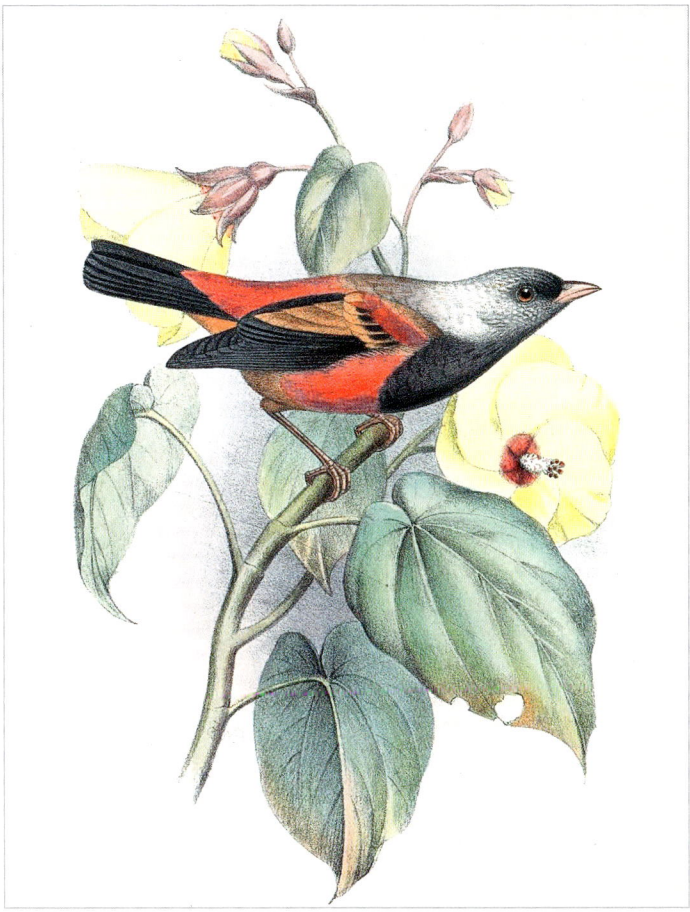

*Ula-ai-hawane in Wilson's* Aves Hawaiienses *(1890)*

## Ula-ai-hawane

This bird with a fabulous name is a single species in the Fringillidae and, like so many Hawaiian birds, is now extinct. In a 1964 book entitled *Arts and Crafts of Hawaii*, P. H. Buck describes a bright red garment called the *ahu'ula*:[463]

> The Hawaiian male nobility wore feather cloaks and capes for ceremonies and battle. Such cloaks and capes were called 'ahu'ula, or "red garments." Across Polynesia the color red was associated with both gods and chiefs. In the Hawaiian Islands, however, yellow feathers became equally valuable, due to their scarcity. They consisted of olona (*Touchardia latifolia*) fibre netting made in straight rows, with pieces joined and cut to form the desired shape. Tiny bundles of feathers were attached to the netting in overlapping rows starting at the lower edge. The exterior of this example

is covered with red feathers from the 'i'iwi bird (*Vestiaria cocchinea*), yellow feathers from the 'o'o (*Moho nobilis*), and black feathers also from the 'o'o.

The first part of the name undoubtedly comes from the association of the red plumage of the bird and the red cape. The *hāwane* is the palm *Pritchardia schattaueri*, endemic to the Hawaiian Islands. So, the name might translate as a "red bird that eats *hāwane* nuts." However, the relationship between the size of the bird's bill and the size of the palm nuts makes it very unlikely that this was the case. It's more likely that they fed on the flowers and ripe flesh of the fruits.

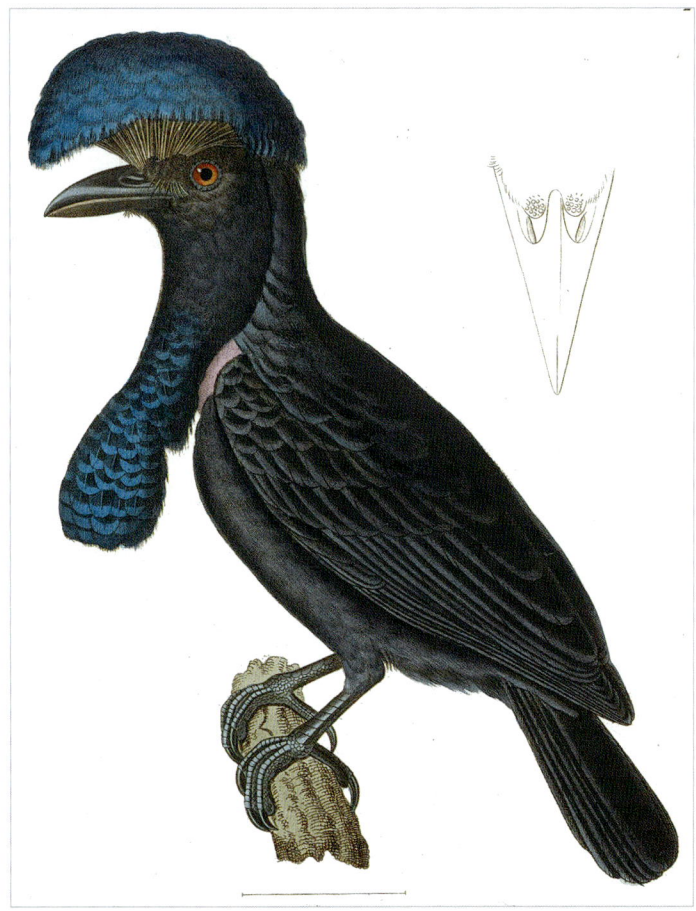

*Amazonian Umbrellabird by Nicolas Huet in Temminck's*
Nouveau recueil de planches coloriées d'oiseaux *(1838)*

# Umbrellabird

These amazing cotingas are three species in the genus *Cephalopterus*. When first observed, their most striking feature is the umbrella-like crest on the head, the obvious derivation of the name. In the words of the wonderful *The Pictorial Museum of Animated Nature* from 1844:[464]

This rare and beautiful bird is distinguished by a crest of full outspreading plumes which tower above its head and fall over the beak, reminding us of the crest of a Grecian helmet.

# V IS FOR VANGA

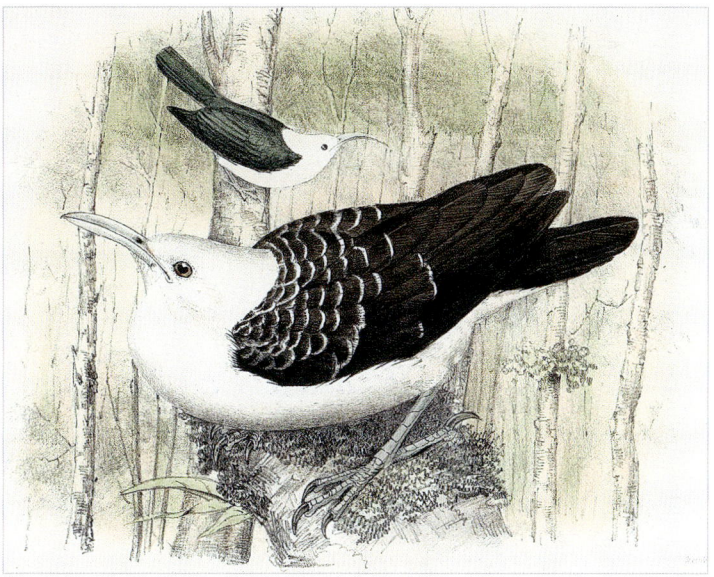

*Sickle-billed Vanga in Pollen's* Recherches sur la faune de Madagascar et de ses dépendances *(1868)*

## Vanga

There are sixteen species represented by the name vanga in the Vangidae, the vangas, helmetshrikes, and allies. In the literature of the 1800s through to the early 1900s, it's stated a number of times that it comes from the Malagasy name *vanga* for the Hook-billed Vanga *Vanga curvirostris*. Latham mentions the name in *A General Synopsis of Birds* in 1781 in his discussion of "Hook-billed Shrike *Lanius curvirostris*":[218]

> This is a native of Madagascar, where it goes by the name of Vanga; it feeds on fruits, and is said to whistle well.

Then, in 1863, in "Notes of a Second Visit to Madagascar," Edward Newton gave the local names for *Vanga curvirostris* as "Voron-bang" and "Vanga."[557] But in *Malagasy-English/English-Malagasy: Dictionary and Phrasebook* (2001), Janie Rasoloson, a native of Madagascar and a doctor of linguistics, lists a number of birds, including vangas.[558] The word "vanga" appears in Malagasy only for small birds, such as Common Stonechat *vangamanga*, Dark Newtonia *vangasoratra*, Wedge-tailed Jery *vangamena*, and White-throated Oxylabes *vangamaintiloha*. The Malagasy name for Hook-billed Vanga, the supposed original vanga, is *todikarasoka* (*todika* is a root word for "turned back" or "turned around," as in a hooked bill).[231]

## Veery

The Veery is just one of twelve species of New World *Catharus* thrushes in the Turdidae. The bird's single-word name comes from its downward-veering call. Ingersoll wrote in 1937:[9]

> "Veery" is an expressive label to the distinctive character of its sweetly rolling music.

Thomas Nuttall tried to transcribe the song in *A Manual of the Ornithology of the United States and of Canada* in 1832:[465]

> This curious whistling note sounds like *'velm 'v'rehu 'v'rehu' v'rehú*, and sometimes *'veã veã 'vrehã 'vrehã vehú*, running up the notes till they become shrill and quick at the close, in the first phrase; but from high to low, and terminating slender and slow, in the latter; another expression seems to be, *'ve ved vehurr*, ascending like a whistle.

## Velvetbreast

The Mountain Velvetbreast, a species of hummingbird of the humid montane forests of the Neotropics, was given its common name by Gould in his *Monograph of the Trochilidæ*.[6] Interestingly, in his discussion of the appearance of the bird, the only reference to velvet is of the plumage of the belly:

> Throat, breast, upper part of the abdomen and flanks very dark shining grass-green; lower part of the abdomen velvety-black.

## Verdin

In the small Remizidae, the penduline-tits, of eleven species there is a single species with a one-word name, the Verdin. It is also the only New World representative of the family. Although for a while the name Yellow-headed Bush-Titmouse held sway, fortunately, the name Verdin gained favor over the long term. Although some, such as Ingersoll, claim that the appellation derives from the Spanish for "green," it is actually an Americanization of the French word for a group of small green birds, the Asian Leafbird (Chloropsidae).[6] The root of the word is *vert*, "green," and the diminutive -*on* suffix.

## Violetear

Five species of violetears are found in the genus *Colibri*, all of which have a patch of elongated violet feathers from below the eye to the auriculars, on the sides of the head, hence the English name. Gould appears to have been the first to use the name "Violet-Ear" in his *Monograph of the Trochilidæ*.[6]

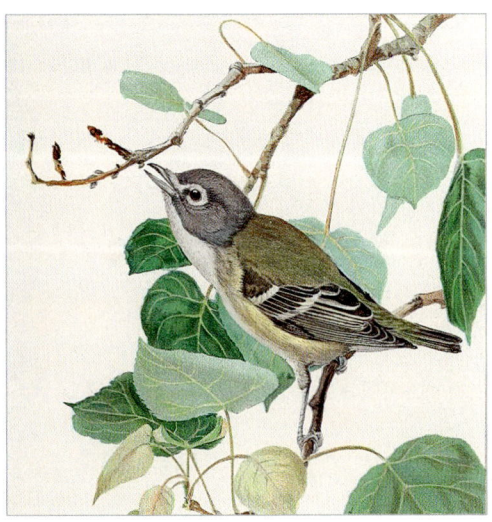

*Blue-headed Vireo in the* Annual Report of the New York (State) Forest, Fish and Game Commission *(1902)*

# Vireo

In the Vireonidae, the vireos, shrike-babblers, and Erpornis, thirty-seven species are called by the name vireo. Pliny first used the word for small green birds, possibly the European Greenfinch and possibly the female Eurasian Golden Oriole, from the Latin for "to be green," *virere*. Ingersoll stated, in hindsight incorrectly (see Greenlet):[9]

> The word is New Latin for greenness, and attempts have been made to popularize it as "green-let," but they have met with little literary success.

# Visorbearer

According to Gould in Volume 1 of his *Monograph of the Trochilidæ*, the two species of hummingbirds in the *Augastes* genus that go by this unusual name[6]

> are very beautiful species, and have had the trivial name of Vizor-bearers applied to them, from the very peculiar manner in which their entire faces are covered with shining metallic feathers, giving the birds the appearance of being masked.

# Vulture

There are two groups of vultures, the Cathartidae, or New World vultures, with five species, and the Accipitridae, or hawks, eagles, and kites, with twelve species, called vultures. The word is thought to have come into English in the 1300s from the Latin name for the scavengers, *vultur*. And this can be traced back to the word *vellere*, "to pluck, to tear," which is clearly appropriate. The word would have undergone a process of metonymy (a figure of speech in which one name takes the place of another with which it has a close association) to apply to the vultures' manner of eating by—you guessed it—plucking and tearing dead flesh. There are many examples of early references to the *voltures* in Middle English, including one in *Bartholomaeus's De Proprietatibus Rerum* in the mid-1200s that gets straight to the point:[559]

> Þe fulture is a foule with euele smelle. [The vulture is a fowl with an evil smell.]

Ray and Willughby were rather uncharitable toward the Accipitridae vultures:[16]

> The more generous, called EAGLES . . . The more cowardly and sluggish, called VULTURES.

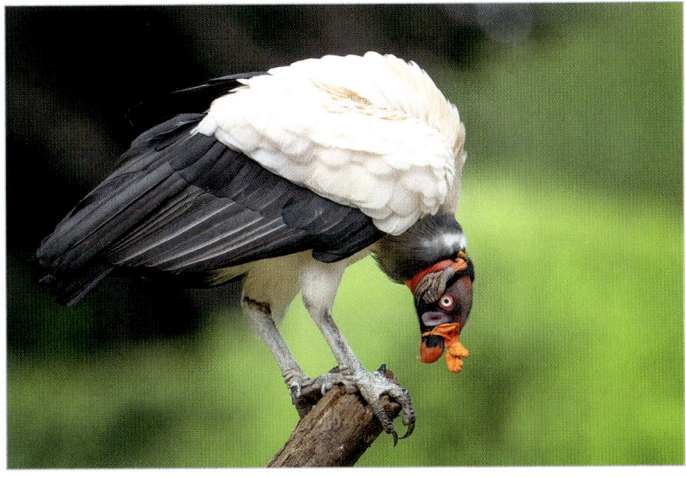

*King Vulture (Costa Rica)*

# W IS FOR WAGTAIL

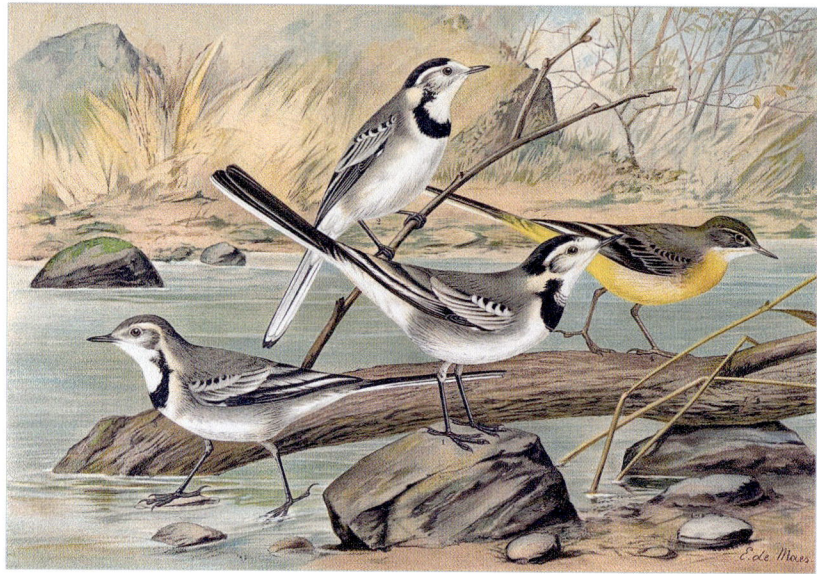

*White and Grey Wagtails in Naumann's* Naturgeschichte der Vögel Mitteleuropas *(1905)*

## Wagtail

The thirteen species of wagtails in the Motacillidae (wagtails and pipits) are all in the *Motacilla*, bar one (Forest Wagtail in *Dendronanthus*). A brief observation of any wagtail will immediately betray the origin of the name—they pump their tails constantly as they walk on the ground in search of their invertebrate prey. In fact, *Motacilla*, the name assigned to the genus by Linnaeus in 1758 means "mobile tail" in Latin. In the 1700s they were referred to in the literature as "wag-tails," with other local names including wagster, wag, waggie, wash-dish, washerwoman, wash-tail, seed, lady, and quake-tail.[15] Forbes wrote in 1905:[466]

> Formerly called "wagstart or wagstert," start, signifying tail.

> This pretty bird is hated in Ireland, and thought always to presage some evil . . . because it plucked away the moss with which the robin had covered and hidden our Saviour from His enemies. In Highlands of Scotland his coming near the doors of houses and among hens, etc., is a sure sign of bad weather.

There is another wagtail, the Willie Wagtail in the Rhipiduridae, the fantails. Another old name for the wagtails familiar to Europeans was Willy Wagtail, which was used for the Pied Wagtail, so it would seem this is a simple case of a name for a familiar bird being transferred to one of somewhat similar appearance by the early English colonists of Australia.[15]

## Wagtail-Tyrant

These two species of tyrant flycatchers in the genus *Stigmatura* have long, cocked tails that they constantly pump and fan, in much the same way as the true wagtails or fantails.

# Wallcreeper

This remarkable bird is the sole member of the Tichodromidae. The Wallcreeper clearly takes its name from its chosen habitat and habits. It is usually found on or near cliff faces of mountain gorges and rocky slopes, where it can be seen foraging for its invertebrate prey using its long, thin bill to probe cracks and crevices. In 1793, Buffon wrote of the "Wall Creeper" in *The Natural History of Birds*:[8]

> All the motions that the preceding [the "Common Creeper"] performs on trees, this performs on walls; it lodges there, and there it climbs, hunts, and breeds: by walls, I mean not only those built by man, but those formed by nature, the huge perpendicular rocks.

Earlier in 1678, Ray and Willughby discussed the "Wall-creeper," giving it another name, the "Spider-catcher," and classifying it as a *Picus* woodpecker.[16]

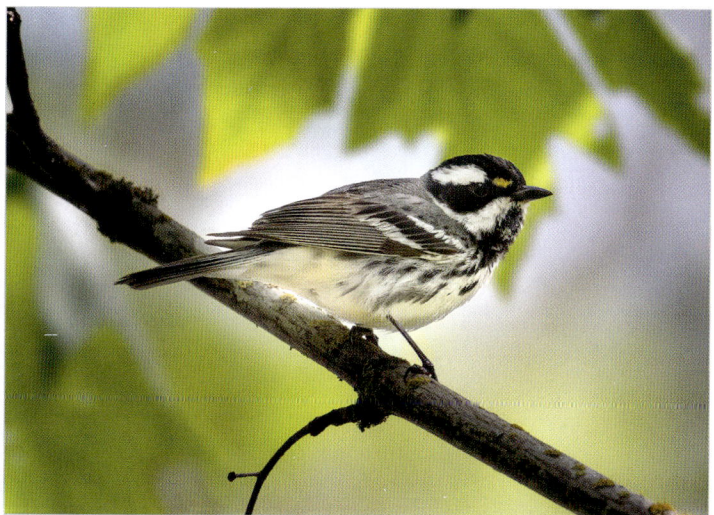

*Black-throated Gray Warbler (Washington State, USA)*

# Warbler

This is one of the most widely used names for birds, whether in its simplest form or hyphenated, as in bracken-warbler, brush-warbler, bush warbler, grass-warbler, ground-warbler, mouse-warbler, plumed-warbler, reed warbler, rufous-warbler, rush-warbler, swamp-warbler, wood-warbler, wren-warbler, and yellow-warbler. Obviously, the warblers are birds that warble, a word that has been used in English since the 1300s, derived from the Frankish word *werbilon* "to whirl." The name for a bird was first coined by Thomas Pennant in 1773 when he wrote of "Wagtails":[281]

> After the example of *Scopoli*, I separate these, the genuine *Motacillae*, from the other soft-bill'd small birds, which he stiles *Sylviae* [warblers]. They are included among the *Motacillae* of *Linnaeus*, and *Ficedulae* of *Brisson*.

The warbler, he said,

> inhabits all parts of the world except the *Arctic*: The most melodious of the smaller *genera*: Insectivorous, seminivorous, delight in woods and bushes. Their pace hopping. *Motacilla* of *Linnaeus*, *Ficedula* of *Brisson*.

The name has since been applied to a host of small songbirds often in an ad hoc manner. As Newton observed:[7]

> "Warbler" has long been used by English technical writers as the equivalent of Silvia, and consequently generally applied to all members of the Family Sylviidae thereon raised, which has since been so much subdivided as to include a vast number of genera, while species almost innumerable have from time to time been referred to it.

## Warbling-Antbird

The seven species of warbling-antbirds in the Thamnophilidae (typical antbirds) are all in the genus *Hypocnemis*. All are known for their loud and varied songs, their "warbling." (See Antbird.)

## Warbling-Finch

The Thraupidae, the tanagers and allies, has fifteen species of warbling-finches. Some, but not all, of these finch-like tanagers have melodious songs, hence the reference to warbling in the name. Sclater described the song of the Black-and-rufous Warbling-Finch in 1888:[22]

> Its voice also, in purity and sweetness of tone, is not unlike that of the Robin; but the song, composed of six unvarying notes, is uttered in a deliberate, business-like manner at regular intervals, and is monotonous.

(See Finch.)

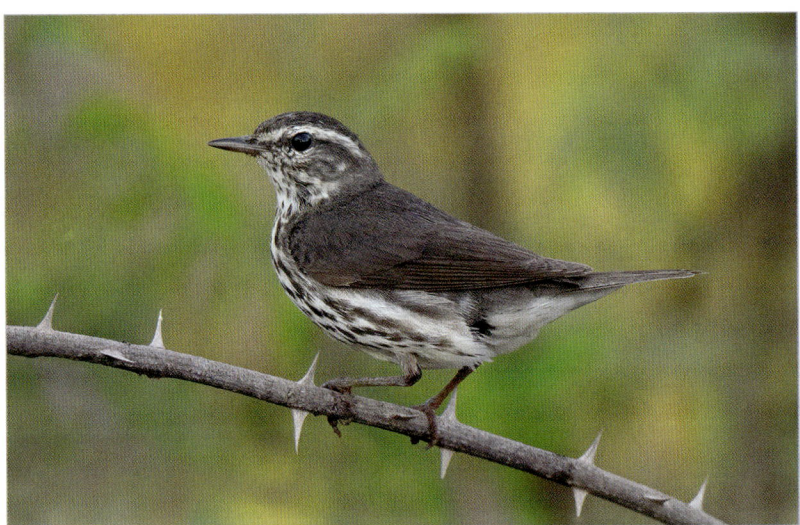

*Northern Waterthrush (Mexico)*

## Waterthrush

Although not thrushes, but New World warblers in the Parulidae, the two waterthrushes are very often closely associated with water, spending much of their time foraging for invertebrates in forest streamsides. The reference to thrushes stems from their cryptic plumage and their loud, resonant song. As Ernest Ingersoll stated in *An Adventure in Etymology* in 1937:[9]

> These are the water-thrushes, a suitable name, for they look like the diminutive thrushes, and are especially fond of woodland water-courses.

# Wattlebird

Although there is a family, the Callaeidae, called wattlebirds, endemic to New Zealand, none of the members now bears the common name wattlebird. In the early literature, such as in Latham's *A General Synopsis of Birds* (1781), one species (the South Island Kokako), at least, was referred to as "Cinereous Wattle-Bird."[218] There are, however, four species of wattlebirds in the genus *Anthochaera* from the Meliphagidae—large, raucous honeyeaters endemic to Australia. Two of them, the Red Wattlebird and the Yellow Wattlebird, have pendulous wattles hanging from the ear coverts, with those respective colors. Despite their names, Little Wattlebird and Western Wattlebird both lack these adornments, the name stemming from their relationship with the aforementioned Red and Yellow Wattlebirds.

# Wattle-eye

There are eleven wattle-eyes in the Platysteiridae, all in the genus *Platysteira* and all found in Africa. These pretty little flycatcher-like insectivores, most representatives of which are black and white, all sport red, black, blue, or green wattles above or around the eye. In the earlier literature, the wattle-eyes were called Platyrynchus, Spectacle Flycatcher [*sic*], or Wattle-eyed Flycatcher.

# Waxbill

These small colorful birds are often referred to as finches, although strictly speaking, they are not. The estrildid family is now usually referred to as the waxbills. Within the family, twenty-one species have the common group name waxbill. Although it was probably a reference to a bird that is now known as Red-billed Firefinch (though still an estrildid), William Hayes wrote in *Portraits of rare and curious birds, with their descriptions* (1794) under the heading "The Wax-Bill," with common names given as Wax-Bill or Wax-Bill Grosbeak:[148]

> This bird derives its name from the colour of the bill being of a bright red, resembling sealing-wax.

Despite this, many of the birds now called waxbill do not have red bills.

# Waxwing

The remarkable waxwings are three quite similar-looking species in the Bombycillidae. Though it is far from their most obvious feature, it is their most unusual one—the keratinous red tips on the trailing edge of the wings where the shafts of some of the feathers extend beyond the barbs. These red appendages on the bird's secondary feathers are said to resemble sealing wax. From *The History of Birds* by the Rev. William Bingley from around 1800:[73]

> The name of Waxwing is given to it from the singular appendages to the secondary quill feathers, bearing much resemblance to a drop of red sealing-wax pressed on the wing.

According to Bingley, another old name for Bohemian Waxwing was the "Waxen Chatterer," the latter word a mistranslation by Ray and Willughby, who rendered Bohemian Chatterer from Gessner's *Garrulus bohemicus*, which was Bohemian Jay.[16] Though long thought to have evolved to protect the bird's feathers, it's now believed the red wax secretions function as plumage enhancements to signal age, maturity, and social status, a useful signal in a species that is often found in large flocks.

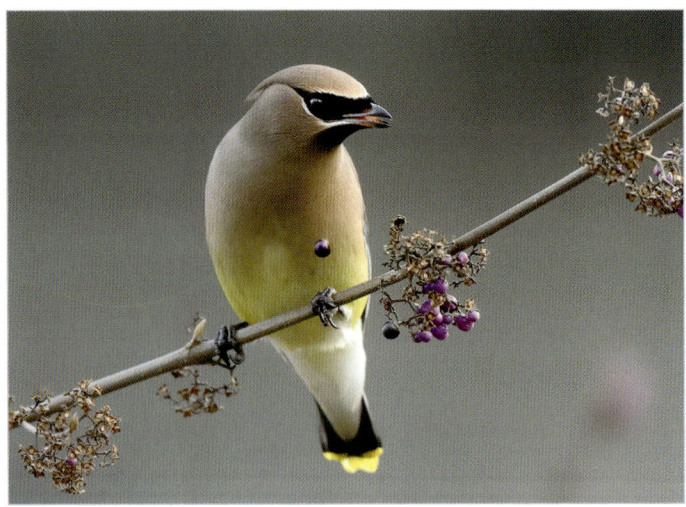

*Cedar Waxwing (Washington State, USA)*

# Weaver

These prolific finch-like birds are a large family, the Ploceidae, and at least eighty species are known by the common group name weaver, either independently or with a hyphenated name. The first parts of these compound names, such as buffalo-weaver, sparrow-weaver, social-weaver, golden-weaver, and masked-weaver, indicate in all cases the group's behavior or appearance. For example, buffalo-weavers are often associated with cattle, while the sociable-weavers build apartment-house-style nests, in which up to 300 pairs build and use separate flask-shaped chambers entered by tubes at the bottom. The birds obviously earned their names from their extraordinary nests intricately constructed of grass, reeds, or sticks. Wood described the nests of the "Mahali Weaver Bird" in 1898:[606]

> The nest of this bird is quite as remarkable [as the Social Weaver Bird] . . . In general shape and size it somewhat resembles the reed-covered bottles which are often to be seen in the windows of wine importers, being shaped somewhat like a flask . . . and being composed of a number of very thick grass stems laid longitudinally, and interwoven in a manner that can hardly be understood without an illustration.

# Wedgebill

Two species of Australian endemic wedgebills belong to the Psophodidae, which also includes the whipbirds. These birds didn't become familiar to Europeans until the 1800s, when Gould, being the first to describe the genus, dubbed one species Crested Wedge-bill in 1848 with the comment:[40]

> The sombre tints of the bird are very like the colour of the earth of the plains it inhabits; and when the nature of its food shall have been ascertained, its wedge-shaped bill will doubtless be found admirably adapted for procuring it.

# Weebill

Australia's smallest bird, the Weebill is a member of the Acanthizidae and is the sole member of the *Smicrornis* genus. The short, stubby bill accounts for the name, but up until the early 1900s, the bird was referred to as the Short-billed Tree-Tit. The earliest use of the present-day name seems to be in an article from a 1924 issue of *The Victorian Naturalist* entitled "An Excursion in South-West Queensland" by Dr. W. MacGillivray:[143]

> The Little Wee Bill, *S. brevirostris*, is searching the box leaves for smaller forms of insect, life, hovering in front of the branchlets or clinging to them when effecting a capture.

# Wheatear

There are twenty-four species of wheatears in the *Oenanthe* in the Muscicapidae and one that previously was classified as such but is now placed in *Mymecocichla*, thanks to molecular studies. The name is an unusual one, which has had many erroneous theories applied to it over the years, but the true story of its etymology is by far the most entertaining one. Smollett wrote in *Travels through France and Italy* in 1766:[467]

> The road to it is agreeable and romantic, lying through pleasant corn-fields, skirted by open downs, where there is a rabbit warren, and great plenty of the birds so much admired at Tunbridge under the name of wheat-ears. By the bye, this is a pleasant corruption of white-a__se, the translation of their French name cal blanc, taken from their colour; for they are actually white towards the tail.

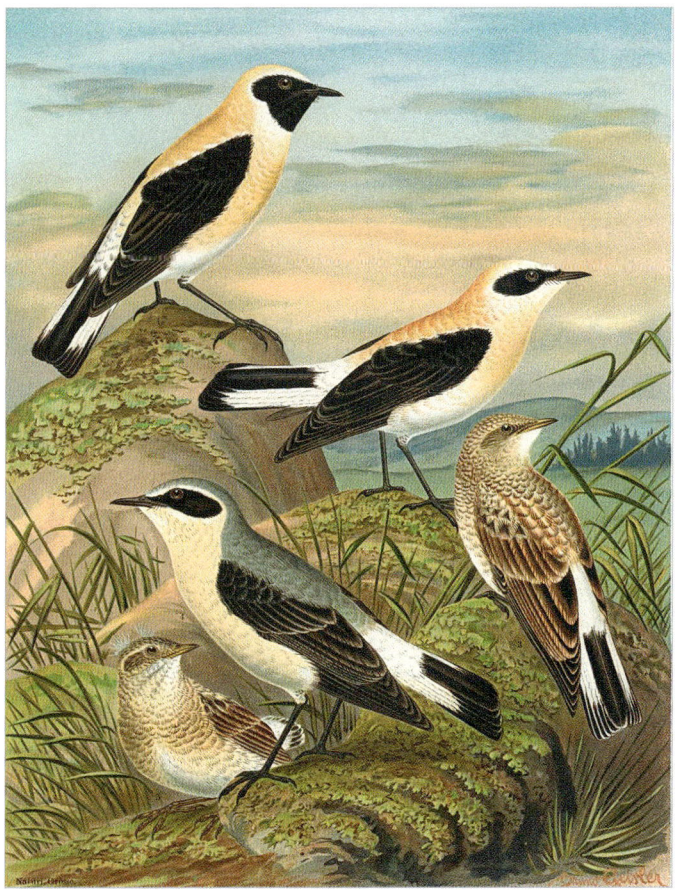

*Western and Eastern Black-eared Wheatears and Northern Wheatear in Naumann's* Naturgeschichte der Vögel Mitteleuropas *(1905)*

The "a__se" entry may be some early form of censorship for the word *arse*, the word arising from the Old English *ærs*, meaning "tail, rump," only later taking on somewhat ruder connotations. And Smythe Palmer wrote in 1904's *The Folk and Their Word-Lore*:[560]

> The word is really a popular corruption of an older form *whit-ers*, compounded of *whit*, *hwit*, white, and *ers* or *ears*, the tail or rump; this distinctive feature giving it also the names of *white-tail*, *white-rump*, French *blanculet* and *cul blanc*, the *whittaile* (Cotgrave), and *saxicola leuc-ura* (i.e. white tail).

# Whimbrel

This curlew, a member of the Scolopacidae, is another single-name bird. The name is an old one that has probably been around since the 1500s. It was first mentioned by Ray and Willughby in 1678:[16]

> Mr. Johnson of Brignal, in his Papers communicated to us, describes this Bird by the name of a Whimbrel thus.

The name is imitative of the bird's far-carrying, whining call, derived from the dialect *whimp*, meaning "to whimper" with the diminutive suffix *-rel*. In 1884, Skeat was quoted in the *Transactions of the Philological Society* in an article entitled "Notes on English Etymology":[561]

> The bird that keeps on uttering a cry imitated by *whim*.

An older version of the name appears in *Proceedings of the Belfast Natural History and Philosophical Society* in an article by R. Lloyd Patterson, Esq. entitled "Some of the Wading Birds frequenting Belfast Lough":[147, 562]

> In the account book of Robert Bennett, bursar of the Priory of Durham, 1530 to 1534, are entered as purchased:—1 1 curlew 3d'; '3 curlews et 1 whympernel (that is whimbrel) 13d.' At that time a barnacle goose was also worth 3d."—(Folkard's Wild Fowler.)

# Whinchat

The Whinchat is a species migratory between Europe and Africa, in the same genus as the stonechats, *Saxicola*. (See Stonechat.) The name stems from its habitat preference, which is meadows, heathlands, swamps, and shrubland. The first part of the compound name comes from the Middle English *whynne*, an old name for gorse, a British and Western European spiny shrub, which is one of the predominant plants in the bird's favored habitat. This word came into English from the Old Norse *hvein*, meaning "gorse, furze or swampy land," ultimately from Proto-Germanic *hwin-*, meaning "swamp or moor." The name probably evolved from the earlier form "Whin Bush-chat" that was used by MacGillivray in his 1840 *Manual of British Ornithology* (he also called the Eurasian Linnet "Whin Linnet."[460] (See Chat.)

# Whipbird

The whipbirds are three species in the small Psophodidae of Australasia. Watkin Tench wrote about the bird around 1790 in *A Complete Account of the Settlement at Port Jackson*, using the name coach-whip, seemingly taking it upon himself to name the bird.[468] He clearly wasn't very familiar with the bird, as his size comparison with "tomtit" was way off the mark.

> To one of them, not bigger than a tomtit, we have given the name of coach-whip, from its note exactly resembling the smack of a whip.

In *Supplement II to the General Synopsis of Birds* (1801), John Latham wrote:[563]

Inhabits New South Wales; native name *Djou*. It has a long single note, not unlike the crack of a coachman's whip, hence called the Coach-whip Bird.

And then later, in 1821, John Latham expanded further on the bird in *A General History of Birds*, quite remarkably referring to it (incorrectly) as a honeyeater:[17]

COACH-WHIP HONEY-EATER.

*Muscicapa crepitans*, Ind. Orn. Sup. li.

Coach-whip Flycatcher, Gen. Syn. Sup. ii. 222.

SIZE of a Thrush. . . .

Inhabits New South Wales, called by the natives *Djou*; has a long, single note, not unlike the crack of a coachman's whip, hence called the Coach-whip Bird . . . I am unable to say of what form the tongue is, as I have only seen the drawings of the bird, but I suspect it from this circumstance to belong to the Honey-Eater Genus.

# Whip-poor-will

As with a number of New World members of the Caprimulgidae, the nightjars, the two species of whip-poor-wills are named for their distinctive vocalizations. Edwards wrote in 1743:[78]

It is called in Virginia, Whip-Poor-Will, from its Cry, which nearly resembles those Words.

He went on to relate correspondence from Catesby, the English naturalist and author of *Natural History of Carolina, Florida and the Bahama Islands*:

The *Indians* imagine these Birds are the Souls of their Ancestors formerly slaughtered by the *English*, and say, that they never appeared in their Country before that Slaughter. Many People here look on them as Birds of Ill-omen.

# Whistler

The Pachycephalidae, whistlers and allies, is essentially an Australasian family, although some species sneak over into the Asian region. There are forty-eight species with the whistler appellation, but, in the English literature, they started life with a very different name—the thick-heads. This is a direct rendering of the family name, from the Greek παχύς *pachys*, "thick,"

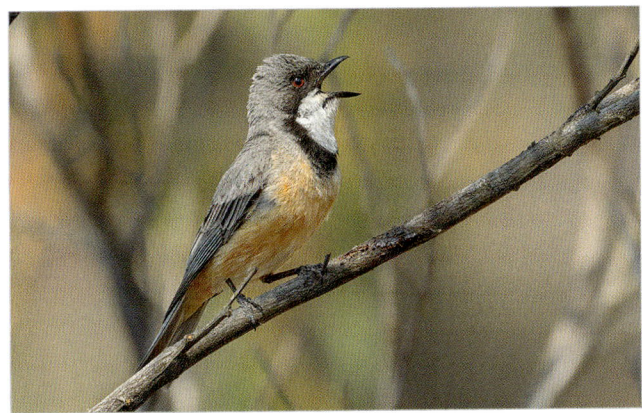

*Rufous Whistler (Victoria, Australia)*

and κεφαλή *kephali*, "head." Of course, this hardly does justice to these beautiful songsters, and, at some stage in the early 1900s, the present-day name gained favor. In the important book *An Australian Bird Book; a pocket book for field use* (1911), John Leach wrote:[144]

> Those badly-named, but often attractive, songsters—the Thick-heads (now called Whistlers)—are placed next. Eighty-eight of these birds are known from the Australian region, though but twenty occur in Australia itself and Tasmania. On account of the difficulty of skinning these birds, they were given the name *Pachycephala*. It is unfortunate that the literal translation—thick head—was the name used by bird people for these beautiful singers. It is now proposed to change the name to Whistler. Strange to say, we have not heard a good local name for these attractive and often gorgeous birds.

# Whistling-Duck

The eight species of whistling-ducks in the genus *Dendrocygna* are members of the Anatidae, although they are often treated in a separate family, the Dendrocygnidae. Despite the common group name, strictly speaking, they are not true ducks. Eyton wrote in *A Monograph on the Anatidae, or Duck Tribe* in 1838:[349]

> The genus appears to form a beautiful connecting link between the true fresh-water ducks and the foregoing genus [*Tadorna*].

In 1847 Jerdon explained the common name:[284]

> The *Dendrocygnæ* are called Whistling Teal by sportsmen in India from the sibilous cry they have, and their Hindustani name of "*Sillee*" also signifies whistler.

# Whistling-Thrush

The *Myophonus* whistling-thrushes are nine species in the Muscicapidae, all found in the Asian region. Like many groups of birds that make their homes near noisy streams and rivers, they have loud, clear, and high-pitched whistle calls, hence the common name. Gould wrote about the bird in *Birds of Asia* (1850–1883):[33]

> It is very impatient of observation, and when intruded on gives utterance to a peculiarly long-drawn plaintive but loud whistling note . . . The natural note of this Thrush, when undisturbed, is very beautiful, and so closely resembles a soft human whistle as to deceive any but practised ears.

# White-eye

The very large Zosteropidae, the white-eyes, yuhinas, and allies, at last count included 114 species of white-eyes. In the early 1800s, publications such as Swainson's *On the Natural History and Classification of Birds* (1836) were using the name "white-eyed warblers."[64] Later, the common name Zosterops, synonymous with the genus name *Zosterops*, came into vogue. Gould used it for three species, Grey-backed Zosterops, Green-backed Zosterops, and Yellow Zosterops in his *Birds of Australia* in 1848.[40] In *A History of the Birds of Ceylon* in 1880, Legge headed his entries for *Zosterops* "white-eye," with alternative names such as "Bush-creepers," "White-eyed Warbler, Latham" and "White-eyed Tit, Jerdon."[564] And by 1896, Alfred Newton made the comment in *Dictionary of Birds*:[7]

> The allusion is to the ring of white feathers round the eyes, which is very conspicuous in many species, and hence by most English-speaking people in various parts of the world the prevalent Zosterops is commonly called "White-eye" or "Silver-eye."

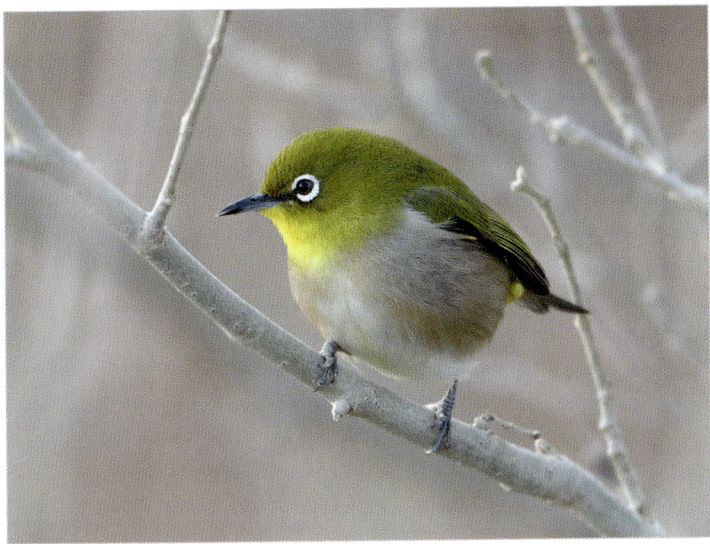

*Warbling White-eye (Japan)*

# Whiteface

These three species of Australian endemics belong to the genus *Aphelocephala* in the Acanthizidae. Not surprisingly, all three have white faces, but they began their journey through English ornithological literature as White-faced Xerophila in Gould's *The Birds of Australia* in 1848 after he found the bird we now know as Southern Whiteface on the streets of Adelaide.[40] At one stage it was also known as White-faced Titmouse, presumably for its perceived similarity to the members of the Paridae (tits, chickadees, and titmice). The first use of the name whiteface seems to be in *The Useful Birds of Southern Australia* by Robert Hall in 1907.[565]

# Whitehead

This single species of small songbird is one of three species in the Mohouidae, the whiteheads, all with one-word names and all endemic to New Zealand. It replaces its very similar congener, the Yellowhead, on the North Island. In earlier literature it was called the "White-head," and Walter Lawry Buller (1888) does refer to its Māori name, Popokatea. He wrote in the *Transactions and Proceedings of the New Zealand Institute* in 1868:[43]

> It would unquestionably be wrong to separate, generically, the two species of Popokatea, *Mohoua albicilla* and *M. ochrocephala*, in the manner proposed, for they are closely allied. In form they resemble each other although their plumage is different, and their habits are precisely the same. They are representatives of each other in the North and South Islands respectively.

# Whitethroat

The Lesser Whitethroat *Sylvia curruca* and Greater Whitethroat *Sylvia communis* are members of the Sylviidae, the sylviid warblers, parrotbills, and allies. Both have white throats, and, especially when signing, the puffed-out plumage is noticeable. Ray and Willighby discussed the bird, which they call "White-throat," in *The Ornithology of Francis Willughby of Middleton in the county of Warwick, esq.* in 1678.[16]

# Whitetip

The two species of hummingbirds in the *Urosticte* genus get their names from the unusual large white spot on the bottom half of the male's central tail feathers. As Gould wrote under the heading White-tip:[6]

> This beautiful species, one of the late discoveries in this lovely tribe of birds, differs in so many particulars from every other member of the family, that I have been constrained to give it a new generic title, and have selected that of *Urosticte* as indicative of the conspicuous white terminations of the four central tail-feathers; in nearly every other instance it is the outer feathers that are thus marked, and not the central ones, and it is the circumstance of the latter being thus decorated in the present bird which renders it so remarkable.

# Whydah

The whydahs, including the longer-tailed paradise-whydahs, are nine species of *Vidua* in the Viduidae, the Indigobirds. In *Cassell's Book of Birds* in 1869, the author discussed the group:[106]

> The Whydah or Widow Birds (*Vidua*) form the group to which we shall next allude, as being most nearly allied to the family of the Weavers. Whether the members of this family have had the name of Widow bird assigned to them by reason of the blackness of their plumage, is a question we shall not attempt to decide; some naturalists affirm that the word Widow is merely a corruption of Whydah, the name of the place from which they were first obtained by the Portuguese.

The first bird to be described in the genus was Yellow-mantled Widowbird *Euplectes macroura* (which is now placed in a different family; see Widowbird). There are theories that both the common and scientific names come from the Latin word *vidua*, meaning "widow." Another theory is that it may also derive from the Portuguese word *viúva*, meaning "widow," a reference to the male's black plumage and long tail. But the more widely accepted view is that it is a corruption of the place name *Whydah*, now *Ouidah*, the slave-trading port in Benin from where the birds were exported to Europe many centuries ago. Finally, it may be a case of a perfect storm in which all of these factors combined to form the name.

# Widowbird

The eight species of widowbirds in the Ploceidae were previously placed in the Viduidae with the whydahs. The two common names share the same etymology with the same root words, be they from the place name Ouidah (Whydah); *vidua*, Latin for "widow"; or *viúva*, Portuguese for "widow."

# Wigeon

The etymology of the name of these three species of Anatidae ducks is a little bit of a mystery. It's an old name in English, albeit with a very different spelling, appearing in recipes for dinner courses from Michaelmas to Christmas in *The Boke of Nurture* (1460) by John Russell:[252]

> In the second course, potage, mortrus, or conyes, or sewe / than roste flesshe, motton, porke, vele, pullettes, chekyns, pygyons, teeles, wegyons [wigeons], mallardes, partryche, woodcoke, plouer, bytture, curlewe, heronsewe / venyson roost, grete byrdes, snytes, feldefayres, thrusshes, fruyters, chewettes, befe with sauce gelopere, roost with sauce pegyll, & other ba\*ke metes as is aforesayde.

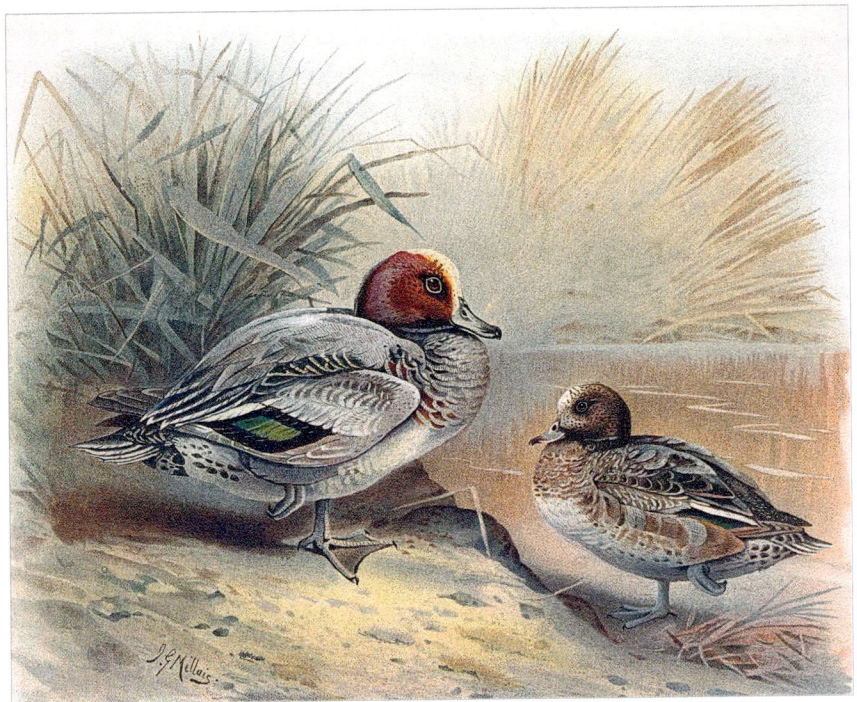

*Eurasian Wigeon in Millais's* The natural history of the British surface-feeding ducks *(1902)*

There's an early reference to "widgeons" by Ogilvy in *America* from 1671:[566]

> In Winter there are great plenty of Swans, Cranes, Geese, Herons, Duck, Teal, Widgeons, Brants, and Pidgeons, with other sorts, whereof there are none in England.

Some believe the name came from the French name for the bird *vigeon, vingeon*, which came from the Old French *vignier*, "to whine or shout" + *-on* (a noun suffix). But the origin of the English word pre-dates the French one, leading others to believe wigeon is merely imitative of pigeon. In what way pigeons bring to mind wigeons is unclear . . . except in the context of food. Could it be that the gourmands of the 1400s are responsible for this unusual name?

# Willet

This American shorebird in the genus *Tringa* bears a unique name given to it by colonists in the United States for its far-carrying vocalizations. Bingley wrote in 1800:[73]

> About the middle of March, however, their lively vociferations of *pill-will-willet, pill-will-willet*, begin commonly to be heard in all the marshes of the sea islands of Georgia and South Carolina.

# Wiretail

Des Murs's Wiretail is the only wiretail and is a member of the Furnariidae found in Chile and Argentina. The name is a reference to the bird's unusual, long, thread-like tail composed of only six filamentous feathers. An older name, from around the 1900s, was Des Murs's Spine-tail.

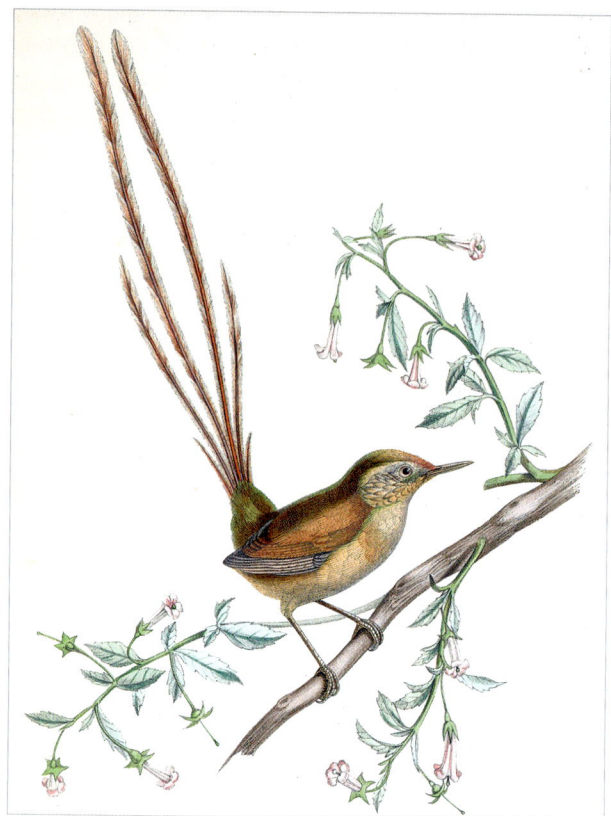

*Des Murs's Wiretail in Athanase's* Iconographie ornithologique *(1849)*

# Woodcock

There are eight species of woodcocks in the genus *Solopax*. The original woodcock, in terms of the English name, was the Eurasian Woodcock, which is widespread in Europe and beyond to East Asia. It featured early on in English literature, not so much for its ornithological merits as for its culinary appeal. It occurs numerous times in *The Boke of Keruynge*, a book of recipes from the 1400s:[483]

> Thye that woodcocke.

> Take a woodcocke, & reyse his legges and his wynges as an henne; this done, dyght the brayne. And here begynneth the feest from Pentecost vnto mydsomer.

The name is a compound from the Old English *wuducoc*, from *wudu*, "wood," and *coc*, "cock." Gurney wrote in *Early Annals of Ornithology* in 1921:[324]

> No bird is oftener alluded to than the Woodcock, whose merit for the spit was well known. This is one of the very few British birds which has not been provided with a string of provincial names. It is par excellence the bird of the woods, and has been so looked upon ever since the Saxons named it *wudecoec* or *wudu-coc*—an appellation which, or its equivalent, is given it in many countries. Especially numerous are the entries in 1548, in which year Mr. le Strange finds that sixty-eight were brought to the house, of which fifty-six were between October 20th and November 1st; probably most of them were caught with horsehair nooses.

# Woodcreeper

This is a large group of furnariid ovenbirds and woodcreepers consisting of fifty-one species in fourteen genera with the woodcreeper moniker. Most of them are so-called for their foraging behavior. In general, these insectivores search for their invertebrate prey by hitching up tree trunks picking and probing for items from the bark. Despite the catch-all name, not all the woodcreepers forage in this manner; many also sally for insects or follow army-ant swarms.

# Woodhaunter

The Striped Woodhaunter is a member of the genus *Automolus*, in which the other eight members are called the foliage-gleaners. This species forages in epiphytes and clumps of dead leaves and is particularly secretive, which could explain the unique name.

# Woodhoopoe

The five species of woodhoopoes from Africa, now in the Phoeniculidae (woodhoopoes and scimitarbills) were previously treated in the same family with the hoopoes, the Upupidae, due to their similar morphology. The name is a compound word referencing this relationship with the hoopoes. The first part of the name is a nod to their preference for woodlands and more arboreal habits compared to the true hoopoes.

# Woodnymph

There are five species of woodnymph hummingbirds in the genus *Thalurania*, all quite similar in appearance. Gould seems to have been the first to use the common name, or "trivial name" as he calls them, in his *A Monograph of the Trochilidæ, or family of humming-birds* in 1861.[6] He used the names in the form "wood nymph" and "wood-nymph." In Greek mythology, a nymph is a spirit of nature in the form of a beautiful maiden inhabiting rivers, woods, and other locations. Presumably, the beauty of these hummingbirds brought to mind fanciful ideas of woodland fairies.

# Wood-Partridge

Three birds are in the genus *Dendrortyx* in the Odontophoridae. The genus name literally translates as "wood quail or partridge" from the Ancient Greek δένδρο *dendro*, "trees," and ὄρτυξ *ortux*, "quail." (See Partridge.)

# Woodpecker

The iconic Picidae, the woodpeckers, consists of 234 species, with 177 of them known by the name woodpecker. The derivation of the name is obvious, as any observer of the birds will soon divine—most spend much of their time probing, chiseling, or hammering the trunks and branches of trees in search of their invertebrate prey. *The Promptorium Parvulorum*, the first English-Latin dictionary, written around 1440 by Dominican friar Anglicus Galfridus, provides a very early mention of the name in the English literature with the simple entry:[261]

Woode hak, or *reyn* fowle: Picus.

Literally, a "wood hacker." There are some equally descriptive older names for woodpeckers. Gurney[324] lists old names including *woodspike* and *woodnawe* that can be found in William Harrison's *List of Birds* from 1577,[324] and *wodake* and the *woodhock* in *Nomina Avium fferorum* from the 1450s.[324] Under the heading of "Peionie" in 1633's *The herball: or, Generall historie of plantes*, John Gerard et al. put an end to some superstitions about the woodpecker:[279]

> Moreover, it is set downe by the said author, as also by Plinie and Theophrasins, that of necessitie it must be gathered in the night; for if any man shall plucke of the fruit in the day time, being seene of the Woodpecker, he is in danger to lose his eies; and if he cut the roote, it is a chaunce if his fundament fall not out. The like fabulous tale hath beene set foorth of Mandrake, the which I have partly touched in the same Chapter. But all these things be most vaine and frivolous: for the roote of Peionie, as also the Mandrake, may be remooved at any time of the yeere, day or hower whatsoever.

Lastly, in *An Adventure in Etymology*, Ernest Ingersoll opined:[9]

> Woodpecker, as a name, is self-evident. In the southern states the absurd rendering "peckerwood" is heard, and there, especially, the giant of the family is a "logcock." Carpintero is the good Spanish name for all these birds in California and southward.

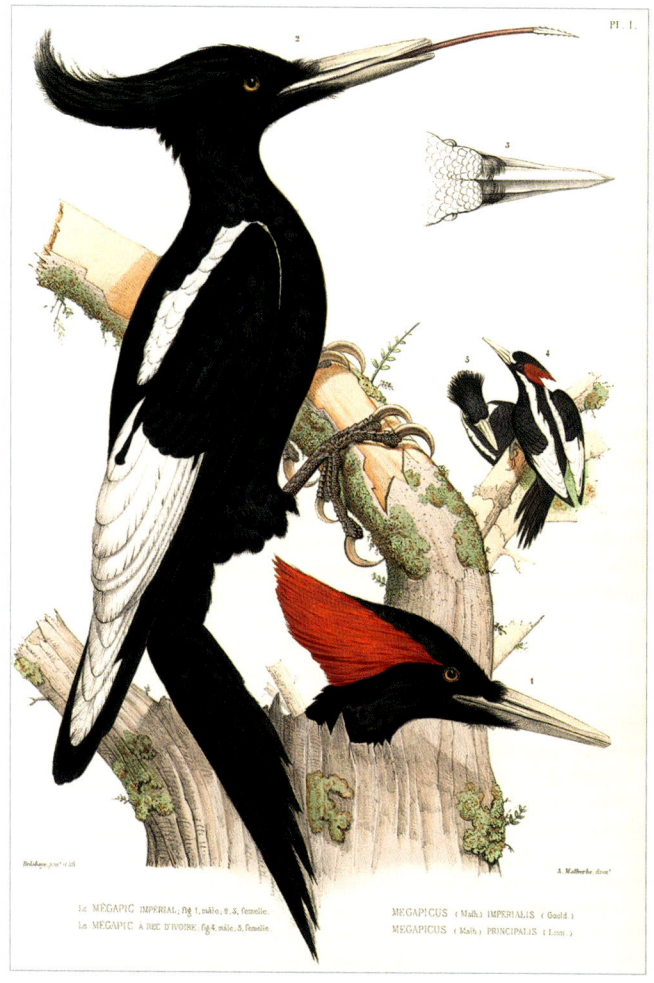

*Imperial and Ivory-billed Woodpeckers in Malherbe's* Monographie des picidées, ou Histoire naturelle des picidés, picumninés, yuncinés ou torcols *(1861)*

# Wood-Quail

These members of the New World quail family, fifteen species in the genus *Odontophorus*, take their name from their preference for dense forest habitats.

# Wood-Rail

The wood-rails are found in two families, the Sarothruridae, the flufftails, and the Rallidae, the rails, gallinules, and coots. The flufftails were, until recently, classified with the Rallidae, and the wood-rails of both families take their names from their preference for undisturbed rain forest.

# Woodshrike

The four species of *Tephrodornis* in the Vangidae go by the name woodshrike. Although the name was once used for a *Lanius* shrike, for example in *British Ornithology* by John Hunt in 1815, by the 1830s it had gained currency for the *Tephrodornis* of Asia.[567] (Early on they were also sometimes referred to as "bush-shrikes," a name now reserved for the African Malaconotidae). Both the birds' woodland habitat and its superficially shrike-like appearance account for the name.

# Woodstar

There are fifteen species of woodstar hummingbirds in eight different genera. When the woodstars were first described, a number were classified in the genus *Calliphlox* from the Ancient Greek *kalliphlox*, meaning "beautiful flame," which is presumably how these stars of the woods got their common names.

# Woodswallow

The Artamidae, the woodswallows, bellmagpies, and allies, contain eleven species of *Artamus* woodswallows. Though unrelated to swallows, these birds do have a superficial resemblance to that group. In *A Description of the Australian Birds*, Vigors and Horsfield (1826) alluded to the origin of the first part of the name when they quoted Caley:[169]

> This species has hence attained the name of Blue-bill among the colonists. It is also called Wood Swallow, as we find in Mr. Caley's* notes. That gentleman further adds:—" I have occasionally seen as many of these birds flying about in some places as I ever did Swallows, which they closely imitate in their mode of flight. This occurred where the ground had been cleared and abandoned. Their resting places were on the stumps of trees which had been felled. I do not think them migratory: if they are so, they depart for no great length of time.

*A collector appointed by Joseph Banks, George Caley arrived in Sydney in 1800 and returned to England in 1808.

In *A Handbook to the Birds of British Burmah* (1883), William Oates referred to the *Artamus fuscus* (Ashy Woodswallow) as "The Swallow-shrike."[112]

# Wood-wren

There are five species of *Henicorhina* called wood-wrens in the Troglodytidae, the wren family. The name was actually previously used for the European Wood Warbler *Phylloscopus sibilatrix*, now called Wood Warbler. Ridgway seems to have been the first to use the name to relate to the *Henicorhina* wren in *The Birds of North and Middle America* in 1904.[506]

# Wren

There are two families of wrens: the Acanthisittidae (New Zealand wrens), with three called wren (the fourth is the Rifleman), and the Troglodytidae (wrens), with eighty-six species bearing the name. The name, originally for the Eurasian Wren *Troglodytes troglodytes*, ironically the only Old World representative of the largely New World Troglyditidae, derives from the Middle English *wrenne* and earlier from the Old English *wrænna*. Whitman wrote in *The Birds of Old English Literature* in 1898:[42]

> Dial[ect] *wran*. ME. [Middle English] *wrenne, wranne*. The literal meaning is the "lascivious bird."

Apparently, the birds' polygynous habits were well known even then. Alternatively, Lockwood traced the name to the Old Norse *rindill*, writing:[84]

> The basic sense is (little) tail, a reference to the perky, cocked up tail, unique among our birds and thus calculated to inspire a name.

The word "wren" appears in the names of many other unrelated groups, including Australasian fairywrens, the Neotropical antwrens, and the Old World wren-babblers, as well as Wrentit and Wrenthrush. To further confuse matters, in the past in Europe, the kinglets were commonly known as "wrens," with the Common Firecrest and Goldcrest known as the "fire-crested wren" and "golden-crested wren," respectively. (See those entries.)

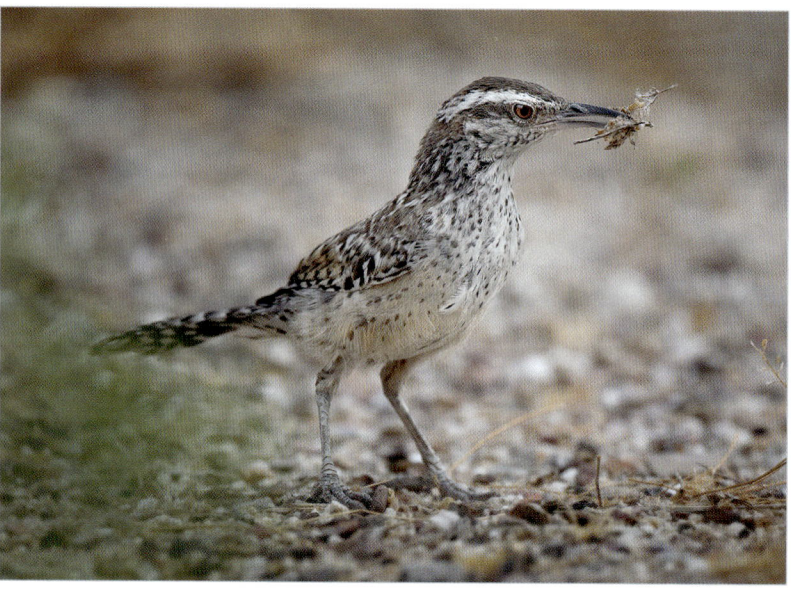

*Cactus Wren (Arizona, USA)*

# Wren-Babbler

The twenty-three species of wren-babblers are found in two families, the Timaliidae, tree-babblers, scimitar-babblers, and allies, with eight species, and the Pellorneidae, ground babblers and allies, with fifteen species. All of them are small, essentially ground-dwelling birds often with cocked tails, many of them with a distinctly wren-like appearance. Previously, all were contained in the babbler family, the Timaliidae, thanks to their "soft, fluffy plumage," and this accounts for the second part of the common generic name. (See Babbler.)

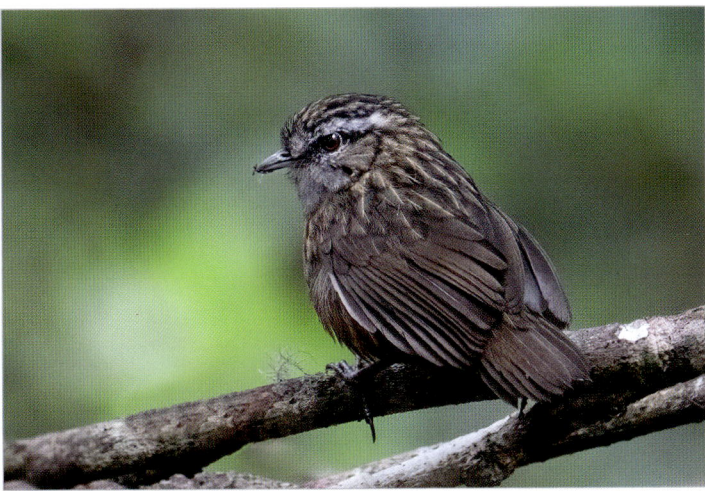

*Mountain Wren-Babbler (Sabah, Malaysia)*

# Wrenthrush

The Wrenthrush is a Central American member of a monotypic family, the Zeledoniidae. Ridgeway brought this species to the attention of Western science in 1888, and, initially, he tentatively placed it in the Turdidae, the thrushes. The name derives from this and the bird's somewhat wren-like appearance, with its short, cocked tail. Ridgeway coined the name "Wren-Thrush" in the 1907 issue of the *Bulletin of the United States National Museum*.[579] It's a shame that such an unusual little bird is known by this rather pedestrian name.

# Wrentit

This sole representative of the Old World Sylviidae, the sylviid warblers, parrotbills, and allies family in North America, is known by the compound of "wren" and "tit" for its diminutive size and wren-like cocked tail.

# Wren-Warbler

Four species of African cisticolids, the wren-warblers bear similarities to both the wrens and the warblers, birds that would have been far more familiar to the early European colonizers of southern Africa. The name was actually already in use in the 1600s for the familiar European bird now known as the Willow Warbler.[17] By the early 1900s, the name was being widely used for the African *Calamonastes* members of the Cisticolidae.

# Wrybill

This shorebird in the Charadriidae is endemic to New Zealand. The unique compound name of the words *wry*, meaning "distorted or contorted," and *bill* recognizes the bird's most outstanding feature, its long bill that curves, always to the right—the only bird in the world to possess such a bill. As Alfred Newton observed in *Dictionary of Birds* in 1896:[7]

> It has its English name from its bill being congenitally bent in the middle and diverted to the right side—a formation supposed to give the bird greater facility in seeking its food, chiefly arthropods that lurk under stones, round which it may be seen running from left to right . . . the wonderful nature of this asymmetry of the bill.

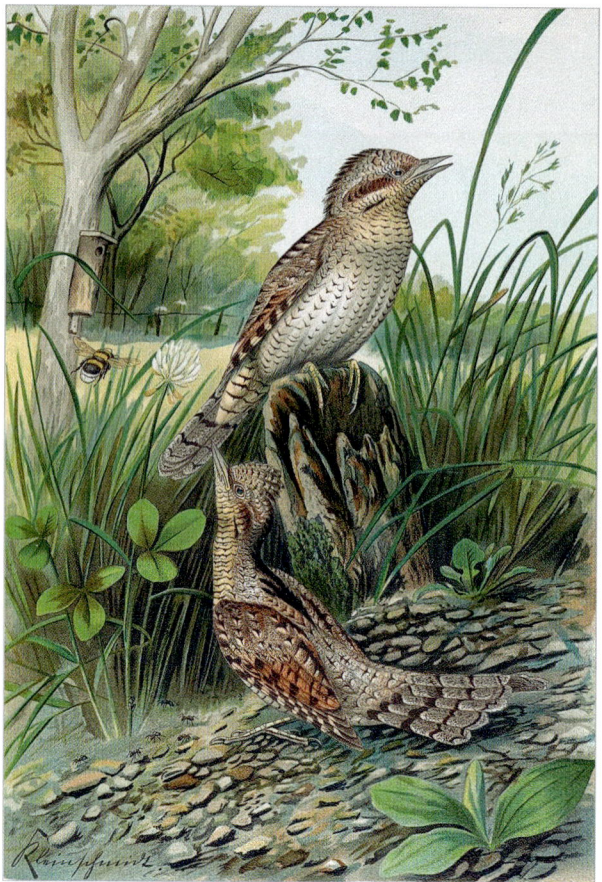

*Wryneck in Naumann's* Naturgeschichte der Vögel Mitteleuropas *(1905)*

# Wryneck

There are two species of *Jynx* wrynecks in the Picidae. The common name is derived from the bird's extraordinary manner of bending and twisting its neck in an almost snake-like manner when threatened. This name originated in the late 1500s, in the compound formation of the words *wry* and *neck*, the word wry meaning "distorted, contorted or twisted to one side." Curiously, this is also where the idea of putting a jinx on someone comes from. The Latin *iynx,* now *Jynx* for the genus name, came from the Ancient Greek name for the wryneck, *iunx.* The Ancient Romans and Greeks traced the bird's mythological origins to a sorceress named *Iynx* who was transformed into a wryneck as punishment for a spell she cast on Zeus. Since the time of the Ancient Greeks, the strange neck twisting along with associated hissing vocalizations was superstitiously believed to be associated with the use of spells and curses in witchcraft. In *Dictionary of Birds* from 1898, Newton wrote the bird is[7]

> so called from its wonderful way of writhing its head and neck, especially when captured . . . in some places it is called "Snake-bird," not only from the undulatory motions just mentioned, but from the violent hissing with which it seeks to repel an intruder from its hole . . . The peculiarity was known to Aristotle, and possibly led to the cruel use of the bird as a love-charm, to which several classical writers refer, as Pindar, Theocritus, and Xenophon.

# X IS FOR XENOPS

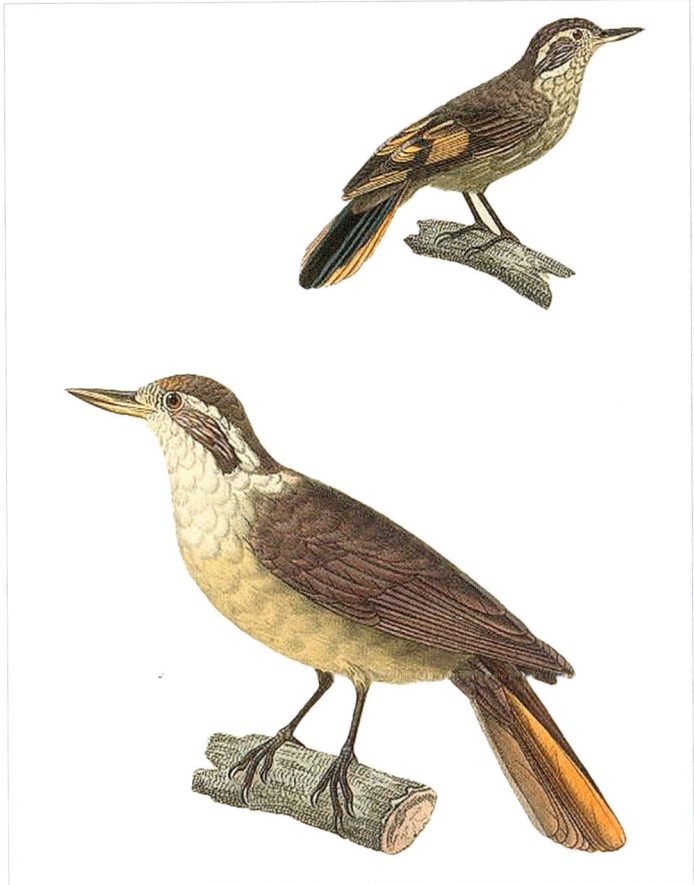

*Plain Xenops in Temminck's* Nouveau recueil de planches coloriées d'oiseaux *(1838)*

## Xenops

The xenops are five species of furnariids (ovenbirds and woodcreepers) in three genera. The name derives from the Ancient Greek ξένος *xénos*, "foreign, strange" and ὤψ *ops* "face, eye," a reference to the unusual shape of the bill in some of these nuthatch-like Neotropical birds.

## Xenopsaris

A single species in the synonymous genus *Xenopsaris* in Tityridae, the tityras and allies, the White-naped Xenopsaris is considered by many to be a most unusual bird due to its nomadism, breeding habits, and taxonomic uncertainties. The name is taken from the Ancient Greek ξένος *xenos*, for "strange," and the now-disused genus name *Psaris*, a synonym for the tityras (used by Georges Cuvier in 1817), based on the Ancient Greek ψάρ *psár*, for "starling."

# Y IS FOR YELLOWHAMMER

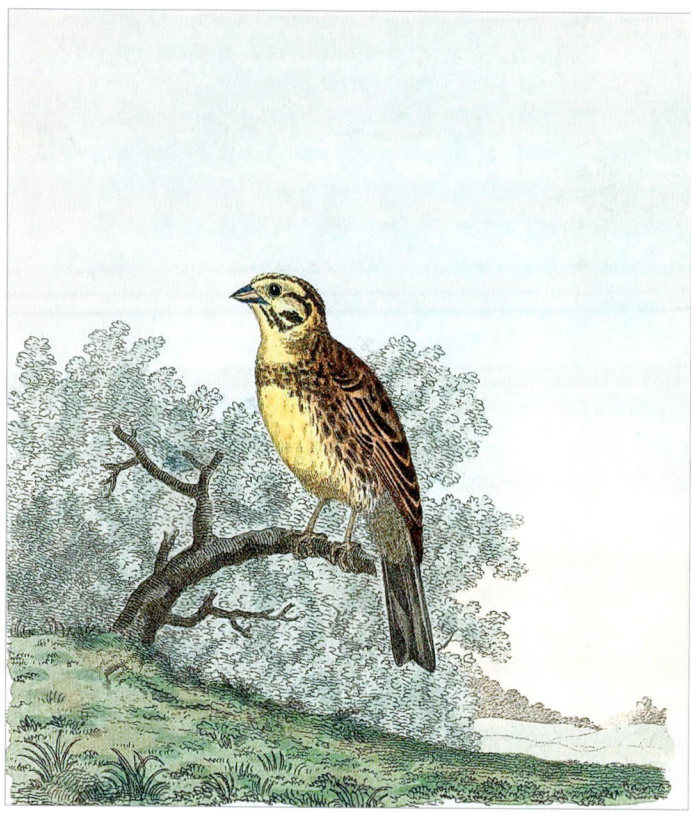

*Yellowhammer in Graves's* British Ornithology: being the history with a coloured representation of every known species of British birds *(1811)*

## Yellowhammer

This little bunting, which is in some ways unremarkable, does have a remarkable name. It is a single species with a single-word name in the Emberizidae, the Old World buntings. The name was probably first used in the mid-sixteenth century when it was called the *yelwambre* from the words *yelwe*, for "yellow," and *ambre*, which comes from the Old German *amaro*, the name for buntings. The modern German word for bunting is, in fact, *ammer*. In the early literature, *ammer* was used for a number of buntings and small birds, but now it persists only in the name of this species.

Interestingly, a colloquial name for the Northern Flicker in some parts of America is Yellowhammer. Ernest Ingersoll wrote, incorrectly, in *An Adventure in Etymology*:[9]

> [The name] would appear at first glance to be simply "yellow hammerer," but really it is a misplaced borrowing of the name given in England to any of several small birds more or less yellow in plumage, combined with the ancient Iclandic word *hamr* for skin—in other words a "yellowbird."

# Yellowhead

This single species of small songbird is one of three species in the Mohouidae, the whiteheads, all with one-word names and all endemic to New Zealand. The color of the bird is the key to the somewhat unimaginative name, although it's unclear why the head was singled out when most of the body is covered in yellow plumage. In the *Manual of the Birds of New Zealand* (1882), Walter Lawry Buller used alternative names "Native Canary. Yellow-head. Popokatea." in the entry for the species.[268] The Māori name *popokatea* was used in the literature until 1910, such as in *Transactions and Proceedings of the New Zealand Institute* in 1868:[43]

> The popokatea (*Mohoua albicilla*) is represented by another species (*M. ochrocephala*), differing in colour, but so closely allied to it that the natives apply the same name to both.

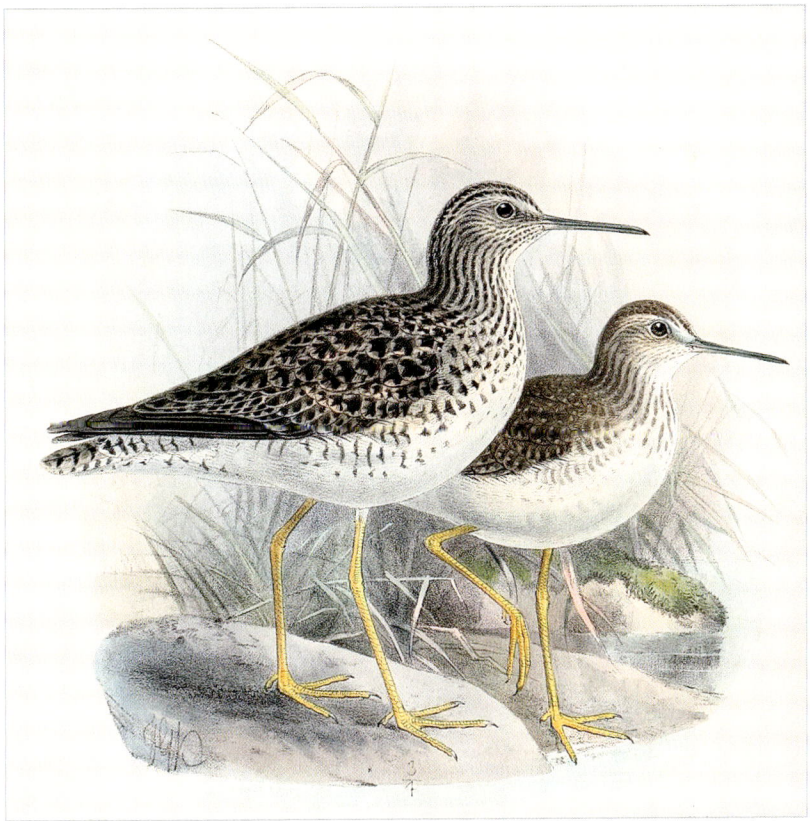

*Lesser Yellowlegs in Dresser's* A History of the Birds of Europe *(1895)*

# Yellowlegs

The genus *Tringa* contains three group names that reference leg color—the redshanks, the greenshanks, and two species of yellowlegs. The former two birds are Old World species with old names, hence the use of the somewhat anachronistic word for legs. Early references to Lesser Yellowlegs refer to the "Yellow Shank," while Shaw used the names "Yellow-shanked Sandpiper" and "Yellow-shank Snipe."[81] But the yellowlegs are New World birds with a relatively more recently coined name, and the term "shank" was dropped for "legs" soon after.

## Yellownape

The yellownapes are two species of superficially similar woodpeckers, and, although they were previously placed in the same genus, they are now recognized to be different enough to warrant placement in different genera. Both have a prominent brushlike yellow crest on the back of the head, around the nape, and this is how they got their names. The common name first given to the bird we know now as Lesser Yellownape was Yellow-crested Woodpecker, as listed in Jerdon's *Catalogue of the Birds of the Peninsula of India* in 1844.[275] Later, both Gould (in 1850's *Birds of Asia*[33]) and Gray (in 1847's *List of the osteological specimens in the collection of the British Museum*[568]) listed the same species as Yellow-naped Woodpecker. When writing about the Greater Yellownape, Gould, calling it the Yellow-naped Woodpecker, also noted:

> I first became acquainted with this noble species of Woodpecker in the year 1833, while engaged in collecting the materials for an intended Monograph of the entire group . . . I then assigned to it the specific name of *flavinucha*, as indicative of the yellow flowing feathers which adorn the occiput.

Even as late as 1985's *A Guide to the Birds of Nepal*, Inskipp and Inskipp used the main headings Lesser Yellow-naped Woodpecker (with alternative names Small Yellow-naped Woodpecker and Lesser Yellownape) and Greater Yellow-naped Woodpecker (with alternative names Large Yellow-naped Woodpecker and Greater Yellownape).[569] And to this day, Indian ornithologists favor the names Small Yellow-naped Woodpecker and Greater Yellow-naped Woodpecker.

## Yellowthroat

There are nine species of *Geothlypis* in the New World warblers family, the Parulidae. It's easy to discern the origins of their names, as all of them possess prominent yellow throats. The bird was first described by Linnaeus in 1766 from specimens collected in Maryland, thus they were called Maryland Yellow-throat in the early literature, such as in *Arctic Zoology* by Thomas Pennant in 1784.[19] By 1866, the bird was being referred to as Maryland Yellowthroat in publications such as *Review of American Birds in the Museum of the Smithsonian Institution* by Baird.[570]

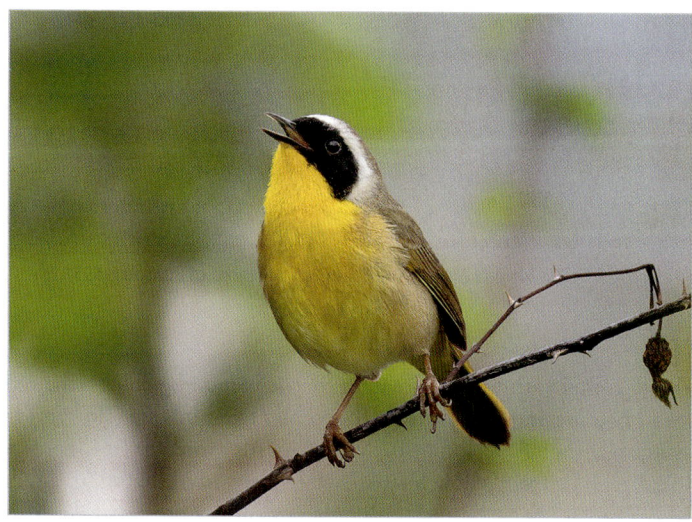

*Common Yellowthroat (Washington State, USA)*

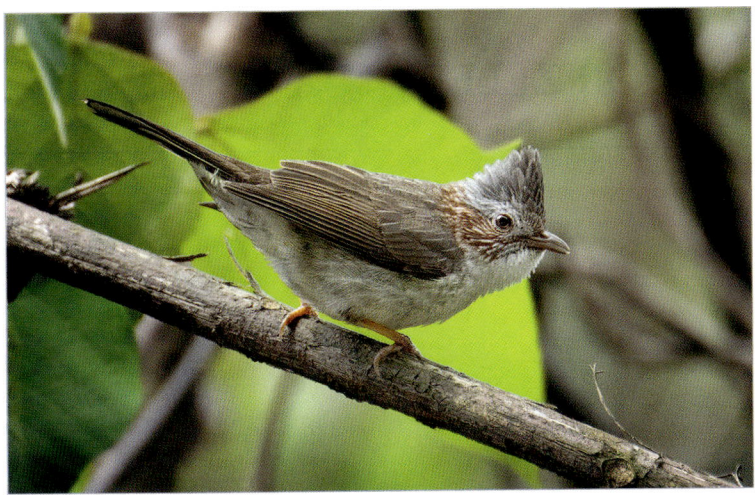

*Indochinese Yuhina (Vietnam)*

# Yuhina

The yuhinas are eleven species now placed in the Zosteropidae, the white-eyes, yuhinas, and allies. These tiny, crested songbirds can often be seen in large, mobile flocks in the canopies of the Asian montane forests. The name is a Nepali-language name युहिना *yuhinā*, assigned to the genus by the British naturalist Brian Houghton Hodgson in 1836. In that year he described three species of yuhinas, new to Western science, but used *yuhin* as the common name. In an 1837 issue of *The Journal of the Asiatic Society of Bengal*, Hodgson wrote about his "new species" with the comment "*Yuhin* of the Nipalese" [*sic*].[584] However, the present-day Nepalese name for the yuhinas is जुरेचरा *Jurichara*, and the Nepali pronunciation of the word could easily be rendered as "yuhina" by the unpracticed European ear.

Later, in 1843, a letter entitled "Series of Propositions for rendering the Nomenclature of Zoology uniform and permanent" appeared in *The Annals and Magazine of Natural History, including Zoology, Botany, And Geology* (1841) signed by twelve men, including one C. Darwin with the following (dare one say, amusing) passage:[571]

> *Names of harsh and inelegant pronunciation.*—These words are grating to the ear, either from inelegance of form, as *Huhua, Yuhina, Craxirex, Eschscholtzi*, or from too great length, as *chirostrongylostinus, Opetiorhynchus, brachypodioides, Thecodontosaurus*, not to mention the *Enaliolimnosaurus crocodilocephaloides* of a German naturalist. It is needless to enlarge on the advantage of consulting euphony in the construction of our language. As a general rule it may be recommended to avoid introducing words of more than five syllables.

# Z IS FOR ZELEDONIA

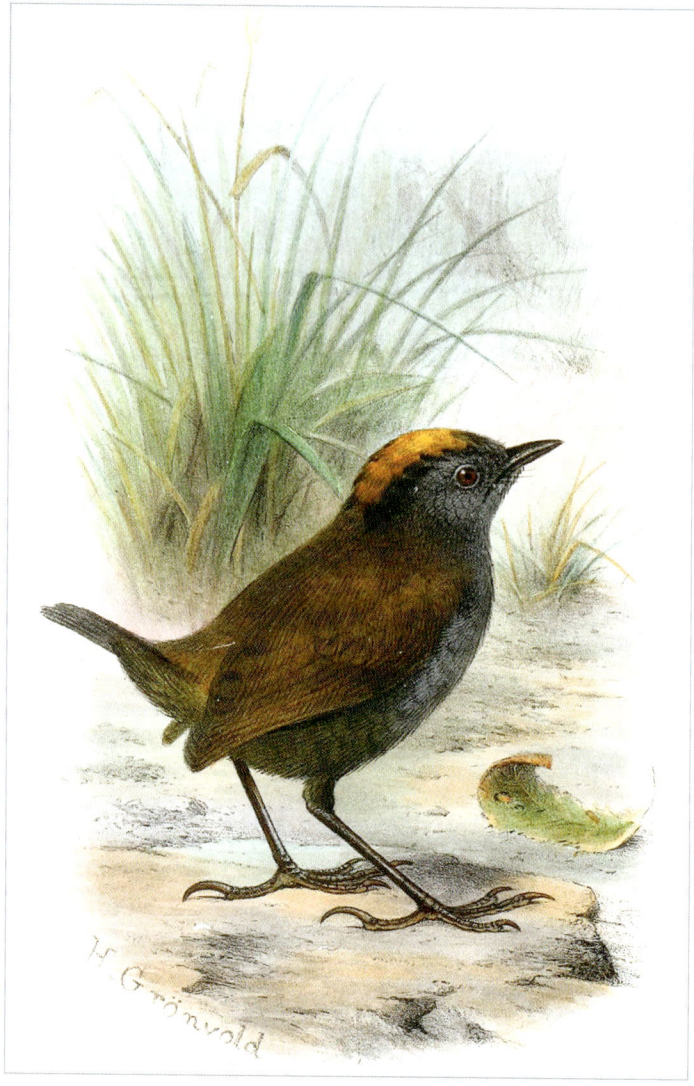

*Zeledonia in* The Ibis *Vol. 5 (1905)*

## Zeledonia

The Wrenthrush (see entry) is sometimes referred to colloquially by its genus name, *Zeledonia*. This is an example of an eponym, a name formed from a person's name. The genus name was established by Ridgway in 1888 in honor of the highly regarded Costa Rican ornithologist José Castulo Zeledón, who lived from 1846 to 1923. *Zeledonia* is also the name of the journal of the Asociación Ornitológica de Costa Rica (Ornithological Association of Costa Rica).

# Glossary

**Aphetism**: a word in which the initial vowel or syllable has been dropped.

**Australian Aboriginal Languages**: between 300 and 400 languages belonging to an estimated 28 language families, spoken by Aboriginal Australians of mainland Australia and a few nearby islands.

**Cognate**: a word having the same linguistic derivation as another; words from the same original word or root (for example, English *bittern* and French *butor* from the Latin *būtiō*).

**Congener**: a living organism belonging to the same genus as another.

**Convergent evolution**: an evolutionary process in which unrelated organisms independently evolve similar traits as a result of having to adapt to similar environments or ecological niches, for example, meadowlarks and longclaws.

**Etymon**: the antecedent form of a word—an earlier form of a word in the same language or an ancestral language (the Old English *fugel* is the etymon of Modern English *foul*); a word in a foreign language that is the source of a particular loanword (the Greek *ornithos* is the etymon of the English *ornithology*).

**Middle English** (ME): a form of English spoken after the Norman conquest (1066) until the late fifteenth century. During this period, writing conventions varied widely as the more standardized Old English language became fragmented and localized and was increasingly being improvised. Significant changes in many Old English grammatical features occurred, with simplification of many nouns, adjectives, and verbs. There was a widespread adoption of the Norman French vocabulary, especially in the areas of politics, law, science, the arts, and religion, while more conventional English vocabulary remained, primarily drawing on Germanic in its sources, with Old Norse influences becoming more apparent.

**Old English** or **Anglo-Saxon**: the earliest recorded form of the English language, spoken in England and southern and eastern Scotland in the early Middle Ages. It was brought to Great Britain by Anglo-Saxon settlers in the mid-fifth century, and the first Old English literary works date from the mid-seventh century.

**Philology/Philologist**: the study of the structure, historical development, and relationships of a language or languages. It involves the study of oral and written literary texts and oral records with elements of textual criticism, literary criticism, history, and linguistics, especially with reference to etymology. Philology involves the establishment of the authenticity and original forms of texts with a determination of their meaning.

**Polyphyletic**: a group of organisms derived from more than one common evolutionary ancestor or ancestral group and therefore not suitable for placing in the same taxon.

**Portmanteau**: a word blending the sounds and combining the meanings of two others, for example, *motel* (from "motor" and "hotel"), and *brunch* (from "breakfast" and "lunch") or, in ornithological terms, *flyrobin* (from "flycatcher" and "robin") and *goosander* (from "goose" and "gander").

**Proto-Germanic**: the reconstructed proto-language of the Germanic branch of the Indo-European languages, thought to have been spoken in southern Scandinavia and north-central Europe from the fifth century BC to fifth century AD. There are no existing texts of the language, which has been reconstructed using a technique called the *comparative* *method*, whereby a feature-by-feature comparison of two or more languages with common descent from a shared ancestor is made and from there extrapolated backward to infer the properties of that ancestor.

**Proto-Indo-European** (**PIE**): the theorized common ancestor of the Indo-European language family. Its proposed features have been derived by linguistic reconstruction from documented Indo-European languages. No direct record of Proto-Indo-European exists.

**Protonym**: the first legitimate name of a taxon, on which the currently accepted name is based.

**Species complex**: a group of closely related organisms that are so similar in appearance that the boundaries between them are often unclear.

**Rectrices**: collectively, a bird's tail feathers. The word *rectrix* (plural *rectrices*) is derived from Latin and means "helmsman."

**Rictal bristles**: bristles around the jaws and face of a bird, thus called because they are located at the *rictus*, the gape of a bird's mouth.

**Tupi-Guarani**: The Tupian language family comprises about seventy languages spoken in South America, of which the best known are Tupi proper and Guarani. The majority of these languages are now extinct, although Guarani is one of the official languages of Paraguay and is one of the most widely spoken American languages. Old or classical Tupi is an extinct Tupian language that was spoken by the aboriginal Tupi people of Brazil, most of whom inhabited the coastal regions in South and Southeast Brazil. In the early colonial period, Tupi was the *lingua franca* throughout Brazil used by Europeans as well as the aboriginal Americans. It was later suppressed almost to extinction, leaving only Nheengatu as the only Tupian language with any significant number of speakers.

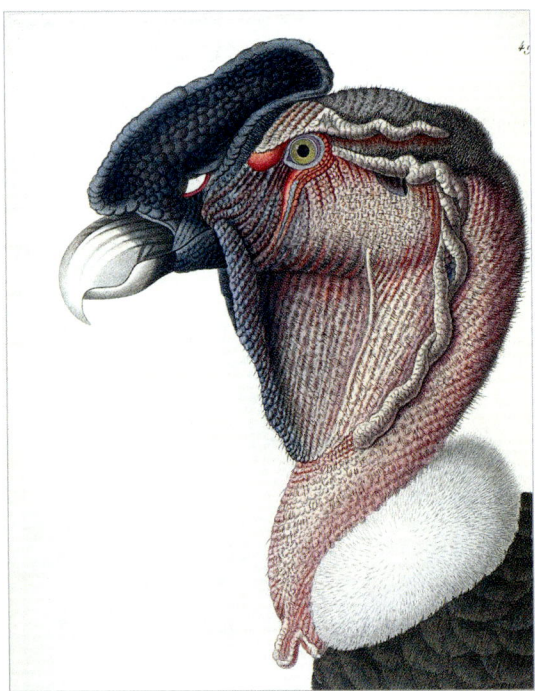

*Andean Condor in Temminck's* Nouveau recueil de planches coloriées d'oiseaux *Vol. I (1838)*

# Illustrations References

Adams, H. G. (1862) *Humming Birds, described and illustrated: with an introductory sketch of their structure, plumage, haunts, habits, etc.* Groombridge and Sons, London. (*Sparkling-tailed Hummingbird, Booted Racket-tail*)

Beal, F. E. L. (1902-1903) *Birds as Conservators of the Forest.* State of New York Forest, Fish and Game Commission, Eighth and Ninth Reports. Albany, N.Y. (*Blue-headed Vireo* by Louis Agassiz Furtes)

Audubon, J. J. (1839-44) *The Birds of America. Vols. 1-7.* J.B. Chevalier, New York (*Common Poorwill, Least Storm-Petrel*)

Barbosa du Bocage, J. V. (1881) *Ornithologie d'Angola.* Ouvrage publié sous les auspices du Ministere de la Marine et des Colonies. Imprimerie Nationale, Lisbonne. (*Bocage's Akalat, Miombo Scrub-Robin* by J. G. Keulemans)

Blackburn, Jane. (1868) *Birds Drawn from Nature.* J. Maclehose, Glasgow. (*Goshawk, Kestrel*)

Bree, Charles Robert (1863) *A history of the birds of Europe, not observed in the British Isles* Groombridge and Sons, London. (*Sacred Ibis* by Benjamin Fawcett)

Brown, P. (1776) *New Illustrations of Zoology.* B. White, London. (*Sungrebe*)

Brown, Thomas (1835) *Illustrations of the American ornithology of Alexander Wilson and Charles Lucien Bonaparte: with the addition of numerous recently discovered species and representations of the whole sylvae of North America.* Frazer & Co., Edinburgh. (*Ocellated Turkey*)

Buller, Walter Lawry (1873) *A History of the Birds of New Zealand.* John van Voorst, London. (*Huia, Stitchbird* by J. G. Keulemans)

Catesby, M. (1729) *The natural history of Carolina, Florida, and the Bahama Islands, Vol. 1.* Printed by the author, London. (*Bald Eagle*)

Conty, H. A. de and Travies, E. (1864) *Types du règne animal. Buffon en estampes.* Paul Dupont, Paris. (*Guianan Cock-of-the-rock, Superb Lyrebird* by Edouard Traviés)

Darwin, C. and Gould, J. (1838) *The Zoology of the Voyage of H.M.S. Beagle ... during the years 1832–1836. Part III Birds.* Smith, Elder & Co., London. (*Bearded Tachuri* by Elizabeth Gould)

Des Murs, Marc Athanase Parfait (1849) *Iconographie Ornithologique: nouveau recueil général de planches peintes d'oiseaux.* Chez Friedrich Klincksieck, Paris. (*Northern Jacana, Des Murs's Wiretail* by Paul Louis Oudard)

Des Murs, O. (1886) *Musée ornithologique illustré: description des oiseaux d'Europe, de leurs oeufs et de leurs nids*, Vol. 2. J. Rothschild, Paris. (*Rock Ptarmigan*)

Descourtilz, J. T. (1854) *Ornithologie Brésilienne, ou, Histoire des oiseaux du Brésil: remarquables par leur plumage, leur chant ou leurs habitudes.* Éditeur, Thomas Reeves, Rio de Janeiro. (*Spot-backed Antshrike, Giant Antshrike, Black-necked Aracari, Curl-crested Aracari, Red-rumped Cacique, Yellow-rumped Cacique, Crested Oropendola, Blue-and-yellow Macaw, Chestnut-fronted Macaw*)

Dresser, H. E. (1895–1896) *Supplement to a History of the Birds of Europe: including all the species inhabiting the Western Palæarctic region, Vol. 9.* Published by the Author, London. (*Desert Finch, Lesser Yellowlegs* by J. G. Keulemans)

Dresser, H. E. (1905) *Descriptions of Three new Species of Birds obtained during the recent Expedition to Lhassa.* Proceedings of the Zoological Society of London, Vol. 1. (*Giant Babax* by H. Grömvold)

Edwards, G. (1743) *A Natural History of Uncommon Birds.* College of Physicians, London. (*Streamertail, Guinea Turaco*)

Elliot, D. G. (1873) *A Monograph of the Paradiseidae.* Printed for the subscibers, by the author, London (*Raggiana Bird-of-Paradise* by J. Wolf)

Fitzinger, Leopold Joseph (1864) *Bilder-atlas zur Wissenschaftlich-populären Naturgeschichte der Vögel in ihren sämmtlichen Hauptformen* K. K. Hof und Staatsdruckerei, Wien. (*Hamerkop, Resplendant Quetzal*)

Frédol, Alfred (1866) *Le Monde de la Mer* L. Hachette & Cie, Paris. (*Wandering Albatross*)

Frohawk F. W. (1890) *The Secretary on Additions to the Menagerie.* Proceedings of the Zoological Society of London. Part 1 containing papers read in January and February. (*Hypocolius*)

Fuertes, Louis Agassiz (1930) *Album of Abyssinian birds and mammals.* Field Museum of Natural History, Chicago. (*Lammergeier*)

Gentry, Thomas G. (1882) *Nests and eggs of birds of the United States.* J. A. Wagenseller, Philadelphia. (*California Quail* by Edwin L. Sheppard)

Gould, J. and Gould, E. (1831) *A Century of Birds from the Himalaya Mountains.* London. (*Bengal Florican, Crested Serpent-Eagle* by Elizabeth Gould)

Gould, J. (1840-1848) *Birds of Australia.* R. and J. E. Taylor, London. (*Australian Darter, Blue-winged Kookaburra, Australian Painted-snipe* by H. C. Richter)

Gould, J. (1849-1861) *Monograph of the Trochilidæ, Vols. 1-5.* Taylor and Francis, London. (*Fiery-tailed Awlbill, Grey-breasted Sabrewing, Marvelous Spatuletail, Fiery Topaz* by H. C. Richter)

Gould, J. (1850-1883) *Birds of Asia, Vols. 1-7.* Taylor and Francis, London. (*Great Argus, Spectacled Barwing, Oriental Bay Owl, Green Cochoa, Grandala, Fire-tailed Myzornis, Golden-mantled Racquet-tail, Gould's Shortwing, Grey-bellied Tesia* by W. M. Hart, H. C. Richter, & J. Wolf)

Gould, J. (1875-1888) *Birds of New Guinea and the Adjacent Papuan Islands.* (*Ribbon-tailed Drongo, Chestnut-backed Jewel-babbler, Greater Lophorina, Trumpet Manucode, Vogelkop Melidectes, Western Parotia, Brehm's Tiger-Parrot* by W. M. Hart)

Grandidier, A. (1876) *Histoire Physique, Naturelle, et Politique de Madagascar, Vol. 13.* L'Imprimerie Nationale, Paris. (*Giant Coua* by J. G. Keulemans)

Graves, George (1811) *British Ornithology: being the history with a coloured representation of every known species of British birds,* Vol. I. Stephen Couchman, London. (*Yellowhammer*)

Gray, George Robert (1844–1849) *The genera of birds: comprising their generic characters, a notice of the habits of each genus, and an extensive list of species referred to their several genera.* Longman, Brown, Green, and Longmans, London. (*Maleo* by D. W. Mitchell)

Guérin-Méneville, M. F. E. (1860) *Revue et Magasin de Zoologie. Series 2, Tome XII.* Bureau de la Revue et Magasin de Zoologie, Paris. (*Kagu*)

Haeckel, Ernst (1904) *Kunstformen der Natur.* Leipzig und Wien, Verlag des Bibliographischen Instituts. (*Hummingbirds*)

Heuglin, Theodor von (1871) *Ornithologie Nordost-Afrika's: der Nilquellen- und Küsten Gebiete des Rothen Meeres und des nördlichen Somal-Landes; Bd 1, Abt. 2.* T. Fischer, Cassel. (*Bare-faced Go-away-bird*)

Holub, E. and Pelzeln, A. von (1882) *Beiträge zur Ornithologie Südafrikas.* A. Hölder, Wien. (*Southern Fiscal*)

Horsbrugh, Boyd Robert (1912) *The Game-birds & Water-fowl of South Africa.* Witherby & Co., London. (*Speckled Pigeon* by C. G. Davies)

Hudson, W. H. (1920) *Birds of La Plata* M. Dent & Sons Ltd., London. (*Crested Gallito* by H. Gronvold)

Huth, G. L. (1768–1776) *Recueil de divers oiseaux étrangers et peu communs qui se trouvent dans les ouvrages de Messieurs Edwards et Catesby.* Chez les Héritiers de Seligmann, Nuremberg. (*Bluethroat, Knot, Northern Mockingbird* by J. M. Seligmann)

Jerdon, T. C. (1871) Supplementary Notes to ' The Birds of India.' *The Ibis,* Vol. 1, Third Series. (*Asian Palm-Swift* by J. G. Keulemans)

Kirby, W. F. and Schubert, G. H. von (1889) *Natural History of the Animal Kingdom for the Use of Young People.* E. & J.B. Young and Co., Brighton. (*Rook*)

Latham, J. (1781–85) *A General Synopsis of Birds.* Benj. White, London. (*Jabiru*)

Le Vaillant, F. (1799) *Histoire Naturelle des Oiseaux d'Afrique.* Chez J. J. Fuchs, Paris. (*Pale Chanting-Goshawk, Southern Tchagra,*)

Lilford, Thomas Littleton Powys, Baron (1885–1897) *Coloured figures of the birds of the British Islands.* R. H. Porter, London. (*Alpine Accentor, Eurasian Griffon, White-throated Needletail* by A. Thornburn, J. G. Keulemans)

Lydekker, Richard (1895) *The Royal Natural History, Vol. 4, section 8.* Frederick Warne & Co., London and New York. (*Somali Ostrich* by W. Kuhnert)

Malherbe, Alfred (1861) *Monographie des picidées, ou Histoire naturelle des picidés, picumninés, yuncinés ou torcols,* Vol. 3. Typ. de J. Verronnais, Metz. (*Imperial and Ivory-billed Woodpeckers*)

Meyer, H. L. (1849) *Coloured illustrations of British birds, and their eggs,* Vol. 6. G. W. Nickisson, London. (*Gadwall*)

Millais, John Guille (1902) *The natural history of the British surface-feeding ducks.* Longmans, Green and Co., London. (*Eurasian Wigeon* by A. Thornburn)

Naumann, Johann Andreas (1897-1905) *Naturgeschichte der Vögel Mitteleuropas.* Fr. Eugen Köhler, Gera-Untermhaus. (*Ruffs and a Reeve, White and Grey Wagtails, Western and Eastern Black-eared Wheatears, Northern Wheatear, Wryneck* by E. de Moues, O. Geisler)

Orbigny, A. D. (1846) *Voyage dans l'Amérique méridionale.* Chez P. Bertrand, Paris. (*Rufous and White-throated Cacholotes*)

Pennant, T. (1812) *British Zoology.,* Printed for Wilkie and Robinson; J. Nunn; White and Cochrane; Longman, Hurst, Rees, Orme, and Brown; Cadell and Davies; J. Harding; J. Booth; J. Richardson; J. Mawman; J. and A. Arch; R. Baldwin, London. (*Common Loon*)

Pollen, François P. L. (1868) *Recherches sur la faune de Madagascar et de ses dépendances, ptie. I.* J. K. Steenhoff, Leyde. (*Sickle-billed Vanga*)

Pycraft, W. P. (1905) "On the Systematic Position of Zeledonia coronata, with some Observations on the Position of the Turdidae " *The Ibis,* Vol. 5. (*Zeledonia* by H. Grönvold)

Robinson, Herbert C. (1911) "On Birds from the Northern Portion of the Malay Peninsula, including the Islands of Langkawi and Terutau; with Notes on other rare Malayan Species from the Southern Districts" *The Ibis,* Vol. 5, Ninth series. (*Giant Ibis* by H. Grönvold)

Rothschild, L. W. R. (1893) *The Avifauna of Laysan and the Neighbouring Islands.* R.H. Porter, London. (*Mamo* by J. G. Keulemans)

Rowley, George Dawson (1878) *Ornithological Miscellany, Vol. 3.* Trübner and Co., London. (*Black-breasted Boatbill* by J. Smit)

Schinz, H. R. and J. Brodtmann (1836) *Naturgeschichte und Abbildungen der Vögel: nach den neuesten Systemen bearbeitet.* Weidmann'sche Buchhandlung, Leipzig. (*Greater Rhea, Eurasian Spoonbill and Roseate Spoonbill* by K. J. Brodtmann)

Sclater, P. L. (1863) "Synopsis of the known Species of Dacnis" *The Ibis*, Vol. 5. (*Scarlet-thighed Dacnis* by J. Jennens)

Sclater, P. L. (1869) *Exotic Ornithology.* Bernard Quaritch, London. (*Chestnut-sided Shrike-Vireo* by J. Smit)

Sharpe, R. B. (1891) *M onograph of the Paradiseidae, or Birds of Paradise and Ptilonorhynchidae, or Bower-birds, Vol. I.* H. Sotheran & Co, London. (*Wallace's Standardwing* by W. Hart)

Studer, J. H. and Jasper, T. (1873) *The Birds of North America.* Jacob Studer, Columbus, Ohio. (*Beach Stone-Curlew* by T. Jasper)

Studer, Jacob Henry (1903) *The Birds of North America: one hundred and nineteen artistic colored plates representing the different species and varieties drawn and colored from nature: including a copious text giving a popular account of their habits and characteristics, based on observations made in the field by the most eminent writers on ornithology: prefaced by a systematic table and index to page, plate and figure.* Published under the auspices of the Natural Science Association of America, New York. (*Gyrfalcon, Peregrine Falcon* by T. Jasper)

Swainson, W. (1820) *Zoological Illustrations.* Printed by R. and A. Taylor for Baldwin, Cradock, and Joy; and W. Wood, London. (*Hooded Berryeater*)

Swainson, William (1841) *A Selection of the birds of Brazil and Mexico: the drawings.* H. G. Bohn, London. (*Black-eared Fairy*)

Temminck, C. J. (1811) *Les pigeons / par Madame Knip, née Pauline de Courcelles; le texte par C. J. Temminck.* Chez Mme. Knip, Paris. (*Luzon Bleeding-heart* by Pauline Knip)

Temminck, C. J. (1838) *Nouveau recueil de planches coloriées d'oiseaux* or *Tableau Methodique, Vols. 1-5.* Chez Legras Imbert et Comp, Strasbourgh. (*Black Baza, Bornean Bristlehead, Andean Condor, Spotted Crocias, Scaled Ground-Cuckoo, Black-casqued Hornbill, Painted Quail-thrush, White-necked Rockfowl, Ratchet-tailed Treepie, Amazonian Umbrellabird, Plain Xenops* by J. G. Prêtre, N. Huet)

Vieillot, L. P. (1825) *La Galerie des Oiseaux.* Constant-Chantpie, Paris. (*African Finfoot, Quail-plover, Sunbittern* by P. Oudart)

von Humboldt, A. and A. Bonpland (1833) *Recueil d'observations de zoologie et d'anatomie compar: faites dans l'ocn atlantique, dans l'intieur du nouveau continent et dans la mer du sud pendant les anns 1799, 1800, 1801, 1802 et 180.* J. Smith, Paris. (*Oilbird* by N. Huet)

Wilson, S. B. (1890) *Aves Hawaiienses: the birds of the Sandwich Islands.* R.H. Porter, London (*Akohekohe, Ula-ai-hawane* by F. W. Frohawk)

Wolf, Joseph (1861) *Zoological Sketches, Vol. 1.* Henry Graves, London. (*North Island Brown Kiwi*)

Wolf, Joseph (1862) *Transactions of the Zoological Society of London, Vol. 4.* Zoological Society of London by Academic Press, London. (*Shoebill*)

Wytsman, P. (1905–1914) *Genera Avium* V. Verteneuil & L. Desmet, Brussels. (*Nuthatches* by C. E. Hellmayr)

# Notes

1.  Online Etymology Dictionary https://www.etymonline.com.
2.  Pittie, Aasheesh (2004) https://www.researchgate.net/publication/316681811_A_dictionary_of_scientific_bird_names_originating_from_the_Indian_region.
3.  Winkler, D. W., S. M. Billerman, and I. J. Lovette (2020) *Typical Antbirds* (*Thamnophilidae*), version 1.0. In Birds of the World (S. M. Billerman, B. K. Keeney, P. G. Rodewald, and T. S. Schulenberg, Editors). Cornell Lab of Ornithology, Ithaca, NY, USA. https://doi.org/10.2173/bow.thamno3.01.
4.  Johnsgard, P. (1983) *Cranes of the World: Origins of Scientific and Vernacular Names of Cranes* University of Nebraska, Lincoln.
5.  *A Dictionary of South African English on Historical Principles* (1996) Oxford University Press, Oxford; Etimologiewoordeboek van Afrikaans (electronic edition, 2003, Stellenbosch).
6.  Gould, John (1849–1861) *A monograph of the Trochilidæ, or family of humming-birds* Taylor and Francis, London.
7.  Newton, Alfred (1896) *Dictionary of Birds* Adam and Charles Black, London.
8.  Buffon, Count de (1793) *The Natural History of Birds. From the French of the Count de Buffon. Illustrated with engravings; and a preface, notes, and additions, by the translator. In Nine Volumes.* Strahan & Cadell, London.
9.  Ingersoll, Ernest (1937) "An Adventure in Etymology" *The Scientific Monthly*, Vol. 45, No. 3, pp. 233–249.
10. British Birds https://archive.org/stream/britishbirds1011unse/britishbirds1011unse_djvu.txt.
11. Liberman, Anatoly (2012) *Oh, what lark!* https://blog.oup.com/2012/12/lark-word-origin-etymology/.
12. Winkler, D. W., S. M. Billerman, and I. J. Lovette (2020) Limpkin (*Aramidae*), version 1.0. In Birds of the World (S. M. Billerman, B. K. Keeney, P. G. Rodewald, and T. S. Schulenberg, Editors). Cornell Lab of Ornithology, Ithaca, NY, USA. https://doi.org/10.2173/bow.aramid1.01.
13. Thompson, D'Arcy Wentworth (1895) *A Glossary of Greek Birds* Clarendon Press, Oxford.
14. Jobling, James A. (2009) *The Helm Dictionary of Scientific Bird Names* Christopher Helm Publishers, London.
15. Hett, Charles Louis (1902) *A Glossary of Popular, Local and Old-fashioned Names of British Birds* Henry Sotheran & Co., London.
16. Ray, John and Francis Willughby (1678) *The Ornithology of Francis Willughby of Middleton in the county of Warwick, esq.* A.C. for John Martyn, London.
17. Latham, John (1821–1828) *A General History of Birds*, Vols. 1–10. Jacob and Johnson for the author, Winchester, England.
18. "The Story of 'Ravenous' and 'Ravishing'" https://www.merriam-webster.com/words-at-play/ravenous-and-ravishing-word-history.
19. Pennant, Thomas (1784–1785) *Arctic Zoology* Henry Hughs, London.
20. Morris, E. E. (1898) *Austral English: A Dictionary of Australasian Words, Phrases and Usages* Project Gutenberg eBook.
21. Baird, Spencer Fullerton (1858) *Birds* Beverly Tucker, Washington.
22. Sclater, Philip Lutley (1888–1889) *Argentine ornithology. A descriptive catalogue of the birds of the Argentine Republic*, Vols. 1 and 2. R. H. Porter, London.
23. Ballance, Alison (2001) "*Takahe: The bird that twice came back from the grave*" in Lee, W. G.; Jamieson, I. G., eds. *The Takahe. Fifty years of conservation management and research* University of Otago Press, Dunedin, New Zealand.
24. Allen, J. A. (1910) "Collation of Brisson's Genera of Birds with those of Linnæs" *Bulletin of the American Museum of Natural History* Vol. 28.
25. Chiaradia, Clóvis (2008) *Dicionário de Palavras Brasileiras de Origem Indígena* [Dictionary of Brazilian Words of Indigenous Origin] Limiar.

26. Evans, A. H. (1903) *Turner on Birds: a short and succinct history of the prinicpal birds noticed by Pliny and Aristotle. First Published by Doctor William Turner. 1544* University Press, Cambridge.

27. Skutch, Alexander F. (1969) "A Study of the Rufous-fronted Thornbird and Associated Birds Part I & II" *The Wilson Bulletin*, 81:5–43.

28. Wilson, Alexander (1808–13) *American Ornithology, Vols. 1–6.* Bradford and Inskeep, Philadelphia.

29. Winkler, D. W., S. M. Billerman, and I. J. Lovette (2020) Thrush-Tanager (*Rhodinocichlidae*), version 1.0. *Birds of the World* (S. M. Billerman, B. K. Keeney, P. G. Rodewald, and T. S. Schulenberg, Editors). Cornell Lab of Ornithology, Ithaca, NY, USA. https://doi.org/10.2173/bow.rhodin1.01.

30. Courtz, Hendrik (1956) *A Carib Grammar and Dictionary* Magoria Books, Toronto.

31. Bosworth, Joseph (1838) *A Dictionary of the Anglo-Saxon Language* University of Michigan Press, Ann Arbor.

32. "Standard English words which have a Scandinavian Etymology" http://www.viking.no/e/england/e-viking_words_2.htm.

33. Gould, John (1850–1883) *Birds of Asia, Vols. 1–7.* Taylor and Francis, London.

34. Sclater, Philip Lutley (1869) *Exotic ornithology: containing figures and descriptions of new or rare species of American birds* Bernard Quaritch, London.

35. Pennant, Thomas (1769) *Indian Zoology* publisher not identified, London.

36. Darwin, C. (1838) *The Zoology of the Voyage of H.M.S. Beagle . . . during the years 1832–1836* Smith, Elder & Co., London.

37. Rothschild, Walter (1893–1900) *The Avifauna of Laysan and the Neighbouring Islands: with a Complete History to Date of the Birds of the Hawaiian Possessions* R. H. Porter, London.

38. *Memoirs of the Bernice Pauahi Bishop Museum of Polynesian Ethnology and Natural History* (1899–1903) Museum Press, Honolulu.

39. Orbigny, Charles Dessalines d' (1816–1819) *Dictionnaire d'Histoire Naturelle.* Chez Deterville, Paris.

40. Gould, John (1840–1848) *The Birds of Australia, Vols. 1–7.* R. and J. E. Taylor, London.

41. Dampier, William (1729) *A Voyage to New Holland, &c. in the Year 1699* James and John Knapton, London.

42. Whitman, Charles Huntington (1898) "The Birds of Old English Literature" *The Journal of Germanic Philology*, Vol. 2, No. 2.

43. Buller, Walter (1868) "Essay on the Ornithology of New Zealand" *Transactions and Proceedings of the New Zealand Institute,* Vol. 1.

44. Williams, H. W. (1906) "Māori Bird Names" *Journal of the Polynesian Society*, Vol. 15, No. 4.

45. Brabourne, Wyndham (1912–1917) *The Birds of South America* R. H. Porter, London.

46. Horsfield, Thomas (1821) "Systematic Arrangement and Description of Birds from the Island of Java" *Transactions of the Linnean Society of London.*

47. Wilson, Scott B. (1890–1899) *Aves Hawaiienses: the birds of the Sandwich Islands* R.H. Porter, London.

48. Anonymous (February 1, 1930) "Maui's Bird" *The New Zealand Railways Magazine*, Vol. 4, Issue 10.

49. Hopkinson, E. (1909) "The Birds of Gambia" *Bird Notes: The Journal Of The Foreign Bird Club*, Vol. 8.

50. Swainson, William (1820–1823) *Zoological Illustrations* Baldwin, Cradock, and Joy; and W. Wood, London.

51. Bonaparte, Charles Lucien (1850) *Monographie des Loxiens* Leiden et Chez Arnz & Comp.

52. Bonaparte, Charles Lucien (1850) *Conspectus Generum Avium* E. J. Brill, Lugduni Batavorum, New York.

53. Cracraft, Joel, and Julie Feinstein (2000) "What is not a bird of paradise? Molecular and morphological evidence places Macgregohain the Meliphagidae and the Cnemophilinae near the base of the corvoid tree" *Proceedings of the Royal Society B: Biological Sciences*, Vol. 267, No. 1440, pp. 233–41.

54. Audubon, John James (1839) *The Birds of America, from drawings made in the United States and their territories* G. R. Lockwood, New York.

55. Homer *Odyssey* (5.66) http://www.perseus.tufts.edu/hopper/text?doc=Perseus:text:1999.01.0136.

56. Robinson, Tancred (1684–1685) "Some Observations on the French Macreuse and Scotch Bernacle; together with a continuation of the Account of Boyling, and other Fountains" *Philosophical Collections Royal Society of London*, Vol.15, No.167–178. Royal Society of London, London.

57. Eisenmann, E. and Poor, H. H. (1946) "Suggested Principles for Vernacular Nomenclature" *The Wilson Bulletin*, Vol. 58. Wilson Ornithological Club.

58. Jardine, William (1833–43) *The Naturalist's Library* W. H. Lizars, Edinburgh.

59. Robin, V., Vishnudas, C.K., Gupta, P. et al. "Two new genera of songbirds represent endemic radiations from the Shola Sky Islands of the Western Ghats, India" *BMC Evol Biol* 17, 31 (2017). https://doi.org/10.1186/s12862-017-0882-6.

60. Broinowski, G. J. (1890–91) *The Birds of Australia* C. Stuart & Co., Melbourne.

61. Finsch, O. (1873) "On *Lamprolia victoria*, a most remarkable new Passerine Bird from the Feejee Islands" *Proceedings of the Zoological Society of London* 733–735 Academic Press, London.

62. Buller, W. (1870) "On *Zosterops lateralis* in New Zealand, with an Account of its Migrations" *Transactions and Proceedings of the New Zealand Institute.*

63. Cabanis, J. L. and F. Heine (1850–1863) *Museum Heineanum: Verzeichniss der ornithologischen Sammlung des Oberamtmann Ferdinand Heine, auf Gut St. Burchard vor Halberstadt. T. 1–4.* R. Frantz, Halbertstadt.

64. Swainson, W. (1836–1837) *On the Natural History and Classification of Birds* John Taylor, London.

65. Richardson, J., W. S. Dallas, T. S. Cobbold, W. Baird, A. White, and J.B. Holder (1877) *The Museum of Natural History, being a popular account of the structure, habits, and classification of the various departments of the animal kingdom: quadrupeds, birds, reptiles, fishes, shells, and insects, including the insects destructive to agriculture* James S. Virtue, New York.

66. Pennant, T. (1781) *Genera of Birds* B. White, London.

67. Dawson, W. L. (1909) *The Birds of Washington; a complete, scientific and popular account of the 372 species of birds found in the state* The Occidental Publishing Co., Seattle.

68. Vieillot, L. P. (1807) *Histoire Naturelle des Oiseaux de l'Amérique Septentrionale: contenant un grand nombre d'espèces décrites ou figurées pour la première fois* Chez Desray, Paris.

69. Burnaby, A. (1904) *Burnaby's Travels through North America: reprinted from the third edition of 1798* A. Wessels Company, New York.

70. Catesby, M. (1729–1747) *The natural history of Carolina, Florida and the Bahama Islands: containing the figures of birds, beasts, fishes, serpents, insects, and plants: particularly, the forest-trees, shrubs, and other plants, not hitherto described, or very incorrectly figured by authors: together with their descriptions in English and French: to which, are added observations on the air, soil, and waters: with remarks upon agriculture, grain, pulse, roots, &c.: to the whole, is prefixed a new and correct map of the countries treated of* Printed at the expense of the author, London.

71. Andrew, L. (1527?) *The noble lyfe a[nd] natures of man of bestes, serpentys, fowles a[nd] fisshes [that] be moste knoweu [sic]* Emprented by me Ioh[a]n of Doesborowe, Antwerp.

72. Reichenbach, H. G. L. (1836) *Die Vollständigste Naturgeschichte der Vögel* Expedition der Vollständigsten Naturgeschichte, Dresden.

73. Bingley, W. (1800) *The History of Birds: their varieties and oddities, comprising graphic descriptions of nearly all known species of birds, with fishes and insects, the world over, and illustrating their varied habits, modes of life, and distinguishing peculiarities by means of delightful anecdotes and spirited engravings* Edgewood Publishing Co., Philadelphia.

74. Buller, Walter L. (1888) *A History of the Birds of New Zealand* Walter Buller, London.

75. Swainson, C. (1886) *The Folk Lore and Provincial Names of British Birds* The Folk Lore Society, London.

76. Brisson, M. J. (1760) *Ornithologie, ou, Méthode contenant la division des oiseaux en ordres, sections, genres, especes & leurs variétés: a laquelle on a joint une description exacte de chaque espece, avec les citations des auteurs qui en ont traité, les noms quils leur ont donnés, ceux que leur ont donnés les différentes nations, & les noms vulgaires* Ad Ripam Augustinorum, apud Cl. Joannem-Baptistam Bauche, bibliopolam, ad Insigne S. Genovesae, & S. Joannis in Deserto, Parisiis.

77. Leivur Janus Hansen and Povl Skårup (2016) *Henrik Høyer and Carolus Clusion on birds and sea beans in the Faroe Islands* Fróðskaparrit 63. bók 2016: 117–145.

78. Edwards, G. (1743–1751) *A Natural History of Uncommon Birds: and of some other rare and undescribed animals, quadrupedes, fishes, reptiles, insects, &c., exhibited in two hundred and ten copper-plates, from designs copied immediately from nature, and curiously coloured after life, with a full and accurate description of each figure, to which is added A brief and general idea of drawing and painting in water-colours; with instructions for etching on copper with aqua fortis; likewise some thoughts on the passage of birds; and additions to many subjects described in this work* Imprimé pour l'Autour [*sic*] au College-Royal des Médecin, Londres.

79. Burchell, W. J. (1822) *Travels in the Interior of Southern Africa* Longman, Hurst, Rees, Orme, and Brown, London.

80. Gould, John (1861) *An introduction to the Trochilidae: or family of humming-birds* Taylor and Francis, London.

81. Shaw, George (1800–1826) *General Zoology, or Systematic Natural History* G. Kearsley, London.

82. Figuier, Louis (1873) *Reptiles and Birds. A popular account of the various orders; with a description of the habits and economy of the most interesting* D. Appleton and Co., New York.

83. Brown, Peter (1776) *New Illustrations of Zoology* Imprimé pour B. White, à Londres.

84. Lockwood. W. B. (1984) *The Oxford Book of British Bird Names* Oxford University Press, Oxford.

85. Wood, J. G. (1859–63) *The Illustrated Natural History* Warne and Routledge, London.

86. Glenn, Ian (2018) "Shoot the Messager? How the Secretarybird *Sagittarius serpentarius* got its names (mostly wrong)" *Ostrich* Vol. 89, No. 3, pp. 287–290.

87. Sparrman A., and J.G.A. Forster (1785) *A voyage to the Cape of Good Hope, towards the Antarctic polar circle, and round the world*, Vol. 1. Translated by J.G.A. Forster. Robinson, London.

88. Fry, C. (1977) "Etymology of 'Secretary-Bird'" *The Ibis*, Vol. 119, No. 550.

89. Deutsche Ornithologische Gesellschaft (1859) "Catalogue of Birds collected on the rivers Camma and Ogobai, Western Africa, by Me. P. B. du Chaillu 1858, with noted and descriptions of new species by John Cassin" *Journal für Ornithologie*.

90. Marcgrave, G. (1648) *Historiæ rerum naturalium Brasili* Ioannes de Laet, Antwerp.

91. Nash, R. (1926) *The Conquest of Brazil* Harcourt Brace, New York.

92. David, Armand and E. Oustalet (1877) *Les Oiseaux de la Chine. Atlas.* https://en.calameo.com/read/00021549823f58f09afac

93. Le Vaillant, François (1799–1808) *Histoire Naturelle des Oiseaux d'Afrique* Chez J. J. Fuchs, Paris.

94. Temminck, C. J. (1838) *Nouveau recueil de planches coloriées d'oiseaux: pour servir de suite et de complément aux planches enluminées de Buffon* Chez Legras Imbert et Comp., Strasbourgh.

95. Pittie, Aasheesh (2004) "A dictionary of scientific bird names originating from the Indian region" *Buceros*, Vol. 9, No. 2.

96. Hodgson, B. H. (1836) "Additions to the Ornithology of Nepal" *The Journal of the Asiatic Society of Bengal*, Vol. 5, pp. 770–781.

97. Hodgson, B. H. (1836) "Meruline Group of Birds, with indication of their generic character" *The Journal of the Asiatic Society of Bengal*, Vol. 5, pp. 358–360.

98. Wilson, A., C. L. Bonaparte, and W. Jardine (1870) *American Ornithology; or, The natural history of the birds of the United States*, Vol. 3, pp. 100–107 Cassell, Petter & Galpin, London.

99. Beebe, William (1906) *The Bird; its form and function: American nature series: Group 2* Holt, New York.

100. Lockwood, W. B. (1981) *Etymological Observations on Brambling, Bunting, Fieldfare, Godwit, Wren* Bono Homini Donum.

101. Aldrovandi, Ulisse (1599–1637) *Ornithologiae hoc est de avibus historiae libri XII* (Bologna, 1599) 1637 edition. Apud Franciscum de Franciscis Senensem, MDXCIX-MDCXXXVII, Bononiae.

102. Singh, Vikram Jit (2020) "Wildbuzz: Mr Besra Sparrowhawk and her hubby" *Hindustan Times*, Chandigarh.

103. Gould, John (1865) *Handbook to the Birds of Australia* Published by the author, London.

104. "Borrowings from Australian Aboriginal Languages" Australian National University Dictionary Centre https://slll.cass.anu.edu.au/centres/andc/borrowings-australian-aboriginal-languages.

105. Royal Zoological Society of New South Wales *Proceedings of the Royal Zoological Society of New South Wales for the years 1968–1969* Mosman, New South Wales [etc.] The Society.

106. Brehm, Alfred Edmund, and T. R. Jones (1875?) *Cassell's Book of Birds, Vols. 1–4.* Cassell, Petter & Galpin, London.

107. Ferrarius, Johannes, and William Bavand (1559) *A vvoorke of Ioannes Ferrarius Montanus, touchynge the good orderynge of a common weale wherein aswell magistrates, as priuate persones, bee put in remembraunce of their dueties, not as the philosophers in their vaine tradicions haue deuised, but according to the godlie institutions and sounde doctrine of christianitie. Englished by william Bauande* Imprinted at London, by Ihon Kingston, for Ihon Wight, dewllyng in Poules Churchyarde, London.

108. Cooper, Thomas (1578) *Thesaurus linguæ Romanæ & Britannicæ tam accurate congestus, vt nihil penè in eo desyderari possit, quod vel Latinè complectatur amplissimus Stephani Thesaurus, vel Anglicè, toties aucta Eliotæ Bibliotheca: opera & industria Thomæ Cooperi Magdalenensis. . . . Accessit dictionarium historicum et poëticum propria vocabula virorum, mulierum, sectarum, populorum, vrbium, montium, & cæterorum locorum complectens, & in his iucundissimas & omnium cognitione dignissimas historias* Henry Denham, Impressum Londini.

109. Aristotle (4th C BC) *Historia Animalium [The History of Animals]* Translated by D'Arcy Wentworth Thompson http://classics.mit.edu//Aristotle/history_anim.html.

110. Phillip, Arthur (1789) *The Voyage of Governor Phillip to Botany Bay with an Account of the Establishment of the Colonies of Port Jackson and Norfolk Island.*

111. Hall, B. P., and R. E. Moreau (1962) "A Study of the Rare Birds of Africa" *Bulletin of the British Museum (Natural History) Zoology,* Vol. 8, No. 7, pp. 313–378. British Museum (Natural History), London.

112. Oates, Eugene William (1883) *A Handbook to the Birds of British Burmah, including those found in the adjoining state of Karennee* R. H. Porter, London.

113. Universidad de Chile (1896) "Catálogo de las Aves Chilenas" *Anales de la Universidad de Chile,* Vol. 93, No. 197. Santiago.

114. Whitney, William Dwight (1889–1891) *The Century Dictionary: an encyclopedic lexicon of the English language* Century Co., New York.

115. Harvie-Brown, J. A. (1879) *The Capercaillie in Scotland [with some account of the extension of its range since its restoration at Taymouth in 1837 and 1838]* D. Douglas, Edinburgh.

116. Skutch, Alexander (1954) "Life Histories of Central American Birds, families Fringillidae, Thraupidae, Icteridae Parulidae and Coerebidae" *Ornithological Society Pacific Coast Avifauna,* No 31.

117. Barrows, W. B. (1883) "Birds of the Lower Uruguay" *Bulletin of the Nuttall Ornithological Club,* Vol. 8. The Club, Cambridge, Massachusetts.

118. Boles, W. (1987) "Alias Emu" *Australian Natural History,* Vol. 22, pp. 215–216.

119. Turnerum (1544) *Avium praecipuarum, quarum apud Plinium et Aristotelem mentio est, brevis et succincta historia* Gymnicus.

120. John Kersey (1715) *Dictionarium Anglo-Britannicum; or, A general English dictionary. To which is added, a collection of words and phrases made use of in our ancient statutes &c* Oxford University.

121. Swann, K. H. (1913) *A Dictionary of English and Folk-Names of British Birds* Witherby & Co., London.

122. Merrett, C. (1666) *Pinax Rerum Britannicarum continens Vegetabilia, Animalia, et Fossil* Roycroft, London.

123. White, Gilbert (1795) *A Naturalist's Calendar, with Observations in Various Branches of Natural History* B. and J. White, London.

124. North, Alfred J. (1889) *Descriptive catalogue of the nests and eggs of birds found breeding in Australia and Tasmania* F. W. White, Sydney.

125. Jackson, S. W. M. (1909) "In the Barron River Valley, North Queensland" *Emu,* No. 8, pp. 233–283.

126. Dixon, R. M. W., Bruce Moore, W. S. Ramson, and Mandy Thomas (2006) *Australian Aboriginal Words in English, their Origin and Meaning* Oxford University Press, Oxford.

127. Hume, A. O., and C. H . T. Marshall (1880) *The Game Birds of India Burmah and Ceylon* Vol 2. Hume and Marshall, London.

128. Wolstenholme, Harry (1926) *Official Checklist of the Birds of Australia compiled by the Checklist Committee, Royal Australasian Ornithologists Union* H. J. Green, Govt. Printer, Melbourne, Victoria.

129. La Condamine, Charles-Marie de (1778) *Relation abrégée d'un voyage fait dans l'interieur de l'Amérique Méridionale, depuis la côte de la mer du Sud, jusqu'aux côtes du Brésil & de la Guyane,*

*en descendant la riviere des Amazones, par M. de La Condamine, de l'Académie des sciences: avec une carte du Maragnon, ou de la riviere des Amazones, levée par le même* Chez Jean-Edme Dufour & Philippe Roux, à Maestricht.

130. Fraser, Ian, and Jeannie Gray (2013) *Australian Bird Names: a complete guide* CSIRO Publishing, Australia.

131. *Catalogue of the Birds in the British Museum* (1874–98) British Museum (Natural History). Department of Zoology. [Birds]. London.

132. Wiradjuri and Associated Community of Wagga Wagga (2002) *Wiradjuri Heritage Study: For the Wagga Wagga Local Government Area of New South Wales* Wagga Wagga City Council NSW Heritage Office.

133. Garcia, R. (1913) "Nomes de Aves em Lingua Tupi" *Boletim do Museu Nacional*, Rio de Janeiro.

134. Sibree, J. (1889) "Madagascar Ornithology; Malagasy birds arranged according to the natural orders, with notes on their habits and habitats, and their connection with native folk-lore and superstition (Part 1)" *The Antananarivo Annual and Madagascar Magazine* John Haddon and Co., London.

135. Lewin, William (1795) *The Birds of Great Britain, systematically arranged, accurately engraved, and painted from nature; with descriptions, including the natural history of each bird* J. Johnson, London.

136. Jerdon, T. C. (1862–1864) *The Birds of India: being a natural history of all the birds known to inhabit continental India, with descriptions of the species, genera, families, tribes, and orders, and a brief notice of such families as are not found in India, making it a manual of ornithology specially adapted for India* Military Orphan Press, Calcutta.

137. Wright, T. (1857) *A Volume of Vocabularies, illustrating the condition and manners of our forefathers, as well as the history of the forms of elementary education and of the languages spoken in this island, from the tenth century to the fifteenth* Privately Printed.

138. Orbigny, Alcide Dessalines d' (1835–1847) *Voyage dans l'Amérique méridionale: (le Brésil, la république orientale de l'Uruguay, la République argentine, la Patagonie, la république du Chili, la république de Bolivia, la république du Pérou), exécuté pendant les années 1826, 1827, 1828, 1829, 1830, 1831, 1832, et 1833* Chez Pitois-Levrault, Paris.

139. Shelley, G. E. (1896–1912) *The Birds of Africa, comprising all the species which occur in the Ethiopian region* R. H. Porter, London.

140. DSAE Dictionary of South African English https://dsae.co.za.

141. Thanh-Lan Gluckman, and Nicholas I. Mundy (2013) "Cuckoos in raptors' clothing: barred plumage illuminates a fundamental principle of Batesian mimicry" *Animal Behavior* University of Cambridge.

142. Inskipp, Collar, and Pilgrim (2010) "Species-level and other changes suggested for Asian birds, 2009" *BirdingASIA* No. 14, pp. 59–67.

143. Field Naturalists Club of Victoria *The Victorian Naturalist* Field Naturalists Club of Victoria, Melbourne.

144. Leach, John Albert (1911) *An Australian Bird Book: a pocket book for field use* Whitcombe & Tombs, Melbourne.

145. Nehemiah Bartley (1892) *Opals and Agates Or, Scenes Under the Southern Cross and the Magelhans: Being Memoirs of Fifty Years of Australia and Polynesia: with Nine Illustrations* Gordon & Gotch, Brisbane.

146. Browne, Patrick (1756) *The Civil and Natural History of Jamaica: in three parts. In three dissertations. The whole illustrated with fifty copper-plates: in which the most curious productions are represented of the natural size, and delineated immediately from the objects* Printed for the author, and sold by T. Osborne and J. Shipton in Gray's-Inn, London.

147. Patterson, R. L. (1878–1879) *Some of the Wading Birds frequenting Belfast Lough* (1852–1880) *Proceedings of the Belfast Natural History and Philosophical Society* Alexander Mayne, Belfast.

148. Hayes, W. (1794–1799) *Portraits of rare and curious birds, with their descriptions: from the menagery of Osterly Park, in the county of Middlesex* W. Bulmer and Co., London.

149. Bewick, Thomas (1797–1804) *A History of British Birds* Sol. Hodgson, Newcastle.

150. Ibarra, José Tomás, Julián Caviedes, and Pelayo Benavides (2020) "Winged Voices: Mapuche Ornithology from South American Temperate Forests" *Journal of Ethnobiology*, Vol. 40, No. 1), pp. 89–100. Society of Ethnobiology.

151. Ricardo Rozzi (2010) *Multi-Ethnic Bird Guide of the Sub-Antarctic Forests of South America* University of Magallanes Press, Punta Arenas, Chile.

152. Lesson, R. P. (1828) *Manuel d'Ornithologie, ou Description des genres et des principales espèces d'oiseaux* Roret, Paris.

153. Fisher, Alexander (1821) *A Journal of a Voyage of Discovery to the Arctic Regions: in His Majesty's ships Hecla and Griper, in the years 1819 & 1820* Longman, Hurst, Rees, Orme, and Brown, London.

154. McAtee, W. L. (1923) "Local Names of Migratory Game Birds" United States Department of Agriculture, Washington.

155. Fradkin, A. (1988) *Reconstructing the Folk Zoological World of Past Cultures: the animal semantic domain of the protohistoric Cherokee Indians* University of Florida, Gainesville.

156. Giraud, J. P. (1844) *The Birds of Long Island* Wiley & Putnam, New York.

157. Feeney W. E., Troscianko J., Langmore N. E., and Spottiswoode C.N. (2015) "Evidence for aggressive mimicry in an adult brood parasitic bird, and generalized defences in its host" *Proc. R. Soc. B* 282: 20150795. http://dx.doi.org/10.1098/rspb.2015.0795.

158. Shaw, L. H. De Visme (1905) *Wild-fowl* Longmans, Green, and Co., London.

159. Sclater, Philip Lutley, and W. H. Hudson (1888–89) *Argentine Ornithology. A descriptive catalogue of the birds of the Argentine Republic* R. H. Porter, London.

160. Sundevall, Carl Jakob (1836) *Ornithologiskt System Kungl. Svenska vetenskapsakademiens handlingar* Series 3 (in Latin) 23:89.

161. Kanahele, George S. (1995) *Waikīkī, 100 B.C. to 1900 A.D.: An Untold Story* University of Hawaii Press, Honolulu.

162. Sundevall, Carl Jakob (1850) *Eremomela. Öfversigt Af Kongl. Vetenskaps-akademiens Forhandlingar* 7: 102.

163. Blyth, E. (1844) "Appendix to Mr. Blyth's Report for December Meeting 1842, (Continued from vol. xii, p. 1011)" *Journal of the Asiatic Society of Bengal*, Vol. 13, pp. 381–396.

164. Desmarest, Anselme Gaëtan (1806) *Histoire naturelle des tangaras, des manakins et des todiers* Garnery, Paris.

165. Moorfield, J. C. (2011) The Te Aka Māori-English, English-Māori Dictionary. Pearson, New Zealand. https://maoridictionary.co.nz.

166. Latham, John (1787) *Supplement to the General synopsis of Birds* Leigh & Sotheby, London.

167. Horsfield, Thomas (1824) *Zoological Researches in Java, and the neighbouring islands* Kingsbury, Parbury, & Allen, London.

168. Iredale, T. (1922–1924) "Fairy Wren" *The Australian Zoologist*, Vol. 3. Royal Zoological Society of New South Wales, Sydney.

169. Vigors, N. A., and T. Horsfield (1826) "A Description of the Australian Birds in the Collection of the Linnean Society; with an attempt at arranging them according to their natural affinities" *Transactions of the Linnean Society of London*, Vol. 15.

170. Hall, Robert (1899) *A Key to the Birds of Australia and Tasmania, with their geographical distribution in Australia* Melville, Mullen and Slade, Melbourne.

171. Baker, C. S. (1892) "The Bulbuls of North Cachar" *Journal of the Bombay Natural History Society*, Vol. 7.

172. Oustalet, M. E. (1901) "Revision de Quelques Espèces D'oiseaux de la Chine Occidentale et Méridionale" *Nouvelles Archives du Muséum d'Histoire Naturelle*, Tome, Paris.

173. Schlegel, H. (1872) *De Dierentuin van het Koninklijk Zoologisch Genootschap Natura Artis Magistra te Amsterdam* Van Es., Amsterdam.

174. Ogilvie-Grant, W. R. (1912) "Further Notes on the Birds of the Island of Formosa" *The Ibis*, Vol. 6. Zoological Society of London, London.

175. Hakluyt, Richard (1583) *The voyage made by M. John Hawkins Esquire, and afterward knight, Captaine of the Jesus of Lubek, one of her Majesties shippes, and Generall of the Salomon, and other two barkes going in his companie, to the coast of Guinea, and the Indies of Nova Hispania, begun in An. Dom. 1564. The Principal Navigations, Voyages, Traffiques, and Discoveries of the English Nation Made by Sea or Overland to the Remote & Farthest Distant Quarters of the Earth at any time within the compasse of these 1600 Yeares* E. P. Dutton & Co., New York.

176. Chubb, Charles (1916–1921) *The Birds of British Guiana: based on the collection of Frederick Vavasour McConnell* Bernard Quaritch, London.

177. Pennant, Thomas (1776–1777) *British Zoology, Vol. 2.* William Eyres, London.

178. Gessner, Conrad (1551–1587) *Historiæ Animalium* Apud Christ. Froschouerum, Tiguri, Zurich.

179. Finsch, O. and Hartlaub, G. (1870) *Die Vogel Ost-Afrikas. Baron Carl Claus von Deer Decken's Reisen in Ost-Africa. Bd. 4* C. F. Winter'sche Verlagshandlung, Leipzig.

180. Ridgely, R; G. Tudor, and W. L. Brown (1994) *The Birds of South America: The Suboscine Passerines* Oxford University Press, Oxford.

181. Adams, Andrew Leith (1867) *Wanderings of a Naturalist in India: the western Himalayas, and Cashmere* Edmonston & Douglas, Edinburgh.

182. Australasian Ornithologists' Union (1905) *The Emu*, Vol. 5: official organ of the Australasian Ornithologists' Union. Walker, May & Co., Melbourne.

183. Du Tertre, Jean-Baptiste (1667) *Histoire générale des Antilles habitées par les François*, Vol. 2. Thomas Joly, Paris.

184. Whitehead, J. (1893) *Exploration of Mount Kina Balu, North Borneo* Gurney and Jackson, London.

185. Stephens, J. F. (1826) *General Zoology: Birds, Vol. 13–Part 1.* J. and A. Arch, etc., London.

186. Latham, R. G. (1866) *A Dictionary of the English Language, Vol. 1, Issue 2.* Longmans, Green & Co., London.

187. McKinlay. J. (1861) *McKinlay's Journal of Exploration in the Interior of Australia (Burke Relief Expedition)* F. F. Bailliere, Publisher in Ordinary to the Victorian Government, and Importer of Medical and Scientific Works, Melbourne.

188. Azara, F. M. de (1805–1809) *Voyages dans l'Amérique Méridionale: depuis 1781 jusqu'en 1801* or *Apuntamientos Para la Historia Natural de los Páxaros del Paraguay Y Rio De La Plata* De La Hija De Ibarra, Madrid.

189. Stresemann, Von E. (1931) *Vorläufiges über die ornithologischen Ergebnisse der Expedition Heinrich 1930–1931* Ornithologische Monatsberichte.

190. Donnegan, J. (1842) *A New Greek and English Lexicon Principally on the Plan of the Greek and German Lexicon of Schneider* Wilkins & Carter, Boston.

191. Johnston, T. H. (1943) "Aboriginal Names and Utilization of The Fauna in the Eyrean Region" *Transactions of the Royal Society of South Australia* Vol. 61. Published and Sold at The Society's Rooms, Adelaide.

192. Duncan-Kemp, A. (1933) *Our Sandhill Country: Nature and Man in South-Western Queensland* Angus and Robertson, Sydney.

193. Porter, S. (1925) "Notes from Rhodesia" *The Avicultural Magazine*, Vol. 3. Avicultural Society, Berkshire.

194. Suolahti, V. H. (1909) *Die Deutschen Vogelnamen: eine wortgeschichtliche Untersuchung* Verlag von Karl J. Trübner, Straßburg.

195. Eulexis-web: Lemmatiser for ancient Greek texts (online app) https://outils.biblissima.fr/en/eulexis-web/index.php.

196. *Reports of Explorations and Surveys, to Ascertain the Most Practicable and Economical Route for a Railroad from the Mississippi River to the Pacific Ocean* Made under the Direction of the Secretary of War, in 1853–6, According to Acts of Congress of March 3, 1853, May 31, 1854, and August 5, 1854. (1858) Vol. 9. A. O. P. Nicholson, Washington.

197. Cory, C. B. (1918) *Of Birds of the Americas and the Adjacent Islands in Field Museum of Natural History and Including All Species and Subspecies Known to Occur in North America, Mexico, Central America, South America, the West Indies, and Islands of the Caribbean Sea, the Galapagos Archipelago, and Other Islands which may properly be included on Account of their Faunal Affinities* Part 2, No. 1. Chicago.

198. Cox, Nicholas (1686) *The Gentleman's Recreation: in four parts, viz. hunting, hawking, fowling, fishing; wherein these generous exercises are largely treated of, and the terms of art for hunting and hawking more amply enlarged than heretofore; whereto is prefixt a large sculpture, giving easie directions for blowing the horn and other sculptures inserted proper to each recreation, with an abstract at the end of each subject of such laws as relate to the same* Freeman Collins, London.

199. Quintilian (ca AD 95) *Institutio Oratoria* English Translation by Harold Edgeworth Butler (1920) Harvard University Press, Cambridge, Massachusetts and William Heinemann, Ltd., London.

200. van Someren, V. G. L. (1939) "Birds of the Chyulu Hills (Part 2)" *The Journal of the East Africa and Uganda Natural History Society*, Vol. 14, Nos. 1–2, pp. 15–129. Longmans, Green, London.

201. Brehm, Alfred Edmund (1874) *Bird-life; being a history of the bird, its structure, and habits, together with sketches of fifty different species* J. Van Voorst, London.

202. Pacheco, J. F., Whitney, B. M., and Gonzaga, L. P. (1996) "A New Genus and Species of Furnariid (Aves: Furnariid Ae) from the Cocoa-Growing Region of Southeastern Bahia, Brazil" *The Wilson Bulletin*, Vol. 108, No. 3, pp. 397–433. The Wilson Ornithological Society.

203. von Berlepsch, H. (1886) "On some interesting Additions to the Avifauna of Bucaramanga, U.S. of Colombia" *The Ibis*, Vol. 4, p. 54, pl. 4.

204. Wetmore, Alexander (1951) "Additional Forms of Birds from Colombia and Panama" Smithsonian Miscellaneous Collections Vol. 117, No. 2. The Smithsonian Institution, Washington, D.C.

205. Wallace, A. R. (1869) *The Malay Archipelago: the Land of the Orang-Utan, and the Bird of Paradise. A Narrative of Travel, with Studies of Man and Nature* Harper & Brother, New York.

206. Hume, A. (1871) "Stray Notes on Ornithology in India" *The Ibis*, Vol. 1, pp. 403–413.

207. Dresser, H. E. (1902–03) *A Manual of Palæarctic Birds* By the author, London.

208. Sharpe, R. B. (1871) "On the Coraciidæ of the Ethiopian Region" *The Ibis*, Vol. 1, pp. 184–289.

209. "Extracts from a MS. Dated '*apud* Eltham, *mense* Jan. 22 Hen. VIII'" (1531) *Archaeologia: Or, Miscellaneous Tracts Relating to Antiquity*, Vol. 3, 1775. Society of Antiquaries of London, London.

210. McGregor, Richard C. (1909) *A Manual of Philippine Birds* Bureau of Printing, Manila.

211. Austin, T. ed. (1888) *Two Fifteenth-Century Cookery-books: Harleian MS. 279 (ab 1430), & Harl. MS. 4016 (ab. 1450), with extracts from Ashmole MS. 1439, Laud MS. 553, & Douce MS. 55* Early English Text Society, Original Series 91.

212. Victorin, J. F. (1858) Zoologiska Anteckningar Under En Resa I Södra Delarne Af Caplandet Aren 1853—1855 Ur Den Aflidnes Papper Samlade Och Ordnade J. W. Grill. K. Vot. Akad. Handl. B. 2.  N:O 10. in Konglika Svenska Vetenskaps-Akademiens Handlingar.

213. Buckley, Francine G. (1968) *Behavior of the Blue-crowned Hanging Parrot* Loriculus galgulus *with Comparative Notes on the Vernal Hanging Parrot* L. vernalis Ibis, Vol. 110, Issue 2, pp. 145–164.

214. Hogg, John (1845) "A Catalogue of Birds observed in South-eastern Durham and in North-western Cleveland" *The Zoologist*, Vol. 3, pp. 1049–1063.

215. Willughby, Francis, John Ray, and Lady Emma Willughby (1676) *Willughbeii: De Middleton in agro Warwicensi, Armigeri, e Regia Societate, Ornithologiae libri tres* Impensis Joannis Martyn, Regiae Societatis Typographi, ad insigne Campante in Carmeterio D. Pauli, Londini.

216. Cayley, Neville W. (1931) *What Bird is That? A guide to the birds of Australia* Angus & Robertson Ltd., Sydney.

217. Kirby, Mary, and Elizabeth Kirby. (1874) *Humming-birds* T. Nelson, London.

218. Latham, John (1781–1785) *A General Synopsis of Birds* Benj. White, London.

219. Cotgrave, R. (1660) *A French and English Dictionary, composed by Mr. Randle Cotgrave; with another in English and French*; ed.J. Howell. Wm. Hunt, in Pye-corner, London.

220. Skeat, W. W. (1888) *An Etymological Dictionary of the English Language* Oxford Clarendon Press, Oxford.

221. Sparrman, Andreas (1777) "An Account of a Journey into Africa from the Cape of Good-Hope, and a Description of a New Species of Cuckow. By Dr. Andreas Sparrman, of the Royal Academy of Stockholm, in a Letter to Dr. John Reinhold Forster, F. R. S." *Philosophical Transactions of the Royal Society of London*, Vol. 67, pp. 38–47. The Royal Society, London.

222. Littré, Emile (1883) *Dictionnaire de la Langue Française* Hachette, Paris.

223. Linné, Carl von (1788–1804) *The Animal Kingdom, or zoological system, of the celebrated Sir Charles Linnæus. containing a complete systematic description, arrangement, and nomenclature, of all the known species and varieties of the mammalia, or animals which give suck to their young Class I Mammalia (and Class II, Birds). Being a translation of that part of the Systemæ Naturæ as lately published, with great improvements, by Professor Gmelin of Goettingen. Vols. 1–3.* A. Strahan, and T. Cadell, London, and W. Creech, Edinburgh.

224. Wakefield, Edward Jerningham (1845) *Adventure in New Zealand from 1839 to 1844 with Some Account of the Beginning of the British Colonization of the Islands, Vol. 1.* www.enzb.auckland.ac.nz.

225. Szabo, M. (1993) "Huia, the Scared Bird" *New Zealand Geographic*, Issue 020.

226. Swinhoe, Robert (1863) "The Ornithology of Formosa, or Taiwan" *The Ibis*, Vol. 5, pp. 377–435.

227. Gould, Elizabeth, and John Gould (1831) *A Century of Birds from the Himalaya Mountains* London.

228. Heine, F. (1859) "Catalogue of Birds collected on the rivers Camma and Ogobai, Western Africa, by Mr. P. B. duChaillu in 1858, with notes and descriptions of new species by John Cassin" *Journal für Ornithologie*, Vol. 7, No. 42, p. 430.

229. Britton, A. S. (2005) *Guaraní Concise Dictionary Guaraní-English, English-Guaraní* Hippocrene Books, New York.

230. Coleção História viva Glossário Tupi Guarani incluindo nomes indígenas de pessoas e cidades. LeBooks Editora.

231. Richardson, J. (1885) *A New Malagasy-English Dictionary* London Missionary Society, London.

232. Cory, C. B. (1918) *Catalogue of Birds of the Americas*, Vol. 13. Field Museum of Natural History Publication 197, Chicago.

233. Clapp, R. B. (1975) Special scientific report (U.S. Fish and Wildlife Service) Vol. 193. The Service, Washington, D.C.

234. Pukui, M. K., and S. H. Elbert (1986) *Hawaiian Dictionary* University of Hawaiʻi Press, Honolulu.

235. Te Ara—the Encyclopedia of New Zealand https://teara.govt.nz/en.

236. Dyer, T. F. T. (1883) *Folk-lore of Shakespeare*.

237. Richter, Joannes (2017) "The Kingfisher" http://docshare04.docshare.tips/files/3076/30767423.pdf.

238. Wada, Yoko (2013) The poem known as *Satire* from London, British Library, MS Harley 913: a new interpretation. 関西大学東西学術研究所紀要 46, pp. 83–100 2013年4月1.日

239. Polynesian Lexicon Project Online https://pollex.shh.mpg.de.

240. Mathews, R. H. (1904) "The Wiradyuri and other languages of New South Wales" *The Journal of the Anthropological Institute of Great Britain and Ireland*, Vol. 34.

241. Lesson, R. P. (1831) *Tableau Méthodique des Ordres, Sous-ordres, Familles, Tribus, Genres, Sous-genres et Races d'Oiseaux* Chez F. G. Levrault, Bruxelles.

242. Perkins, R. C. L. (1895) "Notes on some Hawaiian Birds" *The Ibis*, Vol. 1, p. 117.

243. Sharpe, R. B. (1908) "On further collections of birds from the Efulen District of Camaroon, West Africa, Part V" *Ibis*, Vol. 46, Issue. 9, pp. 117–129 [119].

244. Perkins, R. C. L. (1899–1913) *Fauna Hawaiiensis Vol. 1 Part IV. Vertebrata* Cambridge University Press, London.

245. Henshaw, H. W. (1902) *Birds of the Hawaiian Islands; being a complete list of the birds of the Hawaiian possessions, with notes on their habits* H. T., T. G. Thrum, Honolulu.

246. Andrews, L. (1865) *A Dictionary of the Hawaiian Language*.

247. Wikiaves: bird watching and citizen science for all. www.wikiaves.com.br.

248. Vieillot, Louis P. (1816) *Analyse d'une Nouvelle Ornithologie Élémentaire* Deterville, Paris. 1–70.

249. Clusius, 1605. *Caroli Clusii Atrebatis [of Arras] Exoticorum Libri Decern: Quibus Animalium, Plantarum, Aromatum aliorumque peregrinorum Fructuum historiae describuntur* 1 vol. folio, Leyden.

250. Yule, H. (1886) *Hobson-Jobson: being a glossary of Anglo-Indian colloquial words and phrases and of kindred terms etymological, historical, geographical and discursive* John Murray, London.

251. Baird, Spencer Fullerton (1874) *A History of North American Birds: Land Birds* Little, Brown, Boston.

252. Russell, John (1315) *The Boke of Nurture, following Englondis gise* www.gutenberg.org.

253. Hodgson, B. H. (1843) "Additions to the Catalogue of Nepal Birds" *The Journal of the Asiatic Society of Bengal*, Vol. 12, pp. 447–450.

254. Mundy, Peter (2017) "Griffon—what's in a name?" *Vulture News*, Vol. 72.

255. Perrault, M. (1734) "*Le Grifon*" in *Mémoires pour servir à l'histoire naturelle des animaux* pp. 209–215.Académie Royale des Sciences, Paris.

256. Martin, W. C. L. (1835) *An Introduction to the Study of Birds; or, The elements of ornithology, on scientific principles. With a particular notice of the birds mentioned in Scripture* Religious Tract Society, London.

257. Bowdich, T. E. (1821) *An Introduction to the Ornithology of Cuvier: for the use of students and travellers* J. Smith, Paris.

258. Julien, Esprit (1652) *Voyage d'Orient du R. P. Philippe de la Très-Saincte-Trinité, composé et augmenté par luy mesme et traduit du latin, par un religieux du même ordre.*

259. König, Karl (1939) "Überseeische Wörter im Französischen (16.–18. Jahrhundert)" *Beihefte zur Zeitschrift für romanische Philologie*, 91. De Gruyter, Berlin.

260. *Dicionario Ilustrado Tupi Guarani* www.dicionariotupiguarani.com.br.

261. *The Promptorium Parvulorum: The first English-Latin Dictionary* (ca 1440) Edited from the manuscript in the Chapter Library at Winchester (1908) Oxford University Press, Oxford.

262. The Master of Game by Edward, Second Duke of York, eds. (ca 1450) *Forest Laws* W. A. Baillie-Grohman and F. Baillie-Grohman (1904) 241–42.

263. Tradescant, John (1656) *Musaeum Tradescantianum, or, A collection of rarities preserved at South-Lambeth neer London* John Grismond, London.

264. Meyer, A. B., and L. W. Wiglesworth, (1898) *The Birds of Celebes and the Neighbouring Islands* R. Friedländer & Sohn, Berlin.

265. Munro, G. C. (1960) *Birds of Hawaii* Bridgeway Press, Rutland.

266. Gessner, Conrad (1560) *Icones Avium Omnium* Excvudebat C. Froschovervus, Tigvuri, Zurich.

267. Karttunen, Frances (1992) *An Analytical Dictionary of Nahuatl* University of Oklahoma Press, Norman.

268. Buller, Walter Lawry (1882) *Manual of the Birds of New Zealand* G. Didsbury, Wellington, New Zealand.

269. Lacadena, A. (2008) "The *wa1* and *wa2* Phonetic Signs and the Logogram for *WA* in Nahuatl Writing" *The PARI Journal: A quarterly publication of the Pre-Columbian Art Research Institute*, Vol. 8, No. 4.

270. Dawson, William Leon (1923) *The Birds of California: a complete, scientific and popular account of the 580 species and subspecies of birds found in the state* South Moulton Company, Los Angeles.

271. Sims, R. W. (1956) "Birds collected by Mr. F. Shaw-Mayer in the Central Highlands of New Guinea 1950–1951" *Bulletin of the British Museum (Natural History). Zoology*, Vol. 3, p. 387.

272. Lacey, E. (2016) "Birds and Words: Aurality, Semantics, and Species in Anglo-Saxon England" p. 75 in *Sensory Perception in the Medieval West* Book series: Utrecht Studies in Medieval Literacy, 34. Brepols Publishers, Turnhout https://doi.org/10.1484/M.USML-EB.5.109505.

273. S. Ward, B. S. (1973) "The Breeding of the Spotted Morning Warbler *Cichladusa guttata*, at the Winged World" *The Avicultural Magazine*, Vol. 79, pp. 182–183. Avicultural Society, Ascot, Berkshire.

274. Phillott, D. C., and G. L. Pandit (1908) "Hindustani-English Vocabulary of Indian Birds" *Journal and Proceedings of the Asiatic Society of Bengal* Asiatic Society, Kolkata, India.

275. Jerdon, T. C. (1844) *Catalogue of the Birds of the Peninsula of India, arranged according to the modern system of classification: with brief notes on their habits and geographical distribution, and description of new, doubtful and imperfectly described species*

276. Trumbull, G. (1888) *Names and Portraits of Birds which Interest Gunners, with descriptions in languages understood of the people* Harper & Brothers, New York.

277. Ferguson-Lees, J., and D. A. Christie (2001) *Raptors of the World* Houghton Mifflin Co., Boston.

278. Sclater, P. L. (1873) "Characters of new species of Birds discovered in New Guinea by Signor d'Albertis" *Proceedings of the Zoological Society of London*, pp. 690–710.

279. Gerard, J., R. Davyes, T. Johnson, R. Priest, R. Dodoens, and the Katherine Golden Bitting Collection On Gastronomy (1633) *The herball: or, Generall historie of plantes* London, Printed by Adam Islip, Joice Norton and Richard Whitakers, anno. [Pdf] Retrieved from the Library of Congress, https://www.loc.gov/item/44028884/.

280. Gregory, G. (1806–1807) *A Dictionary of Arts and Sciences* Richard Phillips, London.

281. Pennant, T. (1773) *Genera of Birds* Balfour and Smellie, Edinburgh.

282. von Chun, C. (1902) "Aus den Tiefen des Weltmeeres. Von Carl Chun. Schilderungen vonder deutschen Tiefsee-Expedition. Mit 6 Chromolithographieen, 8 Helio gravüren, 32 als Tafeln gedruckten Vollbildern, 1900" *The Edinburgh Review*, Longmans, Green and Co. London.

283. Blyth, E. (1849) *Catalogue of the birds in the Museum Asiatic Society. Asiatic Museum, Calcutta, India* J. Thomas, Calcutta.

284. Jerdon, T. C. (1847) *Illustrations of Indian Ornithology: containing fifty figures of new, unfigured and interesting species of birds, chiefly from the south of India* P. R. Hunt, American Mission Press, Madras.

285. Staub, F. (1995?) "Dodo and Solitaires, Myths and Reality" *Proceedings of the Royal Society of Arts and Sciences* http://www.potomitan.info/dodo/c32.php.

286. Fuller, Errol (2002) *Dodo—From Extinction to Icon* Harper Collins, London.

287. Reischek, A. (1886) "Ornithological Notes" *Transactions and Proceedings of the New Zealand Institute*, Vol. 19. New Zealand Institute, Wellington.

288. Gill, F., and M. Wright (2006) *Birds of the World: Recommended English Names* Princeton University Press, Princton, New Jersey.

289. Roosevelt, Robert Barnwell (1884) *Florida and the Game Water-birds of the Atlantic Coast and the Lakes of the United States. With a full account of the sporting along our sea-shores and inland waters, and remarks on breech-loaders and hammerless guns* O. Judd Company, New York.

290. Forster, T. (1817) *Observations of the Natural History of Swallows; with a collateral statement of facts relative to their migration, and to their brumal torpidity: and a table of reference to authors* T. and G. Underwood, London.

291. Victorin, P. (2008) "Du papegau au perroquet: Antonomase et parodie" *Cahiers de Recherches Médiévales et Humanistes | Journal of Medieval and Humanistic Studies*, Vol. 15.

292. Layard, Edgar Leopold (1867) *The Birds of South Africa: a descriptive catalogue of all the known species occurring south of the 28th parallel of south latitude* Juta, Cape Town.

293. Layard, E. L., and R. B. Sharpe (1875) *The Birds of South Africa* Bernard Quaritch, London.

294. Linné, Carl von (1776) *A Catalogue of the Birds, Beasts, Fishes, Insects, Plants, &c. contained in Edwards's Natural History in Seven Volumes* J. Robson, New Bond Street, London.

295. Darwin, Charles (1846) *Journal of Researches into the Natural History and Geology of the Countries Visited during the Voyage of H.M.S. Beagle Round the World: under the command of Capt. Fitz Roy* R. N. Harper & Brothers, New York.

296. Chasen, F. N. (1935) *A Handlist of Malaysian Birds: a systematic list of the birds of the Malay Peninsula, Sumatra, Borneo and Java, including the adjacent small islands* Printed at the Govt. Print. Off., Singapore.

297. Lacépède, B. G. (1799) *Discours d'Ouverture et de Clôture du Cours d'Histoire Naturelle* Plassan, Paris.

298. Knowlton, Frank Hall, and R. Ridgway, ed. (1909) *Birds of the World: A Popular Account* Holt, New York.

299. Chapman, F. M. (1928) "Descriptions of New Birds from Eastern Ecuador and Eastern Peru" *American Museum Novitates*, pp. 11, 12, fig. 1.

300. Chapman, F. M. (1926) "Notes on the Plumage of North American Birds. Seventy-seventh Paper" *Bird-Lore*, Vol. 28, No. 6.

301. Hasenfratz, R. ed. (2000) *Ancrene Riwle or Guide for Anchoresses (early 13th Century)* Medieval Institute Publications, Kalamazoo, Michigan.

302. Arnott, W. G. (2007) *Birds in the Ancient World from A to Z* Taylor & Francis, London.

303. Molina, Juan Ignacio (1782) *Saggio Sulla Storia Naturale del Chili* pp. 249–250. Nella Stamperia di S. Tommaso d'Aquino, Bologna.

304. Gay, Claudio (1847) *Atlas de la Historia Fisica y Politica de Chile: Zoologia. T1* En el Museo de historia natural de Santiago, Chile.

305. Swainson, William (1831–1832) *Zoological illustrations, or, Original figures and descriptions of new, rare, or interesting animals, selected chiefly from the classes of ornithology, entomology, and conchology, and arranged according to their apparent affinities* 2nd series Baldwin & Cradock, London.

306. Herrtage, Sidney John Hervon (1881) *Catholicon Anglicum: an English-Latin wordbook dated 1483* Published for the Early English Text Society by Trübner, London.

307. De Vaan, M. (2008) *Etymological Dictionary of Latin and the Other Italic Languages* Brill, Leiden.

308. Oates, Eugene William (1889) "The Fauna of British India, including Ceylon and Burma" *Birds*, Vol. 1. Taylor and Francis, London.

309. Beebe, William (1921) *A Monograph of the Pheasants* Published under the auspices of the New York Zoological Society by Witherby & Co., London, England.

310. Oken, Lorenz (1832) "Mittheilungen über einige mertwürdige Thiere von Wagler" *Isis von Oken* Expedition der Isis, Jena.

311. Cassin, J. (1859) "Catalogue of Birds collected on the Rivers Camma and Ogobai, Western Africa, by Mr. P. B. Duchaillu in 1858, with notes and descriptions of new species" *Proceedings*

*of the Academy of Natural Sciences of Philadelphia*, Vol. 11. Academy of Natural Sciences of Philadelphia, Philadelphia.

312. Swainson, W. (1827) "On several Groups and Forms in Ornithology, not hitherto defined" *The Zoological Journal*, Vol. 3, pp. 158–175.

313. Blyth, E. (1847) "Drafts for a Fauna Indica" *The Annals and Magazine of Natural History: Zoology, Botany, and Geology* Taylor and Francis, Ltd, London.

314. Davies, N. B. and Welbergen J. A. (2008) "Cuckoo–hawk mimicry? An experimental test" *Pro. R. Soc. B*. 275: 1817–1822.

315. Blanford, W. T. (1888) *The Fauna of British India, including Ceylon and Burma* Taylor & Francis, London.

316. Johnson, A. W. (1967) *The Birds of Chile and Adjacent Regions of Argentina, Bolivia and Peru* Platt Establecimientos Graficos, Buenos Aires.

317. Diderot and d'Alembert (1765) "Histoire naturelle—Suite du règne animal—Oiseaux" *Encyclopédie ou Dictionnaire raisonné des sciences, des arts et des métiers,* Vol. 6 (plates), Paris.

318. Chavarría-Duriaux, L., Hille, D. C., and Dean, R. (2018) *Birds of Nicaragua* Zona Tropical Publications, Ithaca, New York.

319. Bonyan, G. R. (1853) "Notes on the Raptorial Birds of British Guiana" *The Annals and Magazine of Natural History: Zoology, Botany, and Geology*, Vol. 11. Taylor and Francis, Ltd., London.

320. Haan, W. de, Korthals, P. W., Müller, S., Schlegel, H. & Temminck, C. J. (1839–1844) "Verhandelingen over de natuurlijke geschiedenis der Nederlandsche overzeesche bezittingen" *Natuurkundige Commissie in Oost-Indië (Netherlands)* J. Luchtmans en C. C. van der Hoek, Leiden.

321. Müller. S. (1846) "Ueber den Cliarakier der Thierwelt auf den Inseln des indischen Archipels, ein Beitrag zur zoologischen Geographie" *Archiv für Naturgeschichte, Jarg*, 12. Bd. 1–2. p. 116.

322. Hodgson, B. H. (1841) "Classical Terminology of Natural History" *Journal of the Asiatic Society of Bengal* Vol. 10, pp. 26–29.

323. Howell, S. N. G., and Sophie Webb (1995) *A Guide to the Birds of Mexico and Northern Central America* Oxford University Press, Oxford.

324. Gurney, J. H. (1921) *Early Annals of Ornithology* H. F. & G. Witherby, London.

325. Percy, T. ed. (1770) *The Regulations and Establishment of the Household of Henry Algernon Percy, the Fifth Earl of Northumberland at his castles of Wresill and Lekinfield in Yorkshire begun Anno Domini M.D.XII. (1512)* London.

326. Vasmer, M. (1964–1973) *Etymological Dictionary of the Russian Language* Universitätsverlag Winter, Wiesbaden.

327. Peter Anderson (2008) " 'Thickening, deepening, blackening': starlings and the object of poetry in Coleridge and Dante" *The Coleridge Bulletin* New Series 32.

328. *Nouveau Dictionnaire d'Histoire Naturelle, appliquée aux arts, à l'agriculture, à l'économie rurale et domestique, à la médecine, etc.* (1816–19) Chez Deterville, Paris.

329. Anonymous (1998) "Vernacular Names of the Birds of the Indian Subcontinent" *Buceros*, Vol. 3, No. 1, pp. 53–109.

330. Matisoff, J. (1991) "Sino-Tibetan Linguistics: Present State and Future Prospects" *Annual Review of Anthropology*, Vol. 20, pp. 469–504.

331. Fryer, J. (1698) *A New Account of East India and Persia, in 8 Letters; being 9 years of Travels. Begun 1672. And Finished 1681* Folio, London.

332. Bontius, J. (1631) *Historia naturalis & medicae Indiae Orientalis.*

333. Schouten, W. (1676) *De Oost-Indische Voyagie van Wouter Schouten.*

334. Munro, I. (1780) *A Narrative of the Military Operations, on the Coromendal Coast, against the combined forces of the French, Dutch, and Hyerally Cawn, from the Year 1780 to the peace in 1784, in a series of letters* T. Bensley, London.

335. Hammer, N. (2017) "Etymology of Sanskrit *Kokiláḥ (Eudynamys scolopacea)*: A Bird's-Eye View" *Zeitschrift der Deutschen Morgenländischen Gesellschaft*, Vol. 167, No. 1, pp. 143–152. Harrassowitz Verlag, Wiesbaden.

336. Williams, M. (1873) *A Sanskrit-English Dictionary* Clarendon Press, Oxford.

337. Lockwood, W. B. (1978) "The Philology of 'Auk', and Related Matters" *Neuphilologische Mitteilungen*, Vol. 79, No. 4, pp. 391–397. Modern Language Society.

338. Mullens, W. H. (1921) "Notes on the Great Auk" *British Birds*, Vol. 15.

339. Charleton, G. (1668) *Onomasticon Zoicon Plerorumque Animalium* Jacobum Allestry, London.

340. Wright, W. A., ed. (1909) *Femina* Roxburghe Club Publications, 152.

341. Kurath, H., and S. M. Kuhn (1958) *Middle English Dictionary* University of Michigan Press, Ann Arbor.

342. Pennant, Thomas (1766) *The British Zoology instituted for the promoting useful charities, and the knowledge of nature, among the descendants of the ancient Britons. Class I. Quadrupeds II. Birds* J. and J. March, for the Society, London.

343. Yarrell, W. (1871) *A History of British Birds 4th ed.* John van Voorst, London.

344. Yarrell, W. (1845) *A History of British Birds 2nd ed.* J. Van Voorst, London.

345. Loat, W. L. S. (1898) "Field-notes on the Birds of British Guiana" *The Ibis*, Vol. 4, pp. 558–567.

346. Jerdon, T. C. (1871) "Supplementary Notes to 'The Birds of India'" *The Ibis*, Vol. 1, Third Series.

347. (1842) "Proceedings of Learned Societies. Zoological Society" *The Annals and Magazine of Natural History including Zoology, Botany and Geology*, Vol. 9, pp. 503–518.

348. Bendire, C. (1892) "Life Histories of American Birds with special reference to their breeding habits and eggs" *Smithsonian Contributions to Knowledge* Smithsonian Institution, Washington, D.C.

349. Eyton, T. C. (1838) *A Monograph on the Anatidae, or Duck Tribe* Longman, Orme, Brown, Green, & Longman . . . and Eddowes, Shrewsbury, London.

350. "Extract of a letter from W. Jameson, Esq., Superintendent Botanic Gardens, N.W. Provinces" *Calcutta Journal of Natural History, and miscellany of the arts and sciences in India*, Vol. 7 (1847) Bishop's College Press, Calcutta.

351. Mitchell, Mason (1909) *Birds of Samoa; a manual of ornithology of birds inhabiting these islands* Printed on the London Missionary Society's press, Malua, Samoa.

352. Pratt, G. (1911) *Pratt's Grammar and Dictionary of the Samoan Language* Malua Printing Press, Apia, Western Samoa.

353. Ministry of Natural Resources and Environnment (MNRE) Government of Samoa (2006) *Recovery Plan for the Ma'oma'o or Mao (Gymnomyza samoensis)*.

354. Leach, William Elford (1814–1817) *The zoological Miscellany: being descriptions of new, or interesting animals* B. McMillan for E. Nodder & Son, London.

355. Lewin, J. W. (1822) *A Natural History of the Birds of New South Wales collected, engraved, and faithfully painted after nature* J. H. Bohte, London.

356. Ramsay. E. P. (1886) "List of Western Australian birds collected by Mr. Cairn, and Mr. W. H. Boyer-Bower, at Derby and its vicinity, with remarks on the species" *Proceedings of the Linnean Society of New South Wales*, Vol. 11, pp. 1085–1100.

357. Vieillot, L. P. and P. L. Oudart (1825–1826) *La Galerie des Oiseaux, Vols. 1–3*. Constant-Chantpie, Paris.

358. Clarke, P. A. (2008) *Aboriginal Plant Collectors: Botanists and Australian Aboriginal People in the Nineteenth Century* Rosenburg Publications, Australia.

359. Hall, Robert (1900) *The Insectivorous Birds of Victoria, with chapters on birds more or less useful* The author, Melbourne.

360. Campbell, A. J. (1893) "A Decade in Australian Oology" *The Victorian Naturalist*, Vol. 10.

361. Augustus Earle, A. (1832) *A Narrative of a Nine Months' Residence in New Zealand, in 1827* Longman, Rees, Orme, Brown, Green, & Longman, London.

362. Buller, W. L. (1905) *Supplement to the 'Birds of New Zealand'* The author, London.

363. Holme, Randle (1688) *The Academy of Armory, or, A storehouse of armory and blazon containing the several variety of created beings, and how they born in coats of arms, both foreign and domestick: with the instruments used in all trades and sciences, together with their their terms of art: also the etymologies, definitions, and historical observations on the same, explicated and explained according to our modern language: very usefel [sic] for all gentlemen, scholars, divines, and all such as desire any knowledge in arts and sciences* Printed at Chester by the Author.

364. Huloet, Richard (1552) *Abecedarium Anglico-Latinum, pro* Riddell, London.

365. Morris, Richard, ed. (1776) *Old English Homilies of the Twelfth Century: from the unique ms. B. 14. 52. in the library of Trinity College, Cambridge: Second series, with three thirteenth century hymns from ms. 54 D.4.14 in Corpus Christi College* Early English Text Society, London.

366. Holland, R. (ca 1450) *The Buke of the Howlat* Edited by D. Laing and W. Scott (1823) Bannatyne Club, Edinburgh, Scotland.

367. Ticehurst, N. F. (1923) "Some British Birds in the Fourteenth Century" *British Birds*, Vol. 17.

368. *Extracts from the Account Rolls of the Abbey of Durham, from the original manuscript* (1898) Vol. 1. Published for the Surtees Society Andrews & Co., Durham.

369. Tunstall, M. (1771) *Ornithologia Britannica* Willughby Society, London.

370. Wood, W. (1634) *New England's Prospect: A true, lively, and experimentall description of that part of America, commonly called New England: discovering the state of that Countrie, both as it stands to our new-come English Planters; and to the old Native Inhabitants* [*sic*].

371. Pennant, T. (1766) *British Zoology Vol. II. Class II Birds. Div. II. Water Fowl* p. 523 and pl. 84. William Eyres, London.

372. Richards, C. (2009) "Dotterel *Charadrius moniellus* Migration through Mynydd, Carmathenshire" *Welsh Birds*, Vol. 6, No. 1. Welsh Ornithological Society.

373. Davie, O. (1900) *Nests and eggs of North American birds: with a chapter on ornithological and oölogical collecting (the preparation of skins, nests and eggs for the cabinet)* Musson, Toronto.

374. Stowe Manuscript 57 (ca 1175) *De Natura Jumentorum, Bestiarum, et cunctorum Animalium.*

375. Coues, E. (1903) *Key to North American Birds. Containing a concise account of every species of living and fossil bird at present known from the continent north of the Mexican and United States boundary, inclusive of Greenland and lower California, with which are incorporated General ornithology, an outline of the structure and classification of birds, and Field ornithology, a manual of collecting, preparing, and preserving birds* D. Estes and Company, Boston.

376. Zupitza, Julius, ed. (1844–1895) *The Romance of Guy of Warwick: The first or 14th-century Version* Published for the Early English Text Society.

377. *The Sherborne Missal* (ca 1399–1407) British Library digitized manuscript, p. 384.

378. Belon, Pierre (1555) *L'Histoire de la Nature des Oyseaux: avec leurs descriptions, & naïfs portraicts retirez du natvrel, escrite en sept livres* Gilles Corrozet, Paris.

379. Layamon's *Brut* (aka *The Chronicle of Britain*) (ca 1190–1215) Published for the Early English Text Society by the Oxford University Press, 1963–1978, London and New York.

380. Centre National de Ressources Textuelle et Lexicales. Outils et Ressources pour un Traitement Optimise de la Langue. https://www.cnrtl.fr/etymologie/hobereau.

381. Anonymous (ca 1430) "The Book containing the Treatises of Hawking, Hunting, Coat-Armour, Fishing and Blasing of Arms" *The Harley Lyrics. As printed at Westminster by* Wynkyn de Worde 1496.

382. Salerne, F. (1767) *L'Histoire Naturelle, éclaircie dans une de ses parties principales, l'ornithologie, qui traite des oiseaux de terre, de mer et de riviere, tant de nos climats que des pays étrangers* Chez Debure Père, Paris.

383. Anonymous (2014) "The pagan roots of St Martin's day—11th November" *Atlantic Religion* https://atlanticreligion.com/2014/11/09/the-pagan-roots-of-st-martins-day-11th-november/.

384. Liberman, A. (2012) "Puzzling heritage: The verb 'fart'" *Word Origins . . . And How We Know Them* Oxford University Press Blog https://blog.oup.com/2012/07/word-origin-fart-fist-etymology/.

385. Sayers, W. (2009) "Mackerel and Penguin: International Words of the North Atlantic" *NOWELE* 56–57: 41–52.

386. Liberman, A. (2009) "Penguin" *Word Origins . . . And How We Know Them* Oxford University Press Blog https://blog.oup.com/2011/06/penguin/.

387. Rowley, S. (1605) *When You See Me You Know Me* https://archive.org/details/cu31924013134113/page/n51/mode/2up.

388. Anonymous (1791) *The Natural History of Birds: containing a variety of facts selected from several writers, and intended for the amusement and instruction of children* J. Johnson, London.

389. Lacey, E. (2015) "When is a hroc not a hroc? When it is a crawe or a hrefn: A case-study in recovering Old English folk-taxonomies" *The Art, Literature and Material Culture of the Medieval World* Four Court, Dublin.

390. Warner, R. (1791) *Antiquitates Culinariæ, Or, Curious Tracts Relating to the Culinary Affairs of the Old English, with a Preliminary Discourse, Notes, and Illustrations* R. Blamire, London.

391. Albin, E. (1737) *A Natural History of English Song-birds: and such of the foreign as are usually brought over and esteemed for their singing: to which are added, figures of the cock, hen and egg, of each species, exactly copied from nature* A. Bettesworth and C. Hitch, London.

392. Gould, John (1850–1883) *The Birds of Asia* Vol. 3, p. 44 (plate 41, Blyth's Laughing-Thrush, *Trochalopteron Blythii* [*sic*]) Taylor and Francis, London.

393. Hume, A. (1874) "In Memoriam" Stray Feathers, *Journal of ornithology for India and its dependencies*, Vol. 2.

394. Harvie-Brown, J. A. (1879) *The Capercaillie in Scotland [with some account of the extension of its range since its restoration at Taymouth in 1837 and 1838]* D. Douglas, Edinburgh.

395. Parsons, B. (1959) "Some Makueni Birds" *Journal of the East Africa Natural History Society*, Vol. 23, No. 4, pp. 164–166. Coryndon Memorial Museum, East Africa Natural History Society, Nairobi, Kenya.

396. Walsh, J. F., Cheke, R. A. and Sowah, S. A. (1990) "Additional Species and Breeding Records of Birds in the Republic of Togo" *Malimbus: Journal of the West African Ornithological Society*, Vol. 12, No. 1, p. 13. Ahmadu Bello University Press, Zaria, Nigeria.

397. Fraser, L. (1849) *Zoologia Typica; or, Figures of new and rare animals and birds described in the proceedings, or exhibited in the collections of the Zoological Society of London* Zoological Society of London, London.

398. Gotch, A. F. (1981) *Birds: Their Latin Names Explained* Blanford Press, London.

399. Jeffreys, M. W. (1973) "The Quelea Finch: the origin of the word" *Bokmakierie*, Vol. 252, pp. 46–48.

400. Blench, R (2005) *A Dictionary of Nigerian English* [draft circulated for comment] https://www.rogerblench.info/Language/English/Nigerian%20English%20Dictionary.pdf.

401. Middle English Compendium https://quod.lib.umich.edu/m/middle-english-dictionary.

402. Parker II, T. A. (1982) "Observations of Some Unusual Rainforest and Marsh Birds in Southeastern Peru" *The Wilson Bulletin*, Vol. 94, No. 4, pp. 477–493. Wilson Ornithological Society, Columbus, Ohio.

403. Deggin, C. (2016) "Turkey's Long, Lustrous Love Afair with the Moustache" *Property Turkey Blog* https://www.propertyturkey.com/blog-turkey/turkeys-long-lusterous-love-affair-with-the-moustache.

404. Cassidy, F. G., and R. B. Le Page, eds. (2002) *Dictionary of Jamaican English* University of the West Indies Press, Kingston, Jamaica.

405. Clute, W. N. (1892) "The Avifauna of Broome County, N.Y." *The Wilson Quarterly*, Vol. 4, No. 1, pp. 59–64.

406. Olearius, Adam (1642) *The Voyages and Travels of the Ambassadors from the Duke of Holstein . . . Whereto are Added the Travels of J. Albert de Mandelslo . . . into the East-Indies*, translated by John Davies Dring and Starkey, London.

407. Cuvier, G (1829) *Le Règne Animal distribué d'après son organisation pour servir de base à l'histoire naturelle des animaux et d'introduction à l'anatomie comparée, Second Edition, Volume 1.* Deterville, Paris.

408. Cuvier, G. (1817) *Le Règne Animal distribué d'après son organisation pour servir de base à l'histoire naturelle des animaux et d'introduction à l'anatomie comparée, First Edition,* Chez Déterville, Paris. [Note: actual publication date was December 1816]

409. Stiles, G. (2018) "Proposal (799) to South American Classification Committee: Establish English names for the two species of Schistes" https://www.museum.lsu.edu/~Remsen/SACCprop799.htm.

410. Seton-Thompson, E (1900) "The Origin of Dick Cissel" *Bird Lore*, Vol. 2, No. 3. Audubon Society, The Macmillan Company, New York.

411. Cheke, A. S., and J. P. Hume (2008) *Lost Land of the Dodo: an Ecological History of Mauritius, Réunion and Rodrigues* T. & A. D. Poyser, New Haven and London.

412. Giraud, J. P. (1844) *The Birds of Long Island* Wiley & Putnam, New York.

413. Durham Cathedral (1844) *The Durham Household Book, or, The accounts of the bursar of the monastery of Durham from Pentecost 1530 to Pentecost 1534* J. B. Nichols and Son, London.

414. Gandavo, P. M. (1576) *História da Província de Santa Cruz, a que vulgarmente chamamos Brasil* Typ. da Academia real das sciencias.

415. "John Hunter—journal kept on board the Sirius during a voyage to New South Wales, May 1787—March 1791" Manuscript DL MS 164 https://www.sl.nsw.gov.au/collection-items/john-hunter-journal-kept-board-sirius-during-voyage-new-south-wales-may-1787-march.

416. Guildford, Nicholas de, fl. 1250, supposed author *The Owl and the Nightingale: edited with introduction, texts, notes, translation and glossary* by Atkins, J. W. H. (1922) Cambridge University Press, Cambridge.

417. Buffon, G. L. L. (1770–1785) *Histoire Naturelle des Oiseaux*, Tomes 1–18. De l'imprimerie royale, Paris.

418. Stanley, Edward (1835) *A familiar History of Birds: their nature, habits, and instincts* John W. Parker, London.

419. Burrow, J. A., and T. Turville-Petre (2013) Ch. 5, "Sir Orfeo" *A Book of Middle English* Blackwell Publishing, Oxford.

420. Nozeman, Cornelis, and M. Houttuyn (1770–1829) *Nederlandsche vogelen; vogens hunne huishouding, aard, en eigenschappen beschreven, Vol. 1–5.* J.C. Sepp, Amsterdam.

421. Lilley, R. Sir. (1962–1966) *A Comparative Dictionary of Indo-Aryan Languages* Oxford University Press, Oxford.

422. Revue et Magasin de Zoologie pure et appliquée 2e SÉRIE.—T, XII. p. 441. (1860) *Description d'Oiseaux nouveaux de la Nouvelle-Calédonie et indication des espèces déjà connues de ce pays, par MM. Jules Verreaux et O. des Murs* Bureau de la Revue et Magasin de Zoologie, Paris.

423. Harting, James Edmund (2013) *The Ornithology of Shakespeare. Critically examined, explained and illustrated* John van Voorst, Paternoster Row, London.

424. Phillips, Edward (1658) *The New World of English Words, or, a General Dictionary* London.

425. Minsheu, J. (1617) *Ductor in Linguas* (*The Guide into Tongues*)

426. Hartert, E. (1924) "Notes on Some Birds from Buru" *Novitates Zoologicae: a journal of zoology in connection with the Tring Museum*, Vol. 31, pp. 104–111.

427. Schlegel, H. (1880) "On an Undescribed Bird of the Timalia-Group *Malia grata*" *Notes from the Leyden Museum*, Vol. 2, pp. 165–167.

428. Daudin, F. M. (1802) "Observations Sur les Oiseaux ranges dans le genre Tangara, avec la description d'une espèce nouvelle, trouvée en Afrique" *Annales du Muséum d'Histoire Naturelle*, Vol. 1, pp. 148–151.

429. Bird, I. L. (1876) *The Hawaiian Archipelago: Six months among the palm groves, coral reefs, and volcanoes of the Sandwich Islands* Murray, London.

430. Olson, S. L., and C. Levy (2013) "Eleazar Albin in Don Saltero's Coffee-house in 1736: How the Jamaican mango hummingbird got its name, *Trochilus mango*" *Archives of Natural History*, Vol. 40, No. 2, pp. 340–344.

431. Maximilianus Transylvanus (1523) *Maximiliani Transylvani Caesaris a Secretis Epistola, de admirabili & novissima hispanoru in orientem navigatione, que auriae, & nulli prius accesae regiones sunt, cum ipsis etia moluccis insulis* Published in Cologne.

432. Lesson, R. P. (1838) *Compléments de Buffon* P. Pourrat Frères, Paris.

433. Baker, R. H. (1951) *The Avifauna of Micronesia, Its Origin, Evolution, and Distribution* University of Kansas publications, Museum of Natural History. Vol. 3, pp. 1–359.

434. Rendall, P. (1896) "Notes on the Ornithology of the Barberton District of the Transvaal" *The Ibis*, No. 6.

435. Mathews, G. M. (1927) *Systema Avium Australasianarum—A systematic list of the birds of the Australasian region* British Ornithologists' Union, London.

436. Blyth, E. (1843) "Mr. Blyth's Monthly Report for December Meeting, 1842, with Addenda subsequently appended" *Journal of the Asiatic Society of Bengal*, Vol. 12, pp. 925–1010.

437. Sagra, Ramón de la (1839) *Histoire Physique, Politique et Naturelle de l'Ile de Cuba* Vol. 3, pt. 2 A. Bertrand, Paris.

438. Olphe-Galliard, Léon (1884) *Contributions à la Faune Ornithologique de l'Europe Occidentale: recueil comprenant les espèces d'oiseaux que se reproduisent dans cette région ou qui s'y montrent régulièrement de passage, augmenté de la description des principales espèces exotiques les plus voisines des indigènes ou susceptibles d'être confondues avec elles, ainsi que l'énumération des races domestiques* L. Lasserre, Bayonne.

439. Lesson, R. P. (1835) *Histoire Naturelle des Oiseaux de Paradis et des Epimaques: ouvrage orné de planches, dessinées et gravées par les meilleurs artistes* A. Bertrand, Paris.

440. Rand, A. L., and E. T. Gilliard (1967) *Handbook of New Guinea Birds* Weidenfeld & Nicolson, London.

441. Temminck, C. J. (1820) *Manuel d'Ornithologie, ou Tableau systématique des oiseaux qui se trouvent en Europe: précédé d'une analyse du système général d'ornithologie, et suivi d'une table alphabétique des espèces, Vol. 1.* Chez Gabriel Dufour, Paris.

442. Lowery, Jr., G. H., and D.A. Tallman (1976) "A New Genus and Species of Nine-primaried Oscine of Uncertain Affinities from Peru" *The Auk* Vol. 93, No. 3, pp. 415–428.

443. de Saluces, Thomas (1403–1404) *Le Chevalier Errant* https://gallica.bnf.fr/ark:/12148/btv1b10509668g/f1.item.

444. Baker, E. C. S. (1922) *Birds—Vol. 1. The Fauna of British India, including Ceylon and Burma* Taylor and Francis, London.

445. Whitbourne, Richard (1620) *A Discourse and Discovery of New-found-land: with many reasons to prooue how worthy and beneficiall a plantation may there be made, after a far better manner than now it is* Felix Kyngston for William Barret, London.

446. Littlejohns, Raymond Trewolla, and S. A. Lawrence (19—) *Birds of our Bush; or, Photography for nature-lovers* Whitcombe & Tombs, Melbourne.

447. Hill, John (1752) *An History of Animals: containing descriptions of the birds, beasts, fishes, and insects, of the several parts of the world, and including accounts of the several classes of animalcules, visible only by the assistance of microscopes . . . : illustrated with figures* Printed for Thomas Osborne, London.

448. Hellmayr, Carl Eduard (1932) *The Birds of Chile* Field Museum of Natural History, Chicago.

449. *Wile E. Coyote and the Road Runner* https://en.wikipedia.org/wiki/Wile_E._Coyote_and_the_Road_Runner.

450. Alexander, Boyd (1907) *From the Niger to the Nile, Vols. 1 and 2.* Edward Arnold, London.

451. Barton, Benjamin Smith (1799) *Fragments of the Natural History of Pennsylvania* Way & Groff, Philadelphia.

452. Ridgway, R. (1874) "Catalogue of the Birds ascertained to occur in Illinois" *Annals of the Lyceum of Natural History of New York.*

453. Sharpe, Richard Bowdler, and Claude W. Wyatt (1885–1894) *A Monograph of the Hirundinidae: or family of swallows* Printed for the authors, London.

454. Sclater, W. L. (1903) *The Birds of South Africa, Vol. 3.* R. H. Porter, London.

455. Cotton, John (1849) "A List of Birds which frequent the upper portion of the River Ooulburn, in the district of Port Phillip, New South Wales" *The Tasmanian Journal of Natural Science, Agriculture, Statistics, &c.* Vol. 3, Royal Society of Van Diemen's Land. Dowling, Launceston, Tasmania.

456. Costa, Lucimara Alves da Conceição (2011) *Estudo lexical dos nomes indígenas das regiões de Aquidauana, Corumbá e Miranda no estado de Mato Grosso do Sul: a toponímia rural* (Thesis in Portuguese) Três Lagoas, MS, Brazil.

457. Montes de Oca, Rafael (1875) *Ensayo Ornitológico de los Troquilideos ó Colibríes de México* Escalante, México.

458. Greene, W. T. (1884–1887) *Parrots in Captivity, Vols. 1–3.* George Bell and Sons, London.

459. Mantell, W. B. D. (1850) "Notice of the discovery by Mr. Walter Mantell in the Middle Island of New Zealand, of a living specimen of the Notornis, a bird of the Rail family, allied to Brachypteryx, and hitherto unknown to naturalists except in a fossil state" *Proceedings of the Zoological Society of London*, No. 214.

460. MacGillivray, William (1840–1842) *A Manual of British Ornithology: being a short description of the birds of Great Britain and Ireland* Scott, Webster, and Geary, London.

461. Grandidier, Alfred (1885-1930) *Histoire physique, naturelle, et politique de Madagascar, Vols. 1–39.* Impr. nationale, Paris.

462. Gray, George Robert (1859) *Catalogue of the Birds of the Tropical Islands of the Pacific Ocean, in the collection of the British Museum* British Museum (Natural History), London.

463. Buck, Peter Henry (1957) *Arts and Crafts of Hawaii* Bernice Pauahi Bishop Museum. Bishop Museum Press, Honolulu.

464. Knight, Charles (1844?) *The Pictorial Museum of Animated Nature*, Vols. 1 and 2. C. Cox, London.

465. Nuttall, Thomas (1832–1834) *A Manual of the Ornithology of the United States and of Canada* Hilliard and Brown, Cambridge.

466. Forbes, Alexander Robert (1905) *Gaelic Names of Beasts (mammalia), Birds, Fishes, Insects, Reptiles, Etc in Two Parts: I. Gaelic-English.—II. English-Gaelic* Oliver and Boyd, Edinburgh.

467. Smollett, Tobias (1766) *Travels through France and Italy* Blurb, Inc. repub. 2021.

468. Tench, W. (1793) *A Complete Account of the Settlement at Port Jackson, in New South Wales: including an accurate description of the situation of the colony; of the natives; and of its natural productions.*

469. Bechstein, Johann Matthäus (1793) *Getreue Abbildungen naturhistorischer Gegenstände in Hinsicht auf Bechsteinskurzgefasste gemeinnützige Naturgeschichte des In- und Auslandes: für Eltern, Hofmeister, Jugendlehrer, Erzieher undLiebhaber der Naturgeschichte.* Schneider und Weigelschen Kunst- und Buch- handlung, Nürnberg.

470. Pliny the Elder (AD 23–79) *The Natural History. Book XXXVII. The Natural History of Precious Stones.*

471. MacDowall, Simon (2015) *Catalaunian Fields AD 451: Rome's Last Great Battle* Bloomsbury Publishing, London.

472. Nathan, David (1995) English–Kamilaroi Wordlist http://www.dnathan.com/language/gamilaraay/dictionary.

473. Iredale, T. and Whitley, G. P. (1968) "John Roach, the Budgerigar, and the Unfortunate Officer" *Proceedings of the Royal Zoological Society of New South Wales* Vol. 89, pp. 36–39.

474. Linnaeus, C. (1758) *Systema Naturae*, ed.10, p.124.

475. Lawrence, G. N. (1867) "Catalogue of Birds observed on New York, Long and Staten Islands, and the adjacent parts of New-Jersey" *Annals of the Lyceum of Natural History of New York* Lyceum of Natural History, New York.

476. De Vaan, M. (2008) *Etymological Dictionary of Latin and the Other Italic Languages*. Brill, Leiden.

477. Pliny the Elder (AD 23–79) *The Natural History. Book VI. An Account of the Countries, Nations, Seas, Towns, Havens, Mountains, Rivers, Distances, and Peoples Who Now Exist, or Formerly Existed. Chapter 37: The Fortunate Islands.*

478. Imprensa Nacional (1917) *O Oriente Partuguez*, Vol. 14.

479. Thevet, André (1557) *Singularities de la France Antarctique.*

480. Compiler: Pheifer, J. D. (1974) *Old English Glosses in the Épinal-Erfurt Glossary* Clarendon Press, the University of Michigan.

481. Real Academia de la Lengua Española, de la lengua Española Asociacion de Academias (2015) *Diccionario de la lengua Española RAE 23a.* 23rd edition Planeta Publishing.

482. Raffles, Sir T. (1822) "Descriptive Catalogue of a Zoological Collection, made on account of the Honourable East India Company, in the Island of Sumatra and its Vicinity, under the Direction of Sir Thomas Stamford Raffles, Lieutenant-Governor of Fort Marlborough; with additional Notices illustrative of the Natural History of those Countries" *Transactions of the Linnean Society of London*, Vol. 13.

483. Wynkyn de Worde (1513) *The Boke of Keruynge.*

484. Author unknown (1947) "Appendix 1. A Checklist of the Birds of the Nature Reserves of Singapore" *The Gardens' Bulletin* Singapore. Govt. Print. Off., Singapore.

485. Gould, J. (1838) *Icones Avium, or, Figures and descriptions of new and interesting species of birds from various parts of the globe.* Published by the author, London.

486. Sharpe, R. B. (1887) "Notes on a Collection of Birds made by Mr. John Whitehead on the Mountain of Kina Balu, in Northern Borneo, with Descriptions of new Species" *The Ibis*, Vol. 5.

487. Smythies, B. E. (1964) "The Birds of Mt Kinabalu and their Zoogeographical Relationships" *Proceedings of the Royal Society of Biological Sciences.*

488. Heinrich, F., ed. (1896) *Ein Mittelenglisches Medizinbuch.* British Museum Manuscript (Additional 33,996).

489. Grey, G. (1841) *Journals of Two Expeditions of Discovery in North-West and Western Australia* T. and W. Boone, London.

490. Drayton, M. (1612) *The Poly-Olbion* The Poly-Olbion Project. https://poly-olbion.exeter.ac.uk.

491. The Cornell Lab of Ornithology Clements Checklist. Updates and Corrections: September 2012. https://www.birds.cornell.edu/clementschecklist/updateindex/sep12overview/sept12/.

492. Collins, M., Black, A., Cussans, T., Farndon, J. and Parker, P. (2017) *Remarkable Books: The World's Most Historic and Significant Works* DK Publishing, New York.

493. Exeter Book, Exeter Cathedral Library MS 3501 (late tenth century).

494. Gunning, J.W.B. and A. Haagner (1910) "A Check-list of the Birds of South Africa" *Annals of the Transvaal Museum*, Vol. 2.

495. Taczanowski, P. L. and H. V. Berlepsch (1885) "Troisieme liste des Oiseaux recueillis par M. Stolzmann dans l'Ecuadeur" *Proceedings of the Zoological Society of London*.

496. Woodward, R. B. and J.D.S. (1898) "Further Notes on the Birds of Zululand" *The Ibis* Vol. 4, Seventh Series.

497. Lewin, J. W. (1808) *Birds of New Holland with their Natural History collected, engraved and faithfully painted after nature by John William Lewin* White and Bagster, London.

498. Mandeville, Sir John (1400–50) *English miscellany with Mandeville's Travels, Piers Plowman, catechetical texts, and devotional poems (The Travels of Sir John Mandeville)* Manuscript held by the British Library (Harley MS 3954).

499. Vigors, N. A. (1832) "Observations on a Collection of Birds from the Himalayan Mountains, with Characters of New Genera and Species" *Proceedings of the Committee of Science and Correspondence of the Zoological Society of London*.

500. Dumbacher, J. P., T. F. Spande, and J. W. Daly (2000) "Batrachotoxin alkaloids from passerine birds: A second toxic bird genus (Ifrita kowaldi) from New Guinea" *PNAS* November 21, 2000, 97 (24).

501. A Māori Perspective. Tiritiri Matangi: An education resource for schools. https://www.doc.govt.nz.

502. Conolly, M. J. (2009) Goo-Goor-Gaga the Kookaburra. https://www.kullillaart.com.au/dreamtime-stories/Goo-Goor-Gaga-the-Kookaburra.

503. Vieillot, L. P. (1805) *Histoire Naturelle des Plus Beaux Oiseaux Chanteurs de la Zone Torride* Chez J.E. Gabriel Dufour, Paris.

504. Sclater, P. L. (1883) *A List of British Birds Compiled by a Committee of the British Ornithologists' Union.* John van Voorst, London.

505. Gould, J. (1875–1888) *The Birds of New Guinea and the Adjacent Papuan Islands: including many new species recently discovered in Australia*, Vols. 1–5. Henry Sotheran & Co., London.

506. Ridgway, R. (1901–1919) *The Birds of North and Middle America: a descriptive catalogue of the higher groups, genera, species, and subspecies of birds known to occur in North America, from the Arctic lands to the Isthmus of Panama, the West Indies and other islands of the Caribbean sea, and the Galapagos Archipelago* Government Printing Office, Washington, D.C..

507. Gould, J. (1887) *A Monograph of the Trochilidae, or family of humming-birds. Supplement.* Henry Sotheran & Co., London.

508. Schlegel, H. (1866) "A communication was read from Dr. H. Schlegel, F.M.Z.S., containing the following list of the most remarkable species of Mammals and Birds collected by Messrs. Fr. Pollen and D. C. van Dam in Madagascar, and about to be described in a work entitled 'Recherches sur la Faune de Madagascar et de ses dépendances,' par MM. H. Schlegelet Fr. Pollen" *Proceedings of the Zoological Society of London*.

509. Hodgson, B. H. (1837) "Indication of a New Species of Insessores, tending to connect the Sylviadæ and Muscicapidæ" *India Review and Journal of Foreign Science and the Arts*, Vol. 1.

510. Byron, G. B. A., Baron and R. R. Bloxam (1826) *Voyage of H. M. S. Blonde to the Sandwich islands, in the years 1824–1825* J. Murray, London.

511. Livingston, C. H. (1943) "Osprey and Ostril" *Modern Language Notes*, Vol. 58, No. 2, pp. 91–98. Johns Hopkins University Press, Baltimore.

512. Audubon, J. J. (1839) *A Synopsis of the Birds of North America* Adam and Charles Black, Edinburgh.

513. Gould, J. (1869) *The Birds of Australia, Supplement.* Printed by Taylor and Francis, London.

514. Rafinesque, C. S. (1815) *Analyse de la nature: or, Tableau de l'univers et des corps organisés.* Aux dépens de l'auteur, Palerme.

515. Gould, J. (1836) "Characters of some new Birds in the Society's Collection, including two new genera, Paradoxornis and Actiniodura" *Proceedings of the Zoological Society of London*, Part 4, p 17. Academic Press, London.

516. Stone, W. (1929) "Proper Name of the 'Parauque'" *The Auk*, Vol. 46, No. 3.

517. Skutch, A. F. (1972) *Studies of Tropical American Birds.* The Nuttall Ornithological Club, No. 10. Cambridge, Massachusetts.

518. Fletcher, F. (1577) "Francis Fletcher's Notes (Sloane MS No. 61)," in Penzer, N. M., ed., 1926, *The World Encompassed and Analogous Contemporary Documents Concerning Sir Francis Drake's Circumnavigation of the World with an Appreciation of the Achievement by Sir Richard Carnac Temple* Argonaut Press, London.

519. Thier, Katrin (2000) "Of Picts and Penguins: Celtic Languages in the New Edition of the *Oxford English Dictionary*."

520. D. M. (1893) "A Field Naturalist's Collection" *The Argus* (Melbourne, Victoria: 1848–1957) Sat 25 Mar 1893, p. 4. https://trove.nla.gov.au/.

521. Ray, J. (1713) *Joannis Raii Synopsis Methodica Avium & Piscium: opus posthumum.* Impensis Gulielmi Innys, Londini.

522. Sclater, P. L. (1904) "On a Rare Passerine Bird from New Guinea" *The Ibis*, Vol. 4, Eighth Series, pp. 373–375.

523. Ridgway, R. (1887) "Description of Two New Races of Bonap" *The Auk*, Vol. 4. American Ornithologists' Union, Washington, D.C.

524. Swainson, W. (1837) "The Natural History of the Birds of Western Africa" *Ornithology*, Vol. 7. *The Naturalist's Library edited by Sir William Jardine*. W. H. Lizars, Edinburgh.

525. *Bericht über die (XVII)* "November-Sitzung" *Journal für Ornithologie*, XXVI. Jahrgang. Vierte Eolge, 6. Band.

526. van der Hoeven, J. (1856) *A Handbook of Zoology*, Vol. 1. Cambridge University Press, Cambridge.

527. Reichenbach, L. (1853). "Icones ad synopsin avium No.11. Scansoriae B" *Handbuch der speciellen Ornithologie* Dresden und Leipzig: Expedition Vollständigsten Naturgeschichte. p. 276.

528. Macgillivray, W. (1910–11) "The Region of the Barrier Range: An Oologist's Holiday" *The Emu*, Vol. 10. Royal Australasian Ornithologists' Union, Melbourne.

529. Hearne, Thomas, ed. (1774) *Joannis Lelandi Antiquarii De Rebus Britannicis Collectanea* 6 vols (3rd ed.) London.

530. Percy, T. (1770) *Northumberland, Henry Algernon Percy, Earl of, 1478–1527. The Regulations and Establishment of the Houshold of Henry Algernon Percy, the fifth earl of Northumberland: at his castles of Wresill and Lekinfield in Yorkshire: begun anno Domini M.D. XII.*

531. Wilson, A. (1828) *American Ornithology or Natural History of the Birds of the United States.* Collins & Co., New York.

532. Gould, J. (1869) *The Birds of Australia, Supplement.* Printed by Taylor and Francis, London.

533. Grimes, S. A. (1940) "Scrub Jay reminiscences" *Bird Lore*, Vol. 42, pp. 431–436.

534. North, A. J. (1897) *A List of the Insectivorous Birds of New South Wales* Part II, p. 22. Charles Potter [etc.], Govt. Printer, Sydney.

535. Vosmaer, A. (1769) *Description d'un Oiseau de Proie, Nommé le Sagittaire, tout-à-fait inconnu jusqu'ici; apporté duCap de Bonne Espérance* Pierre Meyer, Amsterdam.

536. Hodgson, B. H. (1837) "Indication of some new forms belonging to the Parianae (Cont.)" *India Review* 2 (2): 87–90.

537. "Sora" *Bird Watcher's Digest* https://www.birdwatchersdigest.com/bwdsite/learn/identification/wading-birds/sora.php.

538. Jardine, W. and J. P. Selby (1837) *Illustrations of Ornithology*, Vol. 4, pl. 4. W. H. Lizars, Edinburgh.

539. Day, Leslie (2007) *Field Guide to the Natural World of New York City* Johns Hopkins University Press, Baltimore.

540. The Cornell Lab, *All About Birds* allaboutbirds.org.

541. West, R. (1619) *The Booke of Demeanor: from small poems entitled the schoole of vertue* Roxburghe Club, London.

542. Gould, J. (1848) "Drafts for a new arrangement of the Trochilidae" *Proceedings of the Zoological Society of London*, part 16.

543. Pallas, P. S. (1781) *Neue nordische Beyträge zur physikalischen und geographischen Erd-und Völkerbeschreibung, Naturgeschichte und Oekonomie.* Bey Johann Zacharias Logan, St. Petersburg.

544. Massachusetts Zoological and Botanical Survey (1839) *Reports on the fishes, reptiles and birds of Massachusetts* Dutton and Wentworth, State Printers, Boston.

545. Hodgson, B. H. (1837) "On three new Genera or sub-Genera of long-legged Thrushes, with descriptions of their species" *The Journal of the Asiatic Society of Bengal*, Vol. 6, pp. 101–104.

546. Field Museum of Natural History (1925) *Catalogue of birds of the Americas and the adjacent islands in Field Museum of Natural History and including all species and subspecies known to occur in North America, Mexico, Central America, South America, the West Indies, and islands of the Caribbean Sea, the Galapagos Archipelago, and other islands which may properly be included on account of their faunal affinities.* Vol. 13, part 4. The Museum (1910–1943), Chicago.

547. North, A. J. (1901) "The Destruction of Native Birds in New South Wales" *Records of the Australian Museum* The Australian Museum, Canberra.

548. Gurney, J. H. (1864) "A Sixth additional List of Birds from Natal" *The Ibis*, Vol. 6.

549. Levins, P. (1923) *Manipulus Vocabulorum: A Dictionary of English and Latin Words, Arranged in the Alphabetical Order of the Last Syllables.* First Printed A. D. 1570.

550. Gould, J. (1875–1888) *The Birds of New Guinea and the Adjacent Papuan Islands: including many new species recently discovered in Australia,* Vols. 1–5. Henry Sotheran & Co., London.

551. Coues, E. (1874) *Birds of the Northwest: a hand-book of the ornithology of the region drained by the Missouri River and its tributaries* Government Printing Office, Washington, D.C.

552. Bulger, G. E. (1866) "List of Birds observed at Wellington, Neilgherry Hills, about 6000 feet above the level of the sea, during the months of April and May, 1866" *Proceedings of the Zoological Society of London.*

553. Beavan, R. C. (1867) "The Avifauna of the Andaman Islands" *The Ibis*, Vol. 3, New Series.

554. Chisholm, K. C. (1924) "The Avifauna Around Tumbarumba, N.S.W." *The Emu*, Vol. 24.

555. Dampier, William (1729) *A New Voyage Round the World: describing particularly the isthmus of America, several coasts and islands in the West Indies, the Isles of Cape Verde, the passage by Terra del Fuego, the South-Sea coasts of Chili, Peru, and Mexico; the Isle of Guam one of the Ladrones, Mindanao, and other Philippine and East-India islands near Cambodia, China, Formosa, Luconia, Celebes, &c. New-Holland, Sumatra, Nicobar Isles; the Cape of Good Hope, and Santa Hellena: their soil, rivers, harbours, plants, fruits, animals, and inhabitants : their customs, religion, government, trade, &c.* James and John Knapton, London.

556. Tobias, J. A, D. J. Lebbin, A. Aleixo, M. J. Andersen, E. Guilherme, P. A. Hosner, and N. Seddon (2008) "Distribution, Behavior, and Conservation Status of the Rufous Twistwing (*Cnipodectes superrufus*)" *The Wilson Journal of Ornithology*, Vol. 120, No. 1, pp. 38–49.

557. Newton, E. (1863) "Notes of a Second Visit to Madagascar" *The Ibis*, Vol. 5.

558. Rasoloson, Janie Noëlle (2001) *Malagasy-English, English-Malagasy Dictionary and Phrasebook* Hippocrene Books.

559. Bartholomaeus Anglicus (ca 1240) *De proprietatibus rerum* [*On the Properties of Things*]

560. Smythe Palmer, Abram (1904) *The Folk and Their Word-lore: An Essay on Popular Etymologies.*

561. Skeat, W. W. (1888–1890) "Notes on English Etymology" *Transactions of the Philological Society*, Vol. 21, Issue 1, p. 22.

562. Folkard, H. C. (1864) *The Wild-Fowler: a treatise on ancient and modern wild-fowling, historical and practical* Longman, Green, Longman, and Roberts, London.

563. Latham, J. (1801 or 1802) *Supplement II to the General Synopsis of Birds* Leigh, Sotheby, & Son, London.

564. Legge, W. V. (1880) *A History of the Birds of Ceylon* Published by the author, London.

565. Hall, R. (1907) *The Useful Birds of Southern Australia, with notes on other birds.* T. C. Lothian, Melbourne.

566. Ogilby, J. (1671) *America: being the latest, and most accurate description of the New World: containing the original of the inhabitants, and the remarkable voyages thither: the conquest of the vast empires of Mexico and Peru, and other large provinces and territories, with the several European plantations in those parts: also their cities, fortresses, towns, temples, mountains, and rivers: their habits, customs, manners, and religions: their plants, beasts, birds, and serpents: with an appendix containing, besides several other considerable additions, a brief survey of what hath been discover'd of the unknown south-land and the Arctick region* Printed by the author, London.

567. Hunt, J. (1815–1822). *British Ornithology; containing portraits of all the British Birds including those of Foreign Origin which have become domesticated* Bacon & Co., Norwich.

568. Gray, John Edward (1847) *List of the Osteological Specimens in the Collection of the British Museum* British Museum (Natural History), London.

569. Inskipp, Carol, and Tim Inskipp (1985) *A Guide to the Birds of Nepal* Christopher Helm, London.

570. Baird, S. F. (1866) *Review of American Birds, in the Museum of the Smithsonian Institution Part 1 North and Middle America* Smithsonian Institution, Washington, D.C.

571. Strickland, H. E., J. S. Henslow, John Phillips, W. E. Shuckard, John Richardson, G. R. Waterhouse, Richard Owen, W. Yarrell, Leonard Jenyns, C. Darwin, W. J. Broderip, and J. O. Westwood (1843) "Series of Propositions for rendering the Nomenclature of Zoology uniform and permanent, being the Report of a Committee for the consideration of the subject appointed by the British Association for the Advancement of Science" *The Annals and Magazine of Natural History, including zoology, botany, and geology* Taylor and Francis, Ltd, London.

572. Teara: The Encyclopedia of New Zealand https://teara.govt.nz/en/biographies/1b46/buller-walter-lawry.

573. Stone, W. (1901) "John Cassin" *Proceedings of the Delaware Valley Ornithological Club of Philadelphia*, No. 5.

574. Taquet, Philippe (2007) "Georges Cuvier" *The Great Naturalists* Thames & Hudson, Australia.

575. Vigors, N. A. and T. Horsfield (1826) "A Description of the Australian Birds in the Collection of the Linnean Society; with an Attempt at Arranging them according to their natural Affinities" *Transactions of the Linnean Society of London*, Vol. 15, Part the First, pp. 170–331.

576. Esposito, A. (2014) *Killing Paradise* Self-published.

577. Rolland, E. (1879) "Faune populaire de la France" T II, *Les Oiseaux Sauvages* pp. 74–75.

578. Ridgway, R. (1888) "Notes on Costa Rican Birds, with Descriptions of Seven New Species and Subspecies and One New Genus" Proceedings of the United States National Museum, Vol. 11, pp. 537–546.

579. Ridgway, R. (1907) "The Birds of North and Middle America" *Bulletin of the United States National Museum*, Part 4, pp. 71–72. Government Printing Office, Washington, D.C.

580. Zenteno, Francisco (?) "Diucón" Las Aves que Viven en Chile (Birds Living in Chile) http://avesvivenchile.blogspot.com/2008/03/diucn.html.

581. Herbert, Thomas (1634) *Description of the Persian Monarchy now beinge' the Orientall Indyes, Iles and other ports of the Greater Asia and Africk* London.

582. Hudson, W. H. (1893) *Birds in Town and Village* E.P. Dutton & Company, New York [1920 reprint].

583. Pliny the Elder (AD 23–79) *The Natural History, Book X, The Natural History of Birds.*

584. Hodgson, B. H. (1837) "On a new Genus of the Sylviadae, with description of three new Species" *The Journal of the Asiatic Society of Bengal*, Vol 6, part 1, pp. 230–232.

585. American Ornithologists' Union Committee on Classification and Nomenclature (1886) *The Code of Nomenclature and Check-list of North American Birds: adopted by the American Ornithologists' Union: being the report of the Committee of the Union on Classification and Nomenclature* New York.

586. Johnston. T. H. (1943) "Aboriginal Names and Utilization of the Fauna of the Eyrean Region" *Transactions of the Royal Society of South Australia*, Vol. 67, pp. 244–312. Published and sold at the Society's rooms, Adelaide.

587. "Proceedings of the Asiatic Society, Report from the Curator" (1842) *The Journal of the Asiatic Society of Bengal*, Vol. 11, part 2.

588. Dalton, H., J. Salo, P. Niemela, and S. Orma (2018) "Frederick II of Hohenstaufen's Australasian cockatoo: Symbol of detente between East and West and evidence of the Ayyubids' global reach" *Parergon*, 35(1), 35–60.

589. *Diccionario Quechua–Español–Quechua/Qheswa–Español–Qheswa* (2005) Academia Mayor De La Lengua Quechua. Cusco, Peru.

590. Cuvier, Georges (1836) *Le Règne Animal distribué d'après son organisation pour servir de base à l'histoire naturelle des animaux et d'introduction à l'anatomie comparée, Third Edition,* Louis Hauman et compe, Bruxelles.

591. Nichols, J. G. (1875) *Pilgrimages to Saint Mary of Walsingham and Saint Thomas of Canterbury by Desiderius Erasmus in 1518* John Murray, London.

592. *Psalterium Davidis Latino-Saxonicum vetus* (1640) Badger, London.

593. Liberman, A. (2021) "Now in the field with a fieldfare" OUPblog: Oxford University Press's Academic Insights for the Thinking World, https://blog.oup.com/2021/06/now-in-the-field-with-a-fieldfare/.

594. Norman, J. A., W. E. Boles, and L. Christidis, (2009) "Relationships of the New Guinean songbird genera Amalocichla and Pachycare based on mitochondrial and nuclear DNA sequences" *Journal of Avian Biology* 40: 640–645.

595. Anonymous (1768) "Of Congo, including the Kingdoms of Congo proper, Benguela and Loango," p. 176. *The wonders of nature and art: being an account of whatever is most curious and remarkable throughout the world, whether relating to its animals, vegetables, minerals, volcanoes, cataracts, hot and cold springs and other parts of natural history, or to the buildings, manufactures, inventions, and discoveries of its inhabitants* Newbery and Carnan, Sons, London.

596. Swinhoe, E. (1877) "On a new Bird from Formosa" *The Ibis*, Vol. 1, Fourth Series.

597. Bonaparte, C. L. (1825–1833) *American Ornithology; or, The natural history of birds inhabiting the United States, not given by Wilson* Carey, Lea & Carey, Philadelphia.

598. Albertus, Magnus (1450-1500) *De animalibus libri XXVI, nach der Cölner Urschrift. Mit unterstützung der Kgl. Bayerischen Akademie der Wissenschaften zu München, der Görresgesellschaft und der Rheinischen Gesellschaft für Wissenschaftliche Forschung* Published by Stadler, Hermann in 1920.

599. Nicholson, W. (1809) *The British encyclopedia, or, Dictionary of arts and sciences: comprising an accurate and popular view of the present improved state of human knowledge* C. Whittingham, London.

600. Albertus, St. Magnus. (ca. 1450–1500) *B. Alberti Magni Ratisbonensis episcopi, ordinis Prædicatorum, Opera omnia ex editione lugdunensi religiose castigata*, Vol. 12. Edited by Émile Borgnet, 1891.

601. Oliveros, C., S. Reddy, and R. Moyle (2012) "The phylogenetic position of some Philippine "babblers" spans the muscicapoid and sylvioid bird radiations" *Molecular Phylogenetics and Evolution*, Vol. 65, pp. 799–804.

602. Isler, M., D. Lacerda, P. Isler, S. Hackett, K. Rosenberg, and R. Brumfield (2009) "Epinecrophylla, a new genus of antwrens (Aves: Passeriformes: Thamnophilidae)" *Proceedings of the Biological Society of Washington*, Vol. 119, pp. 522–527.

603. Swainson, W. (1835) *A treatise on the geography and classification of animals.* Longman, Rees, Orme, Brown, Green & Longman [etc.], London.

604. Sclater, W. L. (1930) *Systema Avium Aethiopicarum: A Systematic List of the Birds of the Ethiopian Region* British Ornithologists' Union.

605. Vinycomb, J. (1906) *Fictitious and Symbolic Creatures in Art with Special Reference to their Use in British Heraldry* Chapman & Hall, London.

606. Wood, J. G. (1898) *Animate creation; popular edition of "Our Living World" a natural history* S. Hess, New York.

607. Buffon, G. L. L. (1808) *Natural History of Birds, Fish, Insects, and Reptiles* H. D. Symonds, London.

608. Brandt, J. F. (1836) *Descriptiones et icones animalium Rossicorum novorum vel minus rite cognitorum Fasciculus I. Aves.* Jussu et sumptibus Academiae Scientiarum, Petropoli.

609. Clayton, Mary (1998) *The Apocryphal Gospels of Mary in Anglo-Saxon England* Cambridge University Press, Cambridge.

610. Isler, M., D. Lacerda, P. Isler, S.nHackett, K. Rosenberg, and R. Brumfield (2006) "Epinecrophylla, a new genus of antwrens (Aves: Passeriformes: Thamnophilidae)" *Proceedings of the Biological Society of Washington*, 119 (4), 522–527.

611. Leichhardt, L. (1847) *Journal of an overland expedition in Australia, from Moreton Bay to Port Essington, a distance of upwards of 3000 miles, during the years 1844–1845* Adelaide.

612. Nash, D. (2010) "The smuggled budgie: case study of an Australian loanblend" *Lexical and Structural Etymology: Beyond Word Histories* De Gruyter, Berlin.

613. Poundly, J. and M. Andrew (1768) *The wonders of nature and art: being an account of whatever is most curious and remarkable throughout the world, whether relating to its animals, vegetables, minerals, volcanoes, cataracts, hot and cold springs and other parts of natural history, or to the buildings, manufactures, inventions, and discoveries of its inhabitants* Newbery and Carnan, London.

614. Ford, E. R. (1930) "On the proper Name of the 'Parauque'" *The Auk*, Vol. 47, No. 2.

615. Steven L. Hilty, S. L., Parker, T. A. and Silliman, J. (1979) Observations on Plush-Capped Finches in the Andes with a Description of the Juvenal and Immature Plumages. *The Wilson Bulletin*, Vol. 91, No. 1 pp. 145–148.